Taphonomy and Palaeoecology of Quaternary Vertebrates: Advances in Fossil and Experimental Studies

Taphonomy and Palaeoecology of Quaternary Vertebrates: Advances in Fossil and Experimental Studies

Guest Editors

Juan Rofes
Janine Ochoa
Emmanuelle Stoetzel

MDPI

Basel • Beijing • Wuhan • Barcelona • Belgrade • Novi Sad • Cluj • Manchester

Guest Editors
Juan Rofes
School of Archaeology
University of the Philippines, Diliman
Quezon City
Philippines

Janine Ochoa
School of Archaeology
University of the Philippines, Diliman
Quezon City
Philippines

Emmanuelle Stoetzel
Histoire Naturelle des Humanités Préhistoriques - UMR 7194 CNRS
Muséum national d'Histoire naturelle, UPVD
Paris
France

Editorial Office
MDPI AG
Grosspeteranlage 5
4052 Basel, Switzerland

This is a reprint of the Special Issue, published open access by the journal *Quaternary* (ISSN 2571-550X), freely accessible at: https://www.mdpi.com/journal/quaternary/special_issues/quaternary_vertebrates.

For citation purposes, cite each article independently as indicated on the article page online and as indicated below:

Lastname, A.A.; Lastname, B.B. Article Title. *Journal Name* **Year**, *Volume Number*, Page Range.

ISBN 978-3-7258-4677-1 (Hbk)
ISBN 978-3-7258-4678-8 (PDF)
https://doi.org/10.3390/books978-3-7258-4678-8

Contents

About the Editors

Juan Rofes

Juan Rofes is an environmental archaeologist and Head of the Zooarchaeology Laboratory at the School of Archaeology, University of the Philippines Diliman. His MA is from Universidad Autónoma de Madrid and his PhD from the University of Zaragoza, Spain. He is interested in the interactions between environmental change and human development and the effects of climate and anthropization on the evolutionary histories of various species of vertebrates. He has expanded interests to any topic related to the zooarchaeology of South East Asia, the applications of AI to archaeology, and archaeology in popular culture.

Janine Ochoa

Janine Ochoa is an archaeologist at the School of Archaeology at the University of the Philippines Diliman. She obtained her BA in Anthropology and MA in Archaeology at the University of the Philippines, and her PhD at the University of Cambridge. Her current research interests include zooarchaeology, plurispecies engagements and tropical biodiversity, collaborative archaeologies, and indigenous ecological knowledge systems.

Emmanuelle Stoetzel

Emmanuelle Stoetzel is a Researcher at the French Center for Scientific Research (CNRS) in the laboratory *Histoire Naturelle des Humanités Préhistoriques*. She obtained her PhD at the Natural History Museum of Paris in 2009. She co-edited three Special Issues and published numerous peer reviewed articles on themes related to the Prehistory and Quaternary Paleoenvironments of Northwestern Africa, and more specifically on the taxonomic, taphonomic and palaeoecological study of fossil and modern microvertebrate assemblages from different geographical regions.

quaternary

MDPI

Editorial

Taphonomy and Palaeoecology of Quaternary Vertebrates: Advances in Fossil and Experimental Studies

Emmanuelle Stoetzel [1,*], Janine Ochoa [2] and Juan Rofes [3,4,5]

1 Histoire Naturelle de l'Homme Préhistorique (HNHP, UMR 7194), CNRS/MNHN/UPVD, Musée de l'Homme-Palais de Chaillot, 17 Place du Trocadéro, 75016 Paris, France
2 Department of Anthropology, University of the Philippines-Diliman, Palma Hall, Roxas Avenue, Quezon City 1101, Philippines
3 School of Archaeology, University of the Philippines-Diliman, Albert Hall, Lakandula St., Quezon City 1101, Philippines
4 Archéozoologie, Archéobotanique: Sociétés, Pratiques et Environnements (AASPE, UMR 7209), CNRS/MNHN, CP56, 55 rue Buffon, 75005 Paris, France
5 National Museum of the Philippines, P. Burgos Drive, Manila 1000, Philippines
* Correspondence: emmanuelle.stoetzel@mnhn.fr

Since the founding work of Efremov in 1940 [1], the discipline called "taphonomy" has developed considerably, and its importance is now unquestionable in terms of revealing the processes of the formation and preservation of fossil assemblages (and the surrounding sedimentary deposits) and the establishment of reliable palaeoenvironmental reconstructions. Several reference books have been published on this subject [2–6]. Hundreds of articles have subsequently strengthened and complemented these seminal works, particularly from a methodological and experimental point of view and with applications to fossil assemblages, leading to the landmark production of the Atlas of Taphonomic Identifications [7]. The continuing strong interest of the international scientific community in this discipline is still visible, notably through two prominent events simultaneously held in June 2022 in Madrid: the 9th International Meeting on Taphonomy and Fossilization (TAPHOS) coupled with the 6th Meeting of the ICAZ Taphonomy Working Group (TWG-ICAZ), in honour of the 80th anniversary of Efremov's proposal of taphonomy as a new branch of palaeontology.

It has long been established that taphonomy and palaeoecology are inseparable. A taphonomic analysis of a fossil assemblage is an essential prerequisite for palaeoecological studies, particularly for identifying the origin of faunal assemblages and potential biases in species and anatomical representation [6,8–10]. In turn, palaeoecology uses fossil data to examine how organisms and environments change throughout time [11–13]. By studying the patterns of evolution and extinction under environmental change, palaeoecologists can examine the concepts of vulnerability and resilience in species and environments at different scales [14–16]. The Quaternary period is well represented in geographically extensive and high-temporal-resolution records and is particularly important to human evolution. Vertebrate remains, whether accumulated by humans or non-human agents, are frequently well preserved in Quaternary palaeontological and archaeological deposits. Recently, the number and methods of taphonomic and palaeoecological analyses on Quaternary vertebrate assemblages have significantly increased and developed, and this Special Issue aims to highlight the recent works illustrating these advances in palaeontology, zooarchaeology, and paleoenvironmental studies, in light of modern taphonomic referentials and experiments.

We are pleased with the success of this Special Issue, which has resulted in the publication of 16 articles with an extensive global, geographic, and temporal scope, presenting seminal research conducted by internationally recognised specialists from Argentina, Mexico, Brazil, USA, Canada, Australia, Israel, Indonesia, Philippines, Thailand, UK, Spain,

Portugal, Belgium, and France. We briefly summarise them below arranged by the nature of their contributions into different domains that are not mutually exclusive.

1. Methodological Development/New Fields of Research

Certain processes and taphonomic alterations are less studied than others, although they really require interest and attention. This is notably the case for insect marks on bone remains, for which few studies are available and modern referentials are lacking. Escosteguy et al. [17] explored the aspect and measurement of insect marks observed on bone remains from La Guillerma in Argentina (ca. 1400–600 years BP). They compared them with a variety of marks described in the literature, using a specific classification of insect-produced modifications, and providing high-quality images of a variety of marks, which can be very useful for future studies. The authors also highlighted the utility of studying this type of trace to better understand bioturbation phenomena and to contribute to palaeoecological information based on the ecological requirements of the trace-maker insects.

In addition to providing information on taphonomic and environmental contexts, the study of microvertebrate assemblages can also shed light on the timing and frequency of occupation of sites by humans and non-human predators. In taking the example of La Roche-à-Pierrot in France (Middle-Upper Palaeolithic), Lebreton et al. [18] showed a negative correlation between the proxies for human occupation and micromammal densities. Stratigraphic layers rich in micromammal remains were thus correlated with periods of abandonment or less intense occupation of the site by humans, while other predators were more frequently present.

The pattern of the fragmentation of micromammal bones is often used to characterise the taphonomic impact of predators on prey remains in modern referentials (owl pellets, faeces) but is rarely applied in fossil contexts. Durocher et al. [19] proposed a new methodology based on geometric morphometric analyses on rodent mandibles, allowing the characterisation of fragmentation patterns in modern and fossil assemblages in a "continuous" and adaptable way, without restriction/resorting to a priori categories based on subjective criteria.

2. Modern Referentials and Experiments

Modern referentials are essential to better understand and disentangle taphonomic processes, characterise the specific marks produced by biotic and non-biotic agents, and generate transpositions to fossil assemblages in reliable and reproducible ways. Several papers in this collection have made methodological contributions by initiating experimental studies and successfully applying these to archaeological contexts.

Fernández-Jalvo et al. [20] designed a trampling experiment using equipment that allows the control of different parameters. They explored and characterised the effects of trampling on modern small mammal bones, raising interesting issues related to site formation processes. The different patterns of breakage observed on modern bones were then applied to the fossil site of Wonderwerk Cave in South Africa. Their comparative analysis showed that trampling was an important factor responsible for the high degree of breakage observed in the assemblage, together with other factors such as predator digestion and excavation procedures.

Significant strides have also been made in the field of traceology in recent years. Okaluk and Greenfield [21] specifically explored chop marks, which were produced on sheep, cattle, pig, and deer bones through an experiment using a variety of implements (chipped stone, ground stone, copper, and bronze axe heads). The macroscopic observations made by the authors, supported by high-quality images, allow the effective discrimination of traces left by various tools and also open the way for future microscopic studies. Here, again, the patterns observed on modern material were applied to an archaeological faunal assemblage (Göltepe in Turkey), and the taphonomic evidence suggests the use of metal tools in the butchery process.

Few studies on modern assemblages produced by avian and mammalian predators opt for a multi-taxa approach, and even fewer are interested in inter-observer variability. Stoetzel et al. [22] applied this integrative approach to the analysis of barn owl pellets from Dominica Island in the French Antilles, in which two observers were used for each considered taxon (rodents, bats, birds, squamates). The authors found high similarities in the taphonomic results obtained for the considered faunal groups, but some differences emerged at several levels: between observers, between taxa, between elements, and even between different parts of the same element. This methodological contribution raises important considerations relating to the reliability and reproducibility of neo- and palaeo-taphonomic analyses.

3. Analysis of Fossil and Sub-Fossil Assemblages

Most studies presented in this Special Issue focus on the new analyses of fossil and sub-fossil faunal assemblages that incorporate taphonomic methods, including investigations on the origin of the assemblages and post-depositional processes and consideration of their cultural and archaeological contexts.

Powley et al. [23] provided a taphonomic analysis of *Stegodon* remains from Mata Menge in Indonesia, a Middle Pleistocene site that has yielded the earliest fossil evidence for *Homo floresiensis*. The authors observed several surface alterations (weathering, fragmentation, trampling, fluvial abrasion), notably highlighting the effects of fluvial transport and long exposure of bones before burial. Although incontrovertible evidence for human modification could not be clearly established, the evidence regarding the formation of the *Stegodon* bonebed provides important additional context for this hominin-bearing layer.

Wattanapituksakul et al. [24] performed a zooarchaeological analysis of the faunal assemblage of Ban Rai Rockshelter in Thailand (Terminal Pleistocene to Middle Holocene). They highlighted the human exploitation of the diversity of species among large and small mammals, birds, reptiles, fish, and arthropods, with differences between the archaeological layers. Although the relative abundance of taxa is influenced by site preservation processes, these differences may be due to an adaptation of human populations to a changing environment during the Pleistocene–Holocene transition and beyond, becoming more focused on arboreal taxa during the Early and Middle Holocene.

Competition between human groups and large carnivores for shelter and prey is also the subject of multiple studies, which span an extensive geographical and time range. Daura et al. [25] delved into this question through the analysis of faunal and human remains, lithic assemblages, and the chronostratigraphic context of Cova del Gegant in Spain (Middle Palaeolithic). The detailed analysis of the large mammals, birds, and microvertebrates revealed a high level of faunal diversity at the site, as well as the intervention of both human and non-human predators in the accumulations. The site was mainly used by carnivores such as hyenas, but also by Neanderthals as a brief stopping place, within the context of high human mobility during the Middle Palaeolithic.

Rofes et al. [26] explored the potential of micromammals to reconstruct human activities, biostratigraphy, and palaeoenvironments at El Portalón de Cueva Mayor in Spain (Early Late Chalcolithic). The authors demonstrated the differences in the small-mammal spectrum among different areas and phases of human activity (e.g., burials, prepared floors, activity floors, and fumiers) and provided the first Holocene sub-fossil record of multiple rodent, shrew, and bat species in the Sierra de Atapuerca. They also showed the presence of the specimens of Late Pleistocene age into the Late Chalcolithic prepared floors, arguing for pollution from allochthonous sediment and advising caution regarding paleoenvironmental conclusions for this unit.

Royer et al. [27] focused on the question of the origin of the small mammal assemblage from Ittenheim in France (Late Saalian–Late Weichselian). They compared the fossil assemblage of Ittenheim with two modern accumulations produced by a red fox and a badger, alongside data from the literature. In light of the data on anatomical representation, the presence of tooth marks, and the intensity of digestion, the authors concluded that a small mammalian carnivore was the main accumulator of the Ittenheim assemblage. They also pointed out that the taphonomic impact of small carnivores is still too little known compared with that of owls, and that differences are observable according to the considered prey, i.e., medium-sized prey such as European hamster vs. small prey such as small microtines.

Tilby et al. [28] provided a preliminary taphonomic analysis of the micromammal assemblage from Shanidar Cave in Iraqi Kurdistan (Middle Palaeolithic levels). They especially considered the fragmentation of bones and teeth, and surface alterations such as digestion, black staining, and root etching. The authors concluded that, in most respects, the taphonomic processes remained constant throughout the sequence and micromammals were accumulated by category 1 or 2 predators. They discussed the potential predators that could have been involved and linked the slight increase in Manganese staining and root etching in the lower layers to the relatively wet and warm conditions at this time.

Compared with large and small mammals, other taxa such as amphibians, reptiles *s.l.*, and birds have received less attention in taphonomic studies. Two works published in the present Special Issue focus on these poorly known taxa.

Rufà and Laroulandie [29] presented the taphonomic history of the bird assemblage of Grotte Vaufrey in France (Middle Pleistocene). The faunal spectrum, as well as anatomical representation, digestion traces, and tooth/beak/claw marks, pointed to a non-human accumulation for most of the assemblage. Moreover, the origin may differ considering the species, and some of them may have frequented and died in the site naturally. Some examples, however, displayed clear anthropic action (such as burning and cutting marks), arguing for the occasional consumption and exploitation of some birds by Neandertals.

Finally, Rubinatto Serrano et al. [30] provided an interesting archaeological and ethnographic overview of sub-fossil assemblages dominated by anuran remains in the Tiwanaku sites of Peru (700–1100 CE), exploring the respective roles of humans and other predators in these accumulations, and showing the need for further taphonomic analysis for such peculiar assemblages. The authors discussed the environmental, cultural, and taphonomic explanations related to the anomalously high quantity of anuran remains found in the Tiwanaku sites in comparison to other sites in the area.

4. Palaeoenvironmental Inferences

Two studies published in this Special Issue are especially focused on the palaeoecological implications and palaeoenvironmental models deduced from faunal analyses.

Arroyo-Cabrales et al. [31] considered the entire faunal spectrum (molluscs, amphibians, reptiles, birds, and mammals) of San Josecito Cave in Mexico (45,000–11,000 BP). A revision of the faunal assemblage and site taphonomy, stratigraphy, and geochronology showed the presence of numerous extinct species at the site and the succession of several local faunas throughout the sequence, with high potential for Late Quaternary climatic reconstructions.

Traditionally, paleoenvironmental reconstructions are based on the faunal spectra and ecology of present-day species. However, in recent years, new methods have emerged that can provide valuable information on landscape and climatic changes through time. Uzunidis [32] used morphometric and use-wear analyses on archaeological *Equus* remains from European sites in an attempt to observe changes under climatic pressure and vegetation changes during the Middle Pleistocene, also exploring geographic and seasonal variability.

In conclusion, owing to all these multidisciplinary contributions, this Special Issue provides both new data and new avenues to explore the importance of taphonomic studies and multi-taxa approaches for understanding depositional processes and palaeoenvironmental reconstructions based on faunal remains in archaeological and palaeontological contexts. We are also attentive to the development of new methods, such as biomolecular approaches, which have not necessarily been mentioned here but can provide new proxies and ways to ascertain taphonomic contexts and taxonomic designations. This diverse array of methods ensures the integrity of the assemblages and surrounding deposits and thus helps in understanding biases in palaeoenvironmental reconstructions.

Acknowledgments: We would like to sincerely thank all the participants who responded positively to the call for this Special Issue of *Quaternary*, as well as the three managing editors of the journal who were always supporting us. We also extend our deepest gratitude to all the reviewers who allotted their generous time and efforts to carefully assess the manuscripts included in this volume.

Conflicts of Interest: The authors declare no conflict of interest.

References

1. Efremov, I.A. Taphonomy: A new branch of paleontology. *Pan-Am. Geol.* **1940**, *74*, 81–93.
2. Behrensmeyer, A.K.; Hill, A.P. *Fossils in the Making: Vertebrate Taphonomy and Paleoecology*; University of Chicago Press: Chicago, IL, USA, 1981; p. 338.
3. Brain, C.K. *The Hunters or the Hunted? An Introduction to African Cave Taphonomy*; University Chicago Press: Chicago, IL, USA, 1981; p. 365.
4. Shipman, P. *Life History of a Fossil: An Introduction to Taphonomy and Paleoecology*; Harvard University Press: Cambridge, UK, 1981; p. 223.
5. Andrews, P. *Owls, Caves and Fossils*; Natural History Museum Publications: London, UK, 1990; p. 231.
6. Lyman, R.L. *Vertebrate Taphonomy*; Cambridge University Press: Cambridge, UK, 1994; p. 524.
7. Fernández-Jalvo, Y.; Andrews, P. *Atlas of Taphonomic Identifications*; Vertebrate Paleobiology and Paleoanthropology Series; Springer: Dordrecht, The Netherlands, 2016; p. 359.
8. Andrews, P. Taphonomic effects of faunal impoverishment and faunal mixing. *Palaeogeogr. Palaeoclimatol. Palaeoecol.* **2006**, *241*, 572–589. [CrossRef]
9. Domínguez-Rodrigo, M.; Musiba, C.M. How accurate are palaeoecological reconstructions of early paleontological and archaeological sites? *Evol. Biol.* **2010**, *37*, 128–140. [CrossRef]
10. Fernández-Jalvo, Y.; Scott, L.; Andrews, P. Taphonomy in palaeoecological interpretations. *Quat. Sci. Rev.* **2011**, *30*, 1296–1302. [CrossRef]
11. Belyea, L.R. Revealing the emperor's new clothes: Niche-based palaeoenvironmental reconstruction in the light of recent ecological theory. *Holocene* **2007**, *17*, 683–688. [CrossRef]
12. Nógues-Bravo, D. Predicting the past distribution of species climatic niches. *Glob. Ecol. Biogeogr.* **2009**, *18*, 521–531. [CrossRef]
13. Huntley, B. Reconstructing palaeoclimates from biological proxies: Some often overlooked sources of uncertainty. *Quat. Sci. Rev.* **2012**, *31*, 1–16. [CrossRef]
14. Birks, H.J.B.; Heiri, O.; Seppä, H.; Bjune, A.E. Strengths and weaknesses of quantitative climate reconstructions based on Late-Quaternary biological proxies. *Open Ecol. J.* **2010**, *3*, 68–110. [CrossRef]
15. Sandweiss, D.H.; Kelley, A.R. Archaeological contributions to climate change research: The archaeological record as a paleoclimatic and paleoenvironmental archive. *Annu. Rev. Anthropol.* **2012**, *41*, 371–391. [CrossRef]
16. Lyman, R.L. Paleoenvironmental Reconstruction from Faunal Remains: Ecological Basics and Analytical Assumptions. *J. Archaeol. Res.* **2017**, *25*, 315–371. [CrossRef]
17. Escosteguy, P.D.; Fernandez, A.E.; González, M.I. Insect Marks on Bones from La Guillerma Archaeological Locality (Salado River Depression, Buenos Aires, Argentina). *Quaternary* **2021**, *4*, 45. [CrossRef]
18. Lebreton, L.; Morin, E.; Gravina, B.; Michel, A.; Bachellerie, F.; Beauval, C.; Flas, D.; Laroulandie, V.; Marot, J.; Rougier, H.; et al. The Potential of Micromammals for the Stratigraphy and the Timing of Human Occupations at La Roche-à-Pierrot (Saint-Césaire, France). *Quaternary* **2021**, *4*, 33. [CrossRef]
19. Durocher, M.; Grouard, S.; Nicolas, V.; Maestri, R.; Evin, A. Defining fragmentation patterns of archaeological bone remains without typologies: A landmarks-based approach on rodent mandibula. *Quaternary* **2022**, *5*, 14. [CrossRef]
20. Fernández-Jalvo, Y.; Rueda, L.; Fernández, F.; García-Morato, S.; Marin-Monfort, D.; Montalvo, C.; Tomassini, R.; Chazan, M.; Horwitz, L.K.; Andrews, P. Understanding the Impact of Trampling on Rodent Bones. *Quaternary* **2022**, *5*, 11. [CrossRef]
21. Okaluk, T.R.; Greenfield, H.J. Macroscopic chop mark identification on archaeological bone: An experimental study of chipped stone, ground stone, copper, and bronze axe heads on bone. *Quaternary* **2022**, *5*, 15. [CrossRef]
22. Stoetzel, E.; Bochaton, C.; Bailon, S.; Cochard, D.; Gala, M.; Laroulandie, V. Multi-Taxa Neo-Taphonomic Analysis of Bone Remains from Barn Owl Pellets and Cross-Validation of Observations: A Case Study from Dominica (Lesser Antilles). *Quaternary* **2021**, *4*, 38. [CrossRef]
23. Powley, M.J.; Sutisna, I.; Mikac, K.M.; Wibowo, U.P.; van den Bergh, G.D. The Stegodon Bonebed of the Middle Pleistocene Archaeological Site Mata Menge (Flores, Indonesia): Taphonomic Agents in Site Formation. *Quaternary* **2021**, *4*, 31. [CrossRef]
24. Wattanapituksakul, A.; Shoocongdej, R.; Conrad, C. Terminal Pleistocene to Middle Holocene Hunter-Gatherer Subsistence at Ban Rai Rockshelter, Northwest Thailand. *Quaternary* **2022**, *5*, 51. [CrossRef]

25. Daura, J.; Sanz, M.; Vaquero, M.; López-García, J.M.; Blain, H.A.; Sánchez Marco, A. Neanderthal fossils, mobile toolkit and a hyena den: The archaeological assemblage of Lateral Gallery 1 in Cova del Gegant (NE Iberian Peninsula). *Quaternary* **2022**, *5*, 12. [CrossRef]
26. Rofes, J.; Ordiales, A.; Iriarte, E.; Cuenca-Bescós, G.; Galindo-Pellicena, M.Á.; Pérez-Romero, A.; Carretero, J.M.; Arsuaga, J.L. Human Activities, Biostratigraphy and Past Environment Revealed by Small-Mammal Associations at the Chalcolithic Levels of El Portalón de Cueva Mayor (Atapuerca, Spain). *Quaternary* **2021**, *4*, 16. [CrossRef]
27. Royer, A.; Mallye, J.-B.; Pelletier, M.; Griselin, S. Who Killed the Small Mammals of Ittenheim? An Integrative Approach and New Taphonomic Data for Investigating Bone Assemblages Accumulated by Small Carnivores. *Quaternary* **2021**, *4*, 41. [CrossRef]
28. Tilby, E.; Miracle, P.; Barker, G. The Microvertebrates of Shanidar Cave: Preliminary Taphonomic Findings. *Quaternary* **2022**, *5*, 4. [CrossRef]
29. Rufà, A.; Laroulandie, V. Unravelling the Taphonomic Stories of Bird Bones from the Middle Pleistocene Layer VIII of Grotte Vaufrey, France. *Quaternary* **2021**, *4*, 30. [CrossRef]
30. Rubinatto Serrano, J.; Valejo-Pareja, M.C..; deFrance, S.D.; Baitzel, S.I.; Goldstein, P.S. Taphonomic and Paleoecological Insights from Anurans on Tiwa-naku Sites in Southern Peru. *Quaternary* **2022**, *5*, 16. [CrossRef]
31. Arroyo-Cabrales, J.; Johnson, E.; Cruz, J.A. San Josecito Cave and Its Paleoecological Contributions for Quaternary Studies in Mexico. *Quaternary* **2021**, *4*, 34. [CrossRef]
32. Uzunidis, A. Middle Pleistocene Variations in the Diet of Equus in the South of France and Its Morphometric Adaptations to Local Environments. *Quaternary* **2021**, *4*, 23. [CrossRef]

quaternary

MDPI

Article

Insect Marks on Bones from La Guillerma Archaeological Locality (Salado River Depression, Buenos Aires, Argentina)

Paula D. Escosteguy [1,*], Alejandro E. Fernandez [1] and María Isabel González [2]

1 CONICET, Instituto de Arqueología, Facultad de Filosofía y Letras, Universidad de Buenos Aires, Buenos Aires PC C1002ABE, Argentina; alejandro.fernandez@uba.ar
2 Instituto de Arqueología, Facultad de Filosofía y Letras, Universidad de Buenos Aires, Buenos Aires PC C1002ABE, Argentina; igonzale@filo.uba.ar
* Correspondence: pdescosteguy@uba.ar

Abstract: The La Guillerma archaeological locality is located in the northeast sector of Buenos Aires province (Argentina). Two of its sites (LG1 and LG5), dated between ca. 1400- and 600-years BP, have a great amount of faunal remains including deer, rodents, fish and small birds that are subjected to taphonomic agents and processes (e.g., weathering, manganese, roots). Previous studies have shown osteophagic behaviour in different insects (e.g., Coleoptera, Blattodea). In this paper, we evaluate their incidence on La Guillerma faunal assemblage. We performed an analysis on marks that were identified in bone remains of various taxa and applied the criteria for identifying bone alteration by insects (i.e., by measuring each trace and comparing them with the types of insect marks described in the literature). Fifteen specimens (LG1 = 6 and LG5 = 9) exhibited different types of modifications (e.g., pits with striae in base, pits with emanating striae, striations) that are related to the action of insects. Although the proportion of affected bones is low in relation to the total sample, we highlight our study as the first detailed analysis of insect marks on archaeological bones from Argentina. We also emphasize the significance of addressing insect-produced modifications on Argentinean archaeological sites.

Keywords: taphonomy; faunal remains; pampas; late Holocene; bone damage

1. Introduction

Many taphonomic processes may modify animal carcasses and bones between the time of death and the burial (e.g., [1–3]). Also, some parasites can modify bone tissue during the life of the animal [4,5]. The analysis of surface modifications on bone remains and the recognition of the different agents involved have exponentially grown during the last few decades. This identification is primarily supported by the experience observing fossil and modern sites and by experimental studies (e.g., [1–3]). However, insects have not been considered as a major taphonomic agent until recently [6–10]. In fact, insect damage has often been inferred in the fossil record with no comparative data. Thus, laboratory experiments on this issue require urgent attention [2,11].

Recent studies have shown that analysis of osteophagus insects had great importance in the interpretation of the bioarchaeological record since their damage could be misinterpreted either as cultural modifications or as pathologies [8,12,13]. Other studies emphasize the considerable contribution that insect activity plays in the site formation processes and the alteration or destruction of faunal remains [14].

The aim of this article is to assess the incidence of insects as taphonomic agents that affected the archaeofaunal assemblages of the La Guillerma archaeological locality and to evaluate their importance as a formation process in sites from the Salado River Depression. This area is located in the Argentinean pampas and is drained by the middle and lower course of the river (Figure 1). The chronological dates situate the occupations in the late Holocene, between 2400- and 400-years BP, and the archaeological contexts show a varied repertoire of technological and subsistence remains [15–17]. Based on the use of resources, the settlement pattern and technology, these groups have been characterized as hunter-gatherer-fishermen with pottery [18].

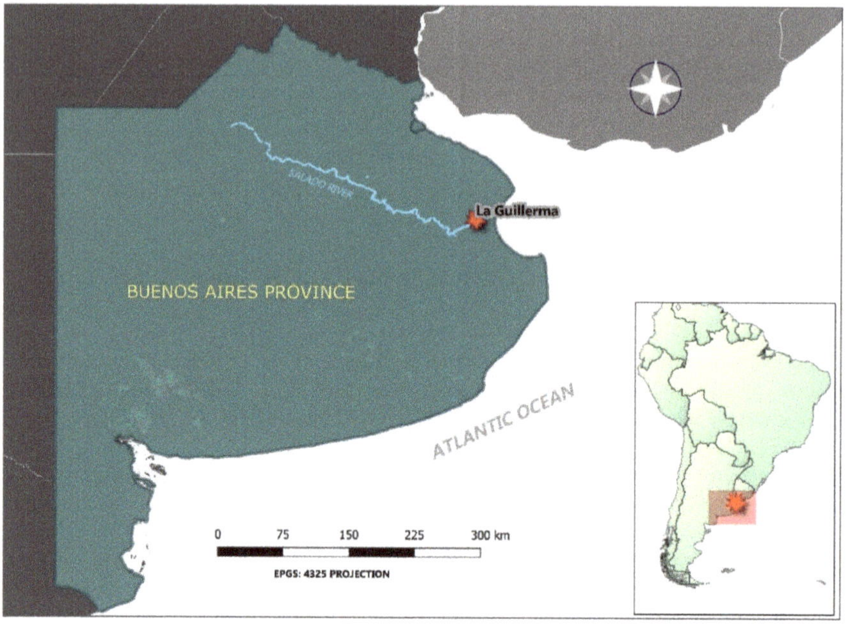

Figure 1. Location of La Guillerma archaeological locality.

2. Background

2.1. Research Area

The Salado River Depression is located northeast of the province of Buenos Aires and has a temperate and humid climate. Its hydrographic network, formed by the Salado River and several shallow lakes, configures a wetland and lentic environment [19]. To the northeast of the river there are soft hills covered by grasslands. Near the northeast edge, there are *talares* (*Celtis tala*) arranged as isolated crescent-shaped archipelagos [20]. Although climatic fluctuations have been detected in several regions of southern South America during the late Holocene, the palaeoenvironmental, palaeoclimatic and palaeoecological reconstruction of the pampas is still tentative and yet not calibrated for interpretation on an archaeological scale [21].

An interdisciplinary study on the Late Quaternary of Salado River middle basin [22], which considered stratigraphical, geomorphological and paleontological data, proposed a model of palaeoenvironmental reconstruction that distinguishes dry conditions between 8000 and 3000 years BP, followed by a gradual return towards wet conditions from 3000 years BP to present time. This last period includes a moment of water deficit conditions circa 1800 years BP, represented by a short depositional hiatus [22]. The data available for the first stages of European arrival indicate more arid conditions, which correspond to the Little Ice Age (approximately 1450–1850 AD). After this, a more humid period followed. During the 19th and 20th centuries, rains and floods became more frequent [23].

The La Guillerma archaeological locality was inhabited during the late Holocene and includes five sites, distributed over a series of elevations along the left margin of the Salado River, at approximately 40 m from its bed. LG1 and LG5 sites (which are 100 m apart) have a great abundance of ceramic sherds, stone artifacts, and plant and animal remains [15,17,

18]. The dates made on these materials showed human occupations between ca. 1200- and 600-years BP for LG1 and ca. 1400- and 660-years BP for LG5 (for further information, see Table 1 and Figure 3 in [15]). The soils show a marked horizonation, with a clay-sandy silt texture, and the pH of the surface horizons is lightly acid. However, LG5 has better internal drainage conditions that favours the bone remains conservation [18]. These are shallow sites, given that archaeological remains are predominantly deposited in the first 30 cm of the soil and are subjected to pedogenetic processes. Among them, a high biological activity including the action of roots, burrowing animals, and microfauna causes the dispersion or concentration of materials [18,24].

Regarding the archaeofaunal record, mammals include deer (*Blastocerus dichotomus* and *Ozotoceros bezoarticus*), rodents (coypu (*Myocastor coypus*), guinea pig (*Cavia aperea*), tucu-tucu (*Ctenomys* sp.), among others) and undetermined armadillos. Among fish, we identified catfish (*Rhamdia quelen*), tiger fish (*Hoplias argentinensis*), mullet (*Mugil platanus*) and black drum (*Pogonias courbina*). Small birds, such as *Eudromia elegans*, *Fulica* sp. and other undetermined Anatidae are also predominant. The estimations of NISP were: LG1 = 2486 and LG5 = 6049 [16,18].

Taphonomic analyses carried out on the archaeofaunal assemblages, following the criteria described by Lyman [3] and Fernandez-Jalvo and Andrews [2], showed the action of different agents and processes that altered bones from both sites but in low proportion. In LG1, roots affected 5.3% (n = 132) whereas the presence of manganese oxide deposits was observed on 5.39% (n = 134). In this site, low stages of weathering (*sensu* Behrensmeyer [25]) were registered on 6.31% (n = 157) of the specimens, being more present on big mammals' bones. In less percentages, we registered the action of rodents (0.68%, n = 17) and carnivores (0.12%, n = 3). In LG5, modifications generated by roots (decolouring and dendrite marks) were observed on 2% of bones (n = 121). In similar rates (2.11%, n = 128), manganese produced dark spots. In greater proportion (3.38%, n = 205), we observed weathering, mainly stages 1 and 2 (*sensu* Behrensmeyer [25]). Only one specimen displayed a puncture made by carnivore (0.01%), and 0.43% (n = 26) of bones exhibited traces generated by rodents. Sometimes, these marks obliterated evidence of anthropic manipulation, which is also in low proportion [18,26–28]. In spite of the recognition of anthills and insect burrows during excavations, along with changes in colour and texture in discrete sectors of the sedimentary context, no insect marks had been identified.

2.2. Insects as Taphonomic Agents

Insects are important soil formation agents, due to their colonizing capacity in a variety of substrates and their ability to rework sediments and use organic matter as raw material to make nests [29]. Ants, termites and other taxa can mix and modify soil and, in the process, affect an archaeological site and its content [30]. In particular, McBrearty [31] noted that termite bioturbation activity in African sites modified the soils' properties and produced pseudofeatures as well as the movement of materials.

Apart from insect remains and associated features, their presence can also be detected by the modifications made on the bone record. Recent studies have shown that analysis of osteophagus insects had great importance in the interpretation of the bioarchaeological record, since their damage could be misinterpreted as cultural modifications or as pathologies [8,12]. Other studies emphasize the contribution that insect activity plays in the site formation processes and the alteration or destruction of faunal remains [14] and its utility to reconstruct past environmental and climatic conditions [6,32].

A great variety of insects can produce damage to bones in many ways. Among these, beetles (Dermestidae, Tenebrionidae and Scarabaeoidea), termites (Kalotermitidae and Termitidae), ants (Formicidae), moths (Tineidae), blowflies (Calliphoridae), mayflies (Polymitarcyidae) and cockroaches (Blattidae) have been registered in Buenos Aires province [33,34].

Dermestid beetles can produce striations, extensive pitting, and edge gnawing [14]. They can also make deep conical perforations when penetrating into the bone [7]. Holden et al. [35] found in an experiment that both dermestid and tenebrionid larvae leave different kinds of damage in different stages of carcass decay, so they conceive that other beetle families may modify bone as well. Also, several researchers have described either archaeological or palaeontological examples of bone alteration by termites [6,10,36,37]. Experimental studies have shown that they produce striations, boreholes, edge gnawing, pits with smooth base, pitting, destruction, smoothed abrasion, polish, surface removal and surface coating. The last two types of damage are more frequent [14]. Cockroaches have been noted as another potential agent of bone alteration, as they colonize corpses. Parkinson [32] made experiments with the species *Periplaneta americana*, which resulted in three types of modification, namely discolouration, destruction and gnawing.

The specific climatic requirements for each kind of insect have also been addressed. The cosmopolitan species *Dermestes maculatus* needs a minimum temperature of 20 °C to feed and is more active during the warmer months [32]. In regards to termites, Kalotermitidae need very little moisture to survive, while Termitidae live underground since the soil has the capacity to hold water for a long period and keep the colony moist [38]. *Trinervitermes trinervoides*, a species of this last family, shows an increased activity during autumn and spring [6]. Cockroaches, on their part, have adapted to highly diverse thermal climatic ranges but prefer a warm and moist environment [32].

Other taxa have been noted to presumably modify bone. Go [39] analyzes a case where a colony of ants nested inside human bone and modified it. Moth larvae tunneling is common on horns and antlers [2], though it has not been registered on bone. Larval blowflies may also produce small bores when they create pupation chambers [6,35]. Finally, mayfly nymphs make U-shaped burrows on wood, and similar traces reported in bone have been attributed to them [7].

3. Materials and Methods

The zooarchaeological assemblages from LG1 and LG5 are curated at the Instituto de Arqueología (Facultad de Filosofía y Letras, Universidad de Buenos Aires), where reference collections of diverse taxa (birds, mammals and fishes) are also available. These have been used for taxonomic and anatomical identification. As stated in the Background section, the samples were LG1 = 2486 and LG5 = 6049 and several taphonomic agents, and processes such as weathering, rodents, and roots of different sizes have modified the faunal assemblages [16,26,27]. However, fifteen specimens (six from LG1 and nine from LG5) displayed traces that had not been recognized previously.

We examined these marks using a high-intensity light from a two-led illuminator and a low-power binocular microscope (Zeiss STEMI 305) with a magnification up to 40×. We took note of their position on each specimen and measured them using the tools provided by ZEN Blue edition software. We measured the diameter of holes, bores and pits. On those specimens where the marks overlapped, we registered the length of the affected area.

We considered the proposals of Pirrone and collaborators [40], Parkinson [11,32] and Backwell and co-authors [6,14] to analyze insect traces. For the description of the general morphology of traces, we distinguished between pits, holes, chambers, tubes/bores, furrows and grooves/striae. To broaden the analysis of traces, we followed Backwell and co-authors' criteria and considered different types of pits: pit with striae in base, pit with smooth base, and pit with emanating striae [14]. We also regarded other modification features such as pitting, etched surface, destruction, dissolution and the presence of surface coating/discolouring [6,14,32].

4. Results

After our analysis, the fifteen bone specimens displayed, at least, one of the traces associated with insect activities (Table 1, Figure 2). These values represent respective low %NISP: LG1 = 0.24% and LG5 = 0.14%.

Table 1. Traces on bones from LG1 and LG5 sites. * Diameter; ** Affected area.

Site	#	Taxa	Bone	Type of Modification	Measure	Observations
LG1	201	Mammalia	Long bone	Pit with emanating striae	4.25 mm *	Figure 10
				Surface residue	–	
				Etched appearance	–	
	5	Mammalia	Long bone	Striations	4.89 mm **	Edge gnawing around nutrient foramen. Figure 8B
	340	Mammalia	Long bone	Pit with striae in base	2.41 mm *	Figure 3C
	4	Cervidae	Long bone	Hole	4.06 mm *	A smooth-based boring. Figure 9B
				Pit with striae in base	3.28 mm *	
	217	*Ozotocerus bezoarticus*	Tibia	Hole	2.66 mm *	A smooth-based boring. Figure 9C
	362	*Ozotocerus bezoarticus*	Phalange 1	Bore	2.93 mm *	Bore hole. Figure 9A
				Etched appearance	–	
LG5	205	Aves	Long bone diaphysis	Pit with striae in base	2.68 mm *	Foraging area. Overlapping of marks. Affected post weathering. Figure 5
				Pit with emanating striae	5.12 mm **	
				Striations	–	
	220	Aves	Femur	Pit with emanating striae	3.52 mm *	Affected post weathering. Figure 4A
				Surface residue	–	
				Striations		
	2005	Mammalia	Tibia	Pit with striae in base	2.90 mm *	Overlapping of marks. Figure 3B
				Etched appearance	–	
	3713	*Cavia aperea*	Femur	Pit with striae in base	2.81 mm *	Overlapping of marks. Figure 3A
				Striations	–	
	1623	*Myocastor coypus*	Humerus	Pit with emanating striae	4.73 mm **	Overlapping of marks. Fissure along the bone. Figure 6
				Pit with striae in base		
				Striations		
	5332	*Myocastor coypus*	Humerus	Pit with emanating striae	3.38 mm *	Striae on base. Figure 7
	5519	*Myocastor coypus*	Radius	Pit with striae in base	3.10 mm *	Overlapping of grooves. Figure 3D
				Striations	–	
	5773	*Myocastor coypus*	Tibia	Pit with emanating striae	2.67 mm *	Figure 4B
				Striations	–	
	14	*Ozotocerus bezoarticus*	Humerus	Striations	3.35 mm **	Overlapping of grooves. Fissure along the bone. Figure 8A

Surface pit is the most frequent type of modification (Figure 2). It was recorded on mammals and bird bones. We distinguished between pits with striae in base (Figure 3) and pits with emanating striae (Figure 4). We identified the former type in two specimens from LG1 and five from LG5, while the latter category was displayed on one specimen from LG1 and five from LG5. The diameter of pits with striae in base ranged from 2.41 mm to 3.59 mm. The pits with emanating striae are formed by multiple overlapping grooves that are radially arranged. Their diameter measured between 2.67 mm and 4.25 mm.

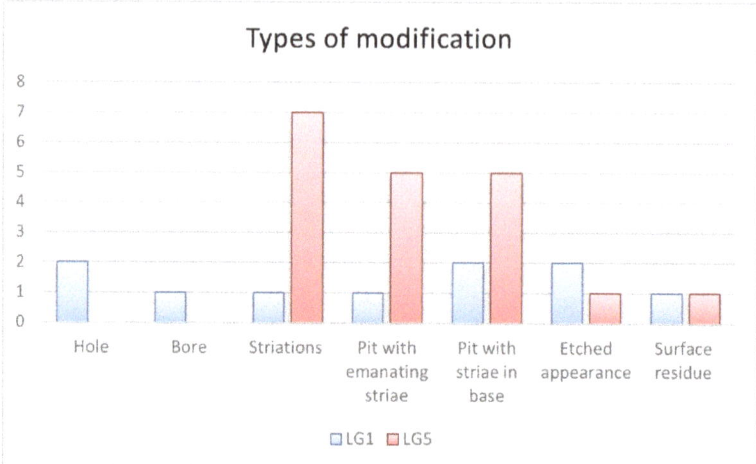

Figure 2. Types of modification on LG1 and LG5 sites.

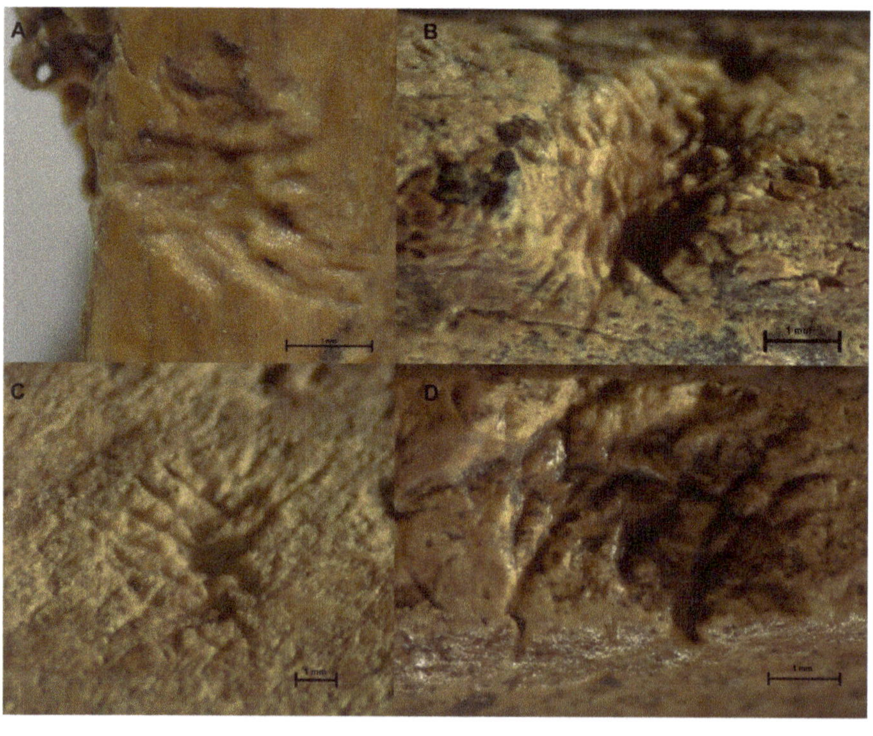

Figure 3. Pits with striae in base. (**A**) Pit on *C. aperea*'s femur (Id. 3713). (**B**) Pit and etched appearance on a mammals' tibia (Id. 2005). (**C**) Pit on a fresh fracture surface (Id. 340). (**D**) Pit on a coypu radius (Id. 5519).

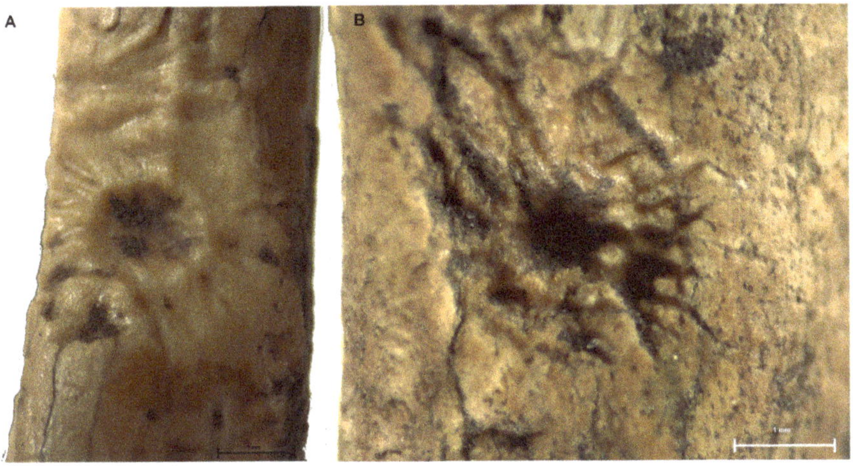

Figure 4. Pit with emanating striae. (**A**) Pit, multiple striations and residue on bird bone (Id. 220). (**B**) Deep pit on a coypus' tibia (Id. 5773).

Two specimens (Id. 205 and 1623) displayed both types of pits. One of them is a long bone of a small-sized bird (Id. 205). This diaphysis exhibits two modified sectors. In one of them, there is an area that displays a shallow pit with striae in base. In addition to this, there are multiple parallel and sub-parallel striations (Figure 5A). The other sector shows overlapping of pits with emanating striae along 5.12 mm (Figure 5B). These marks, composed of multiple grooves radially arranged around a cavity, seem to have been made after the exposure of the bone to aerial conditions; it exhibits weathering on the rest of the cortical surface but not over these marks. The other bone is a coypu's humerus (Id. 1623) that is longitudinally fissured. In its medial shaft, it displays the overlapping of at least a surface pit with striae in base and a pit with emanating striae. Besides these pits, there are clusters of randomly oriented sub-parallel striations within an area of 4.73 mm. We highlight that these marks are previous to the crack, and their deepest part is along it (Figure 6A,B).

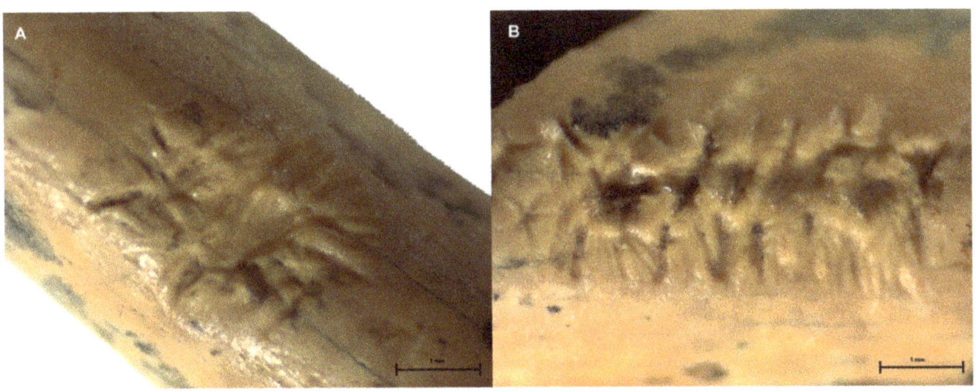

Figure 5. Pits on a bird bone (Id. 205). (**A**) Pit with striae in base and multiple striations. (**B**) Overlapping of pits with emanating striae.

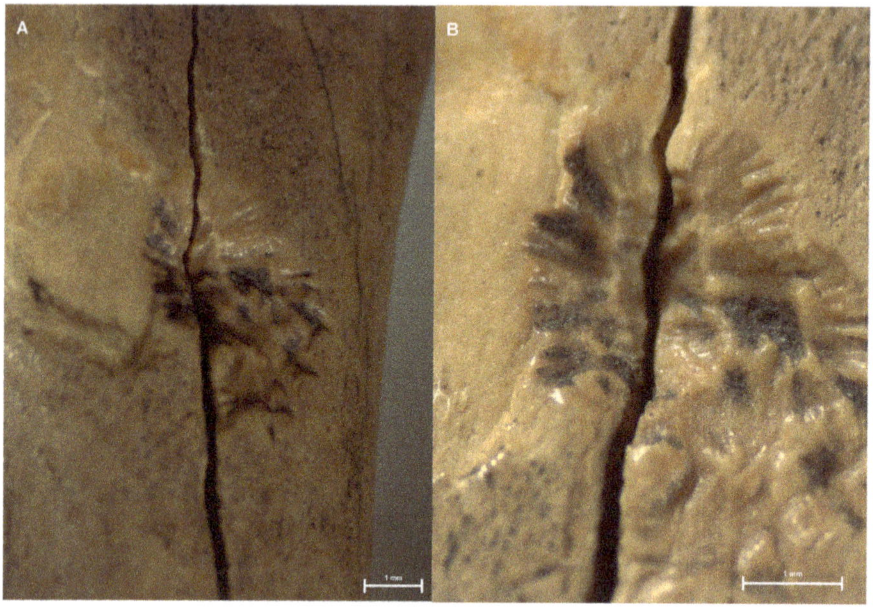

Figure 6. (**A**) Pits with emanating striae over the crack, pit with striae in base, and striations on coypu's humerus (Id. 1623). (**B**) Details of the traces and the crack.

In some cases, the pits penetrate profoundly the cortical bone. On a coypu's tibia (Id. 5773), there is a pit with emanating striae on its distal diaphysis. It is formed by multiple striations radially arranged around a deep center (Figure 4B). Notably, we identified pits that combined attributes, such as the trace displayed on the cortical surface of a coypu humerus (Id. 5332). This is composed of numerous parallel and sub-parallel striations radiating from around the outer circumference of a shallow cavity. The last one has several random grooves that overlap each other (Figure 7A,B).

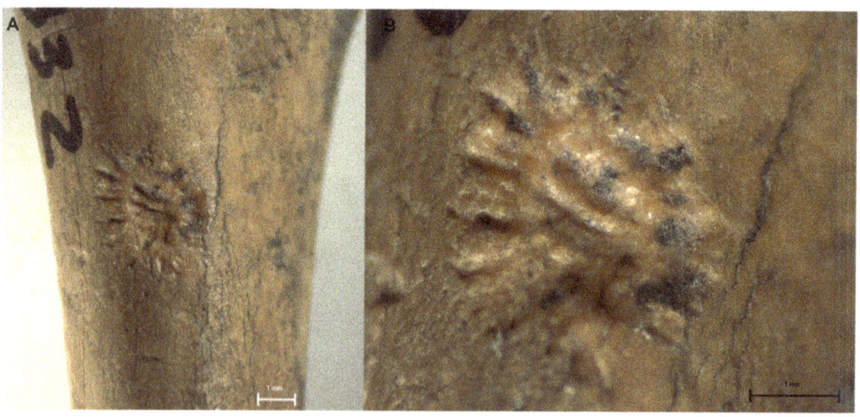

Figure 7. (**A**) Pit with emanating striae and striae in base on coypu's humerus (Id. 5332). (**B**) Details of the traces.

One remarkable issue is the record of this type of traces on weathered bird bones: the previously mentioned (Id. 205) and the shaft of a femur (Id. 220), which displays a pit with a smooth central cavity that has faint striations. Next to this trace, there are numerous sub-parallel striations (a few of them overlapped). It also has brown residue spots on the cortical surface. However, the area affected by the grooves that compose the marks is remarkably lighter, which means that the cracks produced by weathering are previous to the traces (Figure 4A).

Numerous striations were observed, being more frequent in LG5 (LG1 n = 1; LG5 n = 7) (Figure 2). They usually appeared overlapping (Figure 8A); and in specimen Id. 14, the affected area is crossed by a fissure that runs longitudinally along the bone. Besides, on specimen Id. 5 from LG1, grooves are parallelly disposed and transversely orientated around the edge of a nutrient foramen. These marks are visible to the naked eye and the affected sector measured 4.89 mm (Figure 8B). In general, this type of damage occurs associated with other traces such as pits. One example is a femur of guinea pig (*C. aperea*; Id. 3713) that shows a surface pit with overlapped striations in its base and a few parallel striations besides it (Figure 3A), on the proximal sector of the bone's shaft.

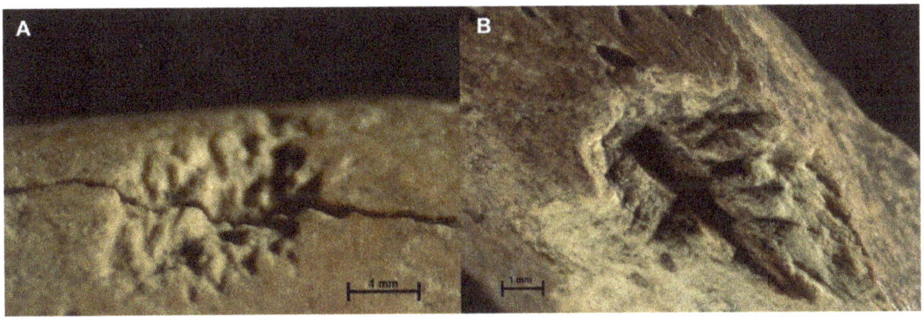

Figure 8. (**A**) Striations on deer's humerus (Id. 14). (**B**) Striations around a nutrient foramen (Id. 5).

We recorded a bore and two circular holes on big mammal bones from LG1 (NISP = 2). The first phalange of pampas deer (*O. bezoarticus*; Id. 362) has a bore (Figure 9A) measuring 2.93 mm in diameter. It is located on the distal end of the phalange and penetrates the bone through to the medullary cavity. Besides, we observed the removal of outer lamellae around its edge and the exposure of the underlying bone structure. The other two holes have smooth bases, and their diameters are 4.06 mm and 2.66 mm (Figure 9B,C). One of them (Id. 4) shows a pit with striae in base next to the deep hole.

The etched appearance is another type of surface damage observed on scarce fragmented bones from La Guillerma sites. We observed it surrounding the aforementioned bore (Id. 362; Figure 9A) and around pits in Id. 2005 and Id. 201, exposing the underlying bone structure (Figures 3B and 10A,B).

Finally, the surface coating was also recorded in association with other traces. In a specimen from LG1, this modification is present as a dark brown coating that coloured the bone and is combined with etching around a pit (Id. 201, Figure 10), while a bird's femur (Id. 220, Figure 4A) has a brown coating that discoloured the cortical surface.

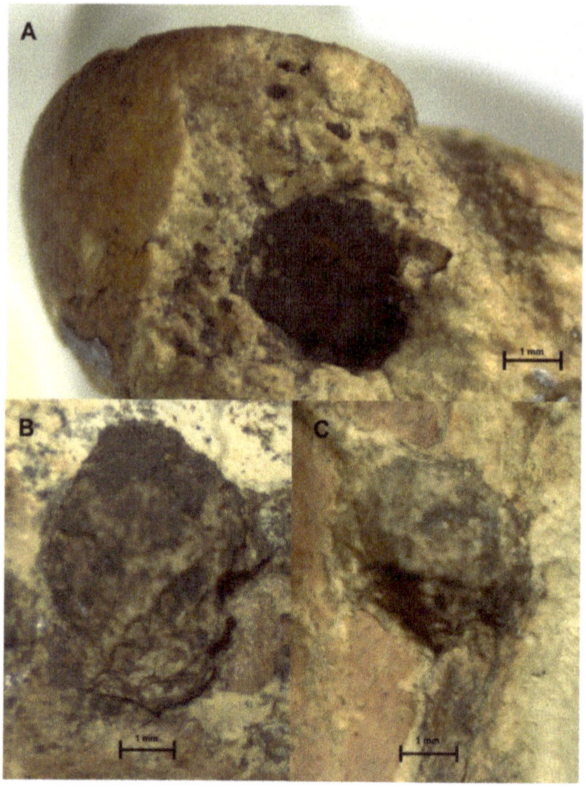

Figure 9. (**A**) Borehole on phalange and etched surface around it (Id. 362). (**B**) Hole on a long bone fragment (Id. 4). (**C**) Hole on a deer's tibia (Id. 217).

Figure 10. (**A**) Pit with emanating striae and surface coating with removal of outer lamellae (Id. 201). (**B**) Detail of the etched appearance.

5. Discussion

As a result of our study, we identified seven types of modifications associated with insect activities: pits with striae in base, pits with emanating striae, bore/tube, holes, striations, etched appearance, and surface coating/discolouring. Now we will discuss each of these types, their possible etiology, and their implications for the integrity of the osteoarchaeological assemblage of La Guillerma. In the first place, both pits with striae in base and with emanating striae were recorded by Backwell and co-authors [14] in their experimental observation of trogid beetles. On the contrary, they were not observed during actualistic research with termite species *Odontotermes badius*. However, the authors suggested that the lack of pits with striae and furrowing could be related to the length and time of exposure; they conducted the experiment in winter when the insects were less active. In this sense, in a previous research project, Backwell and collaborators [6] registered that *Trinervitermes trinervoides* (after six months of trial) had generated surface pits (semi-circular depressions with radially arranged edge gnawing that comprised overlapping individual sub- and parallel striations) and star-shaped marks (several individual grooves radially arranged around a cavity with smooth walls). Parkinson [32] also identified three types of surface pits on bones after a set of experiments with *Dermestes maculatus*: (1.) highly variable shallow depressions with striations radiating around the outer circumference; (2.) semi-circular shallow depressions with randomly orientated striations over them; and (3.) irregular shaped depressions not associated with gnawing striations. Likewise, Holden et al. [35] recorded small and medium circular pits with concave bases that were created by the same insect.

With respect to the traces' depth, one of the deep pits (Id. 1623) had a fissure across it. Pomi and Tonni [37] also recorded insect traces lying over cracks on the bones and proposed that the marks were developed after the opening of the cracks. Rather, in our case, we believe that the insect chewing could have weakened the bone, making it more prone to fragmentation by other taphonomic agent or process (e.g., roots, burrowing animals), and thus generating a deep crack. We support this idea on the basis that the recorded striations seem to be cut by the crack, but they continue on each side. Otherwise, trace's depth is a variable that could be related to the presence of soft or desiccated tissues at the moment of trace's construction [8,11]. Another pit with emanating striae on a coypu bone (Id. 5773) has a deeper centre that could have been produced while anchoring an apical tooth. On this matter, Backwell et al. [6], after observing signs of wear damage on the medial side of the apical tooth of *T. trinervoides* worker mandibles, proposed that "star-shaped marks radiate around a well-defined central cavity that provides anchorage for one half of the mandible, while the other half excavates the bone using an inward gesture" [6] (p. 83). These authors observed that *T. trinervoides*' mandibles are asymmetrical, with a single apical tooth that produces individual striations. In this sense, Britt and collaborators [7] proposed that the parallel grooves were generated by Dermestidae since their mandibles had two widely-spaced apical teeth. Nonetheless, Holden et al. [35] registered that larval Tenebrionidae had similar mandibles with two apical teeth. Therefore, they assert that mandibular marks are difficult to assign since tooth structure may be similar among families and vary between genera and species.

Some authors suggest that different traces represent transitional morphotypes [7,41]. Therefore, if the activity of the insect had continued, pits would have become holes or boreholes. For that matter, Backwell et al. [6] considered that star-shaped marks and surface pits were probably incipient boreholes made by termites searching for lipids. These features were macroscopically visible in their analysis, ranging between 3.41 and 4.20 mm in length and between 2.83 and 2.64 mm in width (with the exception of a borehole that measured 6.09 by 3.06 mm) and holes with diameters between 1.54 and 3.63 mm. For *D. maculatus*, there is a variable register of sizes in different experiments. Those infrequent boreholes recorded by Parkinson [32] tended to be elliptical and small (less than 0.68 mm long). Thus, the author asserted that they probably represent surface pits created by larvae which penetrated into the cortical and trabecular bone, then into the medulla. Their distribution would be related to the exploitation of haversian canals and marrow access and are not connected to the creation of pupal chambers. Holden et al. [35] registered bores that ranged between 2.5 and 3.5 mm in diameter after their live experiments with dermestids. Also, Zanetti et al. [42,43] documented that *D. maculatus* feeds on bone and in the process generates diverse traces (e.g., spherical or ovoid depressions for refuge and pupation). They observed that, in some cases, the largest marks measure more than one centimeter.

Regarding bore and holes, the unique bore recorded in the sample perforates the bone through to the medullae channel on the distal end of a deer phalange. This trace and the two holes range ca. 2.7 to 4.1 mm in diameter. Taking these measures into account, the diameters of bores and holes generated by dermestids and termites overlap. Those recorded on La Guillerma sample have a range of sizes that are consistent with those registered by the other researchers [6,35] for both termites and dermestids, so this data would not allow us to attribute the modification to one or the other of these agents. In relation to the site of emplacement, Backwell and co-authors [6] observed a similar result during the experimental research with termites, as they discerned a predominance of boreholes on proximal or distal ends. However, Parkinson [32] noted that marks produced by dermestids tended to be located either on the epiphysis or close to the diaphysis-epiphysis junction and not on the dense cortical bone areas. Thus, these two types of marks that we have registered in La Guillerma could have been generated by different insects or, as Parkinson [11] remarks, by the same insect in different larval developmental stages. Besides, they could also be a consequence of different behaviours (feeding or pupation), but it would be too speculative to state one of them from such a small sample.

Striations are more frequent in our sample, especially as individually discernible traces. They are considered as the early forms of bone destruction [6]. We highlight the case of a bone fragment from LG1 that displays parallel striations around a nutrient foramen (Id. 5). Britt and co-authors [7] proposed that the concentration of traces around nutrient foramina represent attempts to widen the opening. They interpreted that these areas and articular bones were bored to consume marrow. They also suggested that the insects may have consumed the cartilage and ligaments of articular ends prior to bone consumption. Nevertheless, the modification of bones by dermestids would not be exclusively related to diet. It could be a consequence of the exploration of available ways for the creation of pupation chambers, a frequent behaviour in insects [32]. Holden et al. [35] observed that larvae of both dermestids and tenebrionids expanded foramina. In the case of termites, gnawing around vascular canals would have provided traction for the apical teeth [6].

Additionally, we observed the etched surface appearance on three mammal bones (LG1 = 2; LG5 = 1) that exposed the underlying bone structure. Backwell et al. [6] mentioned, after their experiments with termites, that this acid-etched appearance is a result of their saliva or faecal rich material. But this damage was also recorded on experiments with trogid beetles [14]. We discerned the combination of etched bone with surface residue on a bone fragment from LG1 (Id. 201). However, we distinguished this dark brown coating on another avian bone from LG5 (Id. 220). Noteworthy, we did not recognize the latter as an isolated trace but rather we emphasise the difficulty to identify it on assemblages that are affected by manganese stains or burning that could obliterate this residue or make researchers misinterpret it [6,32]. The discolouration of the bone surface was also stated by Parkinson [32] as an evidence of *P. americana*'s action, although it was not associated with other damages on cortical bones.

We emphasise the presence of patches and lineation of traces (e.g., on two areas of Id. 205). According to Britt et al. [7], the former corresponds to foraging areas and the latter represents galleries. Those authors also mentioned, referring to termites, that "the large pits are composites of small pits related to broad surface foraging/mining" [7] (p. 65);

so, they do not represent the size of the tracemaker. In this sense, within our sample, we frequently recognized the overlapping of multiple traces as grooves or pits.

As insects can access carcasses at different stages of decay, it has also been observed that termites and dermestids alter fresh, dry, weathered and fossilized bones [6,14,32]. Our analysis points to the action of insects on different bone stages: two avian bones distinctly display evidence of light weathering over the cortical surface except on traces. Moreover, Zanetti and co-authors [43] found that dermestids consumed and marked burned tissues. The selection of bones at different diagenetic stages could be related to the consumption of various trace elements (calcium, phosphorus and nitrogen), proteins and lipids [6].

In summary, insect traces recorded on LG1 and LG5 result in both surficial bioerosion and modification of trabecular bone. Additionally, we recognized the two types of bioerosion by insects on bones mentioned by Huchet [44]: mechanical, produced by the action of the mandibles (striations, pits), and chemical (etched surfaces), caused by proteolytic enzymes contained in digestive juices.

Assessing Different Tracemakers

In reference to the tracemaker, diverse studies point out equifinality problems and suggest considering a set of modification types, their association, frequency and distribution patterns e.g., [6,11,32,41]. Although our sample is composed of few specimens with insect traces, we can make some comments in this regard. In this preliminary research we recorded macroscopically visible marks such as pits, holes and striations. A microscopic approach showed more details of traces that are composed by other marks (e.g., pits with striae in base).

To evaluate termites as possible tracemakers, we considered their presence in the study area both in past and present times. In Buenos Aires, there is a register of termites from the late Pliocene to the middle Pleistocene [45,46]. To date, three species have been identified in the northeastern sector of this province: *Rugitermes rugosus*, *Anoplotermes meridianus* and *Anoplotermes cingulatus* [47]. The first one belongs to Kalotermitidae, which nest directly within their food source and consume mainly drywood; the second and third correspond to Termitidae, which are subterranean. *A. cingulatus* feeds on decomposing wood and soil, while *A. meridianus* eats humus and dry grass and leaves [47]. Both locate food sources as they dig a network of tunnels and prefer fine-grained soils since they facilitate excavation [38]. This is consistent with the soils' texture of La Guillerma [18].

Regarding termites' traces, we recognized pits with emanating striae that resembled star-shaped marks, previously associated with termites' action [6,10,44]. The presence of surface coating as a residue (probably made of faecal matter, earth, wood, and saliva [14, 36]) and bone decoloring around it may support the identification of this bonemarker. However, in addition to our small sample, we did not find termite mounds, carton-lined tunnels, chambers or fossil nests in the study area. Laza [46] proposed that the cold climatic trend that occurred by the late Pleistocene would have forced the retraction of termites towards the north of Buenos Aires. Since then, their activity has been limited and that could be the cause of their absence. Nonetheless, the warmer conditions of the late Holocene [22,23] should have reactivated the action of termites, so it cannot be ruled out.

To contemplate dermestids as a potential tracemaker, we considered Parkinson's mention that, "the co-occurrence of surface tunnelling, destruction of bone, infrequent bore holes, surface pits of varying classes and broad areas of gnawing striations can be used as an indicator of *D. maculatus* activities" [32] (p. 139). In this sense, Backwell et al. [14] emphasize that hide beetles produce invasive edge gnawing. We did not identify surface tunnelling or large areas of edge gnawing. Regarding the latter, we only recorded foraging areas (*sensu* Britt et al. [7]) conformed by overlapping of pits.

In the experiment with trogid beetles and *O. badius*, Backwell and collaborators [14] found that the former insects caused more destruction and made more boreholes on the bones. In addition, dermestids' holes and boreholes are frequently described with a tendency to be ovoid or elliptical (e.g., [14,32]). Having in mind that the holes and the bore identified in LG1 specimens are round and their ranges both in size and shape as isolated traces, we could consider termites as the possible causal agent. However, in two cases the marks are associated with other traces: the bore is surrounded by etched surface (a trait that Backwell et al. [14] register in more proportion in termites). However, a hole is clustered with a pit with striae in base (that these same authors only recorded in their experiment with dermestids). Then, based on this inconclusive evidence and the fact that our sample is too small, we cannot propose a unique causal agent.

The attribution of a specific tracemaker permits palaeoecological reconstruction based on geographic distribution and climate requirements. However, this is a difficult task if no comparative material from actualistic experiments is available [35]. In this sense, further research on different carcass-exploiting insects will help infer tracemakers. Although we cannot make thorough interpretations on past ecology and climate, as Farina and Cicchino [34] state, there is a good entomological resilience to flood pulses in the coastal system of Buenos Aires province, which is close to our study area. Thus, these events would not have significantly affected its insects' ecology and population.

There were different moments in which insects could have had access to bone remains. They could consume the bones when they were above ground, still having tissues and medulla. However, they could also mark those remains after having been exposed to environmental conditions. If the agent had subterranean habits (such as Termitidae), it could have affected bones within the soil.

As it was registered on experimental studies (e.g., [6]) and inferred in taphonomic analyses (e.g., [10]), termites can generate preservation biases and, in consequence, alter the representation of skeletal elements and make taxonomic identification more difficult. Also, the modifications produced by *D. maculatus* usually concentrate on epiphysis and the diaphysis-epiphysis junction. Therefore, they can promote increasing separation and fragmentation of elements, which may eventually result in their destruction. Although gnawing of dense bone by dermestids is not prolific, they can destroy small bones (particularly of Aves) within a short period [48]. *P. americana* and *T. trinervoides* also do significant damage to bird bones [6,32]. Hence, insects may at least partially affect the representation of Aves in archaeological and palaeontological sites.

We believe that the actions of insects could have created biases, for instance, if we take into account that the holes and bore are present in large and thicker mammal elements, and that they were not documented neither in bird bones nor in medium to small rodent bones. On the other hand, as referred in one of the cracked elements, the mark could have weakened the bone making it more prone to fracture. We cannot rule out that in smaller or weaker bones the destruction could have been greater.

Since 1988, we have developed different studies on the faunal assemblage from La Guillerma archaeological locality. As a result, we were able to comprehend the role of people and nature as agents on the accumulation and modification of bone remains [16,18,26]. The analysis presented here has allowed us to identify for the first time the action of insects on sites from the Salado River Depression. Our approach contributes to increasing the background of this type of bioerosion in archaeological and palaeontological sites from Argentina (i.e., [12,37]).

Overall, this analysis has shed light on a taphonomic agent that had not been thoroughly addressed and for which there is little background in Argentina. We are still unable to assert which insect or insects were the tracemakers of the bones from La Guillerma sites and therefore to make palaeoecological interpretations. Nonetheless, we have identified a natural formation process that may have introduced variability in the archaeofaunal record and even produce destruction of organic materials such as bone. In this regard, diverse agents and processes were involved in La Guillerma shallow sites' formation. However, concerning the specificity of this kind of sites, we are not able to make statements about the incidence of insects on soil characteristics or to distinguish different deposition events yet. We had previously recorded the action of edaphic fauna and roots that altered the disposition of archaeological remains within the soil and we had also identified weathering and manganese stains on bone surfaces [16,18,24,26,49]. Moreover, the cultural formation processes also played a role as the hunter-gatherer-fisher groups used to resettle periodically in these sites over the centuries [15,16,18]. Hence, the bioturbation produced by insects is one more factor, among others, that contributed to the forming of the shallow archaeological sites of the study area.

Finally, we are aware that experimental analyses are required to improve our knowledge of insect traces and to establish useful criteria to determine causal agents. In this sense, documenting modern analogues will offer unequivocal data to link modification types to specific taphonomic processes and agents [2,11]. Also, a future analysis of our sample through higher magnification equipment will improve the accuracy of our identifications.

6. Final Remarks

The identification of insects (directly or indirectly) in an archaeological context improves the interpretation of bioturbation processes since they can affect as well as displace artifacts or ecofacts from their original context. In addition, insect-produced damage to bone surfaces can obliterate cultural evidence and other formation processes. They are also useful to make palaeoecological interpretations.

Our research provided a better understanding of the faunal assemblages from La Guillerma sites, since it allowed us to identify and analyze a novel taphonomic agent in the area. Despite this type of marks being scant, we should consider potential biases that insects could have produced on bone remains. Moreover, they could also modify the soil structure adding their action to other taphonomic agents such as earthworms [49]. We encourage considering insects in the study of other archaeological sites in the pampas as well as other Argentinian regions.

To improve our interpretation of traces and identification of uncommon taphonomic agents, such as insects, the development of actualistic studies is imperative. For this reason, we are starting experimental research that includes Coleoptera (*D. maculatus*) and local termites (*Cortaritermes cumulans*).

Author Contributions: Conceptualization, P.D.E. and A.E.F.; methodology, P.D.E., A.E.F. and M.I.G.; formal analysis, P.D.E., A.E.F. and M.I.G.; investigation, P.D.E., A.E.F. and M.I.G.; writing—original draft preparation, P.D.E., A.E.F. and M.I.G.; writing—review and editing, P.D.E., A.E.F. and M.I.G.; project administration, M.I.G.; funding acquisition, P.D.E. and M.I.G. All authors have read and agreed to the published version of the manuscript.

Funding: This research was funded by Agencia Nacional de Promoción Científica y Tecnológica, grants PICT-2015-0272 and PICT-2016-0368, and by Secretaría de Ciencia y Técnica, Universidad de Buenos Aires, grant UBACyT-2018-20020170100525BA.

Data Availability Statement: Data are contained within the article and are available on request from the corresponding author.

Acknowledgments: We thank Roxana Mariani, Eduardo Tonni, José Laza, Jean-Bernard Huchet and Jorge Genise for helping us with their knowledge on insect traces; Enrique Laffont and Juan Manuel Coronel for providing us with useful bibliography; Olivia Sokol for drawing the map; Marcelo Zárate for his observations regarding palaeoenvironment; Romina Frontini and Mónica Salemme for their comments on a previous draft. We also thank the three anonymous reviewers for their constructive comments. The authors are the only responsible for the arguments discussed in this paper.

Conflicts of Interest: The authors declare no conflict of interest. The funders had no role in the design of the study; in the collection, analyses, or interpretation of data; in the writing of the manuscript; or in the decision to publish the results.

References

1. Corron, L.; Huchet, J.B.; Santos, F.; Dutour, O. Using Classifications to Identify Pathological and Taphonomic Modifications on Ancient Bones: Do "Taphognomonic" Criteria Exist? *Bull. Mem. Soc. Anthropol. Paris* **2017**, *29*, 1–18. [CrossRef]
2. Fernandez-Jalvo, Y.; Andrews, P. *Atlas of Taphonomic Identifications. Vertebrate Paleobiology and Paleoanthropology*; Springer: Dordrecht, The Netherlands, 2016; pp. 155–166. [CrossRef]
3. Lyman, R.L. *Vertebrate Taphonomy*; Cambridge University Press: Cambridge, UK, 1994.
4. Tomassini, R.L.; Montalvo, C.I.; Ezquiaga, M.C. The oldest record of flea/armadillos interaction as example of bioerosion on osteoderms from the late Miocene of the Argentine Pampas. *Int. J. Paleopathol.* **2016**, *15*, 65–68. [CrossRef] [PubMed]
5. Zonneveld, J.-P.; AbdelGawad, M.K.; Miller, E.R. Ectoparasite borings, mesoparasite borings, and scavenging traces in early Miocene turtle and tortoise shell: Moghra Formation, Wadi Moghra, Egypt. *J. Paléontol.* **2021**, 1–19. [CrossRef]
6. Backwell, L.R.; Parkinson, A.H.; Roberts, E.M.; D'Errico, F.; Huchet, J.-B. Criteria for identifying bone modification by termites in the fossil record. *Palaeogeogr. Palaeoclim. Palaeoecol.* **2012**, *337–338*, 72–87. [CrossRef]
7. Britt, B.B.; Scheetz, R.D.; Dangerfield, A. A Suite of Dermestid Beetle Traces on Dinosaur Bone from the Upper Jurassic Morrison Formation, Wyoming, USA. *Ichnos* **2008**, *15*, 59–71. [CrossRef]
8. Huchet, J.-B.; Le Mort, F.; Rabinovich, R.; Blau, S.; Coqueugniot, H.; Arensburg, B. Identification of dermestid pupal chambers on Southern Levant human bones: Inference for reconstruction of Middle Bronze Age mortuary practices. *J. Archaeol. Sci.* **2013**, *40*, 3793–3803. [CrossRef]
9. Huchet, J.-B.; Greenberg, B. Flies, Mochicas and burial practices: A case study from Huaca de la Luna, Peru. *J. Archaeol. Sci.* **2010**, *37*, 2846–2856. [CrossRef]
10. Matu, M.; Crevecoeur, I.; Huchet, J.-B. Taphonomy and Paleoichnology of Olduvai Hominid 1 (OH1), Tanzania. *Int. J. Osteoarchaeol.* **2017**, *27*, 785–800. [CrossRef]
11. Parkinson, A.H. Traces of Insect Activity at Cooper's D Fossil Site (Cradle of Humankind, South Africa). *Ichnos* **2016**, *23*, 322–339. [CrossRef]
12. Di Donato, R.M. Taphos nomos: El potencial de la tafonomía en la interpretación de contextos funerarios. In *Mamül Mapu: Pasado y Presente desde la Arqueología Pampeana*; Berón, M.A., Luna, L.H., Bonomo, M., Eds.; Libros del Espinillo: Ayacucho, Argentina, 2010; pp. 137–152.
13. Wrobel, G.D.; Biggs, J. Osteophageous insect damage on human bone from Je'reftheel, a Maya mortuary cave site in west-central Belize. *Int. J. Osteoarchaeol.* **2018**, *28*, 745–756. [CrossRef]
14. Backwell, L.; Huchet, J.-B.; Jashashvili, T.; Dirks, P.H.; Berger, L.R. Termites and necrophagous insects associated with early Pleistocene (Gelasian) Australopithecus sediba at Malapa, South Africa. *Palaeogeogr. Palaeoclim. Palaeoecol.* **2020**, *560*, 109989. [CrossRef]
15. Frère, M.M.; González, M.I.; Greco, C. Continuity in the Use of Shallow Sites of the Salado River Basin in the Pampean Region, Argentina. *Radiocarbon* **2016**, *58*, 921–933. [CrossRef]
16. González, M.I.; Escosteguy, P.D.; Salemme, M.C.; Frère, M.M.; Weitzel, C.; Vecchi, R. *Assessing Strategies for Coypu Hunting and Use in the Salado River Depression (Buenos Aires Province, Argentina)*; Springer: Singapore, 2021; pp. 59–81.
17. González, M.I.; Frère, M.M. Río Salado: Espacio de interacción de cazadores-recolectores-pescadores (provincia de Buenos Aires, Argentina). *Rev. Mus. La Plata* **2019**, *4*, 611–632. [CrossRef]
18. González, M.I. *Arqueología de Alfareros, Cazadores y Pescadores Pampeanos. Colección de Tesis Doctorales*; Sociedad Argentina de Antropología: Buenos Aires, Argentina, 2005.
19. López, H.L.; Baigún, C.R.M.; Iwaszkiw, J.M.; Delfino, R.; Padín, O.H. *La Cuenca del Salado: Uso y Posibilidades de sus Recursos Pesqueros. Portal de Libros de la Universidad Nacional de La Plata*; Editorial de la Universidad Nacional de La Plata: Buenos Aires, Argentina, 2001.
20. Quirós, R.; Rennella, A.M.; Boveri, M.B.; Rosso, J.J.; Sosnovsky, A. Factores que afectan la estructura y el funcionamiento de las lagunas pampeanas. *Ecol. Austral.* **2002**, *12*, 175–185.
21. Zárate, M.A. El paisaje pampeano a través del tiempo. In *Mamül Mapu: Pasado y Presente desde la Arqueología Pampeana*; Berón, M.A., Luna, L.H., Bonomo, M., Eds.; Libros del Espinillo: Ayacucho, Argentina, 2010; pp. 19–32.
22. Pommarés, N.N.; Fucks, E.E.; Pisano, M.F.; Luengo, M.S.; Ramos, N.A.; Di Lello, C.V. Late Pleistocene-Holocene paleoenvironments in the middle basin of the Salado river, province of Buenos Aires, Argentina. *J. S. Am. Earth Sci.* **2021**, *105*, 103001. [CrossRef]
23. Fucks, E.E.; Pisano, F.; Carbonari, J.; Huarte, R. Aspectos geomorfologicos del sector medio e inferior de la Pampa Deprimida, provincia de Buenos Aires. *Rev. Soc. Geológica España* **2012**, *25*, 107–118.
24. Zárate, M.A.; de Bonaveri, M.I.G.; Flegenheimer, N.; Bayón, C. Sitios arqueológicos someros: El concepto de sitio en estratigrafía y sitio de superficie. *Cuad. Inst. Nac. Antropol. Pensam. Latinoam.* **2000**, *19*, 635–653.
25. Behrensmeyer, A.K. Taphonomic and ecologic information from bone weathering. *Paleobiology* **1978**, *4*, 150–162. [CrossRef]
26. Escosteguy, P. *Etnoarqueología de Nutrieros. Una Propuesta Metodológica Aplicada al Registro Arqueológico de la Depresión del Salado y del Noreste de la Provincia de Buenos Aires*; Universidad de Buenos Aires: Buenos Aires, Argentina, 2011.
27. Escosteguy, P.D.; Salemme, M.; González, I. Myocastor coypus ("coipo", Rodentia, Mammalia) como recurso en los humedales de la Pampa bonaerense: Patrones de explotación. *Rev. Mus. Antropol.* **2012**, *5*, 13–30. [CrossRef]

28. Escosteguy, P.D.; González, M.I.; Salemme, M.C. Aprovechamiento de venado de las pampas y ciervo de los pantanos en la Depresión del río Salado (región pampeana, Argentina). In *Libro de Resúmenes del IV Encuentro Latinoamericano de Arqueo-Zoología (ELAZ). Homenaje al Dr. Luis A. Borrero*; Cruz, I., Ed.; UNPA Edita: Río Gallegos, Argentina, 2018; pp. 133–134.
29. Genise, J.F.; Mángano, M.G.; Buatois, L.A.; Laza, J.H.; Verde, M. Insect Trace Fossil Associations in Paleosols: The Coprinisphaera Ichnofacies. *Palaios* **2000**, *15*, 49–64. [CrossRef]
30. Wood, W.R.; Johnson, D.L. A survey of disturbance processes in archaeological site formation. In *Advances in Archaeological Method and Theory*; Schiffer, M.B., Ed.; Academic Press: Cambridge, MA, USA, 1978; Volume 1, pp. 315–381.
31. McBrearty, S. Consider the humble termite: Termites as agents of post-depositional disturbance at african archaeological sites. *J. Archaeol. Sci.* **1990**, *17*, 111–143. [CrossRef]
32. Parkinson, A.H. *Dermestes Maculatus and Periplaneta Americana: Bone Modification Criteria and Establishing Their Potential as Climatic Indicators*; University of the Witwatersrand, Johannesburg: Johannesburg, South Africa, 2012.
33. Cigliano, M.M.; Melo, M.C.; Montemayor, S.I.; del Río, M.G.; Dellapé, P.M. BiodAr—Biodiversidad de Insectos de la Argentina y Uruguay. Available online: https://biodar.unlp.edu.ar/ (accessed on 5 April 2021).
34. Farina, J.; Cicchino, A. Una excursión entomológica por la costa atlántica bonaerense. In *La Costa Atlántica de Buenos Aires. Naturaleza y Patrimonio Cultural*; Athor, J., Celsi, C., Eds.; Fundación de Historia Natural Félix de Azara: Buenos Aires, Argentina, 2016; pp. 281–323.
35. Holden, A.R.; Harris, J.M.; Timm, R.M. Paleoecological and Taphonomic Implications of Insect-Damaged Pleistocene Vertebrate Remains from Rancho La Brea, Southern California. *PLoS ONE* **2013**, *8*, e67119. [CrossRef]
36. Huchet, J.; Deverly, D.; Gutierrez, B. Taphonomic evidence of a Human skeleton gnawed by termites in a Mochecivilisation grave at Huaca de la Luna, Peru. *Int. J. Osteoarchaeol.* **2011**, *21*, 92–102. [CrossRef]
37. Pomi, L.H.; Tonni, E.P. Termite Traces on Bones from the Late Pleistocene of Argentina. *Ichnos* **2011**, *18*, 166–171. [CrossRef]
38. Torales, G.J. *Biodiversidad de Artrópodos Argentinos: Una Perspectiva Biotaxonómica*; Ediciones Sur: La Plata, Argentina, 1998.
39. Go, M. A Case of Human Bone Modification by Ants (Hymenoptera: Formicidae) in the Philippines. *Forensic Anthr.* **2018**, *1*, 117–123. [CrossRef]
40. Pirrone, C.A.; Buatois, L.A.; Bromley, R.G. Ichnotaxobases for bioerosion trace fossils in bones. *J. Paléontol.* **2014**, *88*, 195–203. [CrossRef]
41. Odes, E.J.; Parkinson, A.H.; Randolph-Quinney, P.S.; Zipfel, B.; Jakata, K.; Bonney, H.; Berger, L.R. Osteopathology and insect traces in the Australopithecus africanus skeleton StW 431. *S. Afr. J. Sci.* **2017**, *113*, 7. [CrossRef]
42. Zanetti, N.I.; Visciarelli, E.C.; Centeno, N.D. Taphonomic Marks on Pig Tissue Due to Cadaveric Coleoptera Activity Under Controlled Conditions. *J. Forensic Sci.* **2014**, *59*, 997–1001. [CrossRef]
43. Zanetti, N.I.; Ferrero, A.A.; Centeno, N.D. Depressions of Dermestes maculatus (Coleoptera: Dermestidae) on Bones Could be Pupation Chambers. *Am. J. Forensic Med. Pathol.* **2019**, *40*, 122–124. [CrossRef]
44. Huchet, J.B. Man-eating termites: Osteolytic lesions upon human skulls perpetrated by subterranean termites (Insecta: Isop-tera). In Proceedings of the III Meeting of the International Conference in Funerary Archaeoentomology (ICFAE), Bordeaux, France, 5 June 2019.
45. Genise, J.F. A fossil termite nest from the Marplatan stage (Late Pliocene) of Argentina: Paleoclimatic indicator. *Palaeogeogr. Palaeoclimatol. Palaeoecol.* **1997**, *136*, 139–144. [CrossRef]
46. Laza, J.H. Termiteros del Plioceno y Pleistoceno de la provincia de Buenos Aires, República Argentina. Significación paleoambiental y paleozoogeográfica. *Ameghiniana* **2006**, *43*, 641–648.
47. Torales, G.J.; Coronel, J.M.; Laffont, E.R.; Fontana, J.L.; Godoy, M.C. Termite associations (Insecta, Isoptera) in natural or semi-natural plant communities in Argentina. *Sociobiology* **2009**, *54*, 383–437.
48. Kirkland, J.; Bader, K.; Kirkland, J.I.; Bader, K. Insect trace fossils associated with Protoceratops carcasses in the Djadokhta Formation (Upper Cretaceous), Mongolia. In *New Perspectives on Horned Dinosaurs: The Royal Tyrrell Museum Ceratopsian Symposium*; Ryan, M.J., Chinnery-Allgeier, B.J., Eberth, D.A., Ralrick, P.E., Eds.; Indiana University Press: Bloomington, IN, USA, 2010; pp. 509–519.
49. Escosteguy, P.; Fernández, A.E. The Effects of Bioturbation by Earthworms. Preliminary Results of an Actualistic Taphonomy Experiment. *J. Taphon.* **2017**, *15*, 11–17.

quaternary

MDPI

Article

The Potential of Micromammals for the Stratigraphy and the Timing of Human Occupations at La Roche-à-Pierrot (Saint-Césaire, France)

Loïc Lebreton [1,*], Eugène Morin [2,3], Brad Gravina [3], Alexandre Michel [3,4], François Bachellerie [3,5], Cédric Beauval [6], Damien Flas [7,8], Véronique Laroulandie [3], Josserand Marot [9], Hélène Rougier [10], Elise Tartar [8], Dominique Todisco [11] and Isabelle Crevecoeur [3]

1 Histoire Naturelle de l'Homme Préhistorique (HNHP), UMR 7194, Département Homme et Environnement du Muséum National d'Histoire Naturelle, MNHN-CNRS-UPVD, Musée de l'Homme, 75000 Paris, France
2 Department of Anthropology, Trent University, DNA Block C, 2140 East Bank Drive, Peterborough, ON K9J 7B8, Canada; eugenemorin@trentu.ca
3 PACEA, UMR 5199, CNRS, Université de Bordeaux, MC, B2, Allée Geoffroy Saint-Hilaire, CEDEX, 33615 Pessac, France; gravina.brad@gmail.com (B.G.); alex.michel10@laposte.net (A.M.); bachellerie.francois@gmail.com (F.B.); veronique.laroulandie@u-bordeaux.fr (V.L.); isabelle.crevecoeur@u-bordeaux.fr (I.C.)
4 Service Départemental de l'Archéologie, Conseil Départemental de la Dordogne, 2 Rue Paul Louis Courrier, CS11200, CEDEX, 24019 Périgueux, France
5 Archéologie Alsace, 11 Rue Jean-François Champollion, 67600 Sélestat, France
6 Archéosphère, 10 Rue de la Rhode, 11500 Quillan, France; c.beauval@archeosphere.com
7 Service de Préhistoire, Université de Liège, Place du 20-Août 7, 4000 Liège, Belgium; damienflas@yahoo.com
8 TRACES, UMR 5608, CNRS, Université Toulouse Jean Jaurès, Maison de la Recherche, 5 Allées Antonio Machado, CEDEX 9, 31058 Toulouse, France; elise.tartar@cnrs.fr
9 Service Départemental de l'Archéologie, Conseil Départemental de Charente-Maritime, Caserne Brémond d'Ars, Petite Rue du Séminaire, 17100 Saintes, France; marot.josserand@gmail.com
10 Department of Anthropology, California State University Northridge, 18111 Nordhoff St., Northridge, CA 91330-8244, USA; helene.rougier@csun.edu
11 IDEES, UMR 6266, Université de Rouen, 17 Rue Lavoisier, CEDEX, 76821 Mont Saint Aignan, France; dominique.todisco@univ-rouen.fr
* Correspondence: loicleb@hotmail.com

Abstract: As micromammals are highly sensitive to changes in their habitat, variations in species representation are often used to reconstruct local environmental conditions. However, taphonomic aspects of micromammals are often overlooked, despite the fact that they can provide important information for our understanding of archaeological sites. La Roche-à-Pierrot, Saint-Césaire, is a major archaeological site for our understanding of the Middle-to-Upper Palaeolithic transition in Western Europe. Clearly documenting site formation processes, the post-depositional reworking of deposits and the sequence of human occupations is fundamental for providing a secure archaeostratigraphic context of the site. The exceptionally large accumulation of micromammals from recently excavated stratigraphic units at the site makes it possible to track variations in the density of micromammals across the stratigraphic sequence. The taphonomic analysis of micromammals demonstrates these variations are not related to a change in the main accumulation agent or post-depositional phenomena. A negative correlation between small mammal remains and archaeological material suggests that peaks in micromammal densities can potentially be correlated with periods when the site was abandoned or when human occupation was less intense, and therefore provide new data for interpreting the Saint-Césaire stratigraphic sequence.

Keywords: micromammals; taphonomy; density analysis; Late Pleistocene

1. Introduction

The analysis of micromammals has primarily focused on palaeoenvironmental reconstructions, given their high potential for recording climatic fluctuations [1–5]. Most taphonomic analyses of micromammal fossil accumulations tend to concentrate on identifying biases in the representation of species in fossil assemblages and living communities. The frequent documentation of such biases is due to the fact that micromammal accumulations typically result from the feeding practices of one or more predators and post-depositional phenomena, including the fossilisation process. These biases could affect both palaeocommunities and skeletal part representation, as well as distorting the palaeoecological signal [6–10]. Taphonomic studies of small mammals generally focus on the identification of the principal accumulation agent and the characterisation of post-depositional processes [11]. The relationship between micromammals (primarily small rodents and insectivores) and the various categories of archaeological material recovered from a given site has received little attention. Apart from the Chalcolithic site of El Portalón [12], the Upper Palaeolithic sites of Peyrazet [13,14] and Taillis des Coteaux [15] and the Middle Palaeolithic site of Les Pradelles [13], relatively few studies have focused on the correlation between micromammals and archaeological material.

Here we test a simple hypothesis: as raptors are highly sensitive to disturbances in their habitat [16,17], human activity on an archaeological site should be negatively correlated with their presence, and consequently, with the quantity of micromammal accumulated by these predators. As such, tracking shifting densities in micromammal remains can provide complementary data for assessing the timing of human occupations at a given site.

La Roche-à-Pierrot, Saint-Césaire is a key site for understanding the Middle-to-Upper Palaeolithic transition and is perhaps best known for Neandertal remains recovered from a level attributed to the Châtelperronian during excavations [18]. However, a recent techno-typological and taphonomic revision of the lithic material from this level produced no reliable evidence for a Neandertal-Châtelperronian association at the site [19]. A previous analysis of the Saint-Césaire macrofaunal remains documented a negative correlation between the relative abundance of carnivore remains and carnivore damage on bone surfaces. When considered alongside additional evidence for human occupation, such as the proportion of burnt faunal remains, these correlations were interpreted as reflecting changes in the intensity of the site's occupation [20]. Here we use micromammal remains recovered during ongoing excavations at Saint-Césaire to further explore the timing and frequency of the human occupations at the site.

2. Site Presentation

The site of La Roche-à-Pierrot is a collapsed rockshelter located in southwestern France on the commune of Saint-Césaire (Figure 1A). Excavated by F. Lévêque between 1976 and 1987, the site yielded Neandertal remains and a complex Mousterian-Châtelperronian-Aurignacian cultural sequence [21]. Marquet's [22] palaeoecological analysis of rodent remains from Lévêque's excavations highlighted a cold phase reflected in the presence of *Lasiopodomys gregalis* and *Microtus oeconomus*. The large fauna mainly comprises reindeer, bison and horses, with carnivores (spotted hyena, wolf and fox) only sporadically present across the stratigraphic sequence. The presence of small taxa, such as lagomorphs and birds, were interpreted as primarily reflecting background accumulations [20].

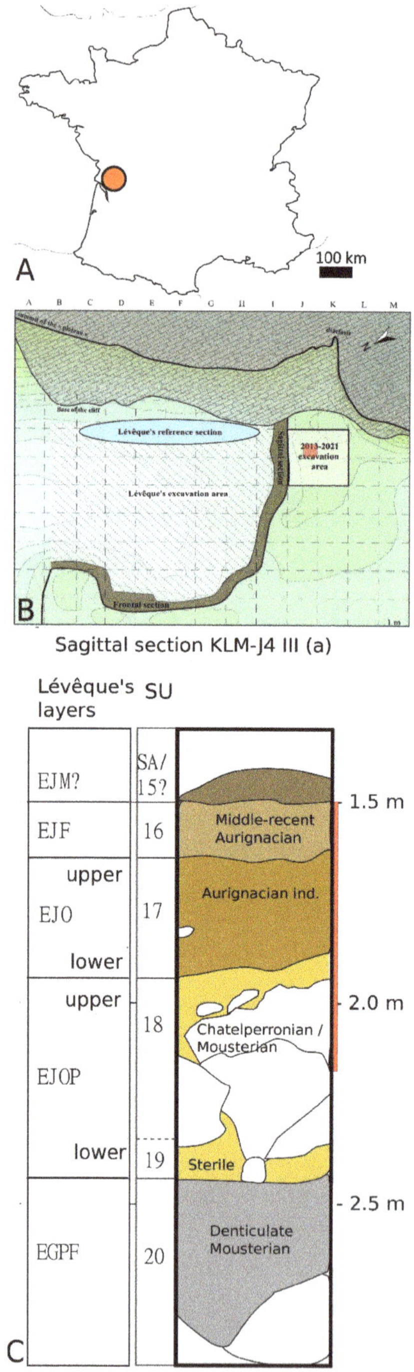

Figure 1. (**A**) Location of La Roche-à-Pierrot, Saint-Césaire (orange circle); (**B**) topographic plan with the excavation grid: the square on the right corresponds to the currently excavated area and the red

inset to the study area; (**C**) stratigraphic log with the correspondence between Lévêque's layers (EJF, EJO, EJOP, EGPF) and the stratigraphic units (SU) identified during the new excavations. The stratigraphic portion studied in the present paper is represented in red.

Excavations at Saint-Césaire were resumed by F. Bachellerie and E. Morin in 2013–2014, and by I. Crevecoeur since 2015. The goal of these new excavations is to produce a better understanding of the site's stratigraphic sequence and associated formation processes [23]. As it was impossible to establish a direct stratigraphic link between the site's sections, stratigraphic units (SU) were—to avoid confusion—numbered from 15 to 19 in the area that was the primary focus of the new excavations. Here, we focus on this series of newly excavated stratigraphic units. The stratigraphic sequence in this area of the site comprises SU 15 to the top of SU 19. Directly underlying a humic deposit referred to as SU 15-humus, the 10 cm thick SU 15 is characterised by dark brown clayey sediments with some evidence for post-depositional reworking. The limited number of lithic artefacts from SU 15 makes it impossible to establish a reliable archaeological attribution for this unit. The 20 cm thick SU 16 is characterised by brown clayey sediment with heterometric limestone fragments and contains a lithic assemblage comprising both Middle and Late Aurignacian lithics. SU 17 is made up of yellow/brown silt-clay sediment with numerous small limestone fragments and contains a still difficult to characterise Aurignacian industry. SU 18 is approximately 20 cm thick and is composed of yellow/orange sandy-silty sediment with numerous small limestone fragments and decimetric blocks. The SU 18 lithic assemblage comprises a mix of Mousterian and Châtelperronian artefacts, which is consistent with a recent revision of material recovered by Lévêque from a broadly equivalent archaeological unit (EJOP upper; [19]). The lowermost SU 19 is a yellow/orange sandy/silty sediment with numerous, sometimes large limestone blocks that seals the underlying Mousterian (Lévêque's EGPF). This unit is about 20 cm thick and contains only a small quantity of bone and lithic artefacts.

3. Material and Methods

The micromammal sample analysed here comes from a succession of 2 cm spits excavated in sub-square J4 III (e.g., spit 176–178 cm, the datum point "0" being situated in the cliff above the stratigraphic sequence) (Figure 1B). This area of the site was selected because it corresponds to a zone where the SUs were the thickest, making it possible to follow a continuous series of spits down to the fallen stone blocks that mark the limit with the relatively sterile level separating the yellow upper sequence (EJ units) from the underlying grey Mousterian (EG units) levels (Figure 1C, and see [21] for more details). All sediments were recovered from successive 2 cm spits and subsequently water-sieved using 4 and 2 mm meshes after the sediment volume was estimated by litre (minus stones with a maximum dimension of 10 cm or more). The mesh size used to sieve the sediments is quite large for micromammal sampling, and this could have had an impact on the representation of some very small taxa (such as *Sicista* sp.) and small bone elements. Sub-square J4 III is divided into four subunits (a, b, c and d), all corresponding to 16ths of a square metre. All objects greater than 2 cm (e.g., lithic, fauna) were piece-plotted in three dimensions with the help of a Total Station. We focused on micromammals, specifically rodents and insectivores, which form the bulk of the small vertebrate accumulation. Avifauna, herpetofauna, ichthyofauna, and lagomorpha are present but only in small proportions.

The taphonomic study of micromammals followed the standard methodology and data available in [6–10]. We focused primarily on the relative abundance, intensity and frequency of digestion marks. The relative abundance of skeletal elements was calculated using the formula provided by Dodson and Wexlar [24] and Andrews [7]:

$$Ri = Ni/(MNI \times Ei)$$

where R*i* is the relative abundance of element *i*; N*i*, the number of elements *i* as measured by MNE (Minimum Number of Elements); MNI, the Minimum Number of Individuals; E*i*, the number of the element *i* in the skeleton. For our analysis, we considered the rodent skeleton to comprise two hemimaxillae; two hemimandibles; four incisors; 12 molars; two humeri; two radii; two ulnae; two femora; two tibiae; two scapulae; two innominates; two tali; two calcanei; 24 ribs; 54 vertebrae; 78 phalanges and metapodia.

Traces of digestion on bone surfaces result from the action of the gastric juices of the predator on the various bone elements. The frequency and intensity of these alterations allows different categories of predators to be identified. For example, nocturnal raptors produce less frequent and less intense bone surface alteration compared to diurnal raptors, and even fewer compared to small carnivores [6–10].

To quantify the micromammal accumulation, we used the MNE as well as the density of remains per litre of sediment ("MNE/L"). We performed a Pearson correlation test using the Past (version 4.06) software to explore whether the density of micromammal remains is negatively correlated with various indicators of human occupations, including the abundance of macrofaunal remains, particularly as they were previously shown to primarily reflect human activity [20,25]. More specifically, proxies considered to reflect human occupation include the number of piece-plotted faunal remains (greater than 2 cm as well as any remain considered identifiable in the field) and the number of lithic artefacts (for simplicity, we only used those that are greater than 2 cm). We also included the same types of material from an adjacent sub-square metre, J4 IV, in order to increase the sample size of proxies for human activity, especially macrofaunal remains. In order to test whether variations result from post-depositional processes or excavation methods, non-piece plotted faunal remains and lithic artefacts recovered from the sieve residue were weighed and subsequently calculated in grams per litre.

4. Results

4.1. Taxonomic Composition

The micromammal accumulation is dominated by *Microtus arvalis/agrestis* and *Lasiopodomys gregalis* (Table 1). We also note the presence of *Arvicola amphibius* and *Microtus oeconomus*, including a few remains attributed to *Spermophilus* sp., *Talpa* sp. and *Sorex* sp. A single lower first molar of *Lemmus/Myopus* sp. was identified in SU 17. Relatively small differences in the taxonomic composition are evident between the SUs. Testing whether these differences reflect sampling error or other factors requires a larger sample of stratigraphic units (Table 1). In comparison with Marquet's data [22], our sample includes fewer remains of *Lasiopodomys gregalis* and higher numbers of *Microtus arvalis/agrestis* remains. In addition, our sample does not include remains of *Sicista betulina*. It is unclear whether these differences could be linked to the mesh size, reflect sampling bias, as Marquet's samples were fairly small, or spatial patterning in the distribution of different micromammal species (Figure 2A–E).

Table 1. Micromammal species diversity at La Roche-à-Pierrot for each stratigraphic unit (SU), expressed in MNI (Minimum Number of Individuals).

Species/SU	15-Humus	15	16	16/17	17	17/18	18	18/19	19
Microtus arvalis/agrestis (Common/field vole)	5	63	33	77	106	26	28	3	1
Lasiopodomys gregalis (Narrow-headed vole)	7	74	40	42	109	32	30	4	2
Arvicola amphibius (European water vole)	0	5	0	0	8	1	1	1	1
Microtus oeconomus (Root vole)	0	4	0	0	2	0	1	0	0
Lemmus/Myopus sp. (Lemming)	0	0	0	0	1	0	0	0	0

Table 1. *Cont.*

Species/SU	15-Humus	15	16	16/17	17	17/18	18	18/19	19
Spermophilus sp. (Ground squirrel)	0	1	0	1	0	1	1	0	0
Sorex sp. (Shrew)	0	1	0	1	0	0	1	0	1
Talpa sp. (Mole)	0	0	0	0	0	0	1	0	0

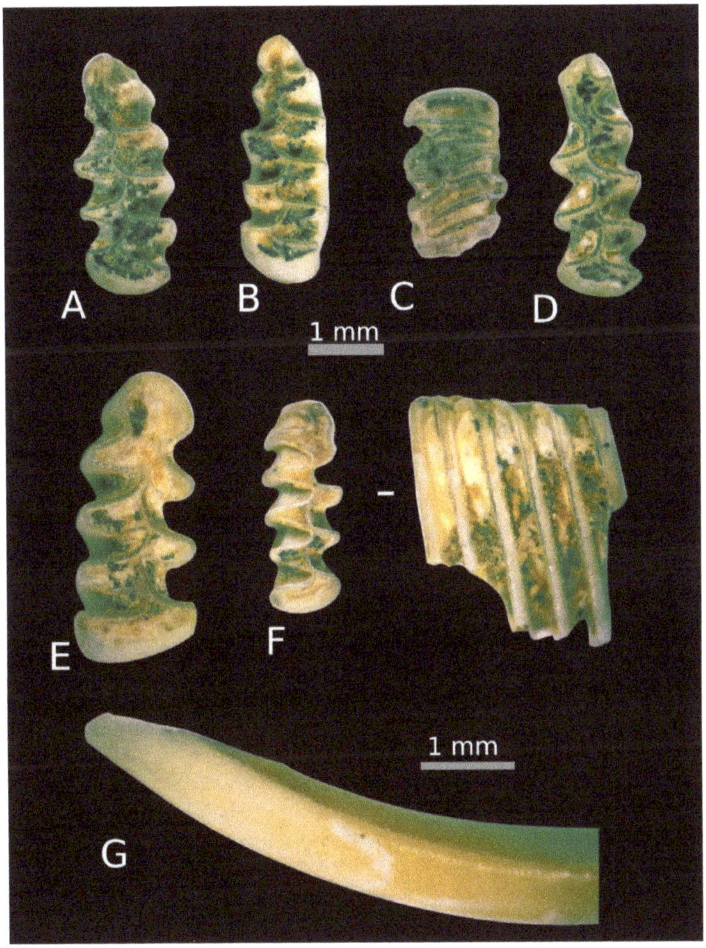

Figure 2. Rodent remains from La-Roche-à-Pierrot. (**A**) First lower molar (left) of *Lasiopodomys gregalis*; (**B**) first lower molar (left) of *Microtus arvalis/agrestis*; (**C**) third upper molar (right) of *Lemmus/Myopus*; (**D**) first lower molar (left) of *Microtus oeconomus*; (**E**) first lower molar (right) of *Arvicola amphibius*; (**F**) first lower molar (right) of *Lasiopodomys gregalis* with signs of moderate digestion (category 2), occlusal and lateral view; (**G**) rodent lower incisor (left) with signs of light digestion (category 1), characterised by the reduction of enamel on the tip (Table 2).

Table 2. Number (N) and percentage (%) of digested Arvicolinae teeth for each SU (Stratigraphic Unit), according to the intensity of the alteration defined by Andrews (1990): 0 = null; 1 = light; 2 = moderate; 3 = strong.

SU	Incisors									Molars								
	0		1		2		3		N Total	0		1		2		3		N Total
	N	%	N	%	N	%	N	%		N	%	N	%	N	%	N	%	
15	82	92	5	6	2	2	0	0	89	62	97	1	2	0	0	1	2	64
16	83	89	6	6	1	1	3	3	93	104	96	0	0	1	1	3	3	108
16/17	95	95	5	5	0	0	0	0	100	177	95	7	4	0	0	3	2	187
17	251	93	18	7	0	0	0	0	269	374	95	12	3	5	1	3	1	394
17/18	66	94	4	6	0	0	0	0	70	92	95	4	4	0	0	1	1	97
18	57	92	3	5	2	3	0	0	62	94	90	5	5	2	2	3	3	104
18/19	1	_	0	_	0	_	0	_	1	6	_	1	_	1	_	0	_	8
19	0	_	0	_	0	_	0	_	0	2	_	1	_	2	_	0	_	5

4.2. Origin of the Micromammal Accumulation

Traces of digestion on Arvicolinae molars vary between 4% and 6% for SU 15–17, and climb to 10% for SU 18, whereas for the incisors, evidence of digestion varies between 5% and 11%. Most teeth exhibit low degrees of digestion, with only a limited number being heavily altered (Table 2). This low digestion rate and low intensity of digestion are consistent with a category 1 predator [7,10], such as the barn owl (*Tyto alba*), short-eared (*Asio flammeus*) or long-eared owl (*Asio otus*). The presence of remains exhibiting moderate to extensive traces of alteration may highlight the limited contribution of another predator (less than 3% of digested teeth) such as diurnal raptors or small carnivores (Figure 2F,G).

All of the SUs are dominated by maxillae and mandibles followed by humeri, femora, incisors and molars. Small and fragile elements (ribs, phalanges, metapodia, vertebrae, radii) are consistently under-represented (Table S1, Figure 3), a pattern likely reflecting a similar post-depositional winnowing or destruction affecting all SUs equally.

4.3. Density Variation

The MNE/L ratio is very high at the top of the sequence (spits 147–150 to 168–170) and includes two density peaks: one at spit 150–152 and a second at spit 166–168 (Figure 4). This ratio is significantly lower towards the bottom of the sequence, except for a peak in spit 176–178. A similar increase in density is also observed between spits 186–188 and 190–192. Density variations observed between smaller subunits (16th of a square metre) are similar to those seen in sub-square J4 III as a whole. The small variations observed between these subunits occur at the interface between SU 17 and 18 and are most likely due to the fact that SU 18 was slightly higher in some places (as in J4 IIIb, spit 190–192 and J4 IIIc, spit 188–190). Additionally, the limited number of micromammal remains in SU 18 is reflected in small variations in specimen counts when SU 17 and 18 were recorded in the same spit during excavations.

The density measures and raw frequencies of micromammals used for the correlation test are provided in Table S2, while Table 3 provides the results of the Pearson correlation tests (Pearson's r) as well as associated p-values. A statistically significant positive correlation is observed between the density and MNE of micromammals, which is related to the small volume difference between spits. Proxies for human occupation (macrofauna and lithics) also show significant positive correlations between them. The significant positive correlation between piece-plotted lithic and faunal material and sieve residues supports variations in these variables reflecting human activity rather than excavation bias.

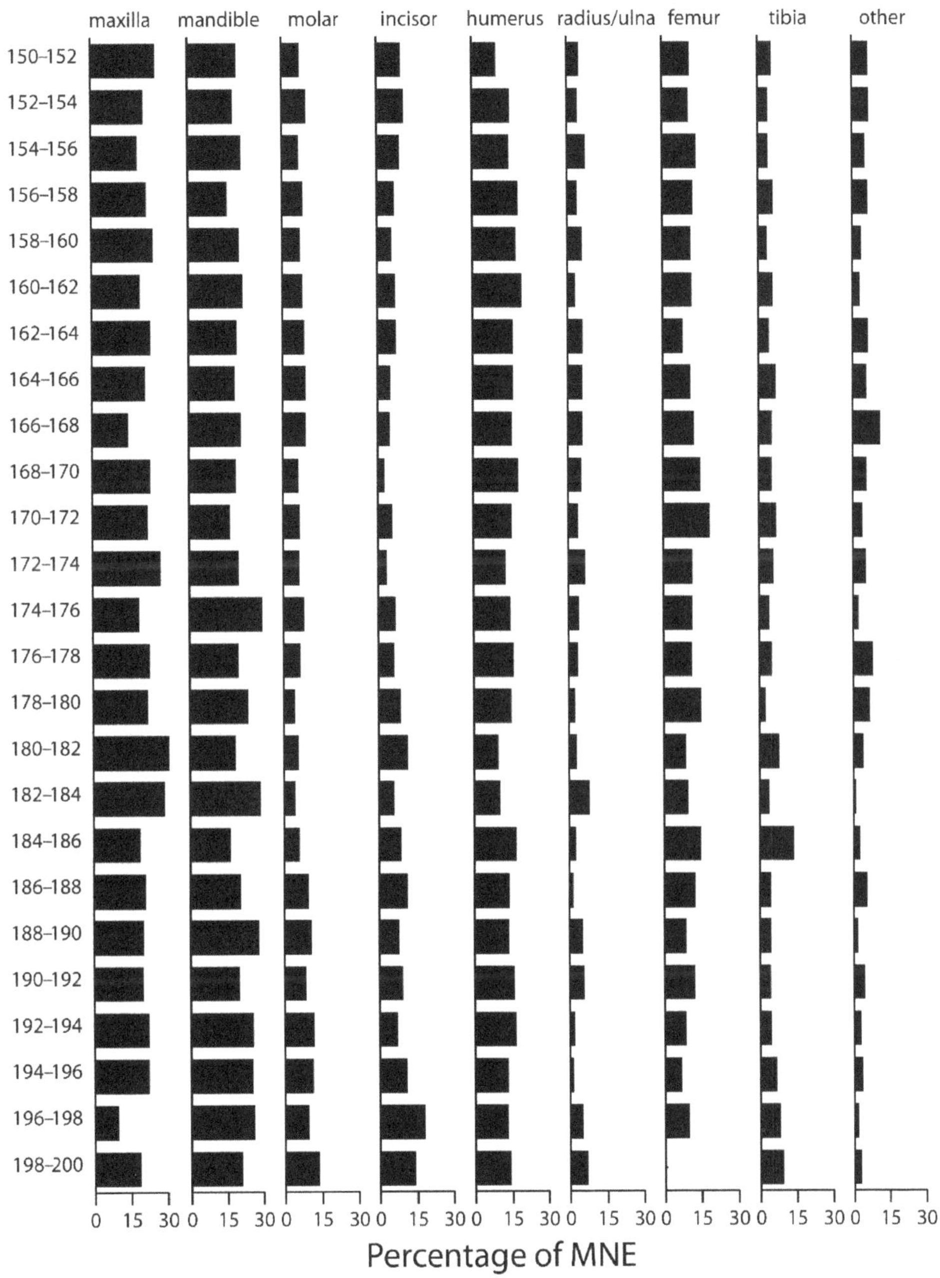

Figure 3. Skeletal part representation of micromammals by spit (cm), expressed in percentage of MNE (Minimum Number of Elements). Spits 200–214 are not represented due to the small size of the samples.

SU	Spit	J4 IIIa	J4 IIIb	J4 IIIc	J4 IIId	J4 III total
15–humus	145–147					54
15	147–150					135
15	150–152					183
15	152–154					144
15/16	154–156					92
16	156–158	104	134	54	92	96
16	158–160	58	78	73	93	76
16/17	160–162	40	55	118	42	64
16/17	162–164	91	27	124	85	82
16/17	164–166	116	77	119	70	95
17	166–168	145	101	126	91	102
17	168–170	111	72	60	65	77
17	170–172	78	26	68	27	50
17	172–174	27	34	39	22	30
17	174–176	117	64	14	25	55
17	176–178	109	48	63	99	80
17	178–180	99	14	19	44	44
17	180–182	34	18	19	18	22
17	182–184	11	5	10	28	14
17/18	184–186	10	13	10	30	16
17/18	186–188	30	67	34	30	40
17/18	188–190	13	24	8	56	25
17/18	190–192	49	2	69	95	54
18	192–194	34	19	24	49	31
18	194–196	13	15	20	24	18
18	196–198	7	14	15	10	11
18	198–200	10	2	16	4	8
18	200–202	6		10	14	10
18	202–204	16		23	14	18
18	204–206	30		6	21	19
18	206–208	2		26	13	13
18	208–210	4		8	19	10
18	210–212	9		4	5	6
19	212–214	7		9	6	7

Figure 4. Density of micromammal remains (MNE/L) along the stratigraphic sequence of the quarter of a square metre J4 III and its smaller subdivisions (a, b, c, and d). Spits highlighted in orange correspond to maximum density peaks. SU: Stratigraphic Unit.

As expected, there is a negative correlation between micromammals (density and MNE micro) and the proxies for human occupation (N fauna and N lithics). All relationships are statistically significant except for the comparison between lithics (J4 III) and the density of micromammals ($r = -0.303$, $p = 0.09$). The lack of statistical significance between these variables may be due to sampling bias, given the small number of lithic artefacts greater than 2 cm recovered from the limited surface area of J4 III.

Table 3. Pearson's *r* correlation coefficient (lower left part of the table) and associated *p*-values (upper right part of the table) for different variables along the spit sequence. Coefficient values in bold are associated with a significant *p*-value at the 5% threshold. Negative correlations in orange, positive correlations in green, the darker shades indicate stronger correlations. N = number of remains and W = weight of faunal remains and lithic artefacts recovered from the sieve residues (g/L). "Density" corresponds to the density of micromammal remains (MNE/L); "MNE micro" refers to micromammals and "fauna" to macrofauna.

	Density	MNE Micro	N Fauna J4 III	N Fauna J4 III-IV	N Lithics J4 III	N Lithics J4 III-IV	W Fauna	W Lithics
Density		1.97×10^{-10}	0.681	0.040	0.091	0.002	0.604	0.717
MNE Micro	**0.864**		0.866	0.003	0.019	3.79×10^{-5}	0.234	0.095
N Fauna J4 III	0.076	0.031		8.55×10^{-5}	0.016	0.789	0.039	0.328
N Fauna J4 III-IV	**−0.364**	**−0.502**	**0.638**		0.001	3.62×10^{-5}	0.0002	0.002
N Lithics J4 III	−0.303	**−0.412**	**0.423**	**0.552**		0.001	0.017	0.088
N Lithics J4 III-IV	**−0.521**	**−0.661**	−0.049	**0.662**	**0.556**		0.012	0.002
W Fauna	0.095	−0.216	**0.366**	**0.621**	**0.417**	**0.441**		8.17×10^{-13}
W Lithics	0.067	−0.300	0.179	**0.531**	**0.306**	**0.530**	**0.907**	

5. Discussion

When compared with data presented by Andrews [7], the accumulation of small mammals at Saint-Césaire is consistent with the action of a category 1 nocturnal raptor. As short-eared (*Asio otus*) and long-eared owls (*Asio flammeus*) do not generally nest in karst contexts [26,27], the assemblage was most likely produced by a barn owl (*Tyto alba*). While this species has not been identified at the site, this does not exclude it from being responsible for the accumulation [7].

The overall accumulation shows a strong over-representation of *M. arvalis/agrestis* and *L. gregalis*, which may be related to the preferences of the main accumulating agent to hunt in open landscapes. As mentioned above, further work is necessary to identify the source of the variations in species frequencies recorded in the stratigraphic sequence. Patterns of skeletal representation do not reveal any major changes in the preservation of skeletal elements that seem to be due to post-depositional processes or shifts in the type of predators. The observed variations in density of micromammal remains therefore appear to be primarily linked to the activities of the accumulating agent rather than external factors, such as variations in micromammal palaeocommunities.

Given that the micromammal remains were most likely accumulated by a nocturnal bird of prey, it is reasonable to assume that phases of human occupation created unfavourable conditions for the presence of these birds at the site. The negative correlations between the proxies for human occupation and the micromammal variables (density measures) are consistent with this conclusion. The density peaks observed along the stratigraphic sequence could therefore correspond to more ephemeral phases of human occupation or periods when the site was abandoned. The high density of remains between depths 162–170, with the density peak in the spit 166–168, could correspond to a period during which human occupations were less frequent or more ephemeral. This portion of the sequence corresponds to the transition between SU 17 (Indeterminate Aurignacian) and SU 16 (Middle/Late Aurignacian). Among the various peaks, the one observed in spit 176–178 is considerably more marked and could correspond to the abandonment of the site by human groups. Moreover, this particular accumulation episode of micromammals within SU 17 distinguishes two portions of this SU 17 that were difficult to identify during fieldwork. The small peak in spit 190–192 coincides with the base of SU 17, which is markedly different from SU 18, which contains a limited number of small mammal remains. This abrupt shift in small mammal density between SU 17 and SU 18 may prove useful in future stratigraphic analyses. Our results confirm that the quantification of micromammal accumulations can be a useful proxy for better documenting stratigraphic sequences, whereas accumulation peaks can provide evidence for phases of site abandonment. Tracking shifts in micromammal densities therefore represents a new complementary tool for characterising the frequency and timing of human occupations at archaeological sites which commonly rely uniquely on macrofauna and lithics. Moreover, with the development of ancient DNA recovered from sediments [28,29], this type of analysis can aid in identifying sampling areas most likely to preserve human DNA.

Our results may be difficult to place in a broader perspective as similar analyses are still rare for Palaeolithic contexts. Nonetheless, they indicate that due consideration should be paid to recovery methods employed during excavations. Among other factors, clast content and sediment volume are particularly critical, as both may produce biases in artefact density estimates. These problems may partly be controlled for by precisely measuring spit volumes or recording spits of constant volume.

6. Conclusions

At Saint-Césaire, higher densities of micromammal remains seem to coincide with phases of limited human occupation whereas lower densities appear to denote periods when the site was only sporadically occupied or abandoned. In the upper part of the archaeological sequence, four accumulation peaks provide new data for (a) securely assigning spits to particular stratigraphic units in inter-unit transitional zones, and (b) identifying different occupation phases within the same stratigraphic unit.

Shifts in micromammal densities may not always produce reliable results due to the impact of post-depositional processes that remobilised archaeological material and/or geochemical alterations that locally affected faunal preservation. Despite these issues, micromammal proxies have considerable potential for documenting changes in patterns of human occupation, particularly if they are coupled with standard taphonomic analyses of microfaunal and macrofaunal remains and when these data are assessed in light of dynamic site formation processes and estimated sedimentation rates. In sum, the development of micromammal analyses should contribute to a more reliable interpretation of archaeological sites as well as the timing and frequency of human occupations.

Supplementary Materials: The following are available online at https://www.mdpi.com/2571-550X/4/4/33/s1, Table S1: Micromammal MNE (Minimum Number of Elements) for each spit and SU, Table S2: Variables taken into account for the correlation test along the spit sequence. N = Number of remains and W = weight of faunal remains and lithic artefacts recovered from the sieve residues (g/L).

Author Contributions: Conceptualisation, L.L. and E.M.; data curation, L.L., E.M., B.G., A.M., F.B., D.F., J.M., H.R. and I.C.; formal analysis, L.L.; methodology, L.L. and E.M.; supervision, I.C.; writing—original draft, L.L.; writing—review and editing, E.M., B.G., C.B., V.L., H.R., E.T., D.T. and I.C. All authors have read and agreed to the published version of the manuscript.

Funding: This research received no external funding.

Institutional Review Board Statement: Not applicable.

Informed Consent Statement: Not applicable.

Data Availability Statement: The data presented in this study are available in this paper.

Acknowledgments: We thank the *Service régional de l'Archéologie* (SRA) *Nouvelle-Aquitaine*, the *Service départemental de l'Archéologie* (SDA) of *Charente-Maritime* and the *Conseil départemental* of *Charente-Maritime* for their financial support to the *Projet Collectif de Recherche* (PCR) La Roche-à-Pierrot directed by I.C. We also want to thank the two anonymous reviewers for their suggestions and comments that improved the final version of the manuscript.

Conflicts of Interest: The authors declare no conflict of interest.

References

1. Royer, A.; Lécuyer, C.; Montuire, S.; Escarguel, G.; Fourel, F.; Mann, A.; Maureille, B. Late Pleistocene (MIS 3-4) climate inferred from micromammal communities and $\delta^{18}O$ of rodents from Les Pradelles, France. *Quat. Res.* **2013**, *80*, 113–124. [CrossRef]
2. Lebreton, L.; Desclaux, E.; Hanquet, C.; Moigne, A.-M.; Perrenoud, C. Environmental context of the Caune de l'Arago Acheulean occupations (Tautavel, France), new insights from microvertebrates in Q-R levels. *Quat. Int.* **2016**, *411*, 182–192. [CrossRef]
3. Stoetzel, E.; Montuire, S. Les rongeurs, indicateurs des paléoclimats: Application aux assemblages de trois sites du nord de la France. *Quaternaire* **2016**, *27*, 227–238. [CrossRef]
4. López-García, J.M.; Blain, H.A.; Sánchez-Bandera, C.; Cohen, J.; Lebreton, L.; Montuire, S.; Stewart, J.R.; Desclaux, E. Multi-method approach using small vertebrate assemblages to reconstruct the Marine Isotope Stage 6 climate and environment of the Lazaret cave sequence (Maritime Alps, Nice, France). *Palaeogeogr. Palaeoclimatol. Palaeoecol.* **2021**, *577*, 110529. [CrossRef]
5. Piñero, P.; Agustí, J.; Blain, H.A.; Laplana, C. Paleoenvironmental reconstruction of the Early Pleistocene site of Quibas (SE Spain) using a rodent assemblage. *C. R. Palevol* **2016**, *15*, 659–668. [CrossRef]
6. Denys, C. Nouveaux critères de reconnaissance des concentrations de microvertébrés d'après l'étude des pelotes de chouettes du Botswana (Afrique australe). *Bulletin du Muséum National d'Histoire Naturelle* **1985**, *7*, 879–933.
7. Andrews, P. *Owls, Caves and Fossils: Predation, Preservation, and Accumulation of Small Mammal Bones in Caves, with an Analysis of the Pleistocene Cave Faunas from Westbury-Sub-Mendip*; University of Chicago Press: Somerset, UK, 1990; 231p.
8. Fernández-Jalvo, Y.; Andrews, P. Small Mammal Taphonomy of Gran Dolina, Atapuerca (Burgos), Spain. *J. Archaeol. Sci.* **1992**, *19*, 407–428. [CrossRef]
9. Fernández-Jalvo, Y.; Denys, C.; Andrews, P.; Williams, T.W.; Dauphin, Y.; Humphrey, L. Taphonomy and palaeoecology of Olduvai Bed-I (Pleistocene, Tanzania). *J. Hum. Evol.* **1998**, *34*, 137–172. [CrossRef] [PubMed]
10. Fernández-Jalvo, Y.; Andrews, P.; Denys, C.; Sese, C.; Stoetzel, E.; Marin Monfort, D.; Pequero, D. Taphonomy for taxonomists, implications in small mammal studies. *Quat. Sci. Rev.* **2016**, *139*, 138–157. [CrossRef]
11. Denys, C. Des référentiels en taphonomie des petits vertébrés: Bilan et perspectives. In *Taphonomie des Petits Vertébrés: Référentiels et Transferts Aux Fossiles*; Laroulandie, V., Mallye, J.-B., Denys, C., Eds.; British Archaeological Reports; International Series 2269; Archaeopress: Oxford, UK, 2011; pp. 7–22.
12. Rofes, J.; Ordiales, A.; Iriarte, E.; Cuenca-Bescós, G.; Galindo-Pellicena, M.Á.; Pérez-Romero, A.; Carretero, J.M.; Arsuaga, J.L. Human Activities, Biostratigraphy and Past Environment Revealed by Small-Mammal Associations at the Chalcolithic Levels of El Portalón de Cueva Mayor (Atapuerca, Spain). *Quaternary* **2021**, *4*, 16. [CrossRef]
13. Royer, A. Intérêts de l'utilisation des décapages lors des fouilles archéologiques pour l'étude des restes de petits vertébrés. *PALEO* **2014**, *25*, 277–286. [CrossRef]
14. Laroulandie, V.; Royer, A.; Mallye, J.-B.; Bailon, S.; Danger, M. Les petits vertébrés de Peyrazet et les chasseurs-collecteurs du Magdalénien supérieur. In *La Grotte-Abri de Peyrazet (Creysse, Lot, France) au Magdalénien. Originalité fonctionnelle d'un Habitat des Derniers Chasseurs de Rennes du Quercy*; Langlais, M., Laroulandie, V., Eds.; Supplément à Gallia Préhistoire 43; CNRS: Paris, France, 2021; pp. 141–152.
15. Rambaud, D.; Laroulandie, V.; Primault, J.; Béarez, P. Les poissons et les oiseaux du Taillis des Coteaux (Antigny, Vienne), niveaux magdaléniens: Origine naturelle ou culturelle? In *Taphonomie des Petits Vertébrés: Référentiels et Transferts Aux Fossiles*; Laroulandie, V., Mallye, J.-B., Denys, C., Eds.; British Archaeological Reports; International Series 2269; Archaeopress: Oxford, UK, 2011; pp. 167–179.
16. Richardson, C.; Miller, C. Recommendations for Protecting Raptors from Human Disturbance: A Review. *Wildl. Soc. Bull. (1973–2006)* **1997**, *25*, 634–638.
17. Martínez-Abraín, A.; Oro, D.; Jiménez, J.; Stewart, G.; Pullin, A. A systematic review of the effects of recreational activities on nesting birds of prey. *Basic Appl. Ecol.* **2010**, *11*, 312–319. [CrossRef]
18. Lévêque, F.; Vandermeersch, B. Découverte de restes humains dans un niveau castelperronien à Saint-Césaire (Charente-Maritime). *C. R. l'Acad. Sci. Paris Sér. D* **1980**, *291*, 187–189.
19. Gravina, B.; Bachellerie, F.; Caux, S.; Discamps, E.; Faivre, J.-P.; Galland, A.; Michel, A.; Teyssandier, N.; Bordes, J.-G. No reliable evidence for a Neanderthal-Châtelperronian association at La Roche-à-Pierrot, Saint-Césaire. *Sci. Rep.* **2018**, *8*, 15134. [CrossRef] [PubMed]
20. Morin, E. *Reassessing Paleolithic Subsistence: The Neandertal and Modern Human Foragers of Saint-Césaire*; Cambridge University Press: New York, NY, USA, 2012; 358p.
21. Lévêque, F.; Backer, A.M.; Guilbaud, M. (Eds.) *Context of a Late Neandertal. Implications of Multidisciplinary Research for the Transition to Upper Paleolithic Adaptations at Saint-Césaire, Charente-Maritime, France*; Monographs in World Archaeology 16; Prehistory Press: Madison, WI, USA, 1993; 131p.
22. Marquet, J.-C. *Paléoenvironnement et Chronologie des Sites du Domaine Atlantique Français d'âge Pléistocène Moyen et Supérieur D'après L'étude des Rongeurs*; Les Cahiers de la Claise, Supplément 2: Tours, France, 1993; 330p.
23. Crevecoeur, I. Reprise des fouilles à La Roche-à-Pierrot, Saint-Césaire. In *Le troisième Homme. Préhistoire de l'Altaï*; Cleyet-Merle, J.-J., Derevianko, A.P., Geneste, J.-M., Gravina, B., Krivoshapkin, A.I., Maureille, B., Turq, A., Shunkov, M.V., Slimak, L., Eds.; Réunion des Musées Nationaux-Grand Palais: Paris, France, 2017; p. 107.
24. Dodson, P.; Wexlar, D. Taphonomic investigations of owl pellets. *Paleobiology* **1979**, *5*, 275–284. [CrossRef]

25. Morin, E. Evidence for declines in human population densities during the early Upper Paleolithic in western Europe. *Proc. Natl. Acad. Sci. USA* **2008**, *105*, 48–53. [CrossRef]
26. Roberts, J.L.; Bowman, N. Diet and ecology of Short-eared Owls *Asio flammeus* breeding on heather moor. *Bird Study* **1986**, *33*, 12–17. [CrossRef]
27. Martínez, J.A.; Zuberogoitia, I. Habitat preferences for Long-eared Owls *Asio otus* and Little Owls *Athene noctua* in semi-arid environments at three spatial scales. *Bird Study* **2004**, *51*, 163–169. [CrossRef]
28. Slon, V.; Hopfe, C.; Weiß, C.L.; Mafessoni, F.; de la Rasilla, M.; Lalueza-Fox, C.; Rosas, A.; Soressi, M.; Knul, M.V.; Miller, R.; et al. Neandertal and Denisovan DNA from Pleistocene sediments. *Science* **2017**, *356*, 605–608. [CrossRef]
29. Rohland, N.; Glocke, I.; Aximu-Petri, A.; Meyer, M. Extraction of highly degraded DNA from ancient bones, teeth and sediments for high-throughput sequencing. *Nat. Protoc.* **2018**, *13*, 2447–2461. [CrossRef] [PubMed]

quaternary

MDPI

Article

Defining Fragmentation Patterns of Archaeological Bone Remains without Typologies: A Landmark-Based Approach on Rodent Mandibula

Marine Durocher [1,2,3,*], Sandrine Grouard [1], Violaine Nicolas [2], Renan Maestri [4] and Allowen Evin [3]

1 Laboratoire Archéozoologie, Archéobotanique: Sociétés, Pratiques, Environnements, Muséum National d'Histoire Naturelle, CNRS, 75005 Paris, France; sandrine.grouard@mnhn.fr

2 Institut de Systématique, Évolution, Biodiversité, Muséum National d'Histoire Naturelle, CNRS, Sorbonne Université, EPHE, Université des Antilles, 75005 Paris, France; violaine.colin@mnhn.fr

3 Institut des Sciences de l'Evolution-Montpellier, UMR 5554-ISEM, CNRS, Université de Montpellier, IRD, EPHE, 34095 Montpellier, France; allowen.evin@umontpellier.fr

4 Instituto de Biociências, Universidade Federal do Rio Grande do Sul (UFRGS), Porto Alegre 90040-060, Brazil; renanmaestri@gmail.com

* Correspondence: marine.durocher@mnhn.fr

Abstract: Fragmentation is a recurring feature of archaeological faunal material, and impacts many aspects of zooarchaeological studies from taxonomical identification to biometric studies. It can result from anthropic and natural actions that occurred respectively before and/or after bone deposit. While several bone fragmentation typologies have been described, they are currently based on both macroscopic observations and researcher subjectivity and lack the universality necessary for inter-study comparisons. To fulfill this need we present a standardized landmark-based protocol for the description and quantification of mandibular fragmentation patterns, using two insular rodents of different sizes as models. The rice rats (Oryzomyini tribe) and the agouti (*Dasyprocta*) from the Lesser Antilles were abundant during the pre-Columbian Ceramic Age (500 BCE-1500 CE). Their mandibles' shapes were quantified using the coordinates of 13 2D-landmarks. We show that landmark-based measurements can be used to:—assess the preservation differences between taxa of the same taxonomic group (e.g., rodents),—estimate the level of preservation of a skeletal part (e.g., mandible),—describe fragmentation patterns without pre-existing typologies and—facilitate the application of geometric morphometric methods to fragmented archaeological material. Our novel approach, leveraging fragmentation analyses and establishing specific fragmentation patterns, frees itself from existing typologies and could be systematically applied to future research.

Keywords: geometric morphometrics; taphonomy; zooarchaeology; mandible; fragmentation; rodents

1. Introduction

Archaeological bones are studied to investigate the relationships between past human societies and animals as well as the environment. Bone fragmentation results from pre- and/or post-depositional taphonomic processes that may be caused by natural or anthropogenic actions [1,2]: butchering and hammering of bones, trampling, climato-edaphic effects, non-human biological agents (e.g., carnivores), and breakage resulting from excavation, transportation, and storage [3]. One common aim of taphonomic studies is to characterize fragmentation patterns throughout the assemblage in order to compare and recognize the taphonomic processes involved [4–12].

Alteration of faunal remains affects preferentially specific bones and specific areas on the bones [13]. Compared to other bones, mandibles of large [14] and small mammals [15] tend to be well preserved in archaeological deposits as a result of their high structural density. However, the bone density differs between the taxa and the bone mineral composition varies according to the age of the animal, which induces different structural and mechanical properties [16]. As a result, within the same group of taxa, e.g., rodents, mandibles do not always have the same mechanical properties [15]. Despite these observations, mandibles tend to show a non-random fragmentation and can be divided into two parts: the anterior part or *corpus* which bears the teeth, and the posterior part or *ramus* that is attached to the skull. The *corpus* is generally better preserved [15,17]. However, the ability to identify to which species or group of species belongs a mandibular fragment depends on its size and the anatomical part represented [18,19].

Landmark-based approaches, such as the ones commonly used in geometric morphometrics [20–22] employ two or three-dimensional coordinates of points to quantify the shape and the size of objects. Many morphometrical and statistical analyses do not accept missing data and require complete specimens. In the case of fragmented material—a common case in archaeology or paleontology—one frequently reduces either the number of specimens [23] or the number of variables [24]. Doing so, a trade-off has to be found between the number of available specimens and the number of available landmarks. Morphometric-based techniques for fossil reconstruction offer the possibility to estimate the location of missing landmarks in order to maximize the number of specimens studied [23,25]. However, such techniques require specimens that can be used as reference—a challenge when the number of available specimens per archaeological site or period is limited and with each specimen displaying only subtle morphometric differences. An alternative approach to studying fragmented shrew mandibles was suggested by Cornette and collaborators (2015) [26]. This approach is based on coordinates of points along the external outline (sliding semi-landmarks) of the mandible from a 2D image of its lateral labial view. Analyses are then performed, without estimating missing parts, but using a pre-established fragmentation typology describing seven fragments ranging from the complete mandible to the single ascending *ramus* [26].

On another hand, other typologies of fragmentation patterns have been described for the mandible [18,27,28] but they vary from one study to the other, relying on the taxa and geographical areas. In this paper, we propose a new, objective, and quantitative method for the analysis of mandibular fragments. With this approach, no zonation typology is required and fragments of all sizes can be studied as long as at least the coordinates of two landmarks can be obtained. This method will also facilitate the selection of the largest number of fragments that can be used in future geometric morphometric studies while retaining the capacity to characterize shape variations, applied here in a case study on modern oryzomyine rodents.

2. Materials

Two rodents were used as a model: the rice rat (Muroidea, Cricetidae, Oryzomyini tribe) and the agouti (Caviomorpha, Dasyproctidae, *Dasyprocta* spp.). These rodents are recovered from Pre-Columbian archaeological sites of the Lesser Antilles. Both rodents were actively consumed by pre-Columbian human populations as evidenced by the presence of butchery cuts and burn marks [29–33].

We analyzed a total of 833 archaeological mandibular fragments among which 767 belong to rice rats and 66 to agoutis. Specimens originate from 17 Lesser Antilleans islands and 45 archaeological sites (Table S1.1). The mandibular fragments studied originate from the same archaeological contexts as the teeth studied elsewhere [34]. Specimens are housed in the collections of the Regional Service of Archaeology (Service Régional de l'Archéologie, Direction des Affaires Culturelles (DAC)) of Guadeloupe and Martinique, at the Muséum national d'Histoire naturelle de Paris (MNHN, France), and at the Florida Museum of Natural History (FLMNH, USA). These remains date from the Ceramic Age (500 BCE–1500 CE) and were part of faunal assemblages excavated and studied heterogeneously. Three categories can be described: the first category (T-1) corresponds to the most recently excavated sites from the 2000–2010s, water sieved with two meshes (1 to 2.7 mm screen) and for which the faunal assemblages have already been studied zooarchaeologically. This category represents 73% of the Oryzomyini and 35% of the Dasyproctidae mandibles included in the present study. The second category (T-2) includes sites from older excavations dated from the 60′ and 90′ where only the "almost-complete" bones were recovered (sometimes with sieving). This corresponds to 11% of the Oryzomyini and 25% of the Dasyproctidae studied. The third category (T-3) includes remains recovered in heavy boxes of sediments and mixed artifacts never properly studied. This last category represents 16% of the Oryzomyini and 40% of the Dasyproctidae. Note that the most fragmented mandibles might not have been recovered and included in our dataset because of the method of study and recovery of the material used for the T-2 and T-3 excavation sites. Conversely, the mandible remains that were recovered from the T-1 category have been studied exhaustively and likely include all types of fragments, due to the good conditions and time dedicated to these zooarchaeological studies.

In addition, 554 complete mandibles of modern rice rats originating from the South American continent (Table S1.2) serve as a basis for comparison in the direct application of the method developed. All these mandibles have already been analyzed [35]. These mandibles belong to 24 genera, are not fragmented, and therefore preserve all the morphological information associated with taxonomic differentiation.

3. Methods

The lateral view of each archaeological mandible fragment was photographed respecting the anatomical positioning of the mandibles. For agoutis, the correct position of the mandibles was checked using a bubble level. In the case of the smaller Oryzomyini mandibles, the position could only be visually inspected. Photographs of all the archeological material were captured by Marine Durocher (M.D.) using a CANON EOS 80D camera, while South American Oryzomyini was captured by Renan Maestri (R.M.) using a Nikon P100 camera [36]. Results of the two data sets were contrasted but not pooled due to methodological and technical interstudy differences (camera and mandible positioning differ). A maximum of thirteen 2D landmarks were digitized by M.D. on each fragment using the TPSdig2 software package [37]. A complete mandible was therefore characterized by 13 2D coordinates while a minimum of two coordinates were recorded on the smallest fragments. The protocol already established for South-American Sigmodontinae [35] was used and adapted to agouti. The protocols for the two rodents differ only by the position of landmark 13 which corresponds to the intersection between the *corpus* and the ventral insertion area of the anterior deep masseter in the agouti and to the most posterior-ventral limit of the mandibular symphysis for Oryzomyini (Table S2).

First, we assessed the number of landmarks present in each specimen (from 2 to 13). Frequency in the number of available landmarks was quantified, and visualized by a pie chart (package 'ggplot 2' v3.1.1; [38]). Then, the frequency of each landmark presence was quantified, and visualized with a bar plot (package graphics v3.5.3; [39]).

The second step was to define all possible fragmentation patterns, i.e., all the combinations of 2 to 13 landmarks according to the fragmentation degree of the mandible (e.g., landmarks #1, #4, and #12; landmarks #2 and #6; . . .). For each number of landmarks (from 2 to 13) the frequencies of the three most common patterns were quantified and visualized with box plots. The R function "printbest" (available upon request) allows us to represent the three most frequent fragmentation patterns in the dataset per number of digitized landmarks. The difference of preservation between the two taxa was assessed with khi2 tests.

In order to assess whether variation in the way remains was recovered from the excavation affects the number of identified fragments and fragmentation patterns, we observed the representation frequency of the most common fragmentation pattern in the assemblage, according to the three types of contexts and for the two rodents.

Finally, in order to determine the characterization capacity of the shape variation of the patterns derived from the methods (second requirement of a geometric morphometrics study), the archaeological patterns, as well as the complete protocol (13 landmarks), were applied on oryzomyine specimens from reference collection. The number of principal components (PCs) that maximize the cross-validation percentage (CVP) was selected. It allowed us to calculate a CVP common to all patterns as well as the complete protocol.

4. Results

4.1. Mandibular Preservation

Only 0.8% (N = 6/767) of the Oryzomyini and none of the agouti mandibles were complete and measured by the full set of landmarks (Figure 1). For Agoutis, a maximum of 10 landmarks could be measured but only on a single specimen (1.9%). In comparison, 10 landmarks could be measured on 6.3% (N = 48/767) of the Oryzomyini mandibles (Figure 1). The same contrast is observed for the smallest fragments with only 0.7% (N = 5/767) of the Oryzomyini against 6% (N = 4/66) of the Dasyproctidae mandibles that are represented by only 2 landmarks (Figure 1). For both taxa, the number of fragments varies depending on the number of landmarks available (Figure 2). Most measured mandibles allow the positioning of six landmarks for the Dasyproctidae (21.6%; N = 14/66) and seven for the Oryzomyini (24.5%; N = 188/767).

All landmarks are not equally preserved but their occurrence displays the same trend for the two taxa (Figure 2). For both taxa, the superior part of the *corpus* of the mandible (quantified by landmarks #1 to #5), as well as the lower part of the *corpus* (quantified by landmarks #12 and #13) are particularly well preserved compared to the *ramus* (landmarks #6 to #11).

For each digitized landmark, the distribution pattern is also quite similar between *Dasyprocta* and Oryzomyini but some variations can be detected. For instance, while landmarks #3 and #12 are the most often present for both species, landmark #9 is never measured on agouti but is present on 20% of the rice rat mandibles. In conclusion, even if the mandibles of the two rodents seem to have similar preservation, the size of the fragments differs between the two (i.e., number of landmarks, Figure 1) as well as the preserved anatomical parts (i.e., location of landmarks, Figure 2). Altogether these results show that mandible preservation deeply differ between taxa (χ^2 = 25.6, df = 12, *p*-value = 0.012). As a consequence, the two taxa were analyzed separately in further analyses.

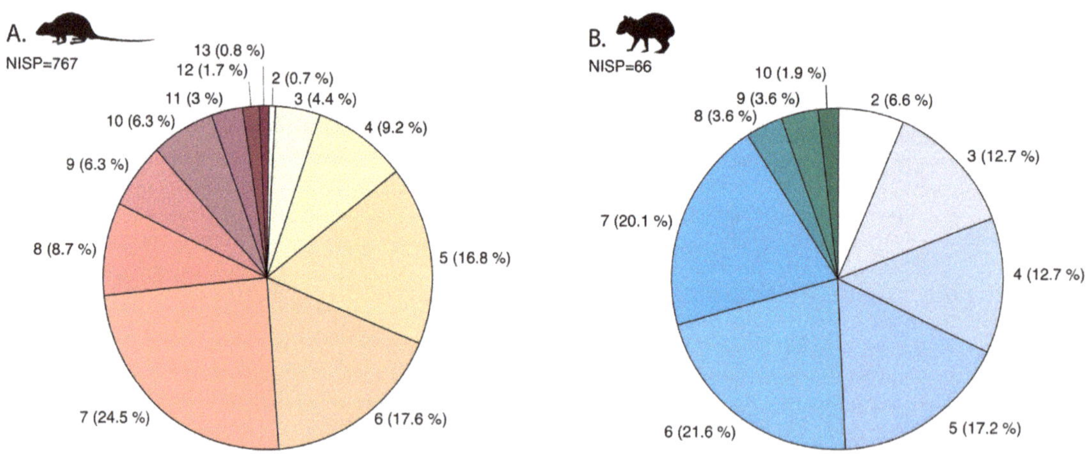

Figure 1. Pie chart representing the preservation for Oryzomyini (**A**) and *Dasyprocta* (**B**) mandibles according to the number of available landmarks (from 2 to 13). See Table S2 for the number of Oryzomyini and *Dasyprocta* specimens (NISP) for each digitalized landmark.

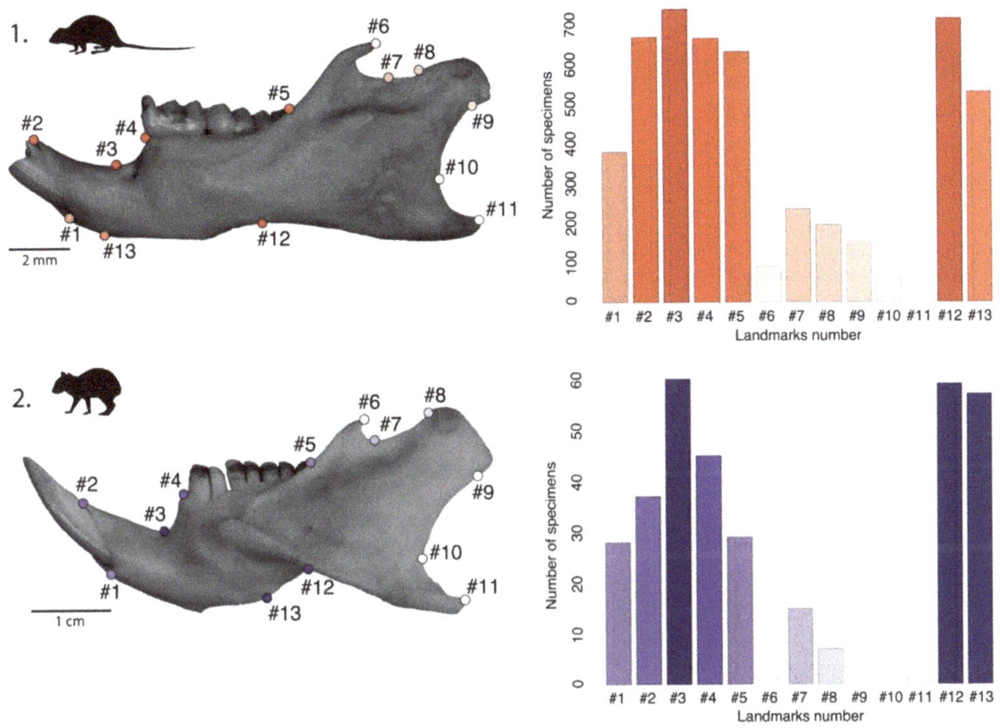

Figure 2. (**Left**) Protocols for positioning the landmarks for the Oryzomyini (**1**) and for the *Dasyprocta* (**2**) colored according to the values of the histogram (**Right**) representing the number of mandibles on which each type of landmark was measured (Oryzomyini NISP = 767, *Dasyprocta* NISP = 66). (See Table S3 for the landmark descriptions).

4.2. Fragmentation Patterns

For each number of measured landmarks (from 2 to 13 for rice rats and from 2 to 10 for agoutis) the three most frequent patterns were identified and visualized (categories P1 to P3, Figure 3). Not surprisingly, the smaller the number of landmarks measured the greater the number of fragments available. On the mandibular *corpus,* up to 7 landmarks can be digitized, which corresponds to the majority of quantified fragments.

The category P1 corresponds to the most represented pattern in the assemblage and allows from 2.09% (13 landmarks) to 89.96% (for 2 landmarks) of the oryzomyine mandibles to be measured. This category allows to measure up to 33% more mandibles than P2 and up to 38% more than P3. Within the Dasyproctidae, P1 allows us to measure from less than 1% (for 10 landmarks) to 98.21% (for 2 landmarks) of the assemblage, which is up to 38% more fragments than P2 and P3 (Figure 3).

These patterns can help to define which anatomical parts of the mandible could be studied with the maximum number of specimens. They are not used here to establish a typology. However, the 11 most present patterns (P1) of the oryzomyine depict an anatomical evolution along the mandible (Table S4), starting from its anterior to its posterior part, in accordance with the morphology and the mechanical properties of the mandible (Figure 3).

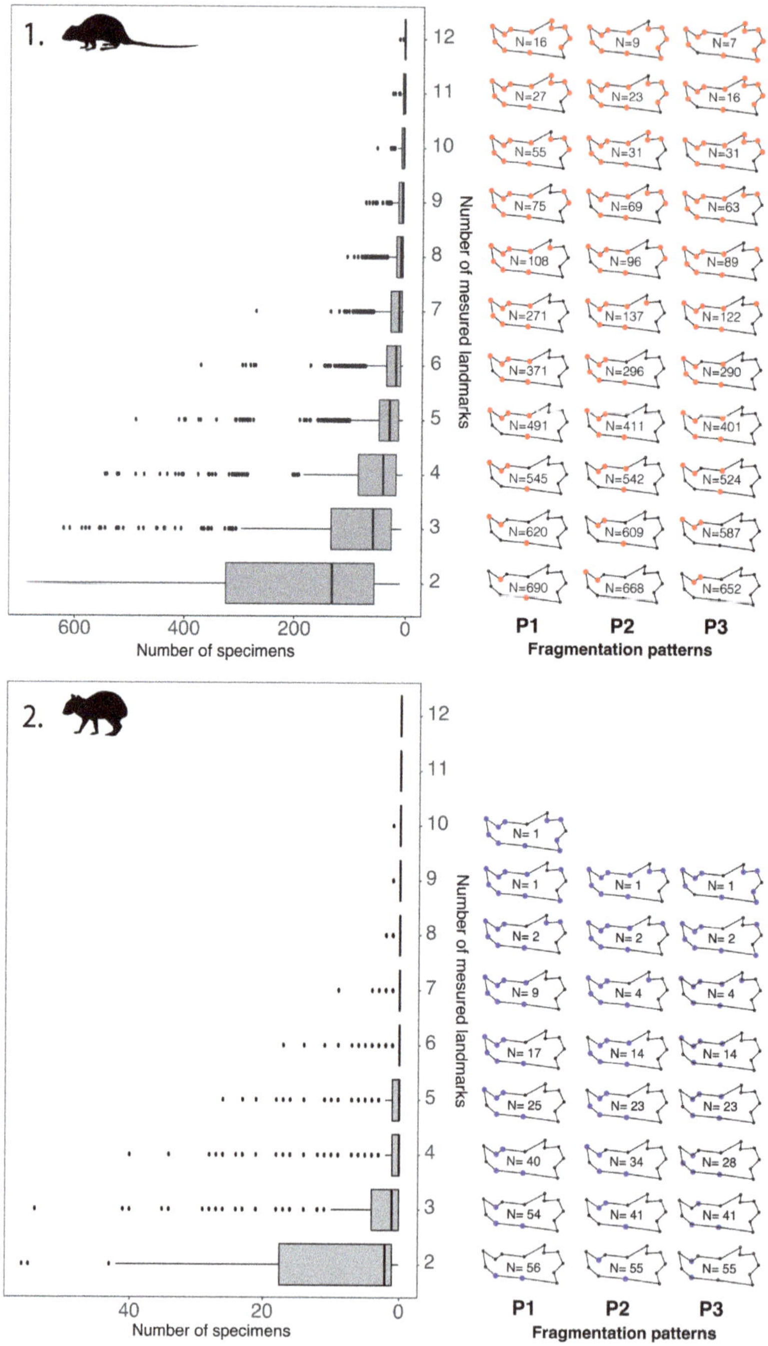

Figure 3. (**Left**) Number of fragmentation patterns for N landmarks on Oryzomyini (**1**) and *Dasyprocta* (**2**) mandibles, (**Right**) categories of the first three most frequent fragmentation patterns for N landmarks (P1, P2, P3).

Because the excavation techniques can have an impact on finding a fragment, the three types of recovery were compared (Figure 4). For both taxa, the three types show the same proportions of remains independently of the size of the fragment (Figure 4). However, a much larger number of fragments come from T-1 suggesting that the recovery methodologies may influence the number but not the size of the fragments identified.

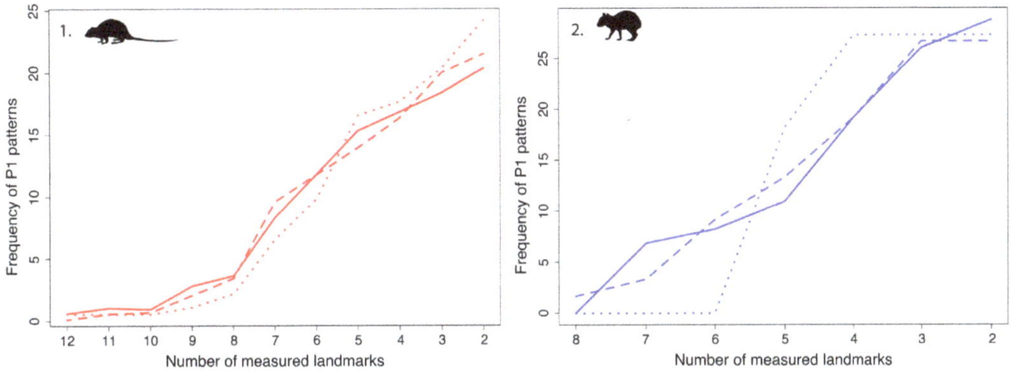

Figure 4. Frequency of the most frequent fragmentation patterns (category P1) in the assemblage, according to the three types of recovery and study of the faunal material (T-1: solid line, T-2: dashed line, T-3: dotted line), for (**1**) Oryzomyini mandibles and (**2**) *Dasyprocta* mandibles.

4.3. Which Pattern to Select for a Geometric Morphometrics Study? The Case of Oryzomyini

When dealing with a geometric morphometrics study numerous criteria must be considered to select the most suitable mandibular fragmentation pattern(s). The largest number of individuals should be targeted. Yet, the patterns should maintain the ability to discriminate groups.

In order to assess the taxonomic information carried by the various patterns, the efficiency to discriminate the Oryzomyini genera was assessed for the most frequent patterns (Figure 3, Table S4). The discriminatory power of the complete mandibula was confronted with the patterns predominantly found in the archaeological record.

With the studied case of Oryzomyini, in order to determine the fragmentation pattern(s) that best meet these two criteria we used a reference sample comprising 554 complete continental oryzomyine mandibles. For each mandible, the taxonomic attribution was used at the genus level to consider the maximal number of individuals. This case study allows us to test the capacity of the different fragmentation patterns to characterize the phenotypic variation between genera.

When studying the 767 archaeological mandibular fragments of Oryzomyini, 27 most frequent fragment patterns are defined according to three counts: the category of frequency of measurement (P1, P2, P3), the number of landmarks measured (between 4 and 12), and the percentage of the archaeological *corpus* that could be measured with each of these patterns (Table S5). This selection meets an important requirement of a geometric morphometrics study, namely the search for the largest number of individuals studied.

To determine the discriminant power of the various fragmentation patterns, the 27 most common fragments in the archaeological records (3 most frequent patterns for 4 to 12 landmarks) were compared to the complete mandible (i.e., including 13 landmarks). The cross-validation percentages decrease with the number of landmarks used to characterize the mandible (Figure 5). While the complete protocol succeeded in correctly identifying at least 79% of the specimens, the protocols including 4 landmarks correctly identify ~49% of the specimens. In all those comparisons the random correct cross-validation turns around 4% (i.e., 100/24 genus). The inter-continental genus cross-validation percentages were calculated (Figure 5) and three patterns provided as good discrimination as the complete protocol. Those protocols are: P1-7LM (size: CVP = 23.3% (CI:19.5–26.8%); shape: 52.3%

(CI:47–57.6%)), P2-6LM (size: 23.3% (CI:19.5–27.6%); shape: 48% (CI:44.2–52. 2%)) and P3-6LM (size: 23.5% (CI: 20.7–27.2%); shape: 44.6% (CI: 39.8–48.9%)). These three patterns, which include 6 or 7 landmarks, therefore carry a strong taxonomic signal and can be applied to the archaeological assemblage.

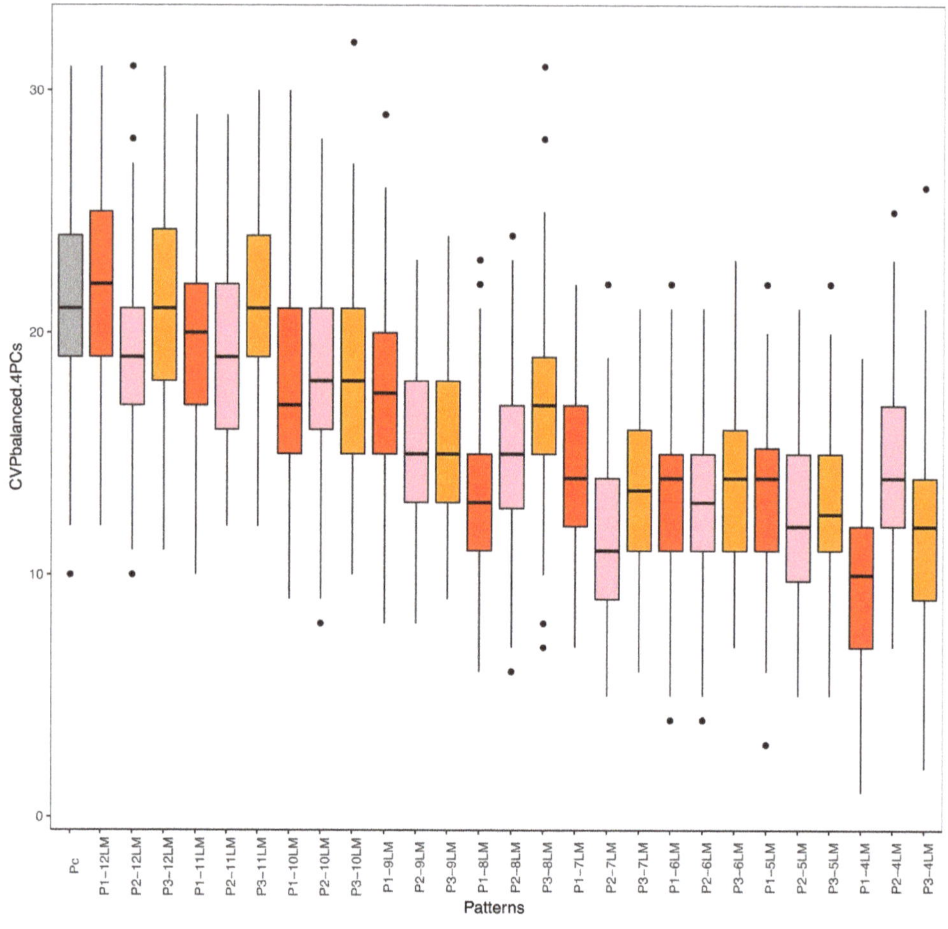

Figure 5. Boxplot representing the distribution of cross-validation percentages (CVP) for 4 PCs of each pattern (between 4 and 12 landmarks), according to the genus of attribution of specimens from the reference collections. Grey: Pc: complete protocol, Red: P1: category of the first most frequent fragmentation patterns with N landmarks, Pink: P2: second category, Orange: P3: third category.

5. Discussion

5.1. Impact of Mandibular Properties on Preservation

Our analyses revealed better preservation of the anterior part of the mandible for both taxa which likely results from the mechanical properties of the mandibles leading to better preservation of the *corpus* than the *ramus* [15,17]. Located on the posterior part of the mandible, the *ramus* has a complex shape with three processes (coronoid, condyloid, and angular), and a large flat and thin area, the masseteric fossa, more frequently fragmented. The angular process is thin and elongated, more developed in agoutis than rice rats, but it never exceeds the condyloid process. The condyloid process is developed on the top of the collar. It is wide and thicker in its rostral part. The coronoid process is thin, elongated, and curved over the mandibular incision. For the rice rat, the coronoid process is higher than the condyloid process, while it is smaller for the agouti (anatomical descriptions adapted to our taxa [40,41]). These three processes are extended beyond the main shape of the mandible and are fragility points that could explain the poor preservation of the *ramus*. The two most frequently measured landmark classes on the mandibles of both Oryzomyini and *Dasyprocta* (Figure 2) are representative of these mechanical properties, with the anterior part of the mandible (*corpus*) having much better preservation than the *ramus*. The degree of preservation of the different parts of the mandible can also be quantified. For example, in archaeological rice rats, the degree of preservation of the coronoid process is much lower than that of the incisor canal (12% versus 76%).

5.2. Application to Zooarchaeological Studies

In the patterns defined here by a methodology not based on diagnostic zones typology, the anterior part of the mandible is well represented. If the mandible is fractured in two pieces, between the *corpus* and the *ramus*, one should expect to identify these two parts in similar proportions, which is not the case here. Two factors may explain this under-representation of the posterior part. Firstly, the different shapes of the two parts of the mandible must be considered. As described before, the *ramus* is composed of three processes that contain significant weakness points that will lead to its fragmentation. Indeed, the different patterns obtained clearly showed that the first breaking parts in the most frequent patterns of fragmentation concern these processes. Moreover, in the entire area of the masseteric fossa, the bone thickness is reduced and the bone fragile. On the other hand, the shape of the *corpus* is less complex and tends to maintain its integrity, despite the taphonomy. The second factor is related to the identification of the fragments themselves. As the posterior part of the mandible tends to be more fragmented than the anterior part, the fragments will be both smaller in size and more variable in shape. These fragments will therefore tend to be classified as "unidentified remains" during the zooarchaeological study since these tiny fragments do not allow a taxonomic attribution or even—sometimes—a complete anatomical identification.

Despite these difficulties, the excavation and study methods used to recover the fragments do not seem to have impacted the representativeness of material fragmentation, at least in the case of these rodent mandibles (Figure 4). However, it cannot be excluded that the under-representation of mandibles from archaeological sites (T-2 & 3, both less precisely studied) in our assemblage are symptomatic of the difference in the recovery process of the remains. For simplicity only the three most frequent patterns were presented (Figure 3), but the proposed methodology enables targeting other specific patterns which may not have been represented here. An overview of all existing patterns could allow us to further develop the analysis. For example, differential preservation through time or space could be explored. It might also be possible to associate fragmentation patterns with more specific taphonomic agents and then quantify their ratio over the fragmentation of the archaeological faunal assemblage.

5.3. Interpretation of Differential Preservation between Taxa

The comparison of two taxa belonging to the same taxonomic class, present in the same archaeological deposits, helps to better understand differential preservation. Direct observation shows that agouti mandibles were more fragile than those of the rice rats. Despite the fact that the agouti mandibles are larger, the bone is proportionally thinner and therefore more subject to fracture. With the approach using no *a priori* typology, we were able to establish that the mandibular preservation differs between the two taxa. We suspect that these differences are due to the density of the mandible, and the size of the alveolar tooth row, but this should be further analyzed and quantified in order to better understand the preservation biases that exist within a faunal assemblage. Agoutis are under-represented compared to rice rats in archaeological sites. Classically, this would be interpreted as a lower proportion of agoutis in the diet of pre-Columbian populations or as a differential anthropological treatment (butchery, cooking, taming practices, or bone industry) of the two rodents as it has been described for other taxa [42]. Indeed, these remains come from human occupation sites, and butchery cuts and burn marks present on the bones suggest an anthropic accumulation. However, we cannot exclude that their under-representation could also be, at least partially, linked to poorer preservation of the bone remains of this taxon.

5.4. Future Geometric Morphometrics Application

Finally, this landmark-based approach should simplify the application of geometric morphometrics to fragmented archaeological material (see Section 4.3). Indeed, the use of geometric morphometrics is increasing in zooarchaeological studies. It is used, for example, to determine taxa that cannot be distinguished by classical comparative anatomy [43,44], or to characterize shape and size variations induced by domestication and/or captivity [45–47]. Derivatives of geometric morphometrics, through landmark-based approaches, like the one proposed in this article, are also in development [48]. They offer more objective and quantitative approaches than traditional methodologies.

Geometric morphometrics often requires the use of complete specimens and therefore the landmark coordinate acquisition protocols require to be suitable for characterizing—anatomically—meaningful shape variations while maintaining the largest number of individuals. The method we propose enables researchers to estimate and describe fragmentation without *a priori* and to select among all the existing patterns the most common ones that can be used for further analyses.

The objective and quantitative definition of patterns that can be established by this method also allow us to avoid the usual issues related to the typologies: i.e., the fragmentation may vary depending on the sites studied, their condition of preservation, and the taxa studied. Finally, the use of the exact same protocol to assess the typology for each of the bones will limit the bias induced by the subjectivity of discrete characters.

6. Conclusions

The proposed landmark-based approach enables us to identify the number and location of the landmarks to be used to increase the number of archaeological fragments suitable for a geometric morphometric study. Archaeological material—which by definition has spent several hundred (or thousands) years buried—is frequently subject to fragmentation. This was reflected in the mandibles available for geometric morphometric analysis. With this new method, all sizes and types of fragments can be analyzed, making it easier to obtain a broad overview of the faunal assemblage diversity. So far, most taphonomic and zooarchaeological (fragmentation patterns) studies are both based on discrete and subjective criteria, and the results are difficult to compare from one study to another due to the lack of a common protocol. Using this without-*a priori*-method, we reduce these limitations, and as consequence, help in studying archaeological faunal remains in a more quantitative manner. In this study, we were able to analyze and interpret the response of a specific skeletal part (i.e., the mandible) according to its mechanical properties and their impact on its preservation. We concluded that the mandible shows a systematic dichotomy in its preservation, with a higher *ramus* fragmentation than that of the mandibular *corpus*. However, there are non-negligible variations within each taxon that must be considered in the differential preservation analysis of an archaeological site. Future bioarchaeological research on agouti and rice rats in the Lesser Antilles archipelago will benefit from this method to perform geometric morphometric studies of the spatio-temporal variation of the taxa.

Supplementary Materials: The following supporting information can be downloaded at: https://www.mdpi.com/article/10.3390/quat5010014/s1, Table S1.1: Number of archaeological studied mandibles per taxa; Table S1.2: Number of Oryzomyini from the South American continent studied; Table S2: Number of identified specimens (NISP) of Oryzomyini and *Dasyprocta*; Table S3: Description of the 13 landmarks digitalized; Table S4: Description of the most common fragmentation patterns for Oryzomyini and *Dasyprocta*; Table S5: Oryzomyini most frequent patterns tested according to the three counts.

Author Contributions: Conceptualization, A.E. and M.D.; methodology, A.E. and M.D.; formal analysis, M.D. and A.E.; investigation, M.D., A.E. and R.M.; writing—original draft preparation, M.D., A.E., S.G. and V.N.; writing—review and editing, M.D., A.E., S.G., V.N. and R.M.; funding acquisition, S.G. and A.E. All authors have read and agreed to the published version of the manuscript.

Funding: This work was supported by the European Research Council (ERC) under the European Union's Horizon 2020 research and innovation program (grant agreement No. 852573); the PEPS ECOMOB grant of the CNRS, INEE (project MIGGRANT) and the ATM Blanche grant of the MNHN (project PILORI).

Institutional Review Board Statement: The study did not require ethical approval.

Informed Consent Statement: Not applicable.

Acknowledgments: We would like to acknowledge the curators and staff of the different museums and institutions for kindly providing access to the archaeological collections under their care, Guadeloupe (including Saint-Martin) and Martinique Departments of Archaeology (Regional Office of Cultural Affairs, French Ministry of Culture), Antigua and Barbuda National Parks and the Barbuda Research Complex (BRC), and the Florida Museum of Natural History in Gainesville (FLMNH). Additionally, to the curators and staff who provide access to samples in biological collections of sigmodontine, including the FMNH, AMNH, NMNH, MLP, MNUFRJ, and MZUSP. We are grateful to Alexis Durocher and Anthony Herrel for their help in language editing and revision. We also warmly thank the anonymous reviewers who gave helpful comments to improve the manuscript.

Conflicts of Interest: The authors declare no conflict of interest.

References

1. Behrensmeyer, A.K.; Kidwell, S.M.; Gastaldo, R.A. Taphonomy and Paleobiology. *Paleobiology* **2000**, *26*, 103–147. [CrossRef]
2. Brugal, J.-P. *TaphonomieS*; Archives Contemporaines: Paris, France, 2017.
3. Davis, S.J. *The Archaeology of Animals*; Routledge: London, UK, 2012.
4. Grouard, S. Première Contribution à La Détermination de l'origine Des Assemblages Fossiles de Prolagus Sardus (Sur Le Massif Corso-Sarde). In *Utilisation Des Statistiques En Recherche Taphonomique*; Mémoire de DEA, Université Paris X Nanterre: Paris, France, 1994.
5. Horard-Herbin, M.-P.; Lepetz, S.; Vallet, C.; Clavel, B.; Corbellini, J.-P.; Guintard, C. Des Traces Observées Aux Gestes Anthropiques: Le Projet D. Coupes. *Nouv. Archéologie* **2017**, *148*, 16–22. [CrossRef]
6. Pickering, T.R. Taphonomy of the Swartkrans hominid postcrania and its bearing on issues of meat-eating and fire management. In *Meat-Eating and Human Evolution*; Oxford University Press Oxford: Oxford, UK, 2001; pp. 33–51.
7. Pickering, T.R.; Clarke, R.J.; Moggi-Cecchi, J. Role of Carnivores in the Accumulation of the Sterkfontein Member 4 Hominid Assemblage: A Taphonomic Reassessment of the Complete Hominid Fossil Sample (1936–1999). *Am. J. Phys. Anthropol. Off. Publ. Am. Assoc. Phys. Anthropol.* **2004**, *125*, 1–15. [CrossRef] [PubMed]
8. Richardson, P.R.K. Carnivore Damage to Antelope Bones and Its Archaeological Implications. *Palaeontologia Africana* **1980**, *23*, 109–125.
9. Val, A.; Taru, P.; Steininger, C. New Taphonomic Analysis of Large-Bodied Primate Assemblage from Cooper's D, Bloubank Valley, South Africa. *S. Afr. Archaeol. Bull.* **2014**, *199*, 49–58.
10. Vigne, J. Découpe Du Cerf (Cervus Elaphus) Au Mésolithique Moyen, à Noyen-Sur-Seine (Seine-et-Marne): Analyses Tracéologique et Expérimentale. *Rev. Paléobiol.* **2006**, *24*, 69.
11. Vigne, J.-D.; Marinval-Vigne, M.-C.; De Lanfranchi, F.; Weiss, M.-C. Consommation Du Lapin-Rat (Prolagus Sardus Wagner) Au Néolithique Ancien Méditerranéen Abri d'Araguina-Sennola (Bonifacio, Corse). *Bull. Société Préhistorique Fr.* **1981**, *78*, 222–224.
12. Vigne, J.-D.; Marinval-Vigne, M.-C. Méthode pour la mise en évidence de la consommation du petit gibier. In *Animals and Archaeology: Hunters and Their Prey*; BAR International Series: Oxford, UK, 1983; Volume 163, pp. 239–242.
13. Gautier, A. L'apport de l'étude Des Vestiges Animaux à La Reconstitution Archéologique: Une Introduction à l'archéozoologie. *Rev. Nord* **1988**, *70*, 23–37. [CrossRef]

14. Brain, C.K. Hottentot Food Remains and Their Bearing on the Interpretation of Fossil Bone Assemblages. *Sci. Pap. Namib Desert Res. Stn.* **1967**, *1967*, 1–11.
15. Pavao, B.; Stahl, P.W. Structural Density Assays of Leporid Skeletal Elements with Implications for Taphonomic, Actualistic and Archaeological Research. *J. Archaeol. Sci.* **1999**, *26*, 53–66. [CrossRef]
16. Marean, C.W. Measuring the Post-Depositional Destruction of Bone in Archaeological Assemblages. *J. Archaeol. Sci.* **1991**, *18*, 677–694. [CrossRef]
17. Gifford-Gonzalez, D. Calibrating Bone Durability. In *An Introduction to Zooarchaeology*; Springer: Berlin/Heidelberg, Germany, 2018; pp. 453–474.
18. Andrews, P.; Cook, J. *Owls, Caves and Fossils: Predation, Preservation and Accumulation of Small Mammal Bones in Caves, with an Analysis of the Pleistocene Cave Faunas from Westbury-Sub-Mendip, Somerset, UK*; University of Chicago Press: Chicago, IL, USA, 1990.
19. Denys, C. Palaeoenvironmental and Palaeobiogeographical Significance of the Fossil Rodent Assemblages of Laetoli (Pliocene, Tanzania). *Palaeogeogr. Palaeoclimatol. Palaeoecol.* **1985**, *52*, 77–97. [CrossRef]
20. Bookstein, F.L. *Morphometric Tools for Landmark Data: Geometry and Biology*; Cambridge University Press: Cambridge, UK, 1991.
21. Mitteroecker, P.; Gunz, P. Advances in Geometric Morphometrics. *Evol. Biol.* **2009**, *36*, 235–247. [CrossRef]
22. Rohlf, F.J.; Slice, D. Extensions of the Procrustes Method for the Optimal Superimposition of Landmarks. *Syst. Biol.* **1990**, *39*, 40–59. [CrossRef]
23. Clavel, J.; Merceron, G.; Escarguel, G. Missing Data Estimation in Morphometrics: How Much Is Too Much? *Syst. Biol.* **2014**, *63*, 203–218. [CrossRef] [PubMed]
24. Fisher, A.E. When Is a Wolf a Dog? Combined Geometric Morphometrics and Stable Isotope Analyses for Differentiating Wild from Domestic Canids on the Northern Plains. *Plains Anthropol.* **2019**, *64*, 316–349. [CrossRef]
25. Neeser, R.; Ackermann, R.R.; Gain, J. Comparing the Accuracy and Precision of Three Techniques Used for Estimating Missing Landmarks When Reconstructing Fossil Hominin Crania. *Am. J. Phys. Anthropol. Off. Publ. Am. Assoc. Phys. Anthropol.* **2009**, *140*, 1–18. [CrossRef] [PubMed]
26. Cornette, R.; Herrel, A.; Stoetzel, E.; Moulin, S.; Hutterer, R.; Denys, C.; Baylac, M. Specific Information Levels in Relation to Fragmentation Patterns of Shrew Mandibles: Do Fragments Tell the Same Story? *J. Archaeol. Sci.* **2015**, *53*, 323–330. [CrossRef]
27. Comay, O.; Dayan, T. Taphonomic Signatures of Owls: New Insights into Micromammal Assemblages. *Palaeogeogr. Palaeoclimatol. Palaeoecol.* **2018**, *492*, 81–91. [CrossRef]
28. Dodson, P.; Wexlar, D. Taphonomic Investigations of Owl Pellets. *Paleobiology* **1979**, *5*, 275–284. [CrossRef]
29. Giovas, C.M. Pre-Columbian Amerindian Lifeways at the Sabazan Site, Carriacou, West Indies. *J. Isl. Coast. Archaeol.* **2018**, *13*, 161–190. [CrossRef]
30. Grouard, S. Variation Des Stratégies de Subsistance Des Précolombiens à Hope Estate, Saint-Martin (Petites Antilles), d'après l'analyse Des Restes Des Petits Vertébrés. In *Petits Animaux et Sociétés Humaines*; APDCA: Antibes, France, 2004; pp. 441–467.
31. Newsom, L.A.; Wing, E.S. *On Land and Sea: Native American Uses of Biological Resources in the West Indies*; University of Alabama Press: Tuscaloosa, AL, USA, 2004; ISBN 0-8173-1315-X.
32. Wing, E.S. Native American use of animals in the Caribbean. In *Biogeography of the West Indies: Patterns and Perspectives*, 2nd ed.; CRC Press: Boca Raton, FL, USA, 2001; pp. 481–518.
33. Wing, E.S. The Sustainability of Resources Used by Native Americans on Four Caribbean Islands. *Int. J. Osteoarchaeol.* **2001**, *11*, 112–126. [CrossRef]
34. Durocher, M.; Nicolas, V.; Perdikaris, S.; Bonnissent, D.; Robert, G.; Debue, K.; Evin, A.; Grouard, S. Archaeobiogeography of Extinct Rice Rats (Oryzomyini) in the Lesser Antilles during the Ceramic Age (500 BCE–1500 CE). *Holocene* **2020**, *31*, 433–445. [CrossRef]
35. Maestri, R.; Luza, A.L.; Barros, L.; Maria Hartz, S.; Ferrari, A.; Freitas, T.; Duarte, L. Geographical Variation of Body Size in Sigmodontine Rodents Depends on Both Environment and Phylogenetic Composition of Communities. *J. Biogeogr.* **2016**, *43*, 1192–1202. [CrossRef]
36. Maestri, R.; Monteiro, L.R.; Fornel, R.; Upham, N.S.; Patterson, B.D.; de Freitas, T.R.O. The Ecology of a Continental Evolutionary Radiation: Is the Radiation of Sigmodontine Rodents Adaptative? *Evolution* **2016**, *71*, 610–632. [CrossRef] [PubMed]
37. Rohlf, F.J. TpsDig, Ver. *Acta Ichthyol. Piscat.* **2010**, *43*, 219.
38. Wickham, H.; Chang, W.; Henry, L.; Pedersen, T.; Takahashi, K.; Wilke, C.; Woo, K. R Package 'Ggplot2'v. 3. *Cran R* **2019**, *11*, 154.
39. Murrell, P. *R Graphics*; CRC Press: Boca Raton, FL, USA, 2018.
40. Barone, R. Anatomie Comparée Des Mammifères Domestiques. In *Tome 1: Ostéologie*; Vigot Freres: Paris, France, 1986; p. 761.
41. Cornette, R.; Herrel, A.; Cosson, J.-F.; Poitevin, F.; Baylac, M. Rapid Morpho-Functional Changes among Insular Populations of the Greater White-Toothed Shrew. *Biol. J. Linn. Soc.* **2012**, *107*, 322–331. [CrossRef]
42. Chaix, L.; Méniel, P. *Archéozoologie: Les Animaux et L'archéologie*; Errance: Paris, France, 2001.
43. Cucchi, T.; Barnett, R.; Martínková, N.; Renaud, S.; Renvoisé, E.; Evin, A.; Sheridan, A.; Mainland, I.; Wickham-Jones, C.; Tougard, C.; et al. The Changing Pace of Insular Life: 5000 Years of Microevolution in the Orkney Vole (Microtus Arvalis Orcadensis). *Evolution* **2014**, *68*, 2804–2820. [CrossRef]
44. Evin, A.; Cucchi, T.; Cardini, A.; Vidarsdottir, U.S.; Larson, G.; Dobney, K. The Long and Winding Road: Identifying Pig Domestication through Molar Size and Shape. *J. Archaeol. Sci.* **2013**, *40*, 735–743. [CrossRef]

45. Evin, A.; Flink, L.G.; Bălăşescu, A.; Popovici, D.; Andreescu, R.; Bailey, D.; Mirea, P.; Lazăr, C.; Boroneanţ, A.; Bonsall, C. Unravelling the Complexity of Domestication: A Case Study Using Morphometrics and Ancient DNA Analyses of Archaeological Pigs from Romania. *Philos. Trans. R. Soc. B Biol. Sci.* **2015**, *370*, 20130616. [CrossRef] [PubMed]
46. Neaux, D.; Sansalone, G.; Lecompte, F.; Noûs, C.; Haruda, A.; Schafberg, R.; Cucchi, T. Examining the Effect of Feralization on Craniomandibular Morphology in Pigs, Sus Scrofa (Artiodactyla: Suidae). *Biol. J. Linn. Soc.* **2020**, *131*, 870–879. [CrossRef]
47. Harbers, H.; Neaux, D.; Ortiz, K.; Blanc, B.; Laurens, F.; Baly, I.; Callou, C.; Schafberg, R.; Haruda, A.; Lecompte, F.; et al. The Mark of Captivity: Plastic Responses in the Ankle Bone of a Wild Ungulate (*Sus Scrofa*). *R. Soc. Open Sci.* **2020**, *7*, 192039. [CrossRef] [PubMed]
48. Ameen, C.; Hulme-Beaman, A.; Evin, A.; Germonpré, M.; Britton, K.; Cucchi, T.; Larson, G.; Dobney, K. A Landmark-Based Approach for Assessing the Reliability of Mandibular Tooth Crowding as a Marker of Dog Domestication. *J. Archaeol. Sci.* **2017**, *85*, 41–50. [CrossRef]

quaternary

MDPI

Article

Understanding the Impact of Trampling on Rodent Bones

Yolanda Fernández-Jalvo [1,*], Lucía Rueda [1,2], Fernando Julian Fernández [3], Sara García-Morato [1,4], María Dolores Marin-Monfort [1,5,6], Claudia Ines Montalvo [7], Rodrigo Tomassini [6], Michael Chazan [8,9], Liora K. Horwitz [10] and Peter Andrews [11]

1 Museo Nacional de Ciencias Naturales (CSIC), José Gutiérrez Abascal, 2, 28006 Madrid, Spain; lucia.rueda.dominguez@gmail.com (L.R.); sagarc16@ucm.es (S.G.-M.); dores@mncn.csic.es (M.D.M.-M.)
2 Sciences de la Vie et de l'Environnement Université de Rennes 1, 35000 Rennes, France
3 CONICET-Grupo de Estudios en Arqueometría, Facultad de Ingeniería, Universidad de Buenos Aires (UBA), Av. Paseo Colón 850 (CP C1063ACV), Ciudad Autónoma de Buenos Aires 1063, Argentina; fernandezf77@yahoo.com.ar
4 Facultad de Ciencias Geológicas, Departamento de Geodinámica, Estratigrafía y Paleontología, Universidad Complutense de Madrid, Jose Antonio Novais 12, 28040 Madrid, Spain
5 Departamento de Botánica y Geología, Universidad de Valencia, Burjassot, Valencia, 28006 Madrid, Spain
6 Instituto Geológico del Sur (INGEOSUR), Department of Geology, Universidad Nacional del Sur (UNS)-CONICET, Avenida Alem 1253, Bahía Blanca 8000, Argentina; rodrigo.tomassini@yahoo.com.ar
7 Facultad de Ciencias Exactas y Naturales, Universidad Nacional de La Pampa, Uruguay 151, Santa Rosa 6300, Argentina; cmontalvolp@yahoo.com.ar
8 Department of Anthropology, University of Toronto, Toronto, ON M5S 2S2, Canada; mchazan@chass.utoronto.ca
9 Evolutionary Studies Institute, University of the Witswatersrand, Johannesburg 2000, South Africa
10 National Natural History Collections, Faculty of Life Sciences, The Hebrew University, Jerusalem 9190401, Israel; lix1000@gmail.com
11 The Natural History Museum, Cromwell Road, London SW7-5BD, UK; peterandrews9@icloud.com
* Correspondence: yfj@mncn.csic.es

Abstract: Experiments based on the premise of uniformitarism are an effective tool to establish patterns of taphonomic processes acting either before, or after, burial. One process that has been extensively investigated experimentally is the impact of trampling to large mammal bones. Since trampling marks caused by sedimentary friction strongly mimic cut marks made by humans using stone tools during butchery, distinguishing the origin of such modifications is especially relevant to the study of human evolution. In contrast, damage resulting from trampling on small mammal fossil bones has received less attention, despite the fact that it may solve interesting problems relating to site formation processes. While it has been observed that the impact of compression depends on the type of substrate and dryness of the skeletal elements, the fragility of small mammal bones may imply that they will break as a response to compression. Here, we have undertaken a controlled experiment using material resistance compression equipment to simulate a preliminary experiment, previously devised by one of us, on human trampling of owl pellets. Our results demonstrate that different patterns of breakage can be distinguished under wet and dry conditions in mandibles, skulls and long bones that deform or break in a consistent way. Further, small compact bones almost always remain intact, resisting breakage under compression. The pattern obtained here was applied to a Pleistocene small mammal fossil assemblage from Wonderwerk Cave (South Africa). This collection showed unusually extensive breakage and skeletal element representation that could not be entirely explained by excavation procedures or digestion by the predator. We propose that trampling was a significant factor in small mammal bone destruction at Wonderwerk Cave, partly the product of trampling caused by the raptor that introduced the microfauna into the cave, as well as by hominins and other terrestrial animals that entered the cave and trampled pellets covering the cave floor.

Keywords: experimental taphonomy; bone compression; microfauna; Wonderwerk Cave

1. Introduction

Understanding the different processes that lead to the formation of fossiliferous assemblages can be problematic as the observed taphonomic signatures are often ambiguous. This since the distinction between damage ensuing from taphonomic agents is frequently subtle and subject to equifinality, i.e., that different taphonomic agents can lead to a common end result [1]. Experiments under controlled conditions or long-term naturalistic monitoring (e.g., [2–4]) have been highlighted as the best ways to distinguish between agents since they enable us to track the more refined signatures of modification in fossil assemblages. Such experiments allow us to gain qualitative insights as to how different processes might occur or influence how fossils accumulate during site formation.

Trampling can be defined as agents treading heavily so as to beat down or crush an object. The effect of trampling would probably not have invited special interest in taphonomic studies, but three decades ago researchers reported that trampling marks and cut marks mimicked each other (e.g., [5,6]). The main surface modifications produced by trampling on bone are striations and microstriations [7–10], shiny and polished surfaces [11,12] and even notches on oblique fracture angles [13]. In order to obtain key traits to distinguish trampling damage from cut marks, and so facilitate assessment of human involvement in a given assemblage, bones of large vertebrates have been extensively studied [14,15]. However, even after these three decades, the importance of distinguishing cut marks from trampling marks is still debated and identification continues to face difficulties even by experts [14]. Machine learning has recently been proposed as a potentially successful procedure to resolve this issue [16]. Bone breakage can be an additional modification derived from trampling and in order to distinguish it from the effects of other types of compression, researchers have characterised bone breakage (e.g., [17]). They have analysed the resistance of bones from different vertebrate species to breakage [18–20], or characterised breakage from trampling versus breakage derived from other taphonomic agents [12,21–24].

Compression testing is one of the simplest methods available that have been extensively used to characterise the mechanical properties of different materials. It has been used in ecology and taphonomy to study the response of bite force of carnivores such as hyaenas [25–27], to distinguish trampling marks obtained when bones are wet and compressed against gravels [19], to evaluate bone resistance of large mammals to sudden or slow movements when bones are wet or dry [28] and to characterise traits of compression in vertebrae of fish to distinguish them from abrasion and digestion [24,29]. Apart from digestion and skeletal element representation, breakage can be used to identify the predator that produced small mammal assemblages [11]. The effects of trampling or compression on such assemblages can thus be linked to the predator that previously modified these bones during ingestion. In such cases, broken edges may be rounded by gastric juice corrosion [11]. Thus, post-depositional processes such as trampling may be superimposed on the taphonomic pattern which characterises each predator, causing distortive effects that can be identified through experimental work.

Our experiment started with modern barn owl (*Tyto alba*) pellets, a raptor known to cause minimal modifications by digestive effects and breakage by ingestion [11]. In the case of small mammals, the fragility of their skeletal elements may give the a priori impression that they will not resist compression. This has, however, never been tested aside from observations made on two nest sites of barn owl and European eagle owl (*Bubo bubo*), and a quick trampling experiment that was described by Andrews (p.10, [11]). In the later experiment, the author stepped on owl pellets and concluded that "the pattern that emerges from these results on trampling is one of breakage of skulls, reduction in numbers of maxillae, considerable loss of teeth from the jaws leading to large numbers of isolated teeth, considerable breakage of larger postcranial elements and some degree of loss, but no loss or breakage of smaller elements (calcanei, talli))". In this study we repeated these experiments under high controlled conditions to explore the patterns of breakage caused to small mammal bones under compression. Results of the compression experiments were then compared to small mammal taphonomic analyses of fossil samples from the Earlier

Stone Age levels in Wonderwerk Cave in South Africa [30] that showed extensive breakage exceeding excavation procedures or digestion by the predator.

2. Material and Methods

The experiment presented here tried to reproduce the results obtained by Andrews (1990) on pellets and on isolated small mammal skeletal elements using compression under controlled conditions. Uniaxial compression experiment was conducted using a Zwick/Roell Z5.0TN machine with testXpert II software, held at the Natural Sciences Museum of Madrid (MNCN-CSIC) at the Laboratory of Environmental Analyses and Experimental Taphonomy (https://www.mncn.csic.es/en/investigacion/servicios-cientifico-tecnicos/laboratory-environmental-analyses (accessed on 27 January 2022)). This uniaxial compression machine has a test load standard of 5 kN (~500 Kg), although for the present experimental study we applied a special load cell to restrict the compression forces to 500 N (~51 Kg), this being an intermediate force between previous experimental and observational samples described by Andrews (1990). This restriction was needed to limit any extra-force exerted by mistake when programming the experiment that could completely destroy these specimens. The equipment to compress the small mammal bones used the same force in the identical direction to prevent the influence of other external parameters which could influence the results. Results of damage were obtained from the compression experiments in the form of curves through the testXpert® II software of the equipment (Figure 1).

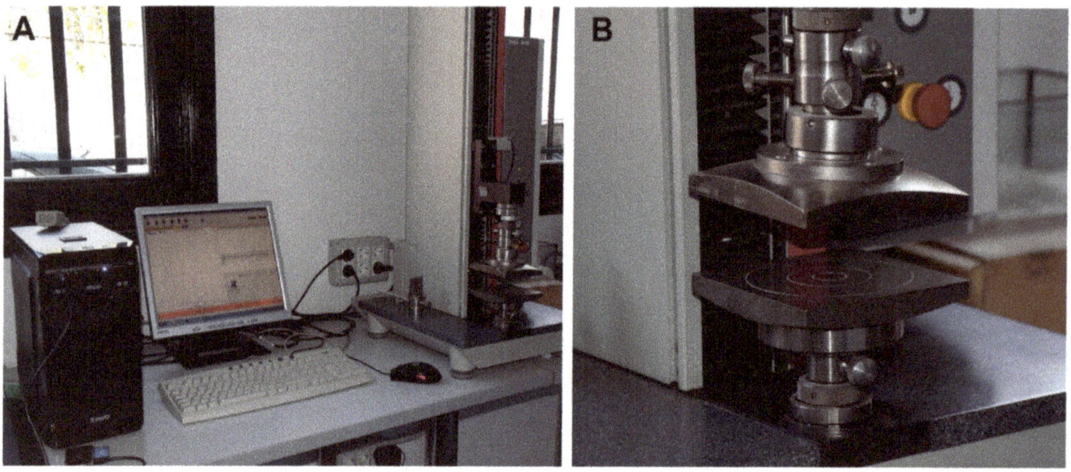

Figure 1. Uniaxial compression (Zwick/Roell Z5.0 TN) equipment. (**A**) Image of the equipment. (**B**) Detail of the load cell.

Small mammal materials used in the experiment derive from a modern collection of pellets of barn owls kept in captivity fed upon laboratory mice (small sized: House mouse, *Mus musculus*; large sized: Brown rat, *Rattus norvegicus*; see Table 1). Barn owl pellets are characterised by a high relative abundance of skeletal elements and low bone and tooth breakage together with a low proportion and degree of digestion on diagnostic elements (e.g., [11,31]. Despite the potential integrity of elements and abundance of skeletal remains, initial compressions performed on four pellets (to test the experiment and observations described by Andrews [11]) resulted in two of them providing only tibiae, ulnae, vertebrae and ribs, without cranial elements, while several of the skeletal elements had been affected by digestion. Eleven pellets were opened, and only five of them provided skeletal elements from cranial (jaws and skulls) and postcranial (long, square and flat bones) elements in sufficient quantity to systematically repeat the experiment several times.

Table 1. The 60 skeletal elements used in the experiments and their position during compression from the 8 sets considered in the experiment.

SETS	State	Size	Skulls	Mandibles	Femora	Humeri	Pelves	Astragali	Calcanei
Set 1	Dry	Large	Dorsal	Lingual	Anterior	Anterior	Ventral	Dorsal Plantar	Absent
Set 2	Dry	Large	Ventral	Buccal	Anterior Posterior	Posterior	Dorsal	Dorsal	Medial
Set 3	Dry	Small	Dorsal	Lingual	Anterior	Anterior	Ventral	Plantar	Lateral
Set 4	Dry	Small	Ventral	Buccal	Anterior Posterior	Anterior	Dorsal	Dorsal	Medial
Set 5	Wet	Large	Dorsal	Buccal	Posterior	Posterior	Ventral	Plantar	Lateral
Set 6	Wet	Large	Ventral	Lingual	Anterior Posterior	Anterior	Dorsal	Dorsal	Medial
Set 7	Wet	Small	Dorsal	Buccal	Posterior Anterior	Posterior	Ventral	Plantar	Lateral
Set 8	Wet	Small	Ventral	Lingual	Posterior	Lateral	Dorsal	Dorsal	Medial

Each selected bone to be compressed was photographed and examined under a binocular light microscope (Leica MZ 7.5). Digested and incomplete bones were discarded to avoid fractures and weaker areas caused by digestion process. Half of selected pellets and skeletal elements were kept dry, whilst the rest were immersed in water for an arbitrary time period of fifteen days in order to examine compression under different environmental conditions. Immersion in water was considered because it could provide seasonal indications due to damp substrates. A total of four complete pellets (two dry and two wet, compressed individually) and 60 isolated skeletal elements (compressed in different sets from 5 opened owl pellets) were used in the experiment. The 60 skeletal elements were: 8 skulls, 8 mandibles, 12 femora, 8 humeri, 8 pelves, 9 astragali and 7 calcanei. The specimens selected were adult and young individuals (<1 kg, both large and small sizes) although not always a large and a small size of the same skeletal element was obtained. Two types of sediment, coarse sands (0.7 mm) and silts (0.065 mm), were chosen for testing. Each set of bones is shown in Table 1.

Pellets, both wet and dry, were pressed directly on the metal surface of the compression plates, in contrast to the isolated anatomical elements that were placed on a sediment tray.

Each set of bones was compressed in three steps as follows:

Step 1: Compression on sand

Anatomical elements selected were placed on a substratum of coarse-medium sand (0.7 mm) to show the protection this sediment type provides to small skeletal elements.

Step 2: Compression on silts

Anatomical elements were placed on silts (0.065 mm) to observe compression on a fine silty sediment typical of caves, as in Wonderwerk Cave which has served as our fossil case study (see below).

Step 3: Compression on silts without skulls

While in nature skulls are associated with other skeletal elements when trampled, the height of the skulls prevented the compression of flatter bones during the experiment. Therefore, the skulls were removed in order to ensure direct compression of the other bones and the whole SET compressed again.

Each step was carried out on four sets of dry skeletal elements and four sets of wet samples of specimens placed in different positions: Dorsal or ventral, lingual or labial, anterior or posterior, plantar or medial (see Table 1 and Figure 2).

For long bones, the resulting breakage was classified following the categories identified by Andrews [11]: Complete, proximal end, shaft and distal end. Classification of long bones was extended to facilitate the identification of breakage types using the categories described by Villa and Mahieu [17] with documentation of incipient fissures or cracks. The breakage categories proposed by Villa and Mahieu [17] were developed for long bones of large mammals, so this nomenclature has been used here without any typological meaning as it is applied to flat and small bones, such as pelves and astragali of microfauna. We have consequently named fracture angles as: Right (R), oblique (O), mixed (M); fracture outline as transverse (T), curved (C), intermediate (I) and fracture edge as smooth (S) or jagged (J). Completeness of the shaft circumference was also evaluated considering the classification proposed by Villa and Mahieu [17]. Circumference completeness includes three categories: Bone circumference less than half of the original ①; circumference more than half of the original ② and complete circumference in at least a portion of the bone length ③ (Figure 3).

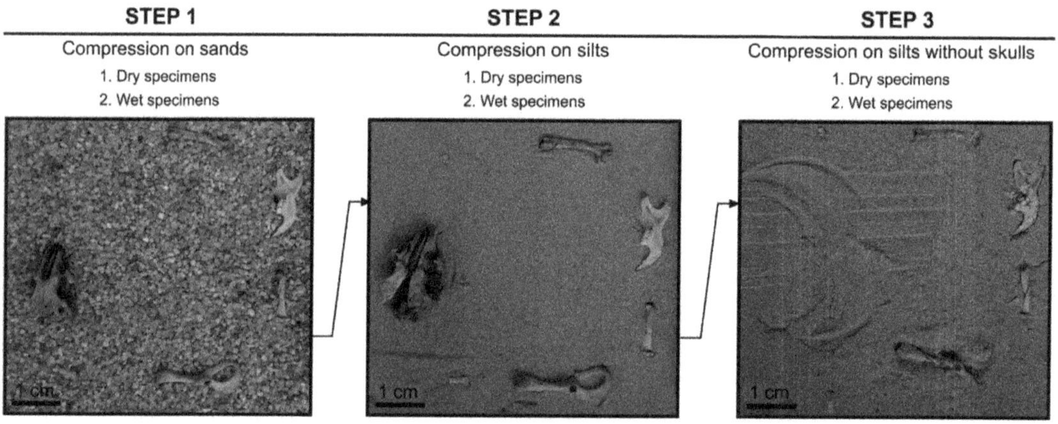

Figure 2. Each set of specimens as shown in this figure was compressed in (1) coarse sand substrate + all the skeletal elements, (2) silty substrate + all the skeletal elements, (3) silty substrate + postcranial elements only, repeated three times and carried out for dry and wet specimens with the skeletal elements in a different position each time.

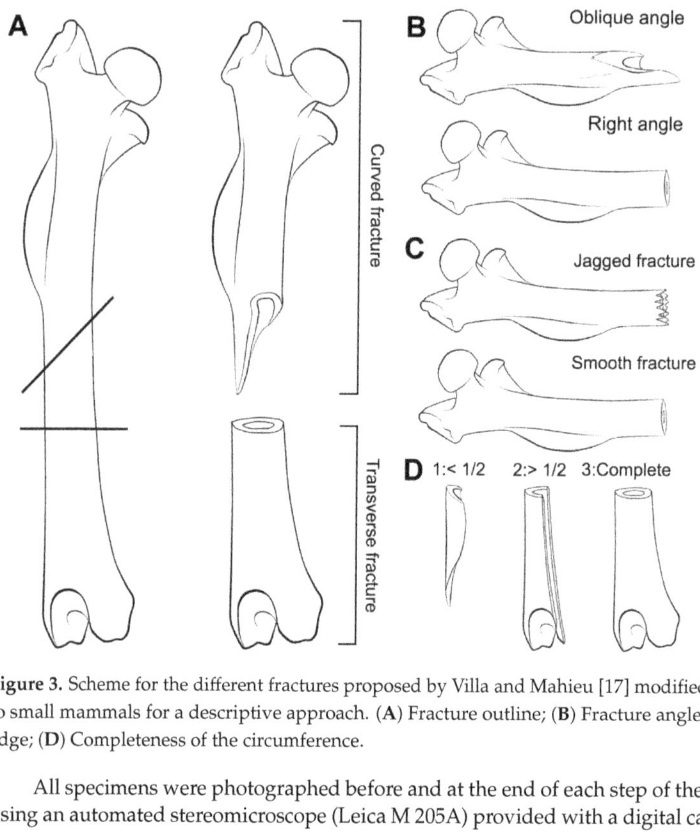

Figure 3. Scheme for the different fractures proposed by Villa and Mahieu [17] modified and applied to small mammals for a descriptive approach. (**A**) Fracture outline; (**B**) Fracture angle; (**C**) Fracture edge; (**D**) Completeness of the circumference.

All specimens were photographed before and at the end of each step of the experiment using an automated stereomicroscope (Leica M 205A) provided with a digital camera (Leica

DFC450) and image software LAS V4.4. This procedure and the control of individual bones allowed us to discard the effects of sifting and sorting of the resulting trampled assemblage or pellet, that potentially could add to detachment of bone fragments or increase breakage.

A Fisher's Exact Test of Independence for numeric data in small samples was applied using R [32] and the library "rcompanion" [33]. This test uses a contingency table and runs an exact procedure especially for small-sized samples [34]. Fisher's Exact Test was considered to be more suitable than other tests, such chi-squared which is better for large samples. Therefore, the "simulate.*p*.value" option was used, which automatically applied a Monte Carlo simulation of the *p*-values based on 2000 replicates. The Monte Carlo approach randomly generates tables to satisfy the null hypothesis for the test and evaluate the test statistic on those tables. A significant *p*-value is considered as $p \leq 0.05$.

Case Study of Wonderwerk Cave

Wonderwerk Cave is a large phreatic cavity (140 m long, 11–26 m wide and 3 to 7 m high) located in the eastern flank of the Kuruman Hills, near the town of Kuruman in central South Africa, (27°50 S, 23°33 E, Figure 4). Archaeological investigations at the site started in the 1940s [35,36], but the cave was most extensively excavated by P.B. Beaumont, A.I. Thackeray and J.F. Thackeray in the 1980s [37–39]. Since 2004, it is being investigated by M. Chazan, L.K. Horwitz and their research team [40,41]. The sediment is mainly silty, sometimes sandy with some dolomite blocks that fell from the roof that are dispersed in the deposit and on the present-day cave surface.

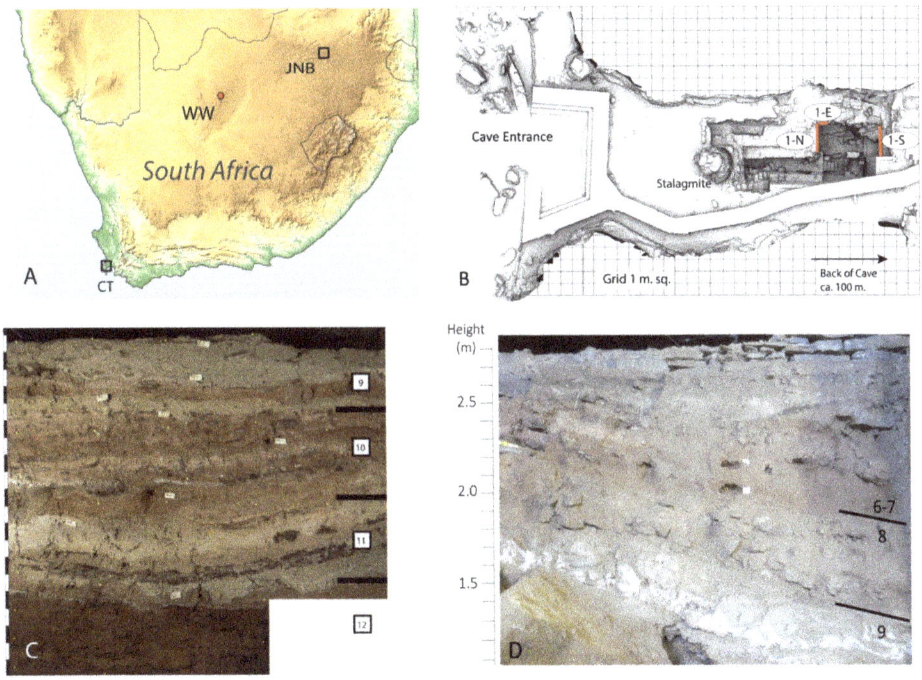

Figure 4. (**A**) Location of Wonderwerk Cave (WW) in South Africa (JNB = Johannesburg; CT = Cape Town). (**B**) Annotated laser scan of Excavation 1 in the cave where 1-N = North Profile, 1-S = South Profile, 1-E = East Profile (scan courtesy of the ZAMANI project, University of Cape Town). (**C**) Stratigraphy of North Profile showing Strata 12 (bottom) to 9 (top) of the sequence. (**D**) Stratigraphy of South Profile showing location of Strata 6/7.

The cave is an exceptional site yielding signs of hominin occupation spanning ca. 2 million years, from the Oldowan through to historic times [41–43]. Small mammal remains are extremely abundant in most of the layers, and in many they formed a background deposit that covered the paleo-cave floor. The physical size of the microfaunal taxa identified at this site can be considered as equivalent to the specimens used in this experiment (for taxon lists see Avery [44,45]). Taphonomic studies of these microfaunal remains have been undertaken on Oldowan and Earlier Stone Age (ESA) strata derived from Excavation 1, which is located ~30 metres from the cave entrance [30,44–47]. These publications have demonstrated the value of research on small mammal fossils at Wonderwork in identifying the predator responsible for their introduction into the cave, and for palaeoenvironmental and palaeoecological information. They have also served as sources of information on past human behaviour.

The small mammal assemblages from the oldest cave strata (St. 12 to St. 10; Oldowan through early ESA) come from the old excavations undertaken by P. B. Beaumont and were recovered by dry sieving using 1 mm mesh. Edentulous jaws and large numbers of loose teeth were the first features that caught the attention to Avery [44] who studied the taxonomy of the P. B. Beaumont collections. She concluded that the extreme rates of breakage may have been influenced by aggressive recovery and preparation procedures. Subsequently, broken edges of these fossils were evaluated by Fernández-Jalvo and Avery [46], and only 20–30% could be attributed to recent breakage. In order to contrast the incidence of recovery procedures, these old samples were compared to new samples from overlying ESA layers, Strata 6 and 7, recovered during the 2018 excavation undertaken by the Chazan and Horwitz team (Table 2). During these excavations, all sediment was collected and processed using a mechanized floatation machine. The recent sampling from Strata 6/7, whose data are reported here (Table 2), combined due to the absence of any significant differences between them, has substantially reduced breakage and loss of small mammal skeletal remains [30], but the main traits observed by Avery [44] are still present in the sample despite processing by floatation.

Table 2. Frequency of long bone (femora and humeri) portions and complete elements from four strata of Wonderwerk Cave site according to Marin-Monfort et al. [30].

STRATA	St 6/7		St10		St11		St12	
FEMORA (Total N)	**N = 1255**	%	**N = 3072**	%	**N = 393**	%	**N = 1451**	%
Complete	353	28%	216	7%	43	11%	72	5%
Proximal (+Prox + 1/2)	650	52%	1953	64%	191	49%	962	66%
Shaft	61	5%	351	11%	76	19%	236	16%
Distal (+Dist + 1/2)	191	15%	552	18%	83	21%	181	12%
HUMERI (Total N)	**N = 1425**	%	**N = 2975**	%	**N = 372**	%	**N = 1185**	%
Complete	446	31%	325	11%	49	12%	85	6%
Proximal (+Prox + 1/2)	212	15%	803	27%	60	15%	104	7%
Shaft	100	7%	505	17%	85	22%	258	18%
Distal (+Dist + 1/2)	667	47%	1342	45%	178	45%	738	51%

3. Results

3.1. *Compression of Pellets*

Compression of pellets have more destructive effects when they are dry than when wet, with some heavily broken bones (see rectangle in Figure 5A) already visible on the pellet's surface. Wet pellets become more plastic, as well as the bones in their interior. After the experiments, the dry pellets recovered their original thickness to some degree, but the wet ones remained deformed and laterally compressed. Only two of the four pellets compressed provided the skeletal elements used here to obtain patterns of breakage (plus some other elements: Vertebrae, scapulae or tibiae). Observations of remains recovered from these two pellets are described in the Supplementary Information. The other two compressed pellets excluded from the Supplementary Information did not yield any of the skeletal elements individually compressed to compare patterns of breakage. Bone remains still inside the wet and dry pellets are more complete than when bones were exposed directly to the force of 500 N. This is partially due to the protection provided by hair and feathers, but also because compression was exerted only once to test the experiment. This test on pellets yielded low numbers of skeletal elements and did not allow us to compare systematically compression effects.

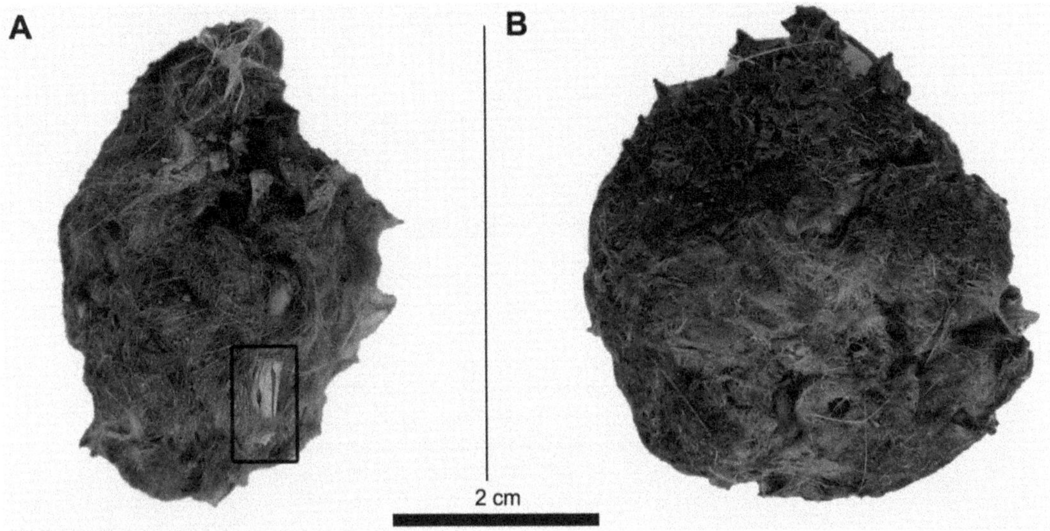

Figure 5. Barn owl pellets compressed under 500 N of force. (**A**) Dry pellet; (**B**) Wet pellet. The square in Figure A shows breakage as result of compression of a long bone (tibia) exposed on the pellet surface.

3.2. Compression of Skeletal Elements

Compression on coarse sands (Step 1) by a 500 N force did not cause significant modifications, except for skulls that opened along their sutures and for long bones that caused removal of the epiphyses, which were not fully fused to the metaphyses. For the rest of the specimens, the skeletal elements were still complete after compression.

Compression of bones on silts delivered consistent patterns, especially for cranial elements. When resting on silts, all remains showed a similar response to compression either in the first (Step 2) or in the second attempt (Step 3), except for two humeri which showed a different response in each compression attempt, even though they were not close to the skulls. These two attempts have been accounted for as two different results given more extensive breakage observed in the second compression.

Most skulls remained deformed with pieces attached together, especially in wet skulls. These pieces, however, will easily detach when the sample is processed (sifted, sorted, moved) as fossil assemblages usually are. The most destructive effects occurred when compression was on the ventral aspect and dry, while dorsal compression mainly separates the skull bones as a result of detachment of the skull sutures (Figure 6). Incisors are recovered in the interior of the alveoli (Figure 6B red arrow), and the M1 is frequently preserved in situ, with or without, the zygomatic arch attached. Frequently the M2 and M3 are detached and isolated from the maxilla (Table 3). These traits should be noted when trampling is tested in a small mammal fossil bone assemblage.

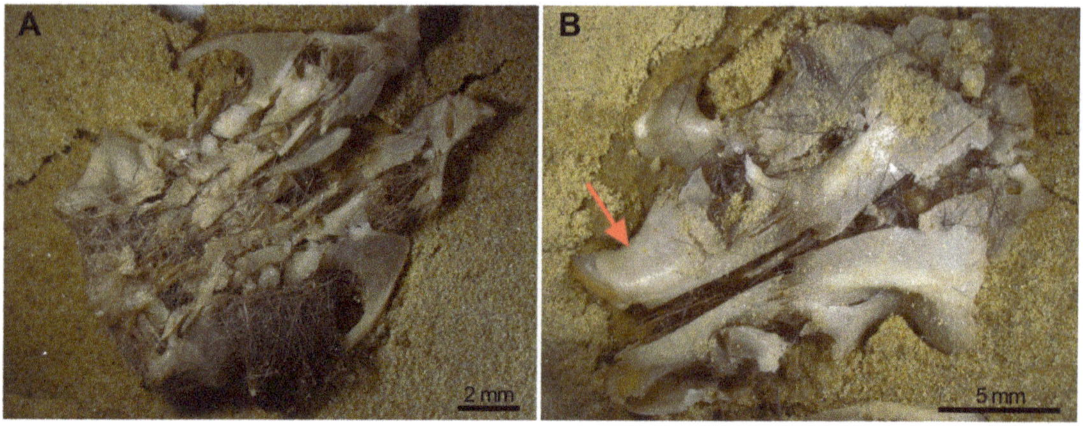

Figure 6. Breakage of the skull is more severe when it is compressed on the ventral aspect (**A**) rather than from the dorsal position (**B**) The red arrow points to the incisors in the interior of their sockets.

Table 3. Representation of cranial parts after the three steps of compression under both dry and wet conditions. Skull fragmentation appears not to be very severe, and most skulls remain attached but deformed (especially in compressed wet skulls), but any further processing of these specimens causes their disintegration into small fragments. Similarly, mandibles compressed wet remain complete. Those compressed dry may have fissures, especially affecting the ascending ramus.

Skulls (n = 8)	N	%
M1 + zygomatic arch	7	88
Incisor in socket	8	100
Isolated molars	8	100
Mandibles (n = 8)		
Complete or longitudinal fissures	4	33
M1 + diastema	7	58
Detached molars	4	33
Incisor in socket	7	58
Detached ascending ramus	4	33

Mandibles show the most regular breakage traits, and these specimens follow the pattern of breakage shown in Figure 7. This pattern consists of: (a) The incisors appear in the interior of their sockets (Figure 7 red arrow), (b) the M1 is attached to the diastema, (c) the ascending ramus appears almost complete but is broken and separated from the mandible, (d) M2 and M3 may be isolated and detached from the mandible or remain in situ, depending on the intensity of compression on the mandible. We, therefore, propose that when trampling aims to be tested, these four observations (Table 3) should be taken into consideration as diagnostic features of this taphonomic agent. When mandibles are wet during compression, breakage is greatly reduced and most of them remain complete. In contrast, dry mandibles show micro-fissures and small breaks compared with deformation for the wet specimens. Even wet mandibles under compression develop cracks and fissures that outline the fragments shown in Figure 7. The original position (lingual or labial) does not influence breakage.

With regard to the postcranial skeleton, pelvis survival is low independent of the position (ventral or dorsal) in which it was compressed. Compression of this element always resulted in small pieces of broken fragments (Figure 8). Amongst these fragments, the acetabulum (the articular cavity) usually survived, although when it was compressed, transversal fissures may be observed (Figure 8A, central arrow). The acetabulum may remain attached to the bones that comprise the pelvis; more frequently it is attached to the illium, less frequently to the pubis (the branches) and never in our experiment is it attached to the ischium. Furthermore, the ischium appears always broken into several pieces (Figure 8B, Table 4).

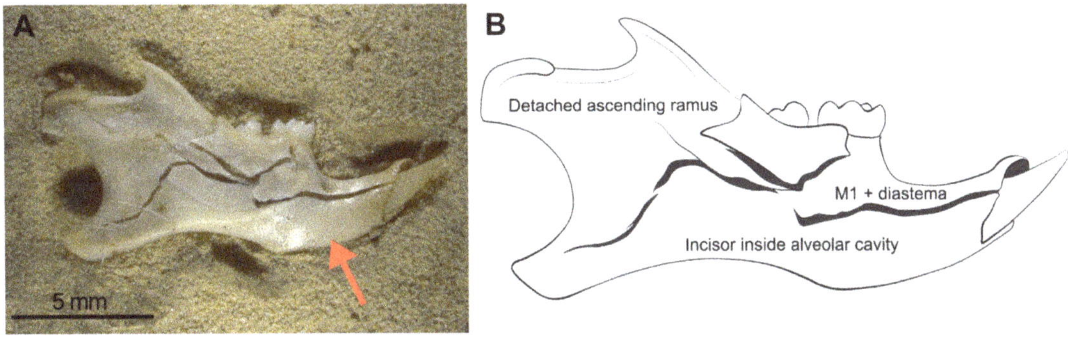

Figure 7. Most mandibles follow this pattern of breakage in which mandible fragments remain joined together, although with minimum movement or processing (sieving, sorting), this specimen will split into pieces. (**A**) Photo of the breakage pattern of mandibles. (**B**) Detailed scheme. The red arrow points to the incisors in the interior of their sockets.

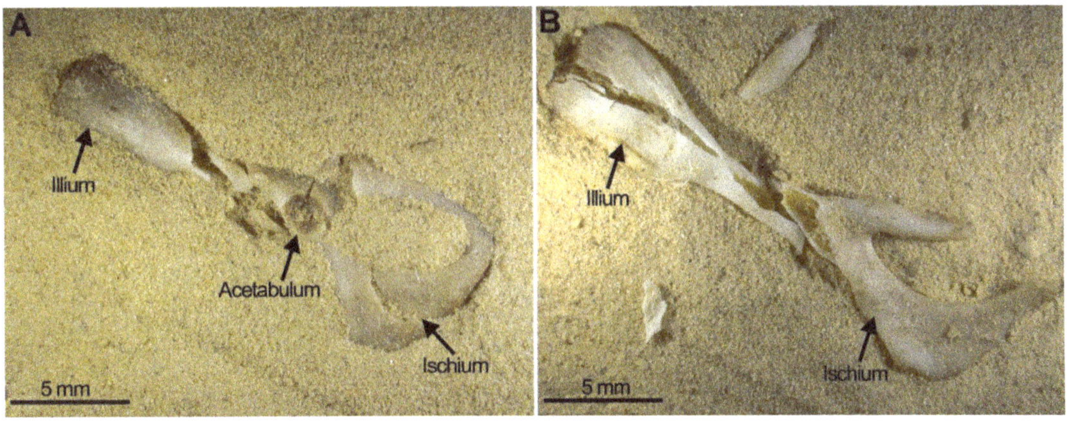

Figure 8. Two examples of pelvis compression: (**A**) Pelvis from SET 3 ventral view with damaged acetabulum (fissure marked by anarrow). (**B**) Pelvis from SET 2 also showing severe destruction leaving small portions of the original bone.

Table 4. Fragments and modifications of the pelvis when compressed, most fragments retain the acetabulum.

Pelvis (n = 8)	N	
Almost Complete	1	13
Articular Cavity (Damaged)	2	25
Cavity + Illium	0	0
Cavity + Ischium	3	38
Cavity + Pubis Branches	2	25

Long bones exhibit a certain degree of variability in breakage patterns resulting from compression (Tables in the Supplementary Information). Most of the breakage types are TRJ (transverse-right-jagged) or COS (curved-oblique-smooth), while most frequently the completeness of the circumference is ③ (complete) and sometimes ① (less than half).

Femora appear more robust and resistant to breakage than humeri, with a higher number of complete elements especially when compression is on wet specimens. Despite this apparent completeness, fissures and cracks are frequent in both femora and humeri (Figure 9, red arrows). These cracks weaken attachments between the bone fragments which potentially may not last long in natural conditions or if the sample is processed by sieving and sorting. To interpret how these cracks and fissures could behave throughout their taphonomic history, we tested how they will be detached and split into two or more fragments following sample processing. Considering the additional breakage that would occur following processing, we created what we call a "compressed + processed" sample (Table 5). We then compared the sample from the compression experiment, the "compressed + processed" sample that we created and samples from the Earlier Stone Age at Wonderwerk Cave.

A Fisher's Exact Test of Independence was applied computing the p-value with a Monte Carlo Simulation. Results from the comparison of each level from Wonderwerk Cave with the "compressed" and "compressed + processed" sample are shown in Table 6. In general terms, St 6/7 is similar to the "compressed" sample while the other stratum, tend to be more similar to the "compressed + processed" sample (see also Figure 10). Only femora from St 12 showed differences in both the "compressed" and the "compressed + processed" sample, due to an anomalous low number of distal parts compared with the rest of the samples (Figure 10).

In our experimental data set, in contrast to all these patterns of breakage to long bones and crania, small square and compact bones, i.e., calcanei and astragali, are almost always unbroken, except in two instances out of 9 astragali (Figure 11). The broken astragali were compressed in dry conditions and broke only when placed in a dorsal position on any type of substrate (sands or silts), that possibly represents a more irregular or weaker surface. Figure 11 shows two astragali compressed in the same compression attempt, using identical force, one close to the other, one broken, the other complete. Calcanei never broke in any instance (Figure 11C).

Table 5. Frequency of long bones (femora and humeri) obtained following compression on current specimens. Several long bones bear cracks or fissures that may result in detachment of parts during sample processing, conforming to what we have called the "compressed + processed sample". * two out of the eight humeri showed a very different response in each compression attempt, and they have been accounted for as two different results given the strong differences observed in the third step of compression, resulting n = 10.

	Experiment Sample	
Femora (n = 12)	Compression experiment (n) %	Compressed + processed (n) %
complete	(7) 41	(3) 13
proximal	(5) 29	(9) 39
shaft	(1) 6	(4) 17
distal	(4) 24	(7) 30
Humeri (n = 10 *)		
complete	(4) 27	(2) 10
proximal	(3) 20	(4) 20
shaft	(2) 13	(6) 30
distal	(6) 40	(8) 40

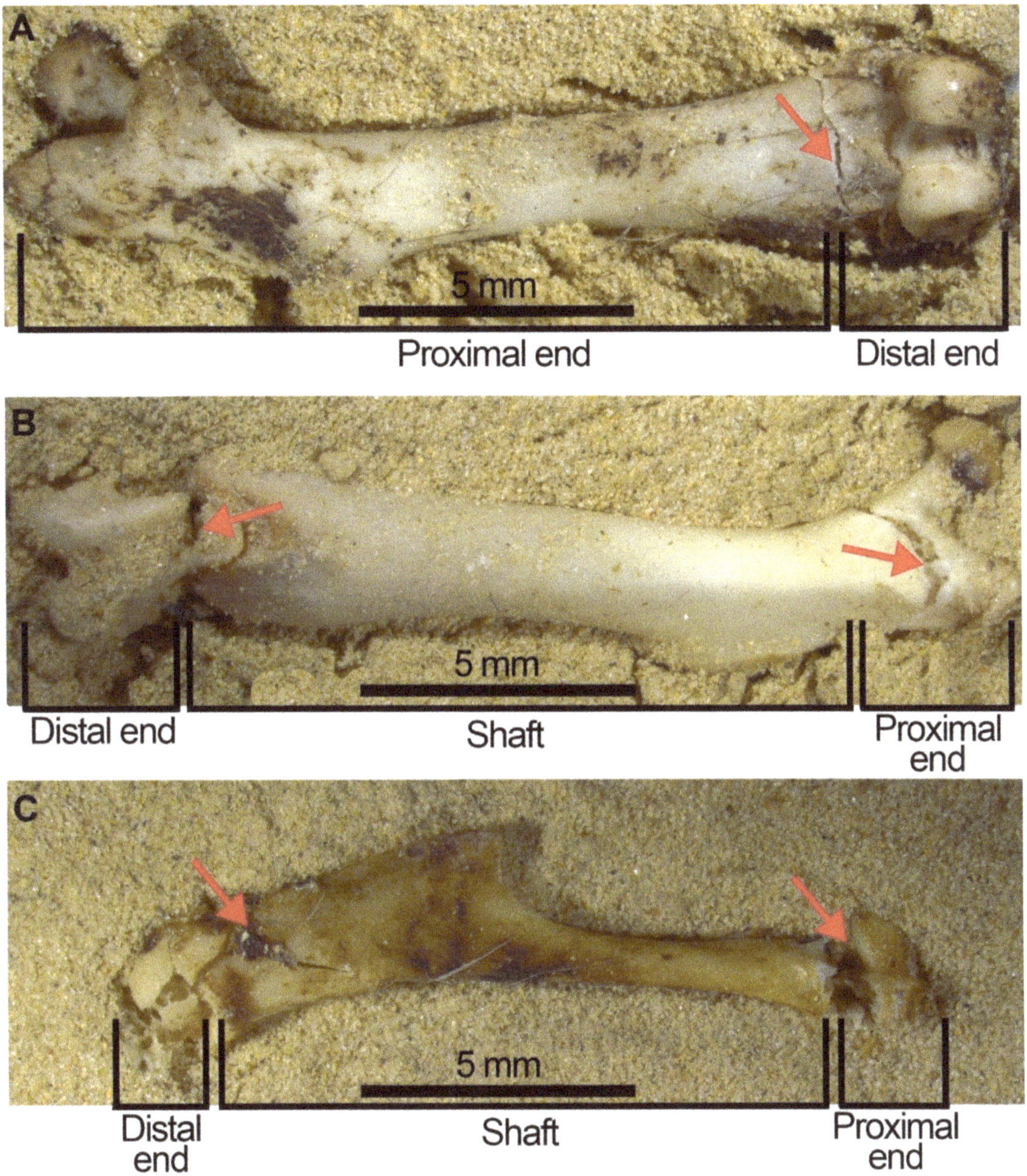

Figure 9. Fissures and open cracks that still join together and constitute complete bones. (**A**) complete femur circumference category ③ with a TRJ fissure on the distal end resulting in the proximal end + epiphysis segment being apart from the distal end in the compressed and processed sample. (**B**) the unfused distal end was detached from the femur's articular end during the compression experiment. The femur remains as a complete element, despite the open crack (category ③ TRJ) shown on the proximal end which will easily cause detachment of the proximal end as well as the distal end from the diaphysis. (**C**) humerus with the distal end broken leaving the proximal end and diaphysis together, although the oblique fissure on the proximal end will cause breakage of the bone into three parts: Proximal end, shaft (almost complete) and distal end.

Table 6. Simulate *p*-values obtained from the Fisher's Exact Test using the data showed in Table 5 compared to Wonderwerk sample assemblages published in Marin-Monfort et al. [30]. Values in bold indicate significant differences (*p*-value ≤ 0.05).

	Femur		Humeri	
Stratum	Compressed	Compressed + Processed	Compressed	Compressed + Processed
St 6/7	0.190	**0.006**	0.085	**0.004**
St 10	**0.001**	0.052	**0.046**	0.518
St 11	**0.005**	0.656	**0.049**	0.753
St 12	**0.001**	**0.010**	**0.002**	0.099

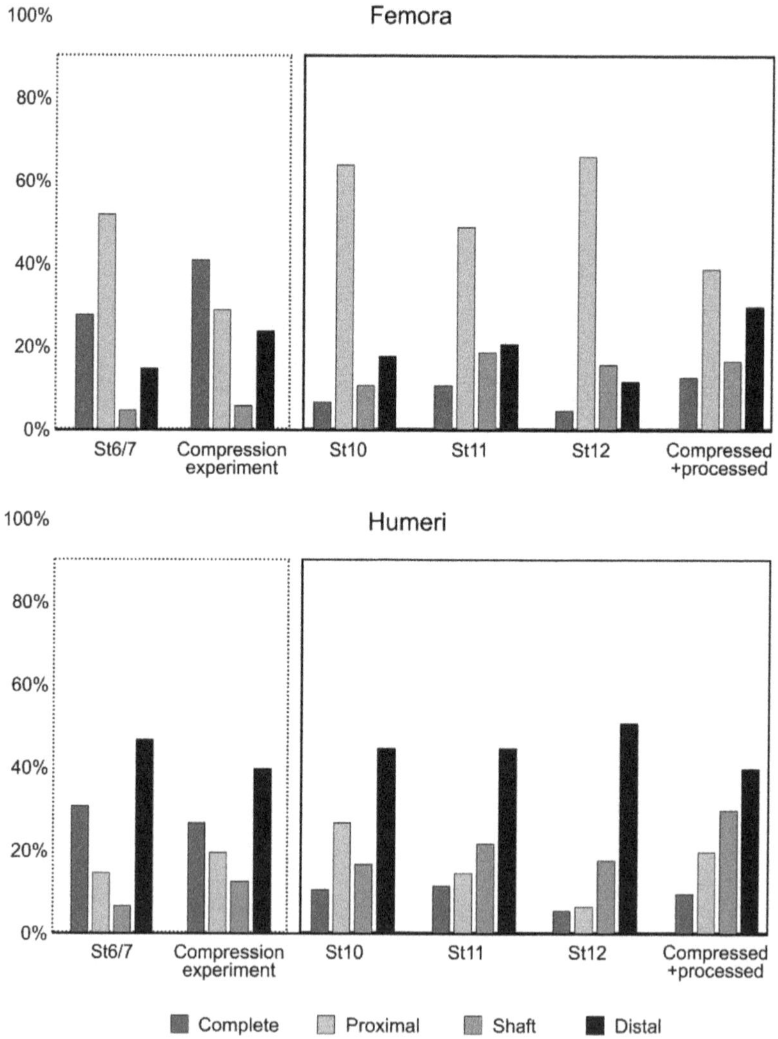

Figure 10. Bar diagrams of skeletal part representation in the Wonderwerk Cave strata (St. 6/7, 10, 11 and 12) compared to the fragments obtained in the compression experiment described here.

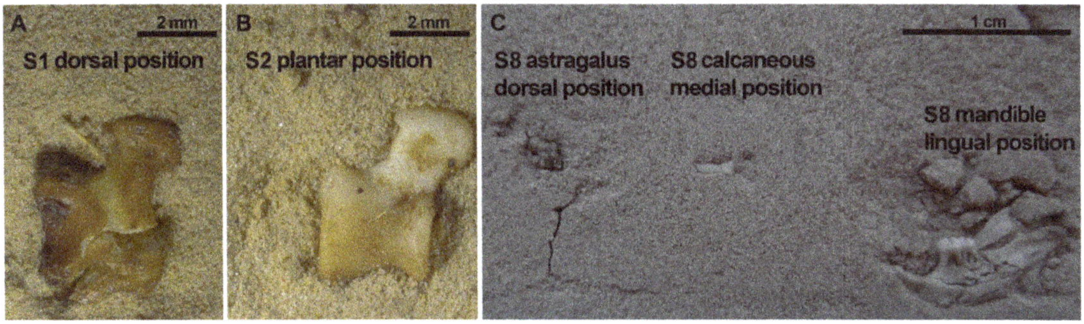

Figure 11. Astragali usually stayed complete after compression (75%) except for two cases in which both were positioned in a dorsal aspect and compressed dry. (**A**,**B**) show astragali from the same SET 1, resulting in one of the astragali breaking while the other remained complete during the 3rd Step of compression. Calcanei were never broken during compression (Table S5). On the right picture (**C**), SET 8, the calcaneus and the astragalus appear complete, in contrast to the mandible that is highly broken despite these elements from SET 8 were immersed in water before compression.

4. Discussion

Frequently, the study of taphonomy is considered to be the study of bias. Probably the reason for such an assumption is due to the inherent comparison with what should be in a site if the faunas and floras represented were complete. The loss of biological information in fossil assemblages is self-evident when observing dead bodies during processes of putrefaction. The lack of muscles and ligaments facilitates the disarticulation of the skeleton, providing important information related to the amount of time passed and weather conditions. A further source of invaluable information is the study of bone breakage and deformation. The "incompleteness" and "imperfection" of fossils, when compared to the original bones, is a source of taphonomic information that is codified in the fossils and needs to be "extracted" in order to identify the past history of assemblages and learn about site formation processes. At all sites, each fossil found can provide information on contemporaneous biotic and abiotic agents, acting either simultaneously or asynchronously on the bone remains, causing the loss of part of their original palaeobiological information and increasing the taphonomic information to be obtained [48].

The current study was initiated as result of uncharacteristically heavy breakage observed in the small mammal remains from Wonderwerk Cave. Most jaws were recurrently edentulous with a high frequency of tooth loss (the latter could easily pass through the mesh of the screens used in the original excavations), and quite extensive modern damage that was incurred due to sifting, resulting in remains that could not be taxonomically identified [44]. Despite the extensive tooth loss [44], other small-sized skeletal elements (calcanei and astragali) appeared in relative high proportions (Figure 12), suggesting that size was not the sole factor determining representation. We tested the possible involvement of highly destructive predators that could cause intense breakage to small mammals, but the predator identified at Wonderwerk Cave is *Tyto alba* (which causes minor destruction according to Andrews, [11]). Thus, this raptor could not explain the extensive destruction evidenced in these small mammal assemblages. Post-depositional abiotic agents (flowing water or falling rocks) could increase breakage in a raptor prey assemblage. However, cave roof spall does not occur in all of the cave strata that were examined here. Moreover, there is no evidence from the sedimentary record of the cave [49] that either of these factors were of any consistent intensity.

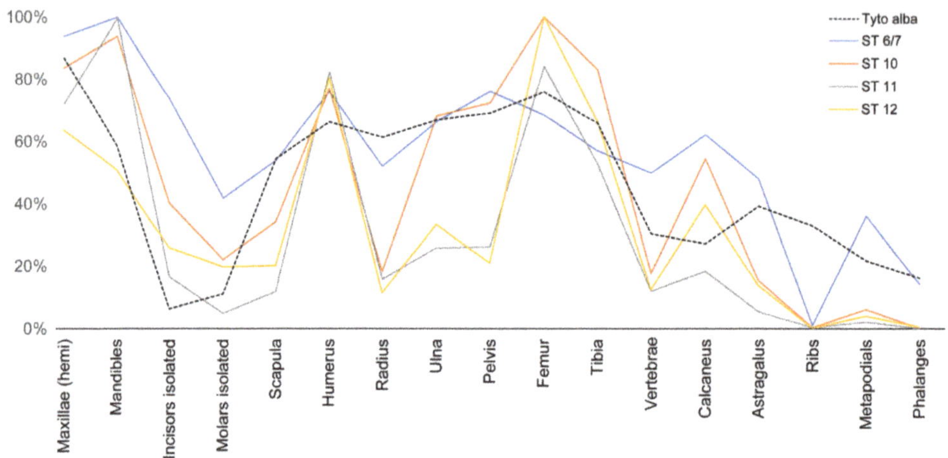

Figure 12. Diagram showing the relative abundance of skeletal elements in individual strata from Wonderwerk Cave (from [30]) compared to a modern barn owl (*Tyto alba*) assemblage [11]. Note that all strata show a high degree of homogeneity and have a relatively high frequency of calcanei and astragali. Note also that all samples have fewer astragali than calcanei, likely because the former incidentally broke during post-depositional taphonomic processes (as seen in our trampling/compression experiment) and possibly got lost during sample processing (i.e., sieving).

The new excavations in the Wonderwerk Cave site provided samples from St. 6/7 that exhibit greater completeness than samples from the old excavations (Table 2), but the frequency of edentulous jaws is still high. The proportion of isolated teeth (detached from jaws) and relative abundance of small-sized elements (astragali and calcanei) are even higher in the new floatation sample. In general, the traits of breakage and destruction observed in the Wonderwerk Cave samples are still higher than is expected in a barn owl assemblage.

Marin-Monfort et al. [30] observed that the destructive traits that characterised the microfaunal samples from Wonderwerk Cave were constant between layers and were associated with the persistent occupation of barn owls in the site during the ESA strata. In fact, barn owls still inhabit the cave today. Thus, the breakage patterns to the small mammal bones were incongruous with the signatures of both biotic and abiotic taphonomic agents that had been identified. We took into account the possibility of an additional factor that had not been considered before, namely that the assemblage may had been subjected to trampling, an overprint on the barn owl signature. To do that, we performed the systematic experiment described in detail in this paper examining breakage in different skeletal elements. Half of these bones were immersed in water and show a more plastic response, in contrast to dry bones that are brittle and so break more easily under compression [19,20,24].

We have confirmed the initial breakage pattern for small mammals described by Andrews [11] but added more detail to the model which can now be defined in three stages:

(A) Breakage of skulls, reduction in numbers of maxillae. In our experiment, we have proved the pneumatic behaviour of the skull which deformed leaving a mass of bones and teeth. The bones of the skull detached into small unidentifiable bone fragments, especially when subjected to water or gravitational movements such as sifting, leaving the zygomatic arches apart from the maxillae which frequently bear the M1 in situ. Incisors were also frequently found in the interior of the alveoli, and both traits have been observed in the Wonderwerk Cave fossil assemblage. Consequently, both observations can be added as taphonomic criteria of trampling in a fossil site.

(B) Considerable loss of teeth from the jaws leading to large numbers of isolated teeth. As seen in Wonderwerk Cave and in the experiment, both mandibles and maxillae showed detachment of molars although in mandibles the M1 remained in situ retained in the diastema portion. The incisors, as observed in maxillae, are also retained in situ (in the alveolar socket of the mandible) and in this type of assemblage it is also common to see the ascending ramus detached from the dental row and the rest of the mandible.

(C) Considerable breakage of larger postcranial elements and some degree of loss, but no loss or breakage of smaller elements (calcanei, talli). High frequency of calcanei and astragali is one of the most representative features of microfaunal assemblages of Wonderwerk Cave site, as well as modern assemblages exposed to trampling. In addition, a high frequency of complete bones was observed when compression was undertaken on wet bones together with jagged edges, fissures and cracks, which were frequently transversal to the length of the bone.

We have obtained clear criteria with which to distinguish trampling on small mammals, by subjecting skeletal elements and pellets to compression. However, in this study our compression of pellets (Supplementary Information) may have been too mild such that the results were less destructive than found by Andrews [11]. Results of compression directly on bones, however, are highly similar. Both experiments respond to the mechanical properties of bone which behaves similar to engineering materials (e.g., ceramic, metals, rocks, building materials) and, therefore, reacts to loading and fracture following the basic principles of mechanics (e.g., [50–55]).

In general, the relevance of detecting trampling in fossil small mammal assemblages facilitates interpretation of the full history of raptor assemblages since, as at Wonderwerk, the barn owl's ingestion and digestion does not produce high bone breakage rates. Identification of trampling may then allow us to better understand how the assemblage formed. Thus, the reconstructed scenario for Wonderwerk Cave is that small mammal bones were probably trampled by the owls themselves while they nested in the cave's interior, although bones still inside the pellets could not be substantially destroyed. The owl pellets could also have been trampled by other cave visitors, including humans, thereby increasing breakage but not totally destroying the small mammal bones unless they were directly exposed to trampling (i.e., not inside a pellet). Bones from disaggregated pellets will be more exposed to post-depositional processes such as weathering and trampling than those bones in the interior of a pellet [11,56]. This scenario was recently documented in the archaeological rock shelter Álvarez 4 (late Holocene) located in the arid and cold environment of northwestern Patagonia of Argentina, where the bones from the pellets were less fractured than bone assemblages isolated in the sedimentary matrix [57].

This paper validates breakage traits obtained from experiments in a compression material testing machine against those from more actualistic experimentsof trampling by humans (e.g., [11]). There are still many experiments that need to be carried out to test other situations of trampling (e.g., pellets of other raptors and excrements of mammalian carnivores, larger and smaller prey, compression by forces around 50 N to simulate breakage by trampling in nests or latrines by predators) and so complement these patterns. There are also modern bones that were collected by one of us (PA) from monitored natural nests that potentially could have been trampled by predators. These were briefly described in Andrews [11] page 8 but as they are of interest, it is planned to study them in greater detail. In addition, it would be useful to increase the size of the experimental sample to better characterise fissures and cracks in postcranial bones and also to process these samples using sieving and floatation methods. Complementary research could extend to monitoring nesting behaviour and studying remains derived from old raptor nests.

5. Conclusions

- There are distinct patterns of bone breakage and anatomical element survival that can be used to recognise the involvement of trampling in small mammal assemblages.

- Several patterns can be proposed based on the results obtained in this and previous experiments. The primary indications of trampling are: (1) Presence of edentulous jaws, (2) high postcranial breakage and (3) high frequency of complete calcanei and astragali.
- Taphonomic categories, especially with reference to cranial elements, reflect other types of breakage patterns due to compression, especially with reference to mandibles and maxillae. Long bones under compression produce characteristic fissures and cracks (transversal to the metaphysis) that may be recorded on complete or almost complete bones which indicate trampling, although these observations need further study and experiments.
- Results from the experiments we have undertaken were compared to those obtained in the actualistic experiment of human trampling (validated here) and compared to the taphonomic results of the Oldowan and Earlier Stone Age small mammal assemblages from Wonderwerk Cave (South Africa) (i.e., Strata 12, 11, 10 and 6/7). Comparison showed that trampling was an important factor responsible for the high degree of breakage observed in these assemblages. Trampling could have been caused by the predators (barn owl) themselves, as well as by other terrestrial animals or humans that visited the cave. The effects of trampling have been augmented by sifting and processing of the samples.

Supplementary Materials: The following supporting information can be downloaded at: https://www.mdpi.com/article/10.3390/quat5010011/s1.

Author Contributions: Conceptualization, Y.F.-J. and P.A.; Data curation, L.R.; Formal analysis, Y.F.-J. and L.R.; Investigation, Y.F.-J.; Methodology, Y.F.-J., F.J.F., S.G.-M., M.D.M.-M., C.I.M. and R.T.; Resources, Y.F.-J.; Supervision, P.A.; Visualization, L.K.H.; Writing—original draft, Y.F.-J., F.J.F., S.G.-M., M.D.M.-M., C.I.M., R.T., M.C. and L.K.H.; Writing—review & editing, L.K.H. All authors have read and agreed to the published version of the manuscript.

Funding: This project was funded by a Leakey Foundation grant to F.-J.Y. and the project CGL2016-79334-P of the Spanish National Program funded by the Ministry of Scientific Research and Innovation and the Spanish Council of Scientific Research (COOPB20287). Funding for field work at Wonderwerk Cave is provided by a grant from the Canadian Social Sciences and Humanities Research Council (SSHRC) to C.M. The Université de Rennes 1 (France) and the Museo Nacional de Ciencias Naturales (CSIC, Spain) provided the Convention de Stage n. 46360 to R.L. to do this work as practical stage of the Master M2 mention Biologie-agro-santé spécialité préhistoire, paléontologie et paléoenvironnememt in 2015. G.-M.S. has a pre-doctoral grant funded by the Universidad Complutense de Madrid (UCM) and Banco Santander (CT42/18-CT43/18).

Data Availability Statement: All data displayed here in tables and Supplementary Information.

Acknowledgments: We are especially grateful to the editors of this volume, Emmanuelle Stoetzel, Janine Ochoa and Juan Rofes for inviting us to participate in this special issue on Taphonomy, and to the reviewers whose comments have greatly improved this paper. All field work and research on the old collections of Wonderwerk Cave is undertaken under permit from the South African Heritage Resources Agency (SAHRA). Thanks also are given to IZIKO museum and curators Thalassa Mathews and Sarena Govender. We offer our warm thanks to David Morris (Head, Department of Archaeology, The McGregor Museum) for facilitating access to the Wonderwerk Cave small mammal collections in his care. Wonderwerk Cave fossil collections are currently hosted at the McGregor Museum of Kimberly (South Africa).

Conflicts of Interest: The authors declare no conflict of interest.

References

1. Lyman, R.L. *Vertebrate Taphonomy*; Cambridge University Press: Cambridge, UK, 1994; p. 524.
2. Andrews, P. Experiments in taphonomy. *J. Archaeol. Sci.* **1995**, *22*, 147–153. [CrossRef]
3. Denys, C. Taphonomy and experimentation. *Archaeometry* **2002**, *44*, 469–484. [CrossRef]
4. Fernández, F.J.; Montalvo, C.I. Actualistic Taphonomy of Small Mammals from Owl Pellets in South America and Its Archaeological Implication. *Glob. J. Arch. Anthropol.* **2017**, *2*, 1–4.
5. Andrews, P.J.; Cook, J. Natural modifications to bones in a temperate setting. *Man* **1985**, *20*, 675–691. [CrossRef]

6. Behrensmeyer, A.K.; Gordon, K.D.; Yanagi, G.T. Trampling as a cause of bone surface damage and pseudo-cutmarks. *Nature* **1986**, *319*, 768–771. [CrossRef]
7. Olsen, S.L.; Shipman, P. Surface modification on bone: Trampling versus butchery. *J. Archaeol. Sci.* **1988**, *15*, 535–553. [CrossRef]
8. Domínguez-Rodrigo, M.; De Juan, S.; Galan, A.B. A new protocol to differentiate trampling marks from butchery cut marks. *J. Archaeol. Sci.* **2009**, *36*, 2643–2654. [CrossRef]
9. Denys, C.; Patou-Mathis, M. *Manuel de Taphonomie (Collection Archéologiques)*; Errance: Paris, France, 2014; p. 248.
10. Courtenay, L.A.; Huguet, R.; Yravedra, J. Scratches and grazes: A detailed microscopic analysis of trampling phenomena. *J. Microsc.* **2020**, *277*, 107–117. [CrossRef]
11. Andrews, P. *Owls, Caves and Fossils*; The Natural History Museum: London, UK, 1990; p. 222.
12. Madgwick, R. What makes bones shiny? Investigating trampling as a cause of bone abrasion. *Archaeol. Anthropol. Sci.* **2014**, *6*, 163–173. [CrossRef]
13. Blasco, R.; Rosell, J.; Fernández-Peris, J.; Cáceres, I.; Verges, J.M. A new element of trampling: An experimental application on the Level XII faunal record of Bolomor Cave (Valencia, Spain). *J. Archaeol. Sci.* **2008**, *35*, 1605–1618. [CrossRef]
14. Domínguez-Rodrigo, M.; Saladié, P.; Cáceres, I.; Hughet, R.; Yravedra, J.; Rodríguez-Hidalgo, A.; Martín, P.; Pinera, A.; Marín, J.; Gené, C.; et al. Use and abuse of cut mark analyses: The Rorschach effect. *J. Archaeol. Sci.* **2017**, *86*, 14–23. [CrossRef]
15. Courtenay, L.A.; Huguet, R.; González-Aguilera, D.; Yravedra, J. A hybrid Geometric Morphometric deep learning approach for cut and trampling Mark classification. *Appl. Sci.* **2019**, *10*, 150. [CrossRef]
16. Cifuentes-Alcobendas, G.; Domínguez-Rodrigo, M. Deep learning and taphonomy: High accuracy in the classification of cut marks made on fleshed and defleshed bones using convolutional neural networks. *Sci. Rep.* **2019**, *9*, 18933. [CrossRef] [PubMed]
17. Villa, P.; Mahieu, E. Breakage patterns of human long bones. *J. Hum. Evol.* **1991**, *21*, 27–48. [CrossRef]
18. Nicholson, R.A. A morphological investigation of burnt animal bone and an evaluation of its utility in archaeology. *J. Archaeol. Sci.* **1993**, *20*, 411–428. [CrossRef]
19. Marin-Monfort, M.D.; Pesquero, M.D.; Fernández-Jalvo, Y. Compressive marks from gravel substrate on vertebrate remains: A preliminary experimental study. *Quat. Int.* **2014**, *330*, 118–125. [CrossRef]
20. Frontini, R.; Morales-Muñiz, A.; Fernández-Jalvo, Y.; Roselló-Izquierdo, E.; Pesquero-Fernández, M.D.; Hernández, A.B.; García, L.A. Evaluation of size-related Salmonid fish vertebrae deformation due to compression. An experimental approach. *Archaeol. Anthropol. Sci.* **2021**, Submitted.
21. Fernández-Jalvo, Y.; Andrews, P. Experimental effects of water abrasion on bone fragments. *J. Taphon.* **2003**, *1*, 147–163.
22. Gaudzinski-Windheuser, S.; Kindler, L.; Rabinovich, R.; Gofen-Inbar, N. Testing heterogeneity in faunal assemblages from archaeological sites. Tumbling and trampling experiments at the early-Middle Pleistocene site of Gesher Benot Ya'aqov (Israel). *J. Archaeol. Sci.* **2010**, *37*, 3170–3190. [CrossRef]
23. Sahle, Y.; El Zaatari, S.; White, T.D. Hominid butchers and biting crocodiles in the African Plio–Pleistocene. *Proc. Natl. Acad. Sci. USA* **2017**, *114*, 13165–13169. [CrossRef]
24. Frontini, R.; Roselló-Izquierdo, E.; Morales-Muñiz, A.; Denys, C.; Guillaud, E.; Fernández-Jalvo, Y.; Pesquero-Fernández, M.D. Compression and digestion as agents of vertebral deformation in fish remains: An experimental study to interpret archaeological assemblages. *J. Archaeol. Method Theory* **2021**, *8*, 1–8.
25. Van Valkenburgh, B. Incidence of tooth breakage among large, predatory mammals. *Am. Nat.* **1988**, *131*, 291–302. [CrossRef]
26. Tanner, J.B.; Zelditch, M.L.; Lundrigan, B.L.; Holekamp, K.E. Ontogenetic change in skull morphology and mechanical advantage in the spotted hyena (*Crocuta crocuta*). *Int. J. Morphol.* **2010**, *271*, 353–365. [CrossRef] [PubMed]
27. Palmqvist, P.; Martínez-Navarro, B.; Pérez-Claros, J.A.; Torregrosa, V.; Figueirido, B.; Jiménez-Arenas, J.M.; Espigares, M.P.; Ros-Montoya, S.; De Renzi, M. The giant hyena *Pachycrocuta brevirostris*: Modelling the bone-cracking behavior of an extinct carnivore. *Quat. Int.* **2011**, *243*, 61–79. [CrossRef]
28. Fernández-Jalvo, Y.; Andrews, P. *Atlas of Taphonomic Identifications: 1001+ Images of Fossil and Recent Mammal Bone Modification*; Springer: Berlin/Heidelberg, Germany, 2016; p. 355.
29. Frontini, R.; Fernández-Jalvo, Y.; Pesquero, M.D.; Vechi, R.; Bayón, C. Abrasion in archaeological fish bones from sand dunes. An experimental approach. *Archaeol. Anthropol. Sci.* **2019**, *11*, 4891–4907. [CrossRef]
30. Marin-Monfort, M.D.; García-Morato, S.; Andrews, P.; Avery, M.D.; Chazan, M.; Horwitz, L.K.; Fernández-Jalvo, Y. The owl that never left!, Taphonomy of Earlier Stone Age small mammal assemblages from Wonderwerk Cave (South Africa). *Quat. Int.* **2022**, *614*, 111–125. [CrossRef]
31. Williams, J. Small Mammal Deposits in Archaeology: A Taphonomic Investigation of *Tyto Alba* (Barn Owl) Nesting and Roosting Sites. Ph.D. Thesis, University of Sheffield, Sheffield, UK, 2001, unpublished.
32. R Core Team. *R: A Language and Environment for Statistical Computing*; R Foundation for Statistical Computing: Vienna, Austria, 2021; Available online: https://www.R-project.org/ (accessed on 27 January 2022).
33. Mangiafico, S.S.; Summary and Analysis of Extension Program Evaluation in R, Version 1.18.8. Monte Carlo Simulation. 2016. Available online: https://rcompanion.org/handbook/ (accessed on 27 January 2022).
34. Kim, H.Y. Statistical notes for clinical researchers: Chi-squared test and Fisher's exact test. *Restor. Dent. Endod.* **2017**, *42*, 152–155. [CrossRef]
35. Malan, B.D.; Cooke, H.B.S. A preliminary account of the Wonderwerk Cave, Kuruman. *S. Afr. J.* **1941**, *37*, 300–312.
36. Malan, B.D.; Wells, L.H. A further report on the Wonderwerk Cave, Kuruman. *S. Afr. J.* **1943**, *40*, 258–270.

37. Humphreys, A.J.B.; Thackeray, A.I. *Ghaap and Gariep: Later Stone Age Studies in the Northern Cape*; South African Archaeological Society: Cape Town, South Africa, 1983; p. 328.
38. Beaumont, P.B.; Vogel, J.C. On a timescale for the past million years of human history in central South Africa. *S. Afr. J. Sci.* **2006**, *102*, 217–228.
39. Beaumont, P.B. The edge: More on fire-making by about 1.7 million years ago at Wonderwerk Cave in South Africa. *Curr. Anthropol.* **2011**, *52*, 585–595. [CrossRef]
40. Chazan, M.; Ron, H.; Matmon, A.; Porat, N.; Goldberg, P.; Yates, R.; Avery, M.; Sumner, A.; Horwitz, L.K. First radiometric dates for the Earlier Stone Age sequence in Wonderwerk Cave, South Africa. *J. Hum. Evol.* **2008**, *55*, 1–11. [CrossRef]
41. Horwitz, L.K.; Chazan, M. Past and present at Wonderwerk cave (Northern Cape Province, South Africa). *Afr. Archaeol. Rev.* **2015**, *32*, 595–612. [CrossRef]
42. Ecker, M.; Brink, J.; Chazan, M.; Horwitz, L.K.; Lee Thorp, J.A. Radiocarbon dates constrain the timing of environmental and cultural shifts in the Holocene strata of Wonderwerk Cave, South Africa. *Radiocarbon* **2017**, *59*, 1067–1086. [CrossRef]
43. Shaar, R.; Matmon, A.; Horwitz, L.K.; Ebert, Y.; Chazan, M. Magnetostratigraphy and cosmogenic dating of Wonderwerk Cave: New constraints for the chronology of the South African Earlier Stone Age. *Quat. Sci. Rev.* **2021**, *259*, 106907. [CrossRef]
44. Avery, D.M. Micromammals as paleoenvironmental indicators of the southern African Quaternary. *Trans. R. Soc. S. Afr.* **2007**, *62*, 17–23. [CrossRef]
45. Avery, D.M. Rodents and other micromammals from the Pleistocene strata in Excavation 1 at Wonderwerk Cave, South Africa: A work in progress. *Quat. Int.* **2022**, *614*, 23–36. [CrossRef]
46. Fernández-Jalvo, Y.; Avery, D.M. Pleistocene micromammals and their predators at Wonderwerk Cave, South Africa. *Afr. Archaeol. Rev.* **2015**, *32*, 751–791. [CrossRef]
47. Fernández-Jalvo, Y.; Tormo, L.; Andrews, P.; Marin-Monfort, M.D. Taphonomy of burnt bones from Wonderwerk Cave (South Africa). *Quat. Int.* **2018**, *495*, 19–29. [CrossRef]
48. Fernández-López, S.R. Taphonomic concepts of or a theoretical biochronology. *Revista Española de Paleontología* **1991**, *6*, 37–49.
49. Goldberg, P.; Berna, F.; Chazan, M. Deposition and diagenesis in the Earlier Stone Age of Wonderwerk Cave, Excavation 1, South Africa. *Afr. Archaeol. Rev.* **2015**, *32*, 613–643. [CrossRef]
50. Currey, J.D. The mechanical properties of bone. *Clin. Orthop. Relat. Res.* **1970**, *73*, 210–231. [CrossRef]
51. Burstein, A.H.; Zika, J.M.; Heiple, K.G.; Klein, L. Contribution of collagen and mineral to the elastic-plastic properties of bone. *J. Bone. Joint. Surg. Br.* **1975**, *57*, 956–961. [CrossRef]
52. Reilly, D.T.; Burstein, A.H. The mechanical properties of cortical bone. *J. Bone. Joint. Surg. Br.* **1974**, *56*, 1001–1022. [CrossRef]
53. Turner, C.H. Bone strength: Current concepts. *Ann. N. Y. Acad. Sci.* **2006**, *1068*, 429–446. [CrossRef]
54. Nyman, J.S.; Roy, A.; Shen, X.; Acuna, R.L.; Tyler, J.H.; Wang, X. The influence of water removal on the strength and toughness of cortical bone. *J. Biomech.* **2006**, *39*, 931–938. [CrossRef] [PubMed]
55. Agnew, A.M.; Bolte, J.H. Bone fracture. Biomechanics and risk. In *Bone Histology: An Anthropological Perspective*; Crowder, C., Stout, S., Eds.; CRC Press: Boca Raton, FL, USA, 2011; pp. 237–256.
56. Terry, R.C. Owl pellet taphonomy: A preliminary study of the post-regurgitation taphonomic history of pellets in a temperate forest. *Palaios* **2004**, *19*, 497–506. [CrossRef]
57. Guillermo, A.A.; Montalvo, C.I.; Fernández, F.J. Palimpsest of micromammal deposits in an archaeological rock shelter (Álvarez 4, late Holocene) from northwestern Patagonia, Argentina. *Archaeol. Anthropol. Sci.* **2021**, *13*, 178. [CrossRef]

quaternary

Article

Macroscopic Chop Mark Identification on Archaeological Bone: An Experimental Study of Chipped Stone, Ground Stone, Copper, and Bronze Axe Heads on Bone

Tiffany R. Okaluk * and Haskel J. Greenfield

Department of Anthropology, St. Paul's College, University of Manitoba, 70 Dysart Road, Winnipeg, MB R3T 2M6, Canada; haskel.greenfield@umanitoba.ca
* Correspondence: tiffanyokaluk@gmail.com

Abstract: This paper presents a new macroscopic method for identifying chop marks on archaeological faunal assemblages and highlights the major differences in the morphology of chop marks created by stone and metal axes. The method provides macroscopic criteria that aid in the identification of both complete and incomplete chop mark types as well as the raw material of the axe. Experiments with modern stone (chipped and ground) and metal (copper and bronze) axes found that the degree of fragmentation within a chop mark is related to both the width and sharpness of the axe and can be classed on a scale from 1–5 using a variety of criteria. The experiments demonstrate that sharp chipped stone axes are fragile (often break upon impact) and do not create clean and well-defined chop marks. Ground stone axes are more durable but tend to create very fragmented chop marks without a clean cut (sheared) surface. Unalloyed copper metal axes can create sheared chopped surfaces; however, the relatively soft metal creates more crushing at the point of entry than bronze axes. In contrast, bronze axes are durable and create chop marks with exceptionally low rates of fragmentation resulting in a clean-cut sheared surface that extends into the bone for more than 3 mm. The method is applied to the faunal assemblage from the Early Bronze Age site of Göltepe, Turkey to determine whether the chop marks on bones were made by stone or metal axes at this early metal processing settlement. The results suggest that many of the chop marks were made by metal implements (e.g., axes). Hence, this method provides another means to monitor the adoption rates of new raw materials at a time when both metal and stone axes coexisted.

Keywords: butchery; chopping tools; axe marks; chop marks; experimental archaeology; ground stone tools; chipped stone tools; copper tools; bronze tools; Early Bronze Age; Anatolia; Near East; innovation; origins of metallurgy

1. Introduction

Animal carcass butchery practices involve a variety of tools, and the resulting actions include bashes, chops, saw marks, and slices. The resulting marks on bone allow zooarchaeologists to map out the process by which the carcass is prepared from slaughter, to secondary stages (skinning, dismemberment, and disarticulation), and final stages (marrow extraction, and filleting) [1–4]. Each of these actions leaves diagnostic marks on the bones. The location and types of marks can also inform on butchering tool preference and raw material choices [5–7], as well as cultural preferences related to food preparation [8–11].

A growing literature exists to define, differentiate, and contextualise slice marks found on archaeological animal bone, e.g., [5,12–20]. Methods for the analysis of slice marks benefit from over 40 years of dedicated experimental, methodological, and theoretical work. Recently, the use of 3D modelling, micro-photogrammetry, and deep learning algorithms has begun to revolutionise the accuracy by which zooarchaeologists recognise and differentiate cut marks by creating more objectivity [18,21–24].

Chop marks, however, have received far less attention and the methods of description and identification are not as advanced on archaeological bone [25,26]. Minimal research has been conducted on chop mark distinguishing characteristics on archaeological bone to identify the raw material choice for the axe and the nature of tool production [20]. Although chop mark studies are not new in the world of zooarchaeology, they are in their infancy when compared to slice marks.

Chopping is an essential part of the butchering process, particularly with respect to dismemberment and disarticulation of limbs and carcasses into portions and for marrow extraction. Bones have been chopped since the Lower Palaeolithic [27–29], and chopping continues to be an important part of carcass preparation in historical [3,30] and modern eras [31]. However, chop marks are not always easy to identify as some are difficult to differentiate from natural breakage, do not leave an easily identifiable chop mark, leave only one side or only a partial mark, or are masked by breakage resulting from marrow processing [32–34]. These types of bone breakage inhibit the identification of chop marks and analysis of this type of butchery mark in detail and our ability to monitor the transition from a stone- to metal-based chopping technology.

In contrast to slice marks, the diagnostic criteria for what constitutes chop marks are not well-defined other than being a somewhat V-shaped impact mark with impact fractures and/or splinters [20,35,36]. However, it is often not possible to apply such criteria if the chop is successful and completely severs the bone. Chopping action on bone creates an incredibly varied range of marks that take multiple macroscopic forms (complete, incomplete, breakage, and shearing). Similar identification problems exists if the tool is roughly formed and/or does not have a sharp and smooth V-shaped edge. This is particularly true with respect to Palaeolithic chipped stone tools, with their differentially shaped ventral and dorsal sides. The lack of specific identification criteria based on experimental studies creates a reciprocal cycle where chop marks are understudied as a line of evidence for butchery practices and tool use.

This lacuna in the study of chop marks is particularly important in the periods when the types of chopping instruments and styles of butchery dramatically evolve over time, particularly with the introduction of new raw materials for axes such as copper, bronze, and iron—in other words, in the Chalcolithic, Bronze, and Iron Ages. While stone tools preserve well, metal tools do not. When they break or are otherwise damaged, they can be melted down and made anew. Consequently, they are much rarer in the archaeological record and cannot be quantitatively used to identify when the transition from a stone to a metal-based chopping technology occurred [7,16].

The existing assemblage of metal items from even later prehistoric and historic sites is only a relatively small percentage of what was originally produced and circulated. Therefore, the higher frequency of chop marks on bones can become a proxy measure for the frequency of tools that have not preserved, and it becomes possible to monitor changes in the frequency of raw material choice, tool form, and tool production over time.

In this paper, we present our recent experimental research to aid in the definition of macroscopic diagnostic differences between several types of chopping tool shapes and raw material choices. Our study demonstrates that the macroscopic morphology of chop marks generally reflects the raw material of the chopping tool (metal versus stone), as well as the nature of tool production (ground versus chipped stone). The experiments focus on chipped stone, ground stone, copper, and bronze axes as these were the tool types available during the transition from a stone to bronze chopping technology during our periods of interest—the Late Chalcolithic (LC) and Early Bronze Age (EBA) of Anatolia [35–37].

This analysis is part of our long-term effort to identify the nature of butchering tools in zooarchaeological assemblages during this crucial period when bronze metallurgy appeared [7,9–11,16,38–47]. Our experimental results are used to differentiate and identify the type of chopping instruments used at the site of Göltepe from central Turkey, at the dawn of the Bronze Age. This experimental study aims to identify and understand the *macroscopic* differences between chop marks made by different types of metal and stone axes potentially used at the site during this crucial period in the evolution of metallurgical technologies.

2. Butchery Chopping Experimental Studies—A Brief Review

2.1. Archaeological Chop Mark Experimentation

Chops are created by sharp wedge-shaped tools that strike the bone with force and are designed to bite into or sever the bone. The limited archaeological literature on chop marks often defines these marks as wide U/V-shaped marks ([2,20], [48] (p. 349), [49]). Alternatively, and more recently, chop marks are sometimes described as a flat/planar surface created by an axe [50]. Further, an experimental study suggests that chop marks made by stone and metal axes have different morphological traits and can be differentiated based on those traits—stone axes left wide U-shaped marks on bone, whereas metal axes left deep V-shaped marks on bone [48].

While pointing the way, these very general criteria do not address chop marks that completely sever the bone. The action of chopping is intended to divide the bone into separate pieces. If the action of chopping results in the division of the bone into multiple pieces, then deep U or V shapes are an unintended consequence of a failed or incomplete chop. Both complete (full separation of the bone) and incomplete (incomplete separation of the bone) chop marks are examined in this study.

2.2. Forensic Chop Mark Experimentation

Forensic and human skeletal researchers have long conducted experiments to identify the effects on bone fragmentation as a function of the type, size, and shape of metal chopping instruments [25,26,51–55]. For the most part, their focus is on the type of metal chopping tool used in potential modern murders. It is clear from these experiments that both the width and sharpness of a metal chopping (axe) or hacking (with a sword/machete) instrument affects how the bone fractures upon impact. A wide and dull metal instrument is more likely to shatter the bone on impact, whereas a thin sharp instrument (sword/machete) is more likely to cut through a significant portion of the bone before fragmentation occurs [56].

A common observation between all such experimental chopping studies is bone fragmentation. High-speed impact from an axe will often result (up to 30% of the time) in complete fragmentation of the bone and can leave very little evidence on the bone material of the chopping event [52]. Even when complete fragmentation does not occur, a clean cut through the bone may only be visible anywhere from a few mm to a few cm into the surface of the bone before the bone fragments/splits due to the force of the tool and stress on the bone due to wedge action. The degree of wedge action is directly related to the width of the tool. Therefore, a thinner metal blade can cut into the bone much deeper than a wide tool.

The problem with directly extrapolating these conclusions from experimental forensic studies for use in archaeological identification is that these studies focus exclusively on modern steel metal tools. It is unknown if the same diagnostic criteria apply to early copper and bronze axes, and how chop marks made by various types of stone axes (chipped and ground) compare to both ancient and modern metal axes. As tool shape, material, and sharpness are all intrinsically related to the morphology of the mark left behind on the bone, it is imperative that all chopping tool types be systematically tested.

3. Materials and Methods

Five replica axes were made and tested in a series of seven separate chopping tests carried out on wood, *Ovis aries* (sheep) crania, *Bos taurus* (cattle) and *Sus scrofa* (pig) ribs, *Sus scrofa* vertebra, and *Odocoileus virginianus* (deer) and *Sus scrofa* long bones. All chops were made by a single individual to control for the relative skill, force, and technique in the butchery process. Each axe was used on a separate bone specimen in every test. The goal of each test was to separate/divide the bone to simulate dismemberment activity. This often-required multiple strikes and was not successful in every test.

3.1. Replica Axe Materials

The five replica axe heads were made from five different types of raw materials: (1) bronze (10% tin: 90% copper), (2) cold hammered pure copper, (3) ground stone (fine basalt), (4) chipped stone (fine grained chert), and (5) chipped stone (Knife River Flint) (Figure 1). The raw materials for the axe heads were selected to represent the range of raw materials used to make axes in LC and EBA Anatolia and also according to what was locally available and feasible within modern contexts for the experiments. Even though arsenical copper is found on the Anatolian plateau during the EBA, it was not included in the experiment since the resulting axe would be as hard and morphologically similar to a tin-bronze axe [57].

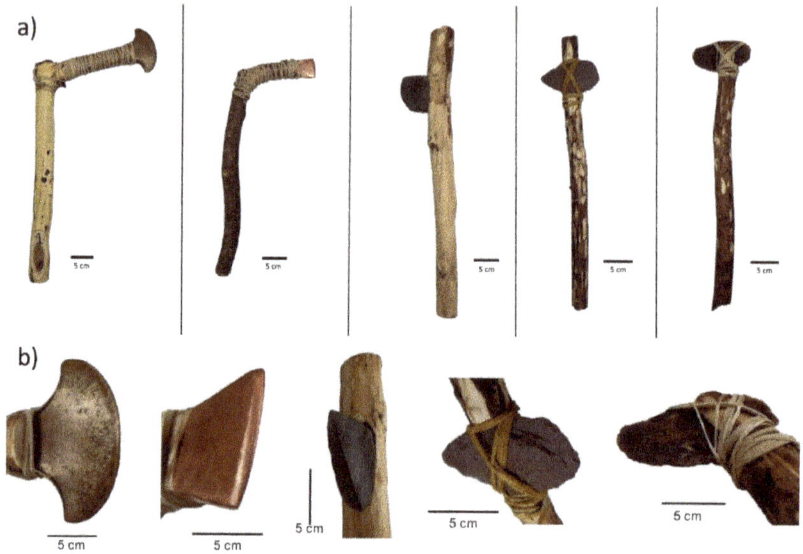

Figure 1. (**a**) The five experimental axes (left to right): bronze, copper, ground stone, chert chipped stone, and Knife River Flint chipped stone; and (**b**) close-up of the same axe heads in the same order as above.

Flint and chert are the raw materials selected for the chipped stone axe heads as these are the most common materials used for these types of axe heads [36,58] (Figure 1). The shape of the chipped and ground stone experimental axes are similar in shape to those found during the LC and EBA. Two sizes and shapes of chipped stone (flint and chert) axe heads are used in the experiment. The chert axe head is robust, and the flint is quite fine and very sharp. Both are bifacially worked, but neither bit is ground down to strengthen the edge since they are modern replicas on loan. The ground stone axes from this region are quite thick [35] with a steep pitch leading to the sharpened edge (Figure 2). The replica is shaped with this in mind. The ground stone axe is made from a fine grain basalt and shaped on a grinding wheel. For detailed description of experimental axe replicas and how they were hafted, see Table 1.

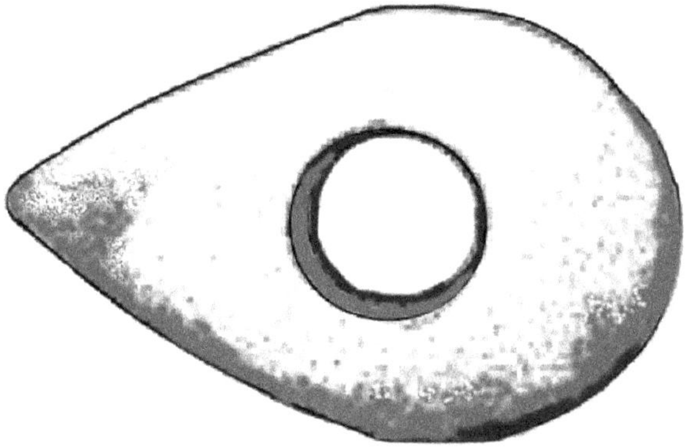

Figure 2. EBII Ground stone axe recovered from Demircihöyük–Sarıket cemetery. Image reproduced from Massa (2014) with permission from Cambridge University Press.

Table 1. Construction materials and techniques used for the experimental axes.

	Bronze	Copper	Ground Stone	Chipped Stone (A)	Chipped Stone (B)
Axe head raw material	90% copper; 10% tin	100% copper (cold hammered)	Fine gained basalt	Fine grained chert	Knife River Flint
Handle raw material	Cherry	Pine	Cherry	Pine	Pine
Haft	Pine resin, sinew, leather cord	Pine resin, sinew, leather cord	Epoxy	Sinew, leather cord	Sinew, leather cord

Although obsidian was available in Anatolia and used for slicing blades, non-votive obsidian axes in the LC and EBA are virtually unheard of. No evidence for the sustained use of obsidian axes is known from the archaeological record in Anatolia or the Levant. Only small ad hoc obsidian flakes were found at Göltepe, and there is no evidence for larger chipped stone tools regardless of the type of stone [59]. Obsidian is a fragile material and is prone to breakage.

Kononenko et al. [60] used a series of 11 experimental obsidian axes to see if the axes could withstand pressures from woodworking. Although the axes could chop through the soft wood, the axes suffered considerable damage. The tools were only useful for between 20 min–1 h before they broke completely and were discarded. The authors of the experimental study concluded that obsidian was only capable of chopping soft wood and is not suitable for harder materials. The materials used for stone axes in Anatolia and the southern Levant are flint and basalt, and these are the materials we chose for use in our study.

3.2. Experimental Bone Sample

Each axe type was tested on several domestic (cattle, sheep, and pig) and wild taxa (white-tailed deer), and osteological elements commonly found in archaeological assemblages (long bones/radius, ulna, tibia, flat bones/rib, irregular bones/cranium and vertebra, etc.). The goal of including multiple element types and animal size classes is to test the chopping capabilities of the different axe types and control for some of the potential variation due to differential bone density (Table 2).

Table 2. Bone sample characteristics.

Bone Sample (Taxon and Element)	Age	Size Class	Amount of Meat (mm) Covering the Bone
Ovis aries crania	Juvenile	Medium	2–5 mm
Bos taurus ribs	Young adult	Large	5 mm
Sus scrofa ribs	Sub-adult	Large	10 mm
Sus scrofa long bones	Sub-adult	Large	20 mm
Sus scrofa vertebrae	Sub-adult	Large	20–25 mm
Odocoileus virginianus long bones	Juvenile	Medium	2–10 mm

All specimens are from older juvenile, sub-adult, and young adult individuals and were purchased from local supermarkets. Only the ossified bone segments were dense enough to survive chopping and leave any diagnostic marks on the bones. Chop marks intersecting unossified bone often separated along ossification/fusion lines during maceration and created problems in identifying and quantifying these chop marks. As a result, these chop marks were not recorded and were discarded from the overall sample resulting in different numbers of chop marks from each axe type as the exact age of the animal was unknown before processing.

In general, meat and skin were not removed prior to chopping to better simulate the dismemberment process. Each bone specimen had at least 2–5 mm (or more) of meat covering the surface of the bone when chopped. The chipped stone axes had difficulty penetrating the bone and sometimes even struggled to cut through the soft flesh. This resulted in significantly fewer chop marks created by the chipped stone axes.

Each axe type was initially tested on a one-inch-thick flat plank of soft wood (pine). Flat pine wood boards provide a relatively even (flat), soft (as to not damage the axe edge), consistent internal structure (avoids the problem of differential bone density). Elsewhere, it has been successfully employed as a control medium for butchering experiments [16]. Second, each axe type was tested on the different bone types from the various taxa. The specifics of each test are described below. Each chopping test was conducted only by the senior author (right-handed, adult female) to control for differential abilities and strength across all tests.

Chops were directed into the bone at both c. 90° and 45° angles to attempt to mimic the angle at which chops are often observed in the archaeological record. It was found that the chop marks were more successful in penetrating the bone surface when directed at a 45° angle.

All axes were tested in both sharp and dull states, and any deformation of the axe bits due to damage was noted before and after each test.

Each chopped bone specimen was boiled (2–3 h), de-fleshed by hand so as not to introduce any unintended tool marks, and re-boiled (2–3 h) to loosen any remaining muscle tissue and extract as much grease as possible. Subsequently, the specimen was left to dry out slowly, and was then labelled, photographed, and the resulting marks were described.

The specimens used in the experiments and from the archaeological site of Göltepe are both curated in the Near Eastern and Biblical Archaeology Lab of St. Paul's College of the University of Manitoba and are available for verification.

3.3. *Recording and Attributes of Investigation*

All chop marks were given a specimen and chop number. Basic zooarchaeological information was recorded for each specimen, including the species, element, location of the mark, and age of the animal. The variables under investigation within this study were: (1) whether the chop was complete or incomplete; (2) the degree of fragmentation within the chop mark on a scale from 1–5, (3) the depth of chop mark into the surface of the bone, (4) the degree of crushing at the location of impact, (5) the angle of chop, (6) the sharpness of the axe, and (7) whether the breakage was smooth due to cutting or irregular (e.g., jagged) due to uneven breakage or crushing. These data are summarised in Table 3.

Table 3. Summary of data for all experimental axe chop marks (cBI—*chopped Butchering Incident*).

cBI	Species	Element	Axe Raw Material	Complete/ Incomplete	Angle of Chop	Sheared/ Not Sheared	Fragmentation Class	Degree of Crushing	Sharp/Dull
1	*Sus scrofa*	Radius	Bronze	Complete	90°	Sheared	5	None	Dull
2	*Sus scrofa*	Radius	Bronze	Complete	90°	Not sheared	3	Light	Dull
3	*Sus scrofa*	Radius	Bronze	Incomplete	90°	Surface	Surface	None	Dull
4	*Sus scrofa*	Radius	Bronze	Complete	45°	Sheared	4	None	Dull
5	*Sus scrofa*	Sternum	Bronze	Complete	45°	Sheared	5	None	Dull
6	*Sus scrofa*	Sternum	Bronze	Complete	45°	Sheared	4	None	Dull
7	*Sus scrofa*	Rib	Bronze	Complete	90°	Not sheared	3	None	Sharp
8	*Sus scrofa*	Rib	Bronze	Complete	90°	Not sheared	2	None	Sharp
9	*Sus scrofa*	Rib	Bronze	Incomplete	45°	Sheared	4	Light	Sharp
10	*Sus scrofa*	Rib	Bronze	Incomplete	90°	Not sheared	3	None	Sharp
11	*Sus scrofa*	Rib	Bronze	Complete	90°	Sheared	5	None	Sharp
12	*Sus scrofa*	Rib	Bronze	Complete	90°	Not sheared	3		Sharp
13	*Sus scrofa*	Rib	Bronze	Complete	90°	Not sheared	3	None	Sharp
14	*Sus scrofa*	Rib	Bronze	Complete	45°	Sheared	5	Light	Sharp
15	*Sus scrofa*	Rib	Bronze	Complete	90°	Sheared	5	None	Sharp
16	*Sus scrofa*	Rib	Bronze	Complete	45°	Sheared	4	Medium	Sharp
17	*Sus scrofa*	Rib	Bronze	Complete	90°	Not sheared	2	Light	Sharp
18	*Sus scrofa*	Rib	Bronze	Complete	45°	Sheared	5	None	Sharp
19	*Sus scrofa*	Rib	Bronze	Incomplete	90°	Surface	Surface	None	Sharp
52	*Bos taurus*	Rib	Bronze	Incomplete	45°	Sheared	5	None	Sharp
21	*Bos taurus*	Rib	Bronze	Incomplete	90°	Not sheared	3	Light	Dull
22	*Bos taurus*	Rib	Bronze	Incomplete	90°	Surface	Surface	No	Dull
23	*Bos taurus*	Rib	Bronze	Incomplete	90°	Surface	Surface	No	Dull
24	*Odocoileus virginiaus*	Tibia	Bronze	Incomplete	45°	Sheared	4	Light	Sharp
25	*Odocoileus virginiaus*	Tibia	Bronze	Complete	90°	Not sheared	3	None	Sharp
26	*Odocoileus virginiaus*	Humerus	Bronze	Complete	45°	Sheared	5	None	Sharp
27	*Odocoileus virginiaus*	Humerus	Bronze	Incomplete	90°	Surface	Surface	None	Sharp
28	*Ovis aries*	Cranium	Bronze	Incomplete	90°	Not sheared	3	None	Dull
29	*Ovis aries*	Cranium	Bronze	Complete	90°	Sheared	4	Medium	Dull
30	*Ovis aries*	Cranium	Bronze	Complete	90°	Not sheared	3	Medium	Dull
31	*Sus scrofa*	Rib	Copper	Complete	90°	Not sheared	3	Light	Sharp
32	*Sus scrofa*	Rib	Copper	Complete	90°	Not sheared	3	Light	Sharp
33	*Sus scrofa*	Rib	Copper	Incomplete	45°	Sheared	4	Light	Sharp
34	*Sus scrofa*	Rib	Copper	Complete	90°	Not sheared	3	None	Sharp
35	*Sus scrofa*	Rib	Copper	Complete	90°	Not sheared	2	Light	Sharp
36	*Sus scrofa*	Rib	Copper	Incomplete	90°	Surface	Surface	Light	Sharp
37	*Sus scrofa*	Rib	Copper	Incomplete	45°	Surface	Surface	None	Sharp
38	*Sus scrofa*	Rib	Copper	Complete	90°	Sheared	4	None	Sharp
39	*Sus scrofa*	Rib	Copper	Complete	90°	Not sheared	3	Light	Sharp
40	*Sus scrofa*	Rib	Copper	Incomplete	90°	Surface	Surface	None	Sharp
41	*Sus scrofa*	Rib	Copper	Complete	45°	Sheared	4	None	Sharp
42	*Sus scrofa*	Rib	Copper	Complete	45°	Not sheared	2	None	Sharp
43	*Sus scrofa*	Rib	Copper	Incomplete	45°	Surface	Surface	None	Sharp
44	*Sus scrofa*	Rib	Copper	Complete	45°	Sheared	5	None	Sharp
45	*Sus scrofa*	Rib	Copper	Complete	45°	Sheared	4	None	Sharp
46	*Sus scrofa*	Rib	Copper	Complete	45°	Not sheared	3	Light	Sharp
47	*Sus scrofa*	Rib	Copper	Complete	45°	Not sheared	2	None	Sharp
48	*Sus scrofa*	Rib	Copper	Complete	45°	Not sheared	3	None	Sharp
49	*Sus scrofa*	Ulna/radius	Copper	Incomplete	45°	Surface	Surface	None	Dull
50	*Bos taurus*	Rib	Copper	Incomplete	90°	Surface	Surface	Medium	Dull
51	*Bos taurus*	Rib	Copper	Incomplete	45°	Surface	Surface	Heavy	Dull
25	*Bos taurus*	Rib	Copper	Incomplete	90°	Surface	Surface	Heavy	Dull
53	*Bos taurus*	Rib	Copper	Incomplete	90°	Surface	Surface	Heavy	Dull
54	*Odocoileus virginianus*	Tibia	Copper	Complete	45°	Not sheared	1	Medium	Sharp
55	*Odocoileus virginianus*	Tibia	Copper	Complete	45°	Not sheared	1	Light	Sharp
56	*Odocoileus virginianus*	Tibia	Copper	Incomplete	45°	Not sheared	3	Medium	Sharp
57	*Ovis aries*	Cranium	Copper	Incomplete	45°	Sheared	4	None	Sharp
58	*Ovis aries*	Cranium	Copper	Incomplete	45°	Sheared	4	Light	Dull
59	*Sus scrofa*	Vertebra	Copper	Complete	90°	Not sheared	2	Light	Dull

Table 3. *Cont.*

cBI	Species	Element	Axe Raw Material	Complete/ Incomplete	Angle of Chop	Sheared/ Not Sheared	Fragmentation Class	Degree of Crushing	Sharp/Dull
60	*Sus scrofa*	Vertebra	Copper	Complete	90°	Not sheared	3	Light	Dull
61	*Sus scrofa*	Rib	Copper	Incomplete	90°	Not sheared	2	None	Sharp
62	*Sus scrofa*	Rib	Copper	Complete	90°	Not sheared	2	Light	Sharp
63	*Sus scrofa*	Rib	Copper	Complete	90°	Not sheared	1	Light	Sharp
64	*Sus scrofa*	Rib	Copper	Complete	45°	Not sheared	1	None	Sharp
65	*Sus scrofa*	Vertebra	Copper	Complete	90°	Not sheared	3	Light	Sharp
66	*Sus scrofa*	Vertebra	Copper	Complete	45°	Sheared	4	None	Sharp
67	*Sus scrofa*	Vertebra	Copper	Complete	45°	Not sheared	3	Light	Sharp
68	*Sus scrofa*	Vertebra	Copper	Complete	90°	Sheared	5	Light	Sharp
69	*Sus scrofa*	Vertebra	Copper	Complete	45°	Sheared	4	Light	Sharp
70	*Sus scrofa*	Vertebra	Copper	Complete	45°	Sheared	4	None	Sharp
71	*Ovis aries*	Cranium	Copper	Incomplete	45°	Not sheared	3	None	Dull
72	*Bos taurus*	Rib	Ground stone	Incomplete	90°	Not sheared	2	Heavy	Sharp
73	*Bos taurus*	Rib	Ground stone	Incomplete	90°	Not sheared	1	Heavy	Sharp
74	*Bos taurus*	Rib	Ground stone	Incomplete	90°	Not sheared	1	Heavy	Sharp
75	*Bos taurus*	Rib	Ground stone	Incomplete	90°	Not sheared	1	Medium	Sharp
76	*Bos taurus*	Rib	Ground stone	Incomplete	90°	Surface	Surface	None	Sharp
77	*Sus scrofa*	Vertebra	Ground stone	Complete	45°	Not sheared	3	Medium	Sharp
78	*Sus scrofa*	Vertebra	Ground stone	Complete	45°	Not sheared	3	Light	Sharp
79	*Odocoileus virginianus*	Radius	Ground stone	Complete	90°	Not sheared	2	Medium	Dull
80	*Odocoileus virginianus*	Radius	Ground stone	Incomplete	90°	Surface	Surface	None	Dull
81	*Odocoileus virginianus*	Radius	Ground stone	Incomplete	90°	Surface	Surface	Light	Dull
82	*Sus scrofa*	Vertebra	Ground stone	Complete	45°	Not sheared	2	None	Sharp
83	*Sus scrofa*	Vertebra	Ground stone	Complete	45°	Not sheared	3	Light	Sharp
84	*Ovis aries*	Cranium	Ground stone	Incomplete	90°	Surface	Surface	Light	Sharp
85	*Ovis aries*	Cranium	Ground stone	Incomplete	45°	Not sheared	3	Light	Sharp
86	*Ovis aries*	Cranium	Ground stone	Complete	90°	Not sheared	3	Light	Sharp
87	*Sus scrofa*	Rib	Ground stone	Complete	45°	Not sheared	2	Light	Dull
88	*Sus scrofa*	Rib	Ground stone	Complete	45°	Not sheared	2	Light	Dull
89	*Sus scrofa*	Rib	Ground stone	Complete	45°	Not sheared	1	Light	Dull
90	*Sus scrofa*	Rib	Ground stone	Complete	45°	Not sheared	1	Light	Dull
91	*Sus scrofa*	Rib	Ground stone	Complete	45°	Not sheared	1	Heavy	Dull

Table 3. *Cont.*

cBI	Species	Element	Axe Raw Material	Complete/ Incomplete	Angle of Chop	Sheared/ Not Sheared	Fragmentation Class	Degree of Crushing	Sharp/Dull
92	*Sus scrofa*	Rib	Ground stone	Complete	45°	Not sheared	1	Light	Dull
93	*Sus scrofa*	Rib	Ground stone	Complete	45°	Not sheared	1	Medium	Dull
94	*Sus scrofa*	Rib	Ground stone	Complete	45°	Not sheared	1	Medium	Dull
95	*Sus scrofa*	Rib	Ground stone	Complete	45°	Not sheared	2	Heavy	Dull
96	*Sus scrofa*	Rib	Ground stone	Complete	45°	Not sheared	1	Heavy	Dull
97	*Sus scrofa*	Rib	Ground stone	Complete	45°	Not sheared	2	Light	Dull
98	*Sus scrofa*	Rib	Ground stone	Complete	45°	Not sheared	1	Light	Dull
99	*Sus scrofa*	Rib	Ground stone	Complete	45°	Not sheared	3	None	Dull
100	*Sus scrofa*	Rib	Ground stone	Complete	45°	Not sheared	3	None	Dull
101	*Sus scrofa*	Rib	Ground stone	Complete	45°	Not sheared	2	Light	Dull
102	*Sus scrofa*	Rib	Ground stone	Complete	45°	Not sheared	1	Light	Dull
103	*Sus scrofa*	Rib	Ground stone	Incomplete	90°	Surface	Surface	None	Dull
104	*Sus scrofa*	Rib	Ground stone	Complete	45°	Not sheared	3	None	Dull
105	*Odocoileus virginianus*	Radius	Chipped stone	Complete	90°	Not sheared	1	Light	Dull
106	*Odocoileus virginianus*	Radius	Chipped stone	Incomplete	90°	Surface	Surface	None	Dull
107	*Odocoileus virginianus*	Radius	Chipped stone	Incomplete	90°	Surface	Surface	Light	Dull
108	*Bos taurus*	Rib	Chipped stone	Incomplete	90°	Not sheared	Surface	None	Sharp
109	*Bos taurus*	Rib	Chipped stone	Incomplete	90°	Not sheared	Surface	None	Sharp
110	*Bos taurus*	Rib	Chipped stone	Incomplete	90°	Not sheared	Surface	None	Sharp
111	*Ovis aries*	Cranium	Chipped stone	Incomplete	90°	Not sheared	3	Medium	Sharp
112	*Ovis aries*	Cranium	Chipped stone	Complete	90°	Not sheared	1	Light	Sharp
113	*Ovis aries*	Cranium	Chipped stone	Complete	45°	Not sheared	2	Medium	Sharp
114		Cranium	Chipped stone	Complete	90°	Not sheared	1	Heavy	Sharp
115	*Ovis aries*	Cranium	Chipped stone	Incomplete	90°	Surface	Surface	Medium	Sharp
116	*Ovis aries*	Cranium	Chipped stone	Complete	45°	Not sheared	2	Light	Sharp
117	*Ovis aries*	Cranium	Chipped stone	Complete	90°	Not sheared	1	Light	Sharp

3.4. Scale of Observation

The data presented are based on macroscopic observations. This scale of analysis was chosen for this experiment for three reasons: (1) Chop mark diagnostics are often large enough to be seen by the naked eye [51]. (2) Macroscopic chop mark diagnostic criteria can be defined to distinguish different kinds of axe shapes, production types, and general raw material type. This allows even field-based analysts without access to advanced instrumentation to collect data. (3) Raw material type identifications using microscopic techniques (e.g., SEM, 3D modelling, and micro-photogrammetry) depend on the creation of a surface where the tool imprints a recognisable mark on the bone.

It is only when there is sufficient direct contact between the bone and the tool where a unique identifying signature is created and the material type can be identified [26]. Unfortunately, chop marks do not always produce such a surface since the tool does not always cut cleanly into the bone so as to produce a diagnostic surface. Surfaces are often crushed, chops are incomplete, or the bone completely fragments upon impact. Therefore, traditional methods of microscopic analysis that hinge on viewing the entire area of contact between the bone and tool are not appropriate for analysing all chop marks within an assemblage. Microscopic analysis is only appropriate for chop marks that exhibit a cut/sheared surface [26]. As will be demonstrated below, EBA stone axes are unlikely to create cut/sheared surfaces on bone, while metal axes create such surfaces >50% of the time.

3.5. Chop Mark Terminology

As with any highly specialised discussion, terminology is important. Below is a list of terms used in our experimental study. The definitions below are compiled from the zooarchaeological and forensics literature on chop marks in conjunction with observations from this experimental study.

- Chop: a butchery mark created by high-speed compression forces inflicted from a sharpened wedge-shaped tool with the intention of severing a bone [48,54].
- Complete chop: a chop that severs the bone into two or more pieces [61].
- Incomplete chop: a chop that does not sever the bone but leaves a mark (deep or shallow) in the surface of the bone, often with an acute angle wedge fracture. This includes deep V-or irregular U-shaped grooves [48,61].
- Axis/kerf line: the final penetration point of the axe into the bone surface [51].
- Fracture: breakage of the bone with no visible chop mark characteristics [52].
- Kerf fracture: fracture or breakage extending from the axis/kerf line due to wedge action [52].
- Obtuse angle of chop: inferior side of chop when directed at 45° (see Figure 3) [56].

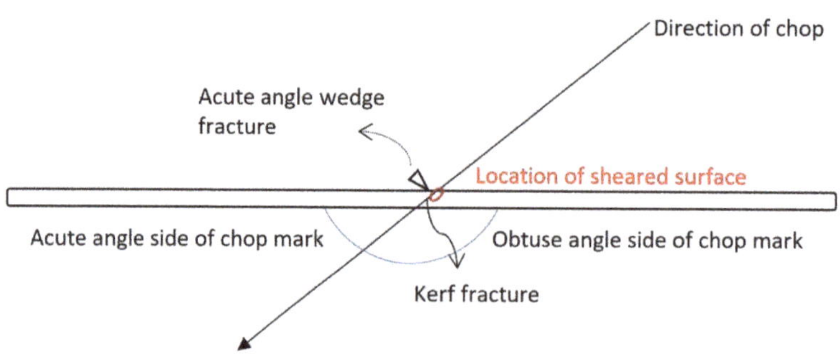

Figure 3. Schematic of chop mark characteristics based on [56].

- Acute angle of chop: superior side of chop when directed at 45° (see Figure 3) [56].
- Bit: The sharpened cutting edge of a wedge-shaped impact tool (axe).
- Sheared surface: a very clean, smooth, cut surface that extends into the bone for at least 3 mm [52].

- Crushing: fractured bone surface pushed into the chop mark by the force of impact [52].
- Acute angle wedge fracture: a small segment of bone on acute side of chop that fractures and is pushed out of chop mark (see Figure 3) [56].

4. Experimental Tests

A total of 117 chop marks were created by the five replica axes. These accumulated because of the following series of controlled chopping tests.

4.1. Test 1

The first test was designed to avoid the issue of variation in chop mark diagnostic criteria due to differential density by bone types and age of individual. Bone density varies depending on the element, age, and taxon. Consequently, each axe type was initially tested on a one-inch-thick flat plank of soft wood (pine). These chop marks did not go all the way through the wood and resulted in preservation of both sides of the chop marks.

Each axe was tested three times at different locations on a pine wood plank. The tests were: (1) a single strike, (2) two strikes in the same location, and (3) multiple strikes in the same location with the aim of severing the board. The marks on each board were labelled, photographed, and examined both macro and microscopically. Chop marks on bone are often only partially preserved due to high rates of fragmentation during the chopping process.

In order to better define the morphology of the chop marks, we chose pine wood as a medium since it would not fragment and both sides of the chop mark would be preserved. We were concerned that the varying morphology, structure, and density of bone would not allow for a clear definition of chop mark morphology (which is borne out by the experiments). While it is recognised that a pine wood plank is not an exact replacement for bone, it is a suitable substitute for this type of experiment.

A pine wood plank is relatively soft and has a more uniform shape and internal structure and density than bone. As such, it is much less susceptible to complete fragmentation upon impact. Chop mark morphology is not affected by the shape and/or density and best preserves the morphology of both the acute and obtuse sides of the mark. Thus, the pine wood chop marks should be considered as "ideal" examples of preserved incomplete chop marks.

The pine wood examples help to highlight specific aspects of axe morphology that leave distinctive and differential impact traces (mark shape, degree of crushing, and apex line morphology) as the entire mark is preserved. Therefore, we suggest caution in directly extrapolating the results of the pine wood test for bone to incomplete chop marks on bones, as the soft and consistent structure of wood preserves these characterises more consistently than bone.

4.2. Test 2

In this experiment, each axe was tested on a separate sheep (*Ovis aries*) cranium. The chops were aimed to divide the cranium into anterior and posterior halves along the coronal plane posterior to the horn core base. This location would expose the brain and is a common location for chop marks within the Göltepe faunal assemblage.

4.3. Test 3

The axes were next tested on cattle (*Bos taurus*) ribs. Cattle ribs were chosen for their dense and relatively consistent structure, particularly in the middle of the shaft. The mid-shaft rib segments were chopped multiple times. None of the axes penetrated the bone when struck at a 90° angle that was perpendicular to the long axis of the bone. The axes were then tested on the rib specimens again, but at an orientation that was parallel to the long axis of the bone.

Once again, each specimen was chopped at 90° and 45° angles, with a minimum of three impact incidences. With this orientation, the edge of each axe blade was able to penetrate the bone's surface, with the exception of the chipped stone axe (Axe 4), which merely bounced off the surface of the bone. Axe 5 (chipped stone) was not used in Test 3 as the axe head suffered significant damage in the second test and was not included in subsequent tests. It is described here to demonstrate the difficulty in using such an instrument for chopping dense bone.

4.4. Test 4

In this test, pig (*Sus scrofa*) ribs were selected for testing to determine if there is a difference between them and cattle ribs (above). Chops were directed only in a perpendicular orientation to the long axis of the bones since they were too small to be tested otherwise. Many of the chops on the pig ribs successfully either severed or shattered the rib bones.

4.5. Test 5 and 6

Tests 5 and 6 are described together since they exhibit similar characteristics and problems. Test 5 was on pig long bones (ulna and radius), while Test 6 was on pig vertebrae. These tests created analytical issues as substantial amounts of meat covered the bone when it was chopped. This prevented the axe from striking the bone in some cases. This was only discovered once the bones were de-fleshed and cleaned. In other words, butchers may sometimes chop carcasses without damaging the bone.

4.6. Test 7

The final series of tests were conducted on a variety of white-tailed deer (*Odocoileus virginianus*) long bones. The four limbs were used to separately test the four remaining axes. These tests were successful in chopping through the flesh and striking the bone, particularly along the middle of the bone shafts.

5. Experimental Chop Mark Classification

5.1. Complete Chops vs. Incomplete Chop Marks

Within the study of chop marks, it is crucial to differentiate between complete and incomplete chop marks as they create distinct but partially overlapping marks that are not always comparable. A complete chop mark successfully severs the bone, whereas an incomplete chop mark does not. The benefit of an incomplete chop is that both sides of the mark are preserved and available for analysis. An incomplete chop is not always a deep mark, and it often presents itself as a slight indentation or imprint of the bit on the bone's surface. Shallow incomplete marks (surface marks) can sometimes resemble slice marks, although they are often deeper and do not have the indicative slicing drag on one side of the mark.

In this study, these very shallow marks are referred to as surface marks and are not included within the larger sample of complete marks. Deeper incomplete marks that penetrate more than a few millimetres into the surface of the bone, but do not sever the bone, leave either a deep V or U shape. Deep incomplete marks often share enough characteristics with complete chops to be included within the complete chopped sample.

For deep incomplete marks to be classed within the complete chop marks, they must exhibit a discernible entry point and deep enough radial cracking to understand how the bone would have broken apart or fractured if more force was applied. This overview of our experimental study focuses primarily on complete chop marks as they constitute the majority of the experimental sample, and illustrate the diagnostic criteria for identifying axe morphology, production type, and raw material more effectively than incomplete marks. An in-depth review of the incomplete marks will be published elsewhere.

5.2. The 5-Point Fragmentation Scale for Complete Chops

The experimental chop marks are incredibly variable. As such, the experimental chop marks are grouped and discussed according to specific characteristics regardless of axe material type. The most common characteristic of all the experimental chops is fragmentation/breakage. This ranges from nil to severe. The purpose of a chop is to separate the bone into two separate pieces by high-speed directed force. The high speed and high force directed into the bone is meant to cause controlled breakage at a specific location. An ideal chop mark, from a functional perspective, creates two separate pieces with no/low fragmentation on either side of the chop.

Understanding fracture patterns is vital as we found that fragmentation took distinct forms that can be classified on a scale from 1–5. The degree of control over how the bones broke apart is discussed according to these groups. Thus, high levels of fragmentation and crushing register lower on the scale (1–2), and low levels of fragmentation and crushing register higher on the scale (4–5). Some marks exhibit extremely low fragmentation, particularly on one side of the mark. These smooth surfaces are referred to as sheared surfaces.

A fragmentation scale is a necessary metric of intensity as it facilitates comparison within a highly variable group of marks by assigning them to defined groups based on shared attributes. The groups are based on the intensity of fragmentation seen in the experimental sample. A total of 88 experimental chop marks were grouped according to this scale.

6. Experimental Results

6.1. Fragmentation Class 1

A Class 1 chop mark on the fragmentation scale occurs when the axe does not penetrate the surface of the bone. Crushing at the place of impact is often the only indicator of the chop (Figure 4). Without the crushing at the edges, this type of chop mark is indeterminable from natural breakage as the bone often fractures into multiple pieces causing uncontrolled breakage. As a consequence, it is easy to misidentify this fragmentation pattern as natural breakage or as non-diagnostic fractures often created by hammerstone percussion [34]. However, Class 1 chop marks do not have similar diagnostic traits as hammerstone percussion as they do not exhibit percussion notches, percussion flakes, percussion marks, shaft cylinders, or impact flakes [34].

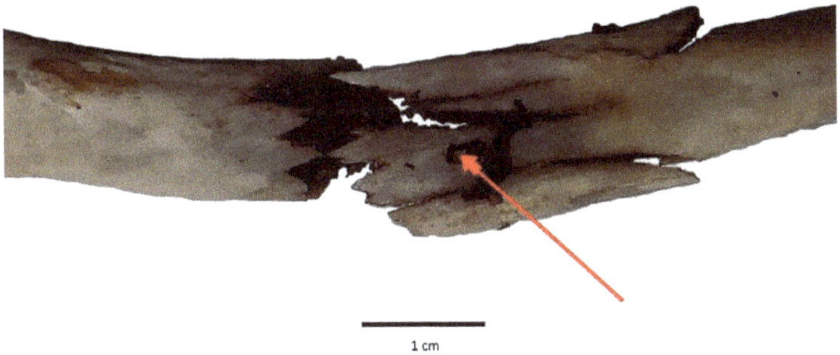

Figure 4. Class 1 experimental chop mark made by the ground stone axe on a pig rib (cBI 91) with uncontrolled breakage and crushing at the place of impact (arrow indicates the place of impact).

This type of ambiguous fracture might also be misidentified as rib peeling. Rib peeling is caused by extreme bending that causes the bone to snap or break. The low and slow pressure of bending force rarely causes complete separation of the bone segments and often leaves small bits of fibrous fresh bone still attached that must then be peeled away. The major distinctive morphological trait of peeling are missing strips of outer cortical bone that radiate from a broken or snapped edge [62,63]. Although some Class 1 chop marks may appear as snap breakage, they do not have the distinctive peeled grooves. No chop marks by any of the experimental axes created marks that mimic rib peeling.

Class 1 chop marks are challenging to identify and could be either misidentified, classed as natural breakage, or missed entirely. Identifying a Class 1 chop mark outside of an experimental setting should be done with caution as there are many natural and cultural agents that can produce seemingly similar fragmentation. Class 1 chop marks created by broad ground stone axes may also be confused with bashes made by hammerstones unless conchoidal fractures are present.

Twenty percent of the complete chops are Class 1. The ground stone axe produced most of the Class 1 marks (67%), chipped stone axes produced 11% of the marks, and copper axes produced 22%. The bronze metal axe did not produce such marks. It is likely that Class 1 chops are underreported because the chop will not be visible if the bone completely fractured upon impact and left no discernible traces. Chopping the bone with meat and periosteum still on the bone often prevents the axe from directly impacting the bone. In the experiment, it was impossible to know if a strike caused complete fragmentation until the bones were de-fleshed and processed. This adds to the difficulty in identifying this mark.

6.2. Fragmentation Class 2

This class of fragmentation is what we call "a somewhat defined chop mark" (Figure 5). Class 2 chop marks are less likely to be confused with natural breakage than Class 1, but still consist purely of breakage rather than cutting. The breakage is slightly more controlled than in a Class 1 mark and the point of impact/straight line where breakage begins is visible. This type of chop can exhibit significant crushing and breakage. Class 2 marks are also difficult to recognise outside of an experimental setting.

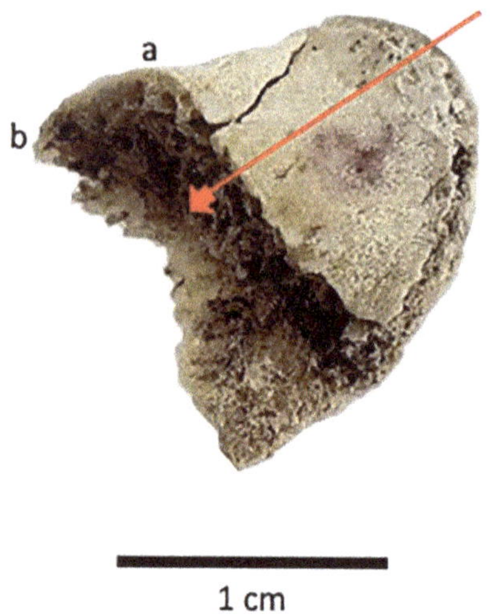

Figure 5. A Class 2 experimental chop mark made by the ground stone axe on a pig vertebra (cBI 82). The arrow indicates the direction of impact, (a) place of impact (relatively straight line where breakage begins), and (b) stepped breakage (somewhat controlled).

Of the experimental chops, 19% were a Class 2, and all axes created at least a few Class 2 marks. The ground stone axe once again produced most of these marks at 47%, but copper also produced a significant 35% of the marks. The bronze axe created far fewer (12% of the marks), while the chipped stone axe only created 6% of the Class 2 marks. An interesting pattern emerges with the metal axes regarding this category. In rare cases when both sides of the chop mark were preserved and it was possible to fit them back together, the acute side of the mark was often identified as a Class 2 mark, while the obtuse side was anywhere between a 2 and a 5. Thus, both sides of the same chop mark do not necessarily present identical fragmentation patterns, and this deviation is often more apparent with the metal axes.

6.3. *Fragmentation Class 3*

A Class 3 chop mark on the fragmentation scale is a well-defined chop mark that clearly represents intentional severing (Figure 6). This chop mark will often have mild crushing at the point of impact and a relatively straight line of breakage below. A Class 3 mark consists of very controlled breakage with a clearly identifiable entry point. Rather than the axe chopping through the bone, the axe produced extremely clean and controlled breakage.

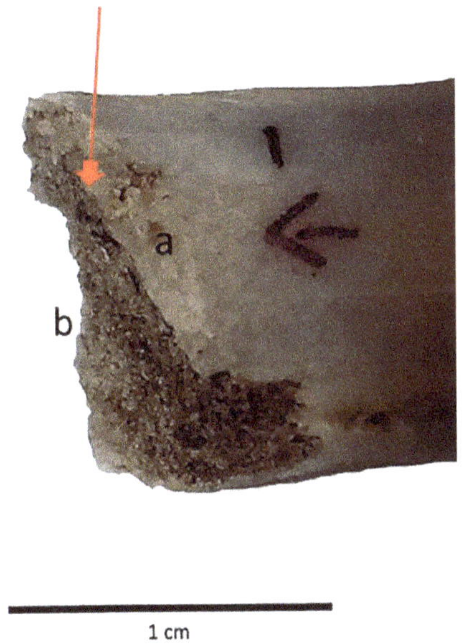

Figure 6. Class 3 experimental chop mark made by the bronze axe on a pig rib (cBI 13); (a) light crushing at place of impact (arrow indicates the place and direction of the impact), and (b) controlled breakage.

Class 3 is the most common mark created by the experimental axes with 33% of all experimental chops placed in this category. The chipped stone axe only produced one mark comparable to a Class 3. The majority of these marks were created by the copper axe (45.5%), followed by equal proportions by the bronze (27%) and ground stone axes (27%). A similar pattern was observed with the metal axes in this category as in the previous category where the two best examples of Class 5 shearing on the obtuse side are mirrored on the acute side with a Class 3. In contrast, the cleanest Class 3 marks created by the ground stone axe are mirrored on the acute side with heavy fragmentation (Class 1). In sum, Classes 1–3 describe breakage patterns due to impact rather than cutting of the bone.

6.4. *Fragmentation Class 4*

A Class 4 chop mark on the fragmentation scale is the first to exhibit shearing. Shearing is the absence of fragmentation, meaning that the axe entered the bone by cutting it and did not cause the chopped surface of the bone to break apart or flake off (except for a kerf fracture). A Class 4 mark can have light crushing, will almost always have a kerf fracture exit point, and will have at least some mild [shearing (a 3–10 mm sheared surface). A sheared surface is clean and smooth with no undulation or macroscopic parallel striations on the planar surface of the chop mark. A sheared surface is extremely identifiable (Figure 7). Class 4 marks have a smaller sheared surface area than a Class 5 because the axe does not penetrate/cut as deep into the bone before kerf fracture occurs.

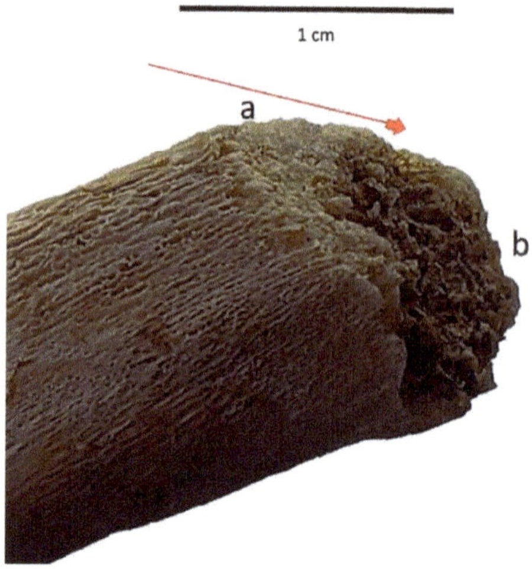

Figure 7. Class 4 experimental chop mark made by the copper axe on a pig rib (cBI 41); (a) beginning of sheared surface, (b) controlled breakage after kerf fracture. The arrow indicates the direction of the chop.

All experimental Class 4 chops were made by metal axes and represent 17% of the total number of complete chop marks. The copper axe produced most (60%) of the Class 4 marks, while the bronze axe created the other 40%.

6.5. Fragmentation Class 5

A Class 5 chop mark on the fragmentation scale exhibited little or no crushing with moderate to extensive shearing (Figure 8). These are highly recognisable marks that have a clean entry point, smooth chopped surface with no macroscopic striations, and an exit fracture. The fracture pattern of a Class 5 chop mark is extremely controlled, and fragmentation is limited to only the kerf fracture at the base of the chop mark. It is also possible for a sheared surface to continue completely through the entire bone. If a sheared surface continues completely through the bone, there will be no kerf fracture and the smooth cut surface will extend the entire length of the chop mark.

Only 11% of the experimental chops are a Class 5. Of these, two marks are made by the copper axe and the remainder by the bronze axe. This is the rarest type of mark found within the experimental sample. Figure 9 shows a breakdown of the fragmentation patterns created by each axe type, while Figure 10 is a schematic representation of chop mark Classes (1–5).

6.6. The Chipped Stone Axes (Axes 4 and 5)

The chipped stone axes produced the most unique marks in comparison to the other axes. The undulating and non-uniform cutting edge produced a 'wavy' chopped surface with uneven macroscopic striations running parallel to the direction of the axe impact. This 'wavy' pattern was also noted by Olsen and Shipman [64] in their experiments with obsidian choppers. Due to the uneven cutting surface of the tool, our chipped stone axes did not create a clean sheared chopped mark on bone. Only rarely did this tool penetrate the surface of the bone further than a few millimetres. The chipped stone axes created only seven complete chops that were clear enough to be analysed after processing, and all of the marks ranged between Classes 1 and 3 on the Fragmentation Scale (Table 4).

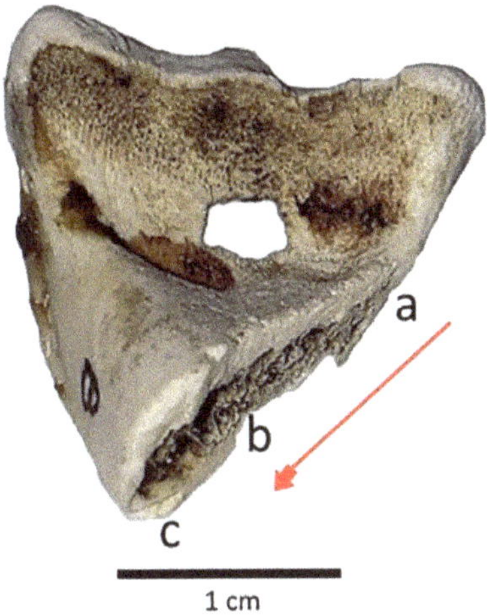

Figure 8. Class 5 chop mark made by the bronze axe on deer humerus (cBI 26); (a) place of impact (beginning of sheared surface), (b) kerf fracture (end of sheared surface), and (c) controlled breakage. The arrow indicates the direction of the axe.

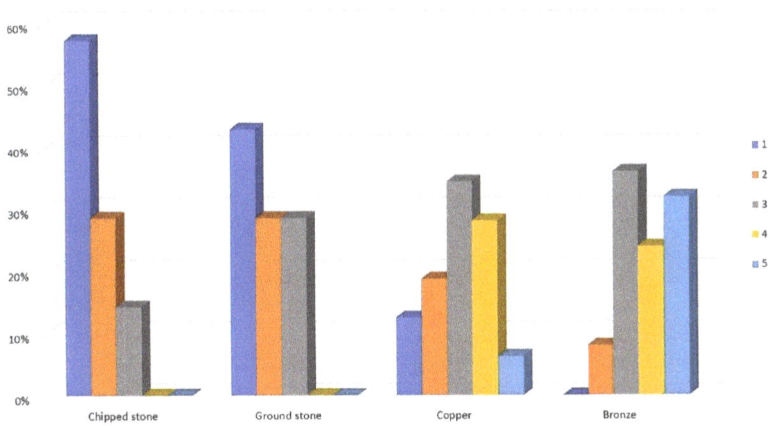

Figure 9. Frequency distribution of experimental axe types (left to right: chipped stone, ground stone, copper, and bronze) by Fragmentation Class (1–5).

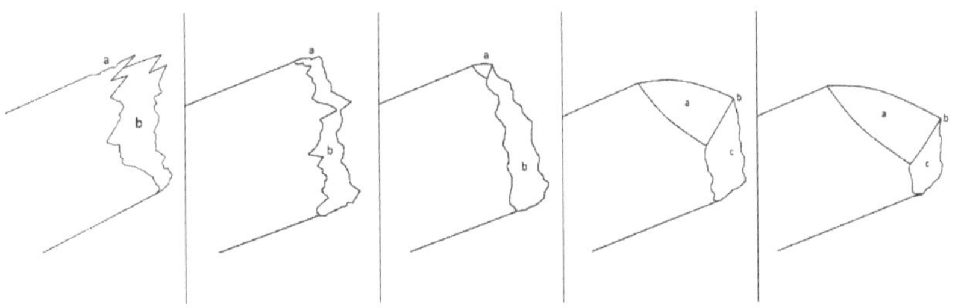

Figure 10. Schematic of Class 1–5 chop marks from left to right. Class 1—(a) severe crushing

(ambiguous point of impact) and (b) uncontrolled breakage. Class 2—(a) moderate crushing and/or visible impact location and (b) uncontrolled breakage. Class 3—(a) clear place of impact with little/no crushing and (b) controlled breakage. Class 4—(a) small (<5 mm) sheared surface with little/no crushing, (b) kerf fracture, and (c) controlled breakage. Class 5—(a) extensive (>10 mm) sheared surface with no crushing, (b) kerf fracture, and (c) controlled breakage

Table 4. Frequency distribution of experimental chipped stone axe chop marks by Fragmentation Class.

Fragmentation Class	Number of Chop Marks
1	4
2	2
3	1
Surface	6
Total	13

The chipped stone axes struggled in the chopping experiments. While they could successfully sever soft (not fully ossified or fused) bone, such as the juvenile *Ovis* skull, and break bone with very little meat covering the bone, they could not sever or break ossified bone with a hard cortex (especially from large mammals) or when a substantial layer of meat covered the bone. Further, such tools are generally ineffective butchery tools since they have fragile cutting edges (bits) that can easily break when applied with high-speed force. We found that chipped stone axes are only capable of successfully chopping softer materials (flesh and unossified bone) and are difficult to use as butchery tools otherwise.

Almost half (46%) of the chipped stone marks are incomplete surface marks that did not result in any fragmentation. The incomplete marks are morphologically unique compared to any other axe type. These marks are punctuated, with various depths along the apex. The marks and their apexes are not straight as they reflect the irregular (wavy) cutting bit of the chipped stone tool. Chipped stone axes require many strikes to the bone before fragmentation occurs, which results in a much higher percentage of incomplete marks compared to all other axe types. The morphology of these incomplete marks is indicative of only this material and manufacture style and resemble "peck marks" (Figure 11). These "peck marks" are more indicative of chopping activity with a chipped stone axe than the actual fragmentation of the bone.

The chipped stone axe heads were in a constant state of remodelling as they fractured, dulled, or re-sharpened themselves during the experiments. Similar to, but less extreme than the experimental results of Kononenko et al. [60], our chipped stone axe bits never produced the exact same signature morphology twice. It is the irregularity and inconsistency of these marks that are the distinctive characteristic of chipped stone axe chop marks.

Stout et al. [65] found very similar marks at the Lower Palaeolithic site of Boxgrove (UK). The marks still have microscopic pieces of flint embedded into the surface and are thought to be made by Acheulean hand axes approximately 500,000 years ago. Our experimental results resemble both the marks from Boxgrove and the experimental marks described by Olsen and Shipman [64].

6.7. The Ground Stone Axe (Axe 3)

The ground stone axe created the least distinct marks of any axe type and only produced marks between a 1–3 on the Fragmentation Scale (Table 5). The ground stone axe produced the heaviest degree of crushing at the point of entry and rarely penetrated the surface further than a few millimetres before fragmentation occurred. A rough chopped surface with no macroscopic parallel striations is diagnostic for these chop marks. Many of the ground stone chop marks are difficult to recognise as chop marks due to their high level of fragmentation. Consequently, they are classed between a 1–3 on the Fragmentation Scale. Only a single mark out of the 32 ground stone chop marks had a relatively smooth point of entry. Since this relatively smooth surface is very small (2 mm) and the remainder of the bone is quite fragmentary, the mark is classed as a Class 3 and not a Class 4.

Figure 11. "Peck mark" created by the chert chipped stone axe on a deer radius (cBI 106).

Table 5. Frequency distribution of experimental ground stone axe chop marks by Fragmentation Class.

Fragmentation Class	Number of Chop marks
1	12
2	8
3	8
Surface	4
Total	32

More extensive crushing is noted on all ground stone chop marks relative to those created by the metal axes. Crushing is further exaggerated on bones with a dense outer cortex (large mammals) as opposed to more delicate elements, such as the vertebrae and crania of smaller individuals. A slight increase in crushing is also noted as the tool dulled slightly after each impact.

All ground stone chops on the pig ribs were directed into the bone at approximately a 45° angle to test the shearing capabilities. Only a single rib bone was divided into two separate sections, while the majority fragmented into three or four pieces, and the most heavily fragmented rib broke into 12 pieces. No sheared surfaces were identified. The chop mark with a relatively smooth entry (mentioned above) was unintentionally directed into the bone at a very low angle (approx. 15°).

A ground stone axe creates significant crushing and fragmentation on both sides of the chop marks and many of the marks would be difficult to identify as chop marks within an archaeological assemblage. When chop marks are more identifiable (a Class 2 or 3), the other side of the mark is often very fragmented and would not be identifiable as a chop mark outside of an experimental setting. Thus, many chops made by ground stone axes are likely to be missed in an archaeological assemblage due to a lack of discernible features.

The ground stone axe produced noticeably fewer surface marks compared to the chipped stone axe. These incomplete marks are much more indicative of the raw material and manufacture than the complete chop marks. Ground stone surface marks are more often a wide U-shape rather than V-shape, with a pitted, relatively straight apex line (see Figure 12 for a comparison of incomplete marks on wood). Identification of the pitted apex line can be aided by examination under a low powered microscope.

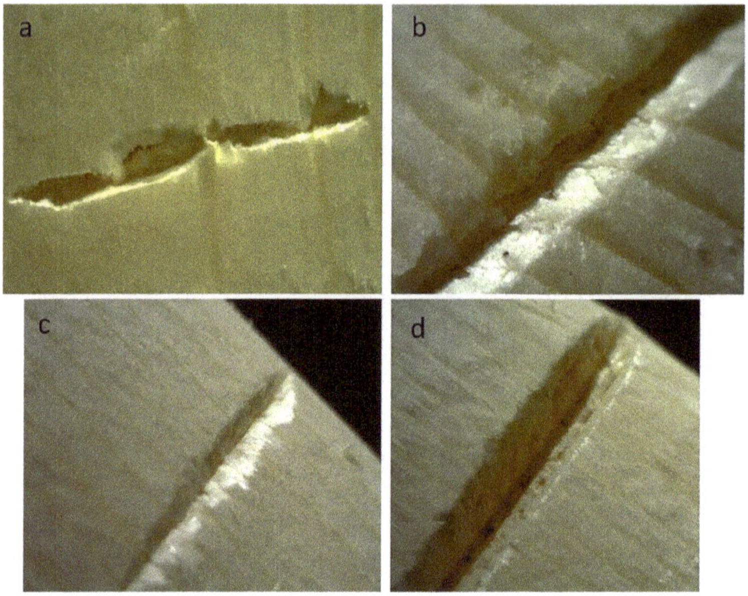

Figure 12. Photographs of incomplete (not fully severed) marks on wood; (**a**) flint chipped stone, (**b**) ground stone, (**c**) copper, and (**d**) bronze.

6.8. The Copper Axe (Axe 2)

The copper axe produced the most variable range of chop marks with multiple chops assigned to each fragmentation class. The copper axe produced more Class 3 marks than any other fragmentation Class. The second highest group was Class 4, and the third most frequent group was Class 2 (Table 6). Most chops demonstrated light crushing at the point of entry, while only a few exhibited heavy crushing.

Table 6. The frequency distribution of experimental copper axe chop marks by Fragmentation Class.

Fragmentation Class	Number of Chop Marks
1	4
2	6
3	11
4	9
5	2
Surface	9
Total	41

Of the 41 copper chop marks, 17% had some level of shearing on the chopped surfaces. The sheared surfaces averaged 5 mm in length, with a single sheared surface closer to 10 mm in length. While the level of shearing was not extensive, it was present and macroscopically recognisable. Shearing only occurred when the bone was struck at a 45° angle, as opposed to a 90° angle.

Most chop marks made by the copper axe were recognisable on both sides of the chopped surface. Only 9.7% of the marks were highly fragmented (Class 1), which suggests that complete chop marks made by copper axes are more likely to be recognised than those created by ground stone axes. The copper axe created significantly less crushing at the place of impact than the ground stone axe. The exception to this is with the dense *Bos taurus* (cattle) ribs where the degree of crushing and the morphology of the marks were similar to the ground stone. The copper axe struggled to cut through dense bone and caused significant damage (crushing) to dense bone.

The wide variety of fragmentation classes created by the copper axe highlights the importance of analysing the entirety of a chopped assemblage as no single macroscopic trait is diagnostic of this material type. The copper axe can produce sheared surfaces but is also capable of creating very fragmented chops and varying degrees of crushing. Copper is a relatively soft material and the edge (bit) continually dulled, deformed, and required resharpening. The inability of copper to consistently retain a sharp cutting edge enabled an increased variability in fragmentation classes and crushing at the point of impact. Therefore, a chopped assemblage (complete marks) created primarily by copper axes will exhibit all types of fragmentation classes (sheared and non-sheared) along with the presence of moderate to mild crushing.

Incomplete surface marks made up 21% of the copper chop marks. The copper axe created V-shaped surface (incomplete) marks. The apex lines were straight and did not exhibit the pitting seen in the marks created by the ground stone axe. The walls of the mark exhibited some fragmentation/crushing and were not as cleanly defined as with examples created by the bronze axe (see Figure 12 for a comparison of incomplete marks on wood).

6.9. The Bronze Axe (Axe 1)

The bronze axe created 30 chop marks, but five (16%) were shallow surface marks. The bronze axe created deep V-shaped surface/incomplete marks. The surface marks varied in depth from 1 mm to almost 1 cm and always had a straight, non-pitted apex similar to those from the copper axe (Figure 12). However, the walls of the bronze axe surface marks were not fragmented and exhibited little to no crushing.

The bronze axe created 25 complete chop marks and was the only axe to produce no Class 1 marks (Table 7). The bronze axe created the least amount of crushing at the point of entry compared to all other axe types, and it was the only axe to produce more sheared than non-sheared marks. The shearing on some of the chop marks was extensive with one sheared surface continuing almost completely through the bone (Figure 8). The fragmentation caused by the bronze axe was more controlled than any other axe type, often resulting in only two fully severed pieces of bone when completely chopped through. The bronze axe was the only axe type capable of cleanly penetrating the hard and dense outer bone surface of the fully ossified *Bos taurus* ribs and *Sus scrofa* long bone shafts without creating crushing at the point of entry.

Table 7. The frequency distribution of experimental bronze chop marks by Fragmentation Class.

Fragmentation Class	Number of Chop Marks
2	2
3	9
4	6
5	8
Surface	5
Total	30

6.10. Morphological Differences between Metal and Stone Chop Marks

Complete (fully severed) chops make up most of the experimental assemblage. These range from a mass of tiny, fragmented pieces to perfectly smooth chopped surfaces where the bone is divided into only two pieces. While the rate of fragmentation within a chopped assemblage is generally associated with the axe material type, all axe materials produced multiple types of fragmentation patterns. The stone axes created more limited patterns of fragmentation than the metal axes. While the stone axes are sometimes able to divide the bone into only two separate pieces, they always create uneven or fragmented breakage on both sides of the chop mark, often with crushing at the entry point. Even with the more controlled ground stone chop marks, the chopped surface always exhibits fragmentation within the mark itself and has no sheared surfaces.

Metal axes produce a wider variety of fragmentation patterns compared with the stone axes. Both the copper and bronze axe created some marks similar to the ground stone axe (Class 1–3) but also created clean sheared surfaces with very low rates of fragmentation (Class 4–5). Metal axes are much more likely to create sheared surfaces when the bone is struck at a 45° angle as opposed to a 90° angle.

In this study, the copper axe produced fewer instances of shearing compared with the bronze axe. This is likely due to the differential hardness between the two materials. The copper axe also produced heavier crushing at the point of impact. The copper axe was never as sharp even though both tools were sharpened with the same metal hand file. The copper axe also did not retain a sharp edge for as long as the bronze axe and required repeated sharpening.

The definitive morphological characteristic of metal axe marks is shearing or the absence of fragmentation on the obtuse chopped surface. Shearing is the result of cutting rather than breakage and is only found on Class 4 and 5 chop marks. None of the stone axes created Class 4 or 5 chop marks, meaning that no sheared surfaces were created by stone axes. While breakage can create a relatively flat surface, it cannot create a smooth/cut surface. The problem with this morphological characteristic is that not all chops made by metal axes create sheared surfaces. Only 27% of the experimental chops created by the copper axe exhibited sheared surfaces, and 47% of the chops created with the bronze axe exhibited sheared surfaces. Shearing was also limited to one side of the chop mark. As noted previously, a sheared mark always produces an opposing side that is never higher than a Class 2 or a 3.

Although shearing did not appear on all chops made by metal axes, it was never associated with chop marks made by stone axes. This is due to the two major variables: width and sharpness [56]. The width of an axe determines how deep the axe can penetrate into the bone tissue before causing the bone to split, either at the kerf or completely [52]. If the axe is quite wide, it does not have the ability to penetrate further than the relatively short portion that was sharpened before it splits the bone as a wedge. This results in breakage at the point of impact rather than cutting through the bone (Figure 13).

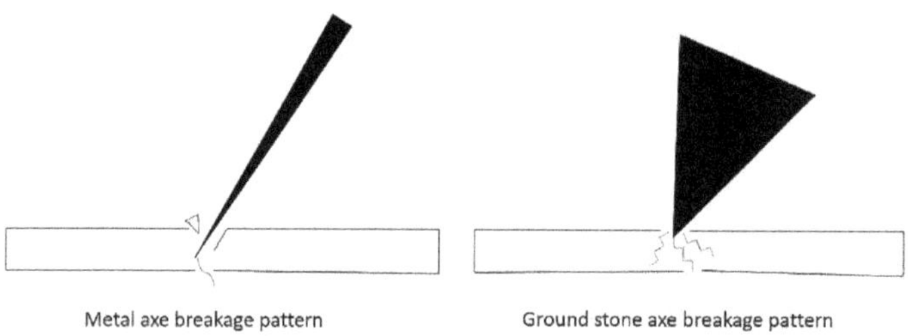

Figure 13. Schematic of metal vs. ground stone fragmentation pattern.

The sharpness of the axe determines how cleanly the axe can enter the surface of the bone. The sharper and thinner the implement, the more concentrated the force that will continue down into the bone instead of spreading out over the surface of the bone. This will create a smoother and deeper chopped surface. Humphrey and Hutchinson [52] found that metal axes, unlike thinner metal tools—such as machetes and cleavers—sometimes push bone material into the chop mark itself. This is defined as crushing, and crushing is more likely when both the width of the tool is increased and the sharpness decreased.

Our experiments found that these criteria are exaggerated in dense bone (large mammal), especially with respect to the crushing and sharpness of the bit. An axe must be thin, very sharp, and strong to cut into the surface of dense (large mammal) cortical bone without creating medium/heavy crushing at the point of impact. In contrast, a thin, moderately sharp and strong axe can cut into medium mammal cortical bone with relatively little crushing. We found the degree of crushing at the place of impact is a function of variables related to sharpness of the axe bit.

The ground stone axe was very successful at splitting the bone; however, it could not cut through the bone without causing splintering and fragmentation on both sides of the chop. A thinner and sharper stone axe does have the ability to penetrate further into the bone as demonstrated by the thin Knife River Flint axe. However, the thin and sharp cutting edge of the Knife River Flint axe head was too fragile to handle the high-speed impact needed for chopping and shattered almost immediately. Although untested, a very thin ground stone axe might also be at risk for significant damage while chopping dense bone material.

Our ground stone axe did not retain its sharp cutting edge for long. The cutting bit consistently chipped off until it reached an equilibrium where the edge was stable enough in relation to the amount of force inflicted upon it. Therefore, the functional form of the tool must be dictated by the structural integrity of the material. Earlier Neolithic ground stone axe forms are considerably thinner than those found in the LC and the EBA. These thinner/sharper forms may have the ability to cut into the bone without causing significant damage to the tool. However, this remains untested.

The fragmentation patterns of the complete chop marks reflect the shape and structural integrity of the tool rather than the raw material. Hence, thick and wide-edged (and consequently dull) axes produce fragmented chops with no shearing. Thin, narrow-edged (and consequently sharp) axes can also produce this type of mark when directed into the bone at a 90° angle. However, when they are used even at a slightly deviating angle, thin sharp-edged axes produce marks with significantly less fragmentation and a sheared surface on the obtuse side. Wenham (1989) also concluded that axe thickness is the major variable for determining the amount of crushing at the surface and the depth the axe can penetrate bone tissue before fracture occurs. As such, it is important to understand tool variability for the period in question before applying the results of this or any experimental study to an archaeological faunal assemblage.

The application of the method is dependent on a sample of chop marks rather than individual chop marks as all axe material types can create multiple fragmentation classes. Chop marks created by metal axes (thin and sharp) cluster between 2–5 on the fragmentation scale, while stone axes (wide and dull) cluster between 1–3 on the fragmentation scale. Individual chop marks should never be considered as indicative, definite evidence of a particular tool material type or be used in place of the overall fragmentation pattern. The fragmentation scale allows for outliers, and this is essential when each material has such a wide range of morphological possibilities.

6.11. Criteria for Identifying Complete Chop Marks

Identifying chop marks that fall lower than a Class 3 on the fragmentation scale is difficult as these marks often do not display a clear point of entry. Without knowing where the axe struck the bone, it is almost impossible to argue that the breakage displayed is in fact created by an axe and not by other depositional or taphonomic processes. A point of entry is defined either by concentrated crushing, a relatively straight line from which breakage begins or the beginnings of a sheared surface.

Any of these criteria indicate the place of impact from a wedge-shaped tool. As this experimental study shows, metal axes create these diagnostic criteria more frequently than stone axes. This means that there can be a clear bias (if one is not careful) toward the identification of metal axe marks as opposed to stone axe marks (especially ground stone) in the archaeological record. Close attention to the placement of high fragmentation could be referenced in relation to ethnographic butchery mark placement see [1]. However, without direct evidence for the butchery marks themselves, this remains only an anecdotal inference.

7. The Zooarchaeological Sample

7.1. Regional Setting

The case study for this experimental project is the Late Chalcolithic/Early Bronze Age site of Göltepe, Turkey (4400–1900 BCE). Göltepe was a mountain village located in the south-eastern corner of the central Anatolian plateau, nestled amidst the Taurus Mountains (Figure 14). Located on the top of a steep hill, the 5 ha village is within sight of the Kestel tin mine located on the neighbouring hill [66]. Early Bronze Age Göltepe was a specialised metal production site, specifically tin (for a detailed description of the tin production process at Göltepe see [67]). There is no evidence for either copper, iron, or gold production at Göltepe during the Early Bronze Age [68].

Figure 14. Map showing the geographic location of Göltepe and the Kestel mine site within modern day Turkey.

It is interpreted as a first-tier metal production site, meaning that Göltepe was a location associated with extraction, grinding, smelting, and the creation of ingots, which would then be traded and or transported to larger centres [69,70]. Tin is an essential component of bronze and is a relatively rare element on earth. Before the discovery of the Kestel tin mine and its associated processing village at Göltepe, it was assumed that tin was imported into Anatolia from the east as it was during the Middle Bronze Age.

Recent geologic, isotopic, and archaeological investigation now indicates that tin was/is locally mined in Anatolia during the Early Bronze Age [70–72]. There is little evidence for the production of finished metal goods at Göltepe apart from a few simple tool moulds for axes, chisels, and ingots. While no finished metal axes were found, the few copper chisels are the only metal tools recovered from at EBA Göltepe [73,74].

7.2. EBA Anatolian Axes

The Early Bronze Age (c. 3000–2000 BCE) in Anatolia is a particularly unique period for axe types as metal (copper, arsenical-copper, and bronze) co-existed with older stone types. As a result, the simple identification of a chop mark on archaeological bone cannot be inherently linked to a specific material type without further experimental work. Four types of axes were common in EBA Anatolia. These are: (1) Chalcolithic style flat axes cast in an open face mould, (2) shaft-hole style battle axes cast using the lost-wax technique, (3) crescent-shaped axes cast in a bivalve mould, and (4) ground stone shaft-hole style axes ([35] (p. 84), [36,58,64], [75] (p. 55)).

While the chipped stone tool industry is often assumed to have declined during the EBA with the introduction of bronze metallurgy, this was not always the case, particularly regarding quotidian activities such as animal butchering. Chipped stone slicing tools continue to be used through the EBA and only begin to decline much later in the Bronze Age. This is particularly true wherever obsidian is abundantly available, including at Göltepe and other sites located near obsidian sources [41,45]. Nonetheless, in general, the range of chipped stone tool types becomes much more restricted and simplistic, and there is a perceived increase in ad hoc/single use stone tools [76]. This pattern is also observed in the southern Levant and often attributed to the increased availability of metal tools [58].

The EBA was a period of non-standardised axes within a society that accepted both new and old forms of this multi-purpose tool. The ratio of stone to metal axes in the EBA is the reverse of the pattern seen in the LC, in that stone axes became the exception rather than the standard. While stone axes, both chipped and ground (votive), disappear completely from the southern Levant after the EB I [58], ground stone axes continued to persist in a limited capacity across Anatolia throughout the EBA [35,36].

7.3. The Chopped Assemblage from Göltepe

The chopped assemblage from EBA Göltepe is comprised of 779 individual chop marks. All primary and secondary context EBA faunal material from Göltepe were examined for evidence of chopping, and 8.86% of the total Number of Identified Specimens (NISP) exhibit one or more chop mark(s). Most chop marks are found on medium mammals (87%), and only 8% are found on large mammals. The sample is primarily comprised of adult (40%) and sub-adult (38%) specimens with smaller frequencies (22%) of chop marks found on juvenile individuals. Chop marks are found on specimens across all EBA phases (EB I, II, and III) at Göltepe (See Yener 2021 for a full description and interpretation of phasing at EBA Göltepe).

Much of the chopped assemblage is comprised of complete chop marks (87%). An incredible 94% of the complete *Chopped Butchering Incidence* (cBI 523) are assigned to Classes 3–5. A butchering incidence is when there is one or more marks made at the same location for the same goal [9,38]. Of these marks, 31.5% (cBI 173) are assigned to Class 3, 37% (cBI 204) are assigned to Class 4, and 26% (cBI 146) to Class 5. (Figure 15) Crushing at the point of entry is minimal for all chop marks considered to be Class 3 and above and is observed in less the 1% of all Class 3 and Class 4 marks. No crushing is observed on any Class 5 marks. (See Figure 15 for fragmentation Class frequencies of complete chop mark incidences).

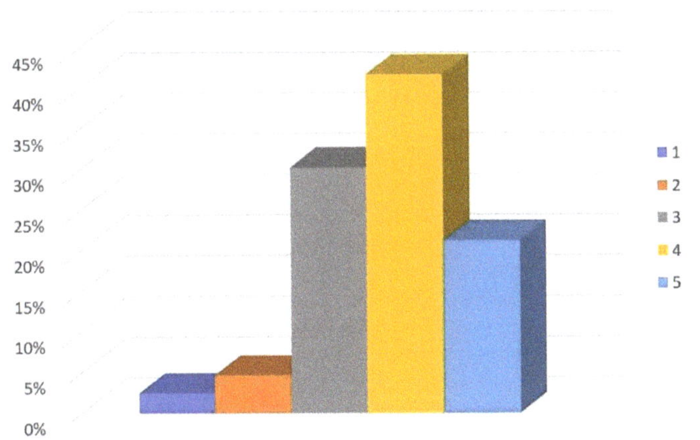

Figure 15. Frequency histogram of all EBA Göltepe chop marks classified by Fragmentation Classes 1–5.

The overall fragmentation pattern does not vary when chop marks are compared according to size class or age. Class 4 chop marks are always the most common with both Class 3 and Class 5 much more frequent than both Classes 1 and 2.

The Class 5 marks within the Göltepe assemblage are remarkable. The depth of the sheared surface before kerf fracture is measurable for 35.5% of the complete Class 5 chop marks, and some of these chop marks are exceptionally deep. The depths ranged from 5 mm up to 30 mm, and averaged 8–15 mm. This is well beyond the depths created by the experimental sample and indicates than an extremely thin and sharp axe created these chop marks. See Figure 16 for examples of chop marks from Göltepe.

Figure 16. Examples of Class 5 complete chop marks from Göltepe (left to right: *Ovis/Capra* vertebra, MRN 2299, bone 13; *Capra hircus* horn core, MRN 3433, bone 35; *Ovis/Capra* vertebra, MRN 2299, bone 12).

Incomplete marks make up only 13% of the chopped assemblage from Göltepe. Many of these marks are quite deep, and due to the depth, as well as the 45° angle rather than a 90° angle, the apex line is only visible on 3.6% of the incomplete marks. Of the visible apex lines, the vast majority are straight and narrow with no macroscopic or microscopic pitting. However, due to weathering and root etching, some incomplete marks are not appropriate for assessment. No "peck marks" were observed on or around any of the identified chop marks.

8. Discussion

Based on comparison with the experimental results from the five axe types, most of the chop marks from EBA Göltepe were likely created by metal axes. This conclusion is based on both our experimental results in conjunction with the specific axe types available during the period of investigation. LC/EBA Anatolia used very wide ground stone axes, and the evidence for chipped stone axes is extremely limited.

The experimental study shows that wide ground stone axes create very fragmented chops and are extremely unlikely to cut through the bone before fracture occurs. While it is possible that ground stone axes created some of the more fragmented chop marks, the overall pattern of chop mark fragmentation from Göltepe closely resembles the fragmentation distribution created by the experimental metal axes, especially the bronze axe. The exceptionally low levels of fragmentation within the chopped assemblage from Göltepe, along with the depth of axe penetration, lack of crushing, and intensity of shearing could not be created by EBA style stone axes (see Figure 2).

The depth of the marks found on the Göltepe chopped assemblage strongly suggests that they may have been made by bronze or possibly arsenical-copper axes. While it may be possible that a heavier pure copper axe might be able to cut into bone deeper than our experimental copper axe did, it does not negate the observation that our copper axe caused much more crushing at the point of impact compared to the bronze axe. The lack of crushing at the point of entry on the majority of the Göltepe chopped assemblage strongly suggests that the axes used were made from a harder alloy material.

Based on the chop marks, the use of metal axes at EBA Göltepe did not increase through time. This suggests that the adoption of metal axes as the preferred tool in butchery dismemberment took place either at or before the Chalcolithic/EB I transition in south central Anatolia. Although Göltepe may have had unique access to metal resources as a metal production site, it is unlikely that the residents of Göltepe existed in a social bubble. Late Chalcolithic/EBA I Göltepe was a seasonal metal production site, and for part of the year, the residents must have resided elsewhere [68]. Since metal axes were used consistently in butchery at Göltepe during this period, metal axes were presumably also in use at other (nearby) sites as well.

The recovered metal finds from Göltepe are not overly impressive, nor do they suggest that metal tools were extensively used in everyday situations. The EBA metal finds mainly consist of small items of personal adornment, such as pins, rings, and bracelets. The only functional items recovered are two metal chisels, an awl, and a needle [73]. The enormous quantity of ground stone tools (mainly grinding stone and pounders) and simple flakes found across the site from all EBA Periods, suggests that stone tools were the preferred utilitarian tool material.

Slice mark analysis on the faunal assemblage supports the theory that metal tools were not extensively used in the EBA, nor were they integrated as a part of everyday butchery activities [45]. The analysis of the chop marks on the faunal material from all EBA phases at Göltepe indicates that this was not the case. While metal knives were not used, metal chopping instruments were extensively used in butchery. Metal chop marks on butchery remains are found across all periods and are not reserved to specific contexts.

Metal chop marks on faunal material are found in almost all contexts, including pit houses, dirt and stone lined pits, outdoor activity areas, Terraced Buildings 1 and 2, stone foundation buildings, stone installations, thermal installations, and middens. This suggests that all butchers had access to metal chopping instruments and that they were an everyday tool available to the general population of EBA Göltepe. Thus, even though metal axes are not physically present in the archaeological record at the site, their residual remains can be identified in the animal bone assemblage.

The extensive use of metal chopping instruments in butchery provides new evidence for how metal functioned and was used within EBA Anatolian society. While metal in the EBA was clearly a prestige material used by the elite classes to differentiate themselves as powerful leaders, metal was not exclusively reserved for the elite class. Metal was used, traded, and consumed by both elite and non-elite individuals. Investigation into the adoption of new technologies and the differential use of metals by elite and non-elite can benefit from butchery analysis, as it may not be self-evident in the material culture.

9. Conclusions

The experimental research presented here is intended as the beginnings of a systematic method (that can be utilised even in field settings) to aid in the recognition of chop marks on archaeological bone. It focuses on and highlights the macroscopic differences between chop marks made by stone and metal chopping instruments.

Chop marks are an incredibly varied grouping of butchery marks. Chop marks can present as U- or V-shaped marks, complete fragmentation, uncontrolled breakage, controlled breakage, and sheared (cut) surfaces. Therefore, it is important to look at the overall fragmentation patterns of the chopped assemblage for variables, such as crushing, degree of fragmentation, depth of chop mark, and shearing. These characteristics generally correspond to the width and sharpness of the cutting edge on the instrument [52,56].

Metal axes create relatively low levels of crushing compared to stone axes and are more likely to cut through the bone instead of breaking the bone. These observations are important as complete chop marks are often defined as a "flat, planar surface" [50]. It is clear from this experimental study that this is a very narrow definition that describes only the obtuse side of some metal chop marks and/or chop marks created by very thin and sharp axes.

Understanding the macroscopic variability within the broad category of chop marks has also revealed the need for further work and experimentation on this subject, especially regarding the microscopic differentiation of chop marks created by different types of metal or stone axes. A more detailed analysis of this topic may require the two overarching material groups (metal and stone) to be microscopically investigated individually, as one material class typically presents as breakage, and the other typically presents as a cut surface.

While our goal is to define macroscopic diagnostic criteria to identify the axe shape, production type, and raw material, which are useful in field settings, subjectivity in descriptive morphological characteristics can be problematic regarding how each researcher interprets the criteria for analysis [21,24]. In this experimental study, we identified key criteria for identifying chop marks and described how these criteria generally present according to material type. Deep Learning algorithms can be programmed to detect these traits more consistently than the human eye and help to remove some subjectivity [22,74,75]. We suggest this as an additional way forward for distinguishing chop marks made by different material types.

Detailed discussions of chop marks on archaeological faunal assemblages are rare but are an important component of butchery studies as demonstrated by the case study from Göltepe. Axes do not necessarily (nor should they be expected to) follow the same technological trajectory as knives. It is a mistake to assume that the replacement of stone tool technologies used in animal butchery occurred with all technologies simultaneously with the advent of the Bronze Age.

Some types of tools were rapidly adopted and replaced older technologies (chopping tools), while others were retained for long periods of time (slicing tools) [7,43,47]. When investigating a time when there was access to diverse types of butchery tools made from different raw materials, it is essential to be aware of how each present as macroscopic marks on bone as more than one type may be in use. In conclusion, chop marks do not always present themselves as flat or sheared surfaces, and a host of attributes must be considered when identifying the raw material and the type of butchering instrument.

Author Contributions: Conceptualization, methods, investigation, writing—T.R.O. and H.J.G.; formal analysis—T.R.O.; curation, funding acquisition, project administration validation—H.J.G. All authors have read and agreed to the published version of the manuscript.

Funding: This research was funded by grants and institutional support from the Social Sciences and Humanities Research Council of Canada, the University of Manitoba, and St. Paul's College to Haskel J. Greenfield.

Acknowledgments: We would like to thank Aslihan Yener and Allan Gilbert for their permission to analyse and publish the data, Tina L. Greenfield for help in sorting out the stratigraphy at the site, and the anonymous reviewers for their comments. Finally, we would like to thank Gord Hill for graciously lending his chert and flint biface tools for the project, Curtis Link for helping to shape the ground stone axe head, and Jon Ross for his help in conducting the butchery experiments. Any errors are the responsibility of the authors.

Conflicts of Interest: The authors declare no conflict of interest.

References

1. Binford, L.R. *Bones: Ancient Men and Modern Myths*; Academic Press: New York, NY, USA, 1981.
2. Lyman, R.L. Archaeofaunas and butchery studies: A taphonomic perspective. *Adv. Archaeol. Method Theory* **1987**, *10*, 249–337.
3. Seetah, K. Meat in history-the butchery trade in the Romano-British period. *Food Hist.* **2004**, *2*, 19–33. [CrossRef]
4. Seetah, K. Butchery as a tool for understanding the changing view of animals: Cattle in Roman Britain. In *Just Skin and Bones? New Perspectives on Human-Animal Relations in the Historical Past*; British Archaeological Reports, International Series 1410; Pluskowski, A., Ed.; BAR: Oxford, UK, 2005; pp. 1–8.
5. Greenfield, H.J. Slicing cut marks on animal bones: Diagnostics for identifying stone tool type and raw material. *J. Field Archaeol.* **2006**, *31*, 147–163. [CrossRef]
6. Seetah, K. Multidisciplinary approach to Romano-British cattle butchery. In *Integrating Zooarchaeology*; Maltby, M., Ed.; Oxbow: Oxford, UK, 2006; pp. 111–118.
7. Greenfield, H.J. "The Fall of the House of Flint": A zooarchaeological perspective on the decline of chipped stone tools for butchering animals in the Bronze and Iron Ages of the southern Levant. *Lithic Technol.* **2013**, *38*, 161–178. [CrossRef]

8. Beller, J.A.; Greenfield, H.J.; Levy, T.E. The butchered faunal remains from Nahal Tillah, an EB I Egypto-Levantine settlement in the southern Levant. In *Archaeozoology of Southwest Asia and Adjacent Areas XIII: Proceedings of the 10th ICAZ-ASWA Conference, Nicosia, Cyprus, 4–10 June 2017*; Daujat, J., Hadjikoumis, A., Berthon, R., Eds.; Lockwood Press: Columbus, OH, USA, 2022; pp. 61–80.
9. Greenfield, H.J.; Brown, A. 'Making the cut': Changes in butchering technology and efficiency patterns from the Chalcolithic to modern Arab occupations at Tell Halif, Israel. In *Bones and Identity: Zooarchaeological Approaches to Reconstructing Social and Cultural Landscapes in Southwest Asia (Proceedings of the ICAZ-SW Asia Conference, Haifa, Israel, 23–28 June 2013)*; Marom, N., Yeshurun, R., Weissbrod, L., Bar-Oz, G., Eds.; Oxbow Press: Oxford, UK, 2016; pp. 273–291.
10. Greenfield, H.J.; Beller, J.A.; Levy, T.E. Butchering technology during the Early Bronze Age I: An examination of microscopic cut marks on animal bones from Nahal Tillah, Israel. In *Tell It in Gath: Studies in the History and Archaeology of Israel. Essays in Honor of A. M. Maeir on the Occasion of his Sixtieth Birthday, Ägypten und Altes Testament. Studien zu Geschichte, Kultur und Religion Ägyptens und des Alten Testaments; Band 90*; Shai, I., Chadwick, J.R., Hitchcock, L., Dagam, A., McKinny, C., Uziel, J., Eds.; Zaphon: Münster, Germany, 2018; pp. 20–40.
11. Greenfield, H.J.; Beller, J.A.; Gaastra, J. Changes in butchering technology and efficiency patterns between the Early and Middle Bronze Ages from Tell Zirā'a, Jordan. In *Tall Zirā'a, The Gadara Region Project (2001–2011) Final Report: Early and Middle Bronze Age (Strata 25–17)*; Vieweger, D., Häser, J., Eds.; Deutsches Evangelisches Institut: Wuppertal, Germany, 2021; Volume 3.
12. Courtenay, L.A.; Yravedra, J.; Aramendi, J.; Maté-González, M.Á.; Martín-Pereaf, D.M.; Uribelarreaf, D.; Baquedano, E.; González-Aguilerac, D.; Dominguez-Rodrigo, M. Cut marks and raw material exploitation in the Lower Pleistocene site of Bell's Korongo (BK, Olduvai Gorge, Tanzania): A geometric morphometric analysis. *Quat. Int.* **2019**, *526*, 155–168. [CrossRef]
13. Domínguez-Rodrigo, M. Distinguishing between apples and oranges: The application of modern cut-mark studies to the Plio-Pleistocene (a reply to Monahan). *J. Hum. Evol.* **1999**, *37*, 793–800. [CrossRef] [PubMed]
14. Domínguez-Rodrigo, M. On cut marks and statistical inferences: Methodological comments on Lupo & O'Connell. *J. Archaeol. Sci.* **2003**, *30*, 381–386.
15. Dominguez-Rodrigo, M.; Barba, R. A study of cut marks on small-sized carcasses and its application to the study of cut-marked bones from small mammals at the FLK Zinj site. *J. Taphon.* **2005**, *3*, 121–134.
16. Greenfield, H.J. The origins of metallurgy: Distinguishing stone from metal cut marks on bones from archaeological sites. *J. Archaeol. Sci.* **1999**, *26*, 797–808. [CrossRef]
17. Greenfield, H.J. Distinguishing metal (steel and low-tin bronze) from stone (flint and obsidian) tool cut marks on bone: An experimental approach. In *Experimental Archaeology: Replicating Past Objects, Behaviors, and Processes. British Archaeological Reports, International Series 1035*; Mathieu, J.R., Ed.; Archaeopress: Oxford, UK, 2002; pp. 35–54.
18. Pante, M.C.; Muttart, M.V.; Keevil, T.L.; Blumenschine, R.J.; Njau, J.K.; Merritt, S.R. A new high-resolution 3-D quantitative method for identifying bone surface modifications with implications for the Early Stone Age archaeological record. *J. Hum. Evol.* **2017**, *102*, 1–11. [CrossRef] [PubMed]
19. Pickering, T.R.; Hensley-Marschand, B. Cutmarks and hominid handedness. *J. Archaeol. Sci.* **2008**, *35*, 310–315. [CrossRef]
20. Shipman, P. Application of scanning electron microscopy to taphonomic problems. *Ann. N. Y. Acad. Sci.* **1981**, *375*, 357–385. [CrossRef] [PubMed]
21. Domínguez-Rodrigo, M.; Saladié, P.; Cáceres, I.; Huguet, R.; Yravedra, J.; Rodríguez-Hidalgo, A.; Patricia, M.; Antonio, P.; Juan, M.; Clara, G. Spilled ink blots the mind: A reply to Merrit et al. (2018) on subjectivity and bone surface modifications. *J. Archaeol. Sci.* **2019**, *102*, 80–86. [CrossRef]
22. Domínguez-Rodrigo, M.; Cifuentes-Alcobendas, G.; Jiménez-García, B.; Abellán, N.; Pizarro-Monzo, M.; Organista, E.; Baquedano, E. Artificial intelligence provides greater accuracy in the classification of modern and ancient bone surface modifications. *Sci. Rep.* **2020**, *10*, 18862. [CrossRef] [PubMed]
23. Yravedra, J.; Maté-González, M.Á.; Palomeque-González, J.F.; Aramendi, J.; Estaca-Gómez, V.; San Juan Blazquez, M.; García Vargas, E.; Organista, E.; González-Aguilera, D.; Arriaza, M.C. A new approach to raw material use in the exploitation of animal carcasses at BK (Upper Bed II, Olduvai Gorge, Tanzania): A micro-photogrammetric and geometric morphometric analysis of fossil cut marks. *Boreas* **2017**, *46*, 860–873. [CrossRef]
24. Merritt, S.R.; Pante, M.C.; Keevil, T.L.; Njau, J.K.; Blumenschine, R.J. Don't cry over spilled ink: Missing context prevents replication and creates the Rorschach effect in bone surface modification studies. *J. Archaeol. Sci.* **2019**, *102*, 71–79. [CrossRef]
25. de Gruchy, S.; Rogers, T.L. Identifying chop marks on cremated bone: A preliminary study. *J. Forensic Sci.* **2002**, *47*, 1–4. [CrossRef]
26. Tucker, B.K.; Hutchinson, D.L.; Gilliland, M.; Charles, T.M.; Daniel, H.J.; Wolfe, L.D. Microscopic characteristics of hacking trauma. *J. Forensic Sci.* **2001**, *46*, 234–240. [CrossRef] [PubMed]
27. Dominguez-Rodrigo, M.; Pickering, T.R.; Semaw, S.; Rogers, M.J. Cutmarked bones from Pliocene archaeological sites at Gona, Afar, Ethiopia: Implications for the function of the world's oldest stone tools. *J. Hum. Evol.* **2005**, *48*, 109–121. [CrossRef] [PubMed]
28. Horwitz, L.K.; Monchot, H. Choice cuts: Hominid butchery activities at the Lower Paleolithic Site of Holon, Israel. In *Archaeozoology of the Near East V: Proceedings of the Fifth International Symposium on the Archaeozoology of Southwestern Asia and Adjacent Areas held at Yarmouk University, Irbid, Jordan in 2000*; Buitenhuis, H., Choyke, A.M., Mashkour, M., Al-Shiyab, A.H., Eds.; ARC: Groningen, The Netherlands, 2002; pp. 48–61.

29. Perez, V.R.; Godfrey, L.R.; Nowak-Kemp, M.; Burney, D.A.; Ratsimbazafy, J.; Vasey, N. Evidence of early butchery of giant lemurs in Madagascar. *J. Hum. Evol.* **2005**, *49*, 722–742. [CrossRef] [PubMed]
30. Maltby, M. Chop and change: Specialist cattle carcass processing in Roman Britain. In *TRAC 2006: Proceedings of the 16th Annual Theoretical Roman Archaeology Conference, the University of Cambridge, Cambridge, UK, 24–25 March 2006*; Croxford, B., Ray, N., Roth, R., Eds.; Oxbow: Oxford, UK, 2007; pp. 59–76.
31. Soulier, M.-C. Exploring meat processing in the past: Insights from the Nunamiut people. *PLoS ONE* **2021**, *16*, e0245213. [CrossRef] [PubMed]
32. Blasco, R.; Domínguez-Rodrigo, M.; Arilla, M.; Camarós, E.; Rosell, J. Breaking bones to obtain marrow: A comparative study between percussion by batting bone on an anvil and hammerstone percussion. *Archaeometry* **2014**, *56*, 1085–1104. [CrossRef]
33. Haynes, G.; Krasinski, K.; Wojtal, P. A study of fractured Proboscidean bones in recent and fossil assemblages. *J. Archaeol. Method Theory* **2020**, *28*, 956–1025. [CrossRef]
34. Pickering, T.R.; Egeland, C.P. Experimental patterns of hammerstone percussion damage on bones: Implications for inferences of carcass processing by humans. *J. Archaeol. Sci.* **2006**, *33*, 459–469. [CrossRef]
35. Massa, M. Early Bronze Age burial customs on the central Anatolian plateau: A view from Demircihöyük-Sarıket. *Anatol. Stud.* **2014**, *64*, 73–93. [CrossRef]
36. Rosen, S.A. Arrowheads, axes, Ad Hoc, and sickles: An introduction to aspects to Lithic variability across the Near East in the Bronze and Iron Age. *Lithic Technol.* **2013**, *38*, 141–149. [CrossRef]
37. Şahoğlu, V.; Tuncel, R. New insights into the Late Chalcolithic of coastal western Anatolia: A view from Bakla Tepe, Izmir. In *Western Anatolia before Troy. Proto-Urbanisation in the 4th Millennium BC*; Horejs, B., Mehofer, M., Eds.; Austrian Academy of Sciences Press: Vienna, Austria, 2014; pp. 65–82.
38. Greenfield, H.J.; Cheney, T.; Galili, E. A taphonomic and technological analysis of the butchered animal bone remains from Atlit Yam, a submerged PPNC site off the coast of Israel. In *Bones and Identity: Zooarchaeological Approaches to Reconstructing Social and Cultural Landscapes in Southwest Asia (Proceedings of the ICAZ-SW Asia Conference, Haifa, Israel, 23–28 June 2013)*; Marom, N., Yeshurun, R., Weissbrod, L., Bar-Oz, G., Eds.; Oxbow Press: Oxford, UK, 2016; pp. 89–112.
39. Saidel, B.; Erickson-Gini, T.; Vardi, J.; Rosen, S.A.; Maher, E.F.; Greenfield, H.J. Egypt, copper, and microlithic drills: The test excavations at Rogem Be'erotayim in western Negev. *Mitkufat Haeven (J. Isr. Prehist. Soc.)* **2006**, *36*, 201–229.
40. Greenfield, H.J.; Greenfield, T.L. Butchering technology in Middle Bronze Age Ashkelon. In *Final Reports of the Leon Levy Expedition to Ashkelon: Ashkelon 6: The Middle Bronze Age Ramparts and Gates of the North Slope and Later Fortifications*; Stager, L.E., Master, D.M., Schloen, J.D., Eds.; Harvard Semitic Museum Publications and Eisenbrauns: University Park, PA, USA, 2018; pp. 509–520.
41. Greenfield, H.J.; Marciniak, A. Retention of old technologies following the end of the Neolithic: Microscopic analysis of the butchering marks on animal bones from Çatalhöyük East. *World Archaeol.* **2019**, *51*, 76–103. [CrossRef]
42. Lev-Tov, J.S.E.; Killebrew, A.E.; Greenfield, H.J.; Brown, A. Puppy sacrifice and cynophagy from early Philistine Tel Miqne-Ekron contextualized. *J. East. Mediterr. Archaeol. Herit.* **2018**, *6*, 1–30. [CrossRef]
43. Greenfield, H.J. Insufficient evidence for metal butchering marks at Tell el-Hesi during the Early Bronze Age: Critique of the analysis of microscopic grooves in 'Cultural Modification Analyses on Faunal Remains in Relation to Space Use and Direct Provisioning from Field VI EBIIIA Tell el-Hesi' by Kara Larson, James W. Hardin, and Sara Cody. *Palest. Explor. Q.* **2021**, *160*, 144–155.
44. Greenfield, H.J.; Brown, A.; De Miroschedji, P. Origins of metallurgy in the southern Levant: Microscopic examination of butchering marks on animal bones at Tel Yarmuth, Israel. In *Themes in Old World Zooarchaeology: From the Mediterranean to the Atlantic*; Albarella, U., Detry, C., Gabriel, S., Ginja, C., Pires, A.E., Tereso, J.P., Eds.; Oxbow Books: Oxford, UK, 2021; pp. 95–107.
45. Greenfield, H.J.; Chaput, T. Butchering technology at the early tin metal producing site of Göltepe. In *Tin Production at Göltepe: Excavations at an Early Bronze Age Mining Town in the Central Taurus Mountains, Turkey*; Yener, K.A., Ed.; Institute for Aegean Academic Press: Philadelpha, PA, USA, 2021; pp. 160–176.
46. Greenfield, H.J.; Marciniak, A. The emergence and transmission of metallurgical technology for subsistence activities in daily life in northern Europe: A microscopic zooarchaeological perspective. *J. Field Archael.* **2021**, *46*, 275–288. [CrossRef]
47. Greenfield, H.J. Monitoring the origins of metallurgy: An application of cut mark analysis on animals bones from the Central Balkans. *Environ. Archaeol.* **2000**, *5*, 119–132. [CrossRef]
48. Olsen, S.L. The identification of stone and metal tool marks on bone artifacts. In *Scanning Electron Microscopy in Archaeology. British*; Archaeological Reports, International Series 452; Olsen, S.L., Ed.; BAR: Oxford, UK, 1988; pp. 337–360.
49. Walker, P.L.; Long, J.C. An experimental study of the morphological characteristics of tool marks. *Am. Antiq.* **1977**, *42*, 605–616. [CrossRef]
50. Gifford-Gonzalez, D.P. *An Introduction to Zooarchaeology*; Springer: Berlin/Heidelberg, Germany, 2018.
51. Lewis, J.E. Identifying sword marks on bone: Criteria for distinguishing between cut marks made by different classes of bladed weapons. *J. Archaeol. Sci.* **2008**, *35*, 2001–2008. [CrossRef]
52. Humphrey, J.H.; Hutchinson, D.L. Macroscopic characteristics of hacking trauma. *J. Forensic Sci.* **2001**, *46*, 228–233. [CrossRef] [PubMed]
53. Knusel, C.; Outram, A.K. Fragmentation: The zonation method applied to fragmented human remains from Archaeological and Forensic contexts. *Environ. Archaeol.* **2004**, *9*, 85–99. [CrossRef]

54. Lynn, K.S.; Fairgrieve, S.I. Macroscopic analysis of axe and hatchet trauma in fleshed and defleshed mammalian long bones. *J. Forensic Sci.* **2009**, *54*, 786–791. [CrossRef] [PubMed]
55. Symes, S.A.; Chapman, E.N.; Rainwater, C.W.; Cabo, L.L.; Myster, S.M.T. *Knife and Saw Toolmark Analysis in Bone: A Manual Designed for the Examination of Criminal Mutilation and Dismemberment*; National Institute of Justice: Rockville, MD, USA, 2010.
56. Wenham, S.J. Anatomical interpretation of Anglo-Saxon weapon injuries. In *Weapons and Warfare in Anglo-Saxon England*; Hawkes, S.C., Ed.; Oxford University Press: Oxford, UK, 1989; pp. 123–139.
57. Hansen, S. Arsenic bronze: An archaeological introduction into a key innovation. *Eurasia Antiq.* **2021**, *23*, 139–162.
58. Rosen, S.A. *Lithics after the Stone Age: A Handbook of Stone Tools from the Levant*; Altamira Press: Walnut Creek, CA, USA, 1997.
59. Yener, K.A. (Ed.) *Tin Production at Göltepe: Excavations at an Early Bronze Age Mining Town in the Central Taurus Mountains, Turkey*; INSTAP (Institute for Aegean Academic Press): Philadelpha, PA, USA, 2021.
60. Kononenko, N.; Torrence, R.; White, P. Unexpected uses for obsidian: Experimental replication and use-wear/residue analyses of chopping tools. *J. Archaeol. Sci.* **2015**, *54*, 254–269. [CrossRef]
61. Lupo, K.D.; Fancher, J.M.; Schmitt, D.N. The taphonomy of resource intensification: Zooarchaeological implications of resource scarcity among Bofi and Aka forest foragers. *J. Archaeol. Method Theory* **2013**, *20*, 420–447. [CrossRef]
62. Pickering, T.R.; Domínguez-Rodrigo, M.; Heaton, J.L.; Yravedra, J.; Barba, R.; Bunn, H.T.; Musiba, C.; Baquedano, E.; Diez-Martín, F.; Mabulla, A. Taphonomy of ungulate ribs and the consumption of meat and bone by 1.2-million-year-old hominins at Olduvai Gorge, Tanzania. *J. Archaeol. Sci.* **2013**, *40*, 1295–1309. [CrossRef]
63. White, T.D. *Prehistoric Cannibalism*; Princeton University Press: Princeton, NJ, USA, 1992.
64. Olsen, S.L.; Shipman, P. Surface modification on bone: Trampling versus butchery. *J. Archaeol. Sci.* **1988**, *15*, 535–553. [CrossRef]
65. Stout, D.; Apel, J.; Commander, J.; Roberts, M. Late Acheulean technology and cognition at Boxgrove, UK. *J. Archaeol. Sci.* **2014**, *41*, 576–590. [CrossRef]
66. Yener, A.K. Background. In *Tin Production at Göltepe: Excavations at an Early Bronze Age Mining Town in the Central Taurus Mountains, Turkey*; Yener, A.K., Ed.; INSTAP (Institute for Aegean Academic Press): Philadelpha, PA, USA, 2021; pp. 1–8.
67. Earl, B.; Özbal, H. Early Bronze Age tin processing at Kestel/Göltepe, Anatolia. *Archaeometry* **1996**, *38*, 289–303. [CrossRef]
68. Yener, K.A. *The Domestication of Metals: The Rise of Complex Metal Industries in Anatolia*; Brill: Leiden, The Netherlands, 2000.
69. Yener, K.A. Summary and Conclusions. In *Tin Production at Göltepe: Excavations at an Early Bronze Age Mining Town in the Central Taurus Mountains, Turkey*; Yener, A.K., Ed.; INSTAP (Institute for Aegean Academic Press): Philadelpha, PA, USA, 2021; pp. 195–206.
70. Yener, K.A.; Kulakoğlu, F.; Yazgan, E.; Kontani, R.; Hayakawa, Y.S.; Lehner, J.W.; Dardeniz, G.; Öztürk, G.; Johnson, M.; Kaptan, E.; et al. New tin mines and production sites near Kültepe, ancient Kanesh in Turkey: A third millennium BC highland production model. *Antiquity* **2015**, *89*, 596–612. [CrossRef]
71. Yener, K.A. *Excavations at Kestel Mine, Turkey: The Final Season: 1996–1997 Annual Report*; The Oriental Institute of the University of Chicago: Chicago, IL, USA, 2012.
72. Yener, K.A. Revisiting Kestel Mine and Göltepe: The dynamics of local provisioning of tin during the Early Bronze Age. In *Ancient Mining in Turkey and the Eastern Mediterranean*; Ozbal, H., Yalcin, U., Pasamehmetoglu, G., Eds.; Atilim University: Ankara, Turkey, 2008; pp. 57–64.
73. Yener, K.A. Small finds. In *Tin Production at Göltepe: Excavations at an Early Bronze Age Mining Town in the Central Taurus Mountains, Turkey, Yener, A.K., Ed.*; INSTAP (Institute for Aegean Academic Press): Philadelpha, PA, USA, 2021; pp. 111–148.
74. Yener, K.A. Horizontal Exposures at Göltepe. In *Tin Production at Göltepe: Excavations at an Early Bronze Age Mining Town in the Central Taurus Mountains, Turkey*; Yener, A.K., Ed.; INSTAP (Institute for Aegean Academic Press): Philadelpha, PA, USA, 2021; pp. 23–74.
75. Massa, M.; McIlfatrick, O.; Fidan, E. Patterns of metal procurement, manufacture and exchange in Early Bronze Age northwestern Anatolia: Demircihüyük and beyond. *Anatol. Stud.* **2017**, *67*, 51–83. [CrossRef]
76. Yakar, J. *Reflections of Ancient Anatolian Society in Archaeology: From Neolithic Village Communities to EBA Towns and Polities*; Homer Kitabevi: Istanbul, Turkey, 2011.

quaternary

Article

Multi-Taxa Neo-Taphonomic Analysis of Bone Remains from Barn Owl Pellets and Cross-Validation of Observations: A Case Study from Dominica (Lesser Antilles)

Emmanuelle Stoetzel [1,*], Corentin Bochaton [2,3], Salvador Bailon [1,4], David Cochard [3], Monica Gala [3] and Véronique Laroulandie [3]

1. HNHP UMR 7194, CNRS/MNHN/UPVD/Sorbonne Universités, Musée de l'Homme—Palais de Chaillot, 17 place du Trocadéro, 75016 Paris, France; salvador.bailon@mnhn.fr
2. Max Planck Institute for the Science of Human History, 10 Kahlaische Strasse, 07745 Jena, Germany; bochaton@shh.mpg.de
3. PACEA UMR 5199, CNRS/Université de Bordeaux/MC, Bât. B2, Allée Geoffroy St-Hilaire—CS 50023, CEDEX, 33615 Pessac, France; david.cochard@u-bordeaux.fr (D.C.); monica.gala@u-bordeaux.fr (M.G.); veronique.laroulandie@u-bordeaux.fr (V.L.)
4. AASPE UMR 7209, CNRS/MNHN/Sorbonne Universités, CP 55, 55 rue Buffon, 75005 Paris, France

* Correspondence: emmanuelle.stoetzel@mnhn.fr

Abstract: Paleo- and neo-taphonomic analyses of bone assemblages rarely consider all the occurring taxa in a single study and works concerning birds of prey as accumulators of microvertebrate bone remains mostly focus on small mammals such as rodents and soricomorphs. However, raptors often hunt and consume a large range of taxa, including vertebrates such as small mammals, fishes, amphibians, squamates and birds. Bone remains of all these taxonomic groups are numerous in many paleontological and archaeological records, especially in cave deposits. To better characterize the predators at the origin of fossil and sub-fossil microvertebrate accumulations and the taphonomic history of the deposit, it is thus mandatory to conduct global and multi-taxa taphonomic approaches. The aim of this study is to provide an example of such a global approach through the investigation of a modern bone assemblage from a sample of pellets produced by the Lesser Antillean Barn Owl (*Tyto insularis*) in the island of Dominica. We propose a new methodology that allows us to compare different taxa (rodents, bats, squamates and birds) and to experiment with a cross-validation process using two observers for each taxonomic group to test the reliability of the taphonomic observations.

Keywords: rodents; bats; lizards; birds; taphonomy; cross-validation; predation; *Tyto insularis*; Caribbean

1. Introduction

The Lesser Antillean Barn Owl, *Tyto insularis* (Pelzeln, 1872), is an endemic Caribbean species characterized by a dark plumage and a small size, with a distribution restricted to the Lesser Antilles (Dominica, St Vincent, Grenadines islands and Grenada) [1–3] and whose taxonomic status as a species or subspecies is still discussed (e.g., [4,5]). Several studies have documented the food habits of the Barn Owl family (Tytonidae) in the Caribbean: *Tyto alba* (Scopoli, 1769) in the Greater Antilles [6–11], *T. insularis* on the island of Dominica [12] and *T. glaucops* (Kaup, 1852) in Hispaniola [2]. However, the taphonomy of the bone assemblages created by these predators in the insular Caribbean has never been investigated. This lack of scientific interest is not restricted to barn owls as this region currently suffers from an almost complete absence of neo-taphonomic investigation conducted on modern bone assemblages. As a result, although the role of barn owls is suspected in the accumulation of some of the known paleontological deposits in the Caribbean [13], the accumulation process of most subfossil bone accumulations has never been investigated. This generalized lack of taphonomic work in the Caribbean makes it difficult to interpret paleobiodiversity data, both for paleoenvironmental reconstructions (e.g., [14,15]) and for the investigation of the evolution of faunal communities over time, under climatic and/or anthropogenic pressure. This is especially damaging as this situation can sometimes lead to a severe misinterpretation of the fossil records. To add to the lack of regional taphonomic comparison points, most available studies conducted in other geographic areas are mostly focused on small mammals such as rodents (e.g., [14,16]), which only account for a very limited part of the insular Caribbean paleobiodiversity. Indeed, native terrestrial small mammals are rare in the Caribbean islands, and commensal rodents (*Dasyprocta* spp., *Rattus* spp., *Mus musculus*) have mostly been introduced throughout the last few millennia by human populations during different waves of settlement [17–20]. Before these rodents arrived and their populations expanded, predators obviously had to exploit the locally available prey species (bats, birds, amphibians and squamates). In this context, it is therefore important to better characterize the taphonomic impact of raptors on the anatomical and faunal representation of the whole diversity of their available prey species. This is mandatory to better understand the history and representativeness of the Caribbean fossil and subfossil microvertebrate assemblages. More generally, multi-taxa taphonomic analyses remain rare [15,21–23] and deserve to be considerably developed.

The present study has several goals: (1) to provide new data on the taphonomic modifications caused by *Tyto insularis* on bone assemblage of Lesser Antillean prey species; (2) to test the effectiveness, reliability and reproducibility of the chosen taphonomic approach by involving several observers who specialize in different taxa (rodents, bats, squamates and birds); and (3) to propose a reliable and simplified methodology for multi-taxa neo-taphonomic analyses that is potentially transposable to fossil and sub-fossil assemblages.

2. Materials and Methods

2.1. Owl Pellets Sampling

For this study, a total of 111 pellets produced by the Lesser Antillean Barn Owl (*Tyto insularis*) were collected in June 2014 (at the beginning of the rainy season) and January 2015 (at the beginning of the dry season) in three coastal localities on the island of Dominica (Figure 1): 34 pellets in a church tower at Salisbury (n = 5 in 2014; n = 29 in 2015), 50 pellets in a small cave near Canefield (n = 43 in 2014; n = 7 in 2015), and 27 pellets in a church tower at Grand Bay (n = 24 in 2014; n = 3 in 2015). Salisbury and Grand Bay are located in relatively rural areas, while Canefield is located in a more urban area.

Whole pellets and bulk material were collected on the ground in nests (Grand Bay, Canefield) or under roosts (Salisbury). Only pellets were considered in the present study. The whole pellets were isolated in a bag marked with a specific number. Each bag was then prepared individually: the pellets were soaked in water and the bones carefully extracted with fine pliers. The bones were then dried and packed in individually marked tubes so that each bone could be cross-referenced to the pellet from which it came.

2.2. Prey Identification

Bat and rodent species were identified through size and morphology of skulls, mandibles and post-cranial elements, thanks to data from the literature [13,24–27] and modern osteological collections (PACEA UMR 5199, Université de Bordeaux; Muséum national d'Histoire naturelle, Paris, France). The body mass was estimated according to [28].

Bird remains were not identified at the species level, mainly due to a lack of suitable reference collection, but also due to the unawareness of osteological characters allowing to identify small Passeriformes, particularly when the preservation state of the material prevents the observation of peculiar anatomical criteria. However, we observed that the majority of the birds belong to small Passeriformes, followed by rare hummingbirds and a single Common Ground Dove, *Columbina passerina*. Bird remains were assigned to size/weight classes (see infra) using some modern skeletons from the PACEA laboratory collection and following the data of body mass from [29].

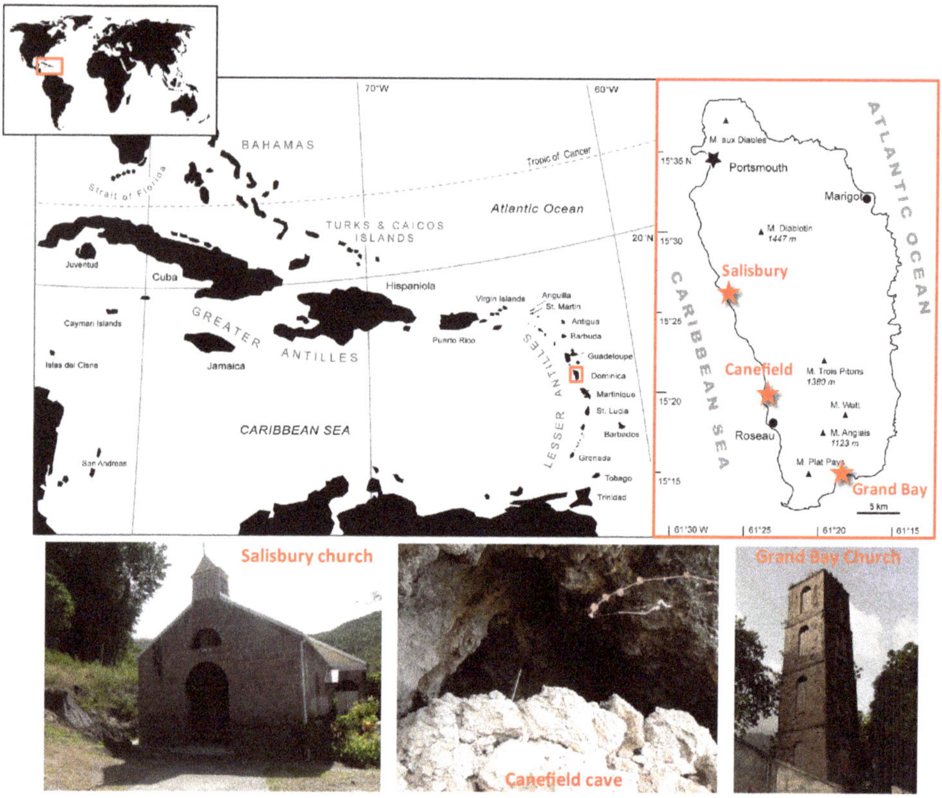

Figure 1. Location of the three collection areas of the studied *Tyto insularis* pellets from the island of Dominica. Pictures: B. Angin.

The identification of squamate species is based on the morphology of both cranial and post-cranial elements. This approach was carried out using comparative skeletons from modern osteological collections (PACEA UMR 5199, Université de Bordeaux; Muséum national d'Histoire naturelle, Paris) and published osteological criteria [30,31].

For all taxa, the Number of Identified Specimens (NISP) was given, and the Minimum Number of Individuals (MNI) was calculated based on the most abundant element for each taxon and for each site (mainly mandibles/dentaries, but also long bones in some cases).

2.3. Taphonomic Analysis

The four faunal groups we considered (bats, rodents, birds and squamates) were well represented in the three studied samples, which allowed us to perform a multi-taxa taphonomic analysis. We tried to adapt the classical methods of analysis so that they were applicable to and comparable for all taxa (for example, we did not consider digestion on teeth, because this is not applicable to birds and because the digestion patterns are very different between mammals and squamates). Only pellets were studied, and the bulk material was not considered. The analyses were performed pellet by pellet, but in this study we present the results as a whole because this is closer to what is found in the fossil and sub-fossil record. To reach a sufficient sample size for the analysis to be relevant we had to pool together the data from the three sites, and we did not consider possible site-specific taphonomic characteristics.

2.3.1. Anatomical Representation

In order to propose a simplified methodology and maximize the analysis efficiency while obtaining informative data on the anatomical representation of the different skeleton parts in each assemblage, only some specific anatomical elements were quantified (Table 1). We considered the same number of bone elements for each taxon. From the skull, mandibles (mammals and birds) and dentaries (lizards) were counted. From the forelimbs, humerus (all taxa) and ulna (rodents and squamates), radius (bats) or carpometacarpus (birds) were studied. From the hindlimbs, femur and tibia were quantified for mammals and lizards, and femur and tarsometatarsus were considered for birds. The choice of these bones allows us to represent the proximal and distal parts of the limbs and concerns elements frequently recovered in the fossil record. Isolated epiphyses were put aside and not considered. We did not explore the presence of anatomical connection because similar observations are rarely transposable to fossil assemblages. To quantify anatomical distributions, several ratios were calculated: cranial/post-cranial, anterior/posterior and stylopodia/zeugopodia (Table 1).

Table 1. Elements considered for the anatomical representation and ratios.

	Anatomical Representation		**Ratios**		
	Cranial Elements	**Post-Cranial Elements**	**Cranial /Post-Cranial ***	**Anterior /Posterior**	**Stylopodia /Zeugopodia**
Aves	Mandible (Md)	Humerus (H), Carpometacarpus (Ca), Femur (F), Tarsometatarsus (Ta)	Md/(H + Ca + F + Ta)	(H + Ca)/(F + Ta)	(H + F)/(Ca + Ta)
Squamata	Dentary (D)	Humerus (H), Ulna (U), Femur (F), Tibia (Ti)	D/(H + U+F + Ti)	(H + U)/(F + Ti)	(H + F)/(U + Ti)
Rodentia	Mandible (Md)	Humerus (H), Ulna (U), Femur (F), Tibia (Ti)	Md/(H + U+F + Ti)	(H + U)/(F + Ti)	(H + F)/(U + Ti)
Chiroptera	Mandible (Md)	Humerus (H), Radius (R), Femur (F), Tibia (Ti)	Md/(H + R+F + Ti)	(H + R)/(F + Ti)	(H + F)/(R + Ti)

* a correction of 8/2 was applied because there are two cranial elements and eight post-cranial elements.

The percentage of anatomical representation (PR) was calculated following the formula: PR = [Fobs/(Fth × MNI)] × 100 (Fobs = observed frequency in the studied assemblage, Fth = theoretical frequency of the bone in an individual, MNI = minimum number of individuals). In rodents and squamates the hemi-mandibles/dentaries are separated, so the frequencies were divided by two for the calculation of the PR in a way that allows comparison with birds and bats for which hemi-mandibles are fused.

2.3.2. Fragmentation

The fragmentation of the remains was only recorded for the long bones considered in the anatomical distribution analyses. We followed the methodology of [14] and identified six fragmentation categories: (0) complete bone, (1) only proximal part, (2) proximal part + shaft, (3) only shaft, (4) distal part + shaft, (5) only distal part.

2.3.3. Surface Modifications

The study of surface alterations focused mainly on the observation of digestion traces on both extremities of long bones and on mandible/dentaries of adult individuals. We did not consider digestion on the distal extremity of rodent ulnae because this fragile part is rarely preserved in both modern and fossil assemblages. For every taxon, we used the following classification of the digestion grades (Figure S1):

- Not digested.
- Slight digestion. On long bones: polished/rounded articular surfaces, cortical bone slightly dissolved and substantia spongiosa slightly visible. On mandibles/dentaries: weak alteration of the anterior part of the bone, but the shape of the bone remains weakly or not modified.
- Moderate digestion. On long bones: substantia spongiosa much more exposed with holes in the cortical bone, but not necessarily on the whole articular surface; the global form of the extremity is preserved. On mandibles/dentaries: strong alteration and partial destruction of the anterior part and alteration of the surface of the *corpus* of the mandible, the shape of the bone and teeth (mammals, lizards) are modified.
- Strong to extreme digestion. On long bones: cortical bone totally dissolved and substantia spongiosa totally exposed or even dissolved; the extremity can be totally destroyed. On mandibles/dentaries: more than one third of the bone is destroyed, teeth (mammals, lizards) are altered.

We are aware that describing a continuous phenomenon by classes has its limits, but such an approach remains relevant to describe global alteration patterns, especially in large assemblages.

2.3.4. Size/Weight Classes of Preys

In order to characterize the size of the exploited prey from different taxonomic groups, we created size/weight classes by taking into account the taxonomic attribution and maturity of bones. This avoids relying exclusively on the overall size of bones, which, depending on the investigated taxa, could be misleading. We created categories based on the weight of each taxon because this is the only variable that allows for comparisons between the different taxa:

- Size category 0: below 10 g (the gecko *Sphaerodactylus* sp.; birds such as small Passeriformes (Parulidae) and hummingbirds (Trochilidae)).
- Size category 1: 10 to 100 g (small lizards of *Anolis* genus; birds such as *Columbina passerina, Salatator albicollis, Mimus gilvus; Mus musculus*, young *Rattus* spp., all identified bat species).
- Size category 2: 100 to 300 g (lizards such as large *Anolis* spp., medium size *Pholidoscelis* spp., *Leiocephalus* sp. and newborn *Iguana* sp.; birds such as *Margarops fuscatus* and largers; adult *Rattus rattus*, young *Rattus norvegicus*).

2.4. *Cross-Validation of Observations*

Because two observers were involved for each taxononomic group, we decided to perform cross-validation of the taphonomic observations. Each observer is a specialist of a specific taxon: M.G. and V.L. for birds, S.B. and C.B. for squamates, D.C. and E.S. for small mammals. For Aves, the two observers made counts together, but taphonomic observations were made independently.

Such an approach remains rare, and, to our knowledge, only one other study has been conducted on observer bias in taphonomy—a blind-test evaluation of digestion on leporid bones [32].

In our approach, we compared the observations made by the two observers at the scale of the complete assemblages, but not bone by bone because this would have been too time consuming and complex to set up. Before starting the study of the material, the different observers met to discuss and agree on the methodology, the definition of digestion grades, etc. Thus, observers worked independently following a detailed methodology and filled in a similar results document that was created prior to the observations. The two resulting datasets for each taxonomic group (mammals, birds and squamates) were then combined and compared. Wherever necessary, Khi^2 tests were performed to assess inter-observer differences (only statistically significant values are shown).

3. Results

In this section, we present the observations made on the different variables used to define the taphonomic properties of bone assemblages. For each variable, we present both the overall results and the inter-observer differences.

3.1. Faunal Spectrum

More than 1470 bone remains were studied, corresponding to around 194 prey individuals. The prey spectrum was large and diverse (Tables 2 and S1), dominated by birds of several sizes, but also included lizards (*Anolis* sp., *Thecadactylus rapicauda*), bats (*Ardops nicholsii, Artibeus jamaicensis, Artibeus schwartzi, Brachyphylla cavernarum, Monophyllus plethodon, Sturnira angelli*) and rodents (*Mus musculus, Rattus rattus*) (Table S1). At Canefield, we observed a larger proportion of rodents (corresponding to commensal mice and rats), which was consistent with the more urban location of this site (Figure 2). Most of the material was composed of prey less than 100 g (Tables 3 and S2), with larger preys at Canefield corresponding mainly to adult rats. These observations did not differ significantly between the observers.

Table 2. Number of Identified Specimens (NISP) and Minimum Number of Individuals (MNI) of the prey taxa found in the three studied pellet samples from Dominica (mean values of the data from the two observers for each taxon; see detailed data in Table S1).

Prey Taxa	Grand Bay		Salisbury		Canefield		TOTAL	
	NISP	MNI	NISP	MNI	NISP	MNI	NISP	MNI
Aves	186.0	23.0	206.0	24.0	295.0	34.0	687.0	81.0
Squamata	75.5	9.5	159.0	20.0	36.5	7.3	271.0	36.0
Rodentia	27.0	5.0	39.0	5.5	207.5	27.5	273.5	38.0
Chiroptera	102.0	18.0	117.0	15.5	21.0	5.0	240.0	38.5
TOTAL	390.5	55.5	521.0	65.0	560.0	73.8	1471.5	193.5

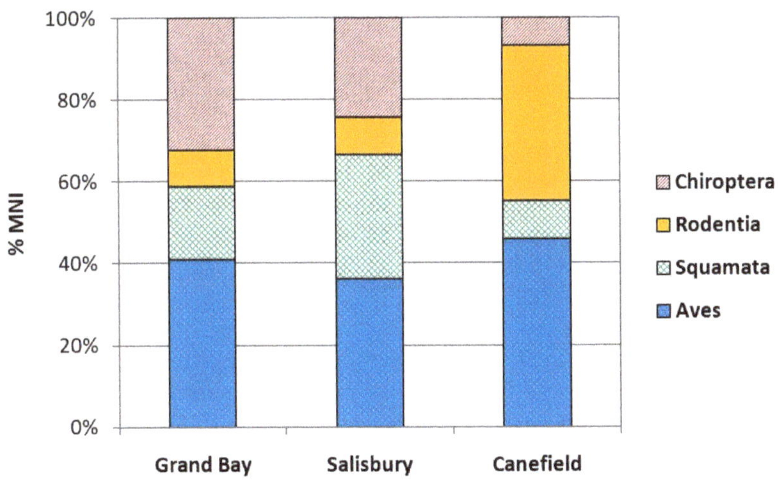

Figure 2. Relative proportions of the considered taxonomic groups in the three studied owl pellet samples.

Table 3. Size categories of the prey taxa found in the three studied pellet samples from Dominica (mean values of the data from the two observers for each taxon; see detailed data in Table S2).

Size Classes	Grand Bay		Salisbury		Canefield		TOTAL	
	NISP	NMI	NISP	NMI	NISP	NMI	NISP	NMI
0 (<10 g)	107.0	13.0	71.5	9.0	119.0	14.0	297.5	36.0
1 (10–100 g)	275.5	40.5	447.5	55.0	431.0	57.0	1154.0	152.5
2 (100–300 g)	8.0	2.0	2.0	0.5	10.0	3.0	20.0	5.5
TOTAL	390.5	55.5	521.0	64.5	560.0	74.0	1471.5	194.0

3.2. Anatomical Representation

On average, the best represented elements among the considered anatomical parts were dentaries/mandibles for squamates and bats, humeri for birds and femora for rodents (Figure 3). In squamates, humeri and femora are well documented, while the radius and tibia are underrepresented compared to the other taxa; this is reflected in the stylopodia/zeugopodia ratio (Table 4). In small mammals, the studied elements are more or less equally found. Posterior elements are slightly more represented than anterior elements in "terrestrial taxa" (lizards, rodents), while the reverse is true in "flying taxa" (bats, birds) (Table 4).

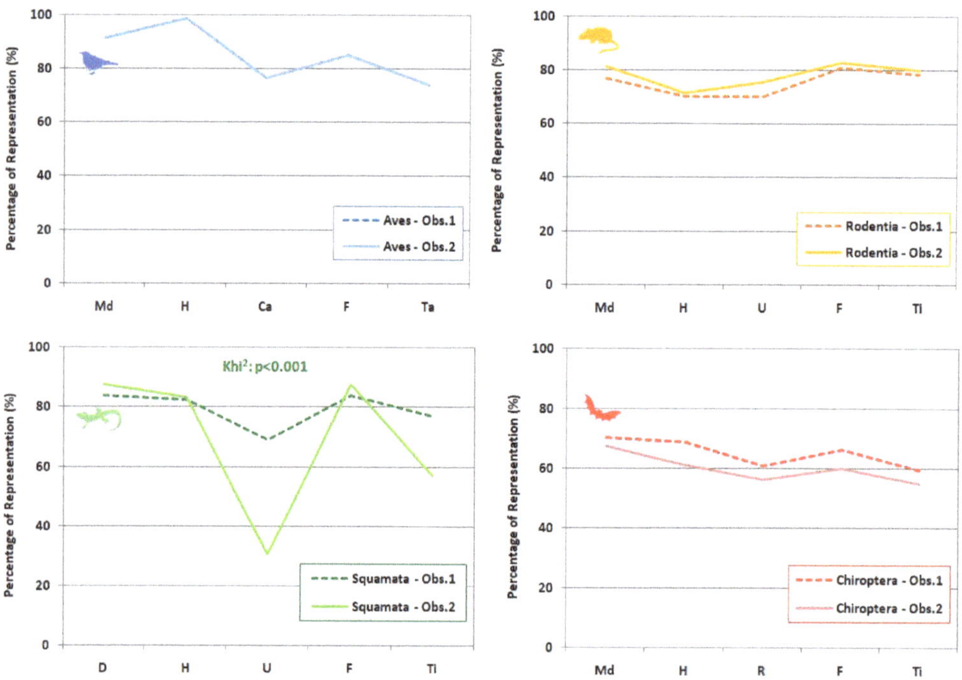

Figure 3. Profiles of anatomical representation for each observer and for each faunal group, all sites combined (detailed data in Table S3). For birds, both curves are identical.

Table 4. Ratios calculated with the considered anatomical elements, all sites and observers combined (detailed data in Table S4). No significant difference was noticed between observers.

	Cranial/Post-Cranial	Anterior/Posterior	Stylopodia/Zeugopodia
Aves	1.09	1.10	1.22
Squamata	1.20	0.87	1.44
Rodentia	1.04	0.89	1.00
Chiroptera	1.13	1.03	1.11

Regarding the values obtained by the different observers for each group (Figure 3; Tables S3 and S4), in most cases there are only a few differences except for squamates, which show huge discrepancies for the ulna and tibia. This is probably due to the identification methodology adopted by both observers: one of them considered and identified smaller and more fragmented elements than the other. However, seeing intra-observer differences in their identification is not surprising since both of these elements are challenging to identify because of their lack of characteristic anatomical features.

3.3. *Fragmentation*

From a global point of view (Table S5, mean values from the two observers), the fragmentation rate is very low, and intact bones represent between 80% and 97% of the material, all taxa combined. Bats show the higher variability in the percentage of fragmentation between the elements. The most fragmented elements are generally the longest and the finest, such as bat radius or bird tarsometatarsus. Conversely, the less fragmented elements are often more robust, such as rodent femora or bird carpometacarpus. Proximal parts (fragmentation categories 1 and 2) are generally better preserved than distal parts (fragmentation categories 4 and 5) (cf. detailed data in Table S5).

Figure 4 shows the percentages of fragmentation obtained by each observer for each faunal group. For birds and rodents, there is little variability between observers, but for squamates and bats the differences are significant (Figure 4). For squamates, this is probably due to the previously mentioned difference in anatomical identification, as observer 1 has identified more numerous elements corresponding to smaller and fragmented bones. For bats, two factors may be responsible for this result: (1) one of the observers considered the quasi-complete elements as complete, and the other as fragmented; (2) the absence of some extremities was attributed either to fragmentation or extreme digestion according to the observers.

3.4. *Modifications of Bone Surface*

Examples of digestion traces observed on the material are given in Figures 5 and 6. The two squamate species (*Thecadactylus rapicauda* and *Anolis* sp.) were separated because they displayed different patterns of digestion. In geckos *s.l.*, including *Thecadactylus* spp., the skeleton may be less mineralized and less robust, and thus generally less preserved and less represented in the deposits. Moreover, as *Anolis* sp. and *Thecadactylus rapicauda* are generally found in different sites, this variation might also be related to differences between several individual barn owls, which would contribute to the formation of the considered samples.

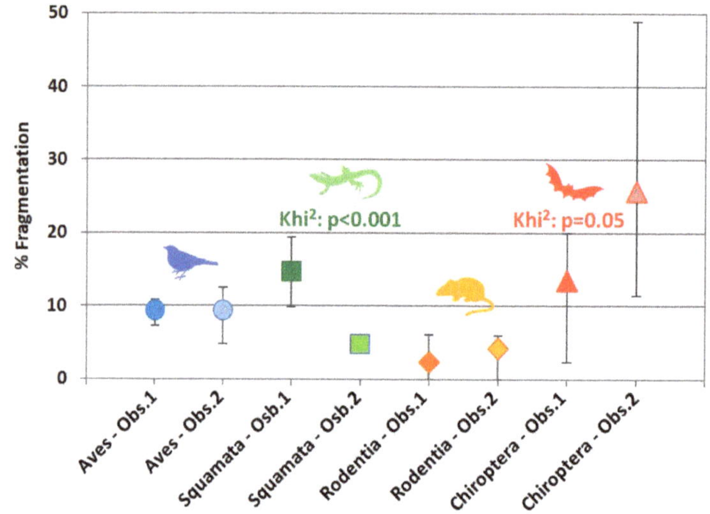

Figure 4. Percentage of fragmentation for each observer and for each faunal group, all sites combined (detailed data in Table S5). The error bars represent the values for the most and the least fragmented elements, which may vary from one group to another.

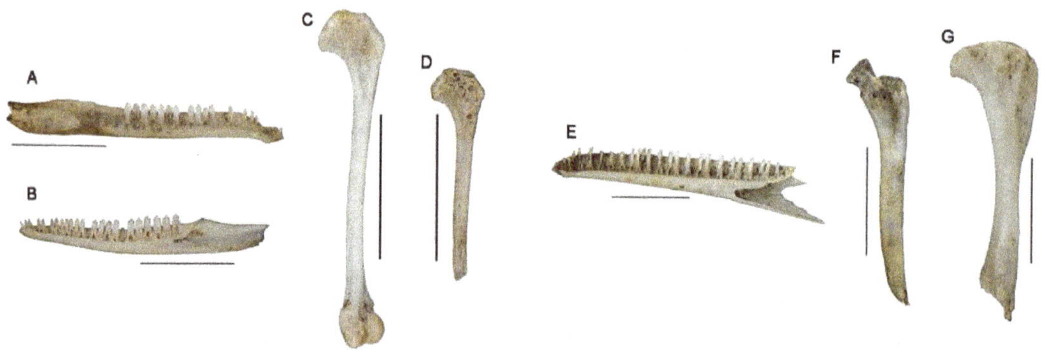

Figure 5. Example of digestion traces observed on squamate remains. (**A–D**) *Anolis* sp. (Salisbury, n°2): (**A**) left dentary with moderate digestion on the extremity and loss of anterior and posterior teeth, (**B**) right dentary not digested, (**C**) complete humerus not digested, (**D**) humerus with moderate digestion on the proximal part and loss of distal extremity. (**E–G**) *Thecadactylus rapicauda* (Canefield, n°30): (**E**) right dentary with light digestion on the anterior extremity, (**F**) femur with moderate digestion on the proximal and distal parts, (**G**) humerus with dissolved distal part. Scale = 5 mm. Pictures: S. Bailon.

For all considered taxa, approximately half of the material is digested, except for *Thecadactylus rapicauda*, which presents higher percentages of digestion (~70%) with a dominance of light digestion, but also with more moderate to strong grades (Table S6). For birds and rodents, most of the digested bones show light to moderate grades. For bats and *Anolis* sp., we observed a higher representation of strong digestion.

We also chose some relevant examples so that we could look at the variability that may exist between elements of the same species and between different parts of the same element, independent of the observer effect.

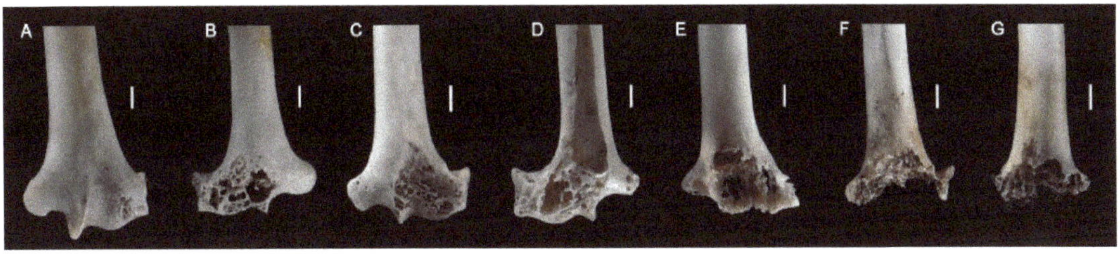

Figure 6. Example of digestion traces observed on bat remains. Posterior view of distal humerus: (**A**) *Brachyphylla cavernarum*, right humerus (Salisbury 2015, n°29) slightly digested; (**B**) *Artibeus jamaicensis*, left humerus (Salisbury 2015, n°12) moderately digested; (**C**) *Ardops nicholsii*, right humerus (Salisbury 2015, n°15) moderately digested; (**D**) *Artibeus jamaicensis*, left humerus (Salisbury 2015, n°32) strongly digested; (**E**) *Brachyphylla cavernarum*, right humerus (Salisbury 2015, n°31) extremely digested; (**F**) *Ardops nicholsii*, right humerus (Salisbury 2015, n°9) extremely digested; (**G**) *Ardops nicholsii*, left humerus (Salisbury 2015, n°15) extremely digested. Scale = 1 mm. Pictures: D. Cochard.

Regarding the differences between observers, we see that for birds and *Thecadactylus rapicauda* results are similar, while for *Anolis* sp., rodents and bats, differences are statistically significant (Figure 7). The main differences in the percentage of digestion are observed in ulna for squamates, mandible, proximal femur and distal tibia for rodents, and proximal humerus for bats (Table S6).

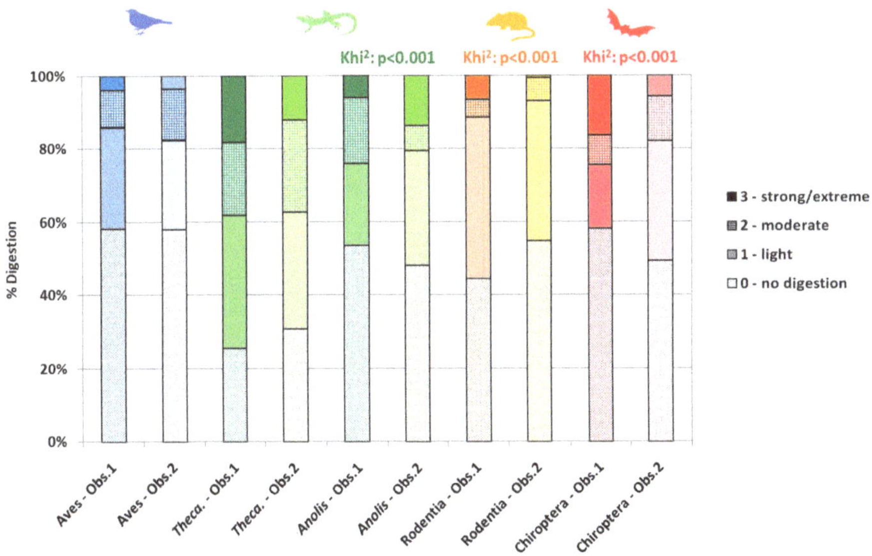

Figure 7. Percentages of digestion grades quoted for each observer and for each faunal group, all sites combined (detailed data in Table S6).

For example, with the same anatomical element, the pattern of digestion may differ according to the species. Such inter-taxa variability was notably observed for mandibles and dentaries (Figure 8). Globally, digestion more strongly affects bird mandibles and lizard dentaries than small mammal mandibles. In contrast, there is no significant difference in the digestion pattern between bird and bat femora for example, because these elements are relatively short and robust with a similar shape and size.

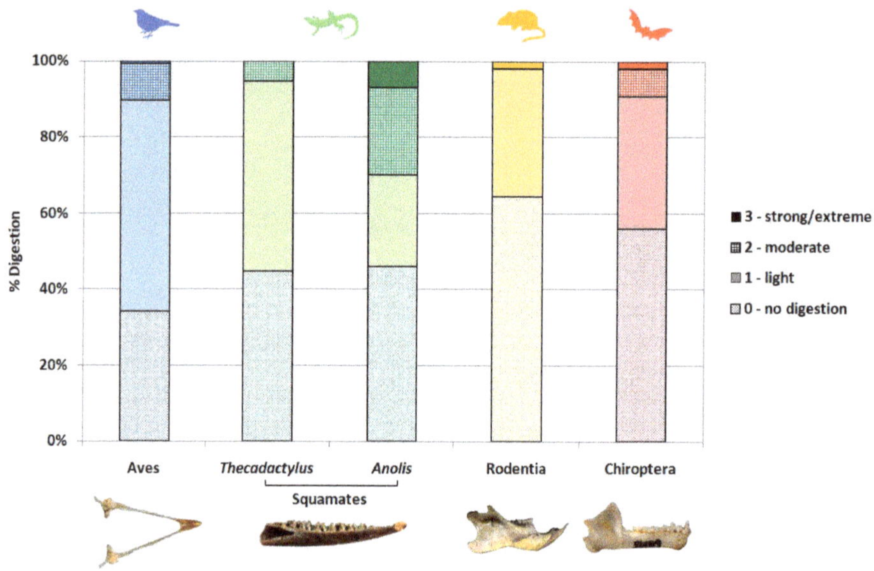

Figure 8. Digestion pattern on mandibles/dentaries (mean values of the data from the two observers for each taxon; detailed data in Table S6).

Regarding differences between elements within the same faunal group (intra-taxa variability), we see, for example, that the bat radius is much more digested than the femur (Figure 9, Table S6). This is probably due to the size and shape of bones: the radius is very long and narrow, and often exceeds the pellet size, making it more exposed to digestion. In contrast, among birds there is no significant difference in the pattern of digestion between the four long bones examined (humerus, femur, carpometacarpus and tarsometatarsus), because these elements all display a relatively short size in the considered species (Figure 10).

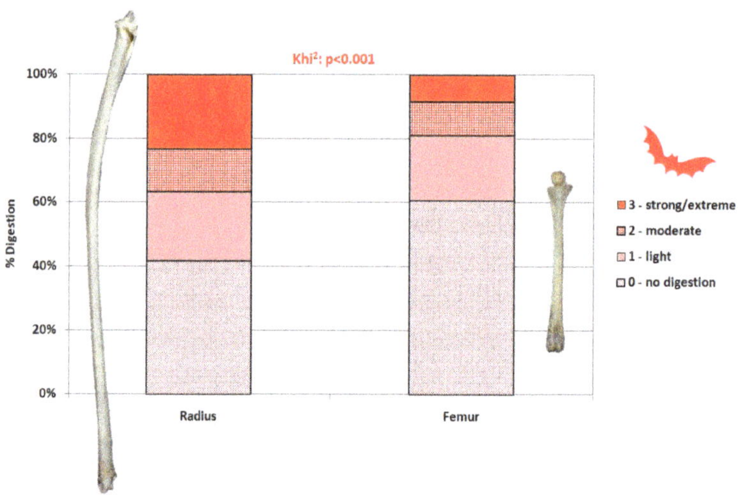

Figure 9. Digestion pattern on bat radius and femur (mean values of proximal and distal parts from the two observers; detailed data in Table S6).

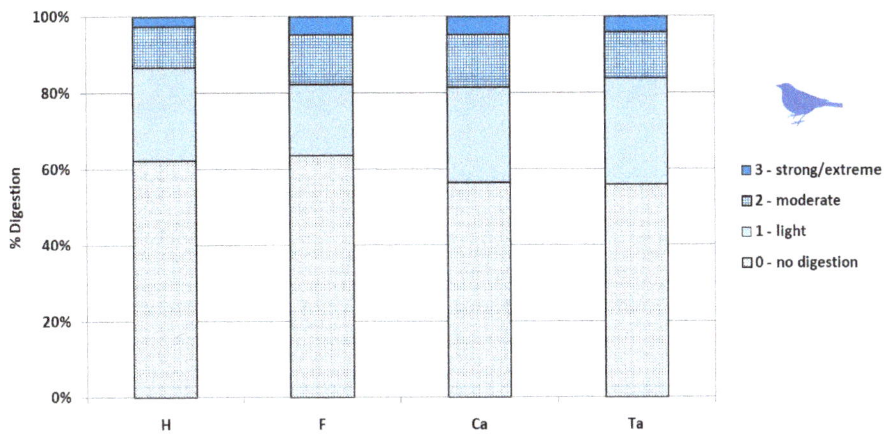

Figure 10. Digestion pattern on bird long bones (mean values of proximal and distal parts from the two observers; detailed data in Table S6).

However, differences can be observed between different parts when comparing the same element. This is the case for bird femora, for which the distal part is more digested than the proximal part (Figure 11). Consequently, within the same bone, the pattern of digestion can differ considerably. The same observation can be made for rodent and bat femora, while the reverse is observed for bird tarsometatarsus and small mammal tibia (proximal part more digested than distal part). In contrast, the digestion pattern is quite similar between proximal and distal parts for rodent humeri, for example (Figure 12).

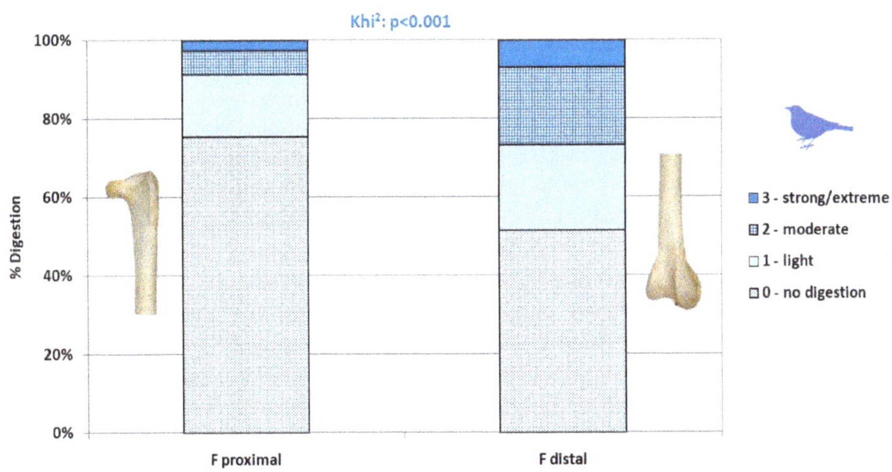

Figure 11. Digestion pattern on proximal and distal bird femur (mean values of the data from the two observers; detailed data in Table S6).

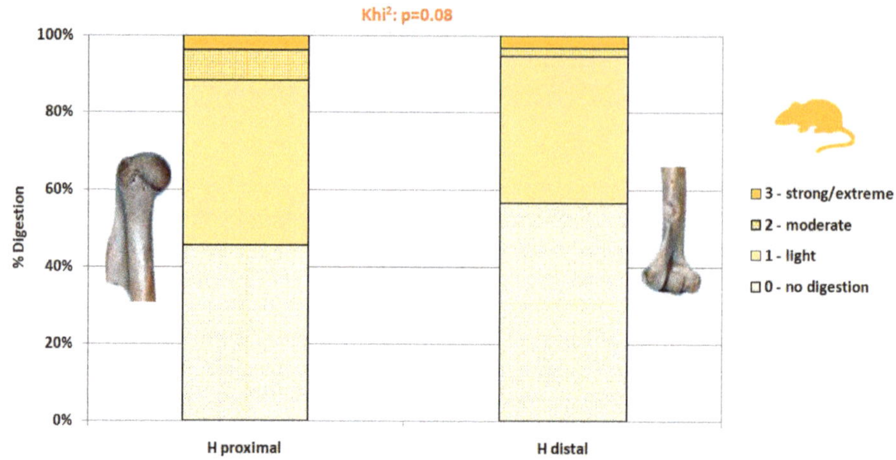

Figure 12. Digestion pattern on proximal and distal rodent humerus (mean values of the data from the two observers; detailed data in Table S6).

4. Discussion

We present here a synthesis and a discussion of our results on the studied *Tyto insularis* assemblages and a comparison with data of the diet from assemblages produced by representatives of the Barn Owl family in the West Indies. As no other taphonomic analysis has been performed on Barn Owl accumulations in the study region, we considered data from diverse geographic areas.

4.1. Diet of Tyto Insularis in Dominica

According to our study, the feeding habits of *Tyto insularis* from the island of Dominica include several bird species (belonging to three size classes), two lizard species, two rodent species and six bat species, with most of the preys weighing less than 100 g. Our observations show that birds are the main prey of *Tyto insularis* in the three studied sites and that a significant difference exists for the other prey taxa. The representation of lizards and bats is higher in "natural" environments (Grand Bay, Salisbury), while a greater representation of rodents is observed in the more anthropized site (Canefield). Moreover, as the samples analyzed in the present study were collected at two different periods (June 2014 and January 2015), a seasonal effect may exist between Grand Bay and Canefield on the one hand, and Salisbury on the other. The expected effect could be a decrease in black rat populations in the dry season, which is very clear in dry forest environments [33]. This seasonality effect may be the same for bats and forest birds and lizards, but it deserves further analyses.

Our results partly differ from the previous studies of the diet of *Tyto alba* and *Tyto glaucops* in the West Indies. Indeed, these studies predominantly show a general pattern indicating that the preferred prey in the Caribbean islands are, from most to least abundant, commensal rodents > birds > squamates/amphibians/bats, with few records of invertebrates such as insects and crustaceans [2,6–8,34,35]. For *Tyto glaucops* in Hispaniola, [2] identified in several pellet samples around 40 vertebrate taxa (including five bat species), and the large pellet sample studied by [35] led to the identification of 125 vertebrates (including 13 bat species). According to [12], a sample composed of 57 pellets produced by *Tyto insularis* in Dominica yielded at least 28 species of vertebrates, including seven bat species, and the most abundant species in terms of MNI (*Anolis* cf. *oculatus*) was not the most important in terms of biomass (*Rattus* cf. *rattus*). Hispaniola is a larger island than Dominica, and the studied material for *Tyto glaucops* is more abundant than for *Tyto insularis,* which may explain the higher species richness in the diet of *Tyto glaucops* compared to *Tyto insularis*, but both of them appear as opportunist predators. Thus, barn owls appear to have a broad spectrum of prey species with a few species dominating, probably depending on environmental conditions (specific diversity and abundance of prey in the environment) and seasonality, and *Tyto insularis* may exhibit a more varied diet than *Tyto alba* in the Caribbean [12].

4.2. *Taphonomic Impact of* Tyto Insularis *on Small Vertebrate Bone Assemblage*

4.2.1. Remarks on the Size/Weight Classes of Preys

For the present study we chose to consider only three size/weight classes according to the taxa we identified in the *Tyto insularis* assemblages. To apply the developed methodology to other predators, additional classes may be added according to their diet range and the size of hunted preys. For example, for the study of eagle-owl pellet assemblages, we can add at least one or two more classes (size category 3: 300 to 500 g including, for example, birds of pigeon size and adult *Rattus norvegicus;* size category 4: more than 500 g including larger birds, squamates, rodents and lagomorphs).

4.2.2. Anatomical Representation

For squamates, the fact that one of the observers considered and identified smaller and more fragmented elements than the other has meant that, compared to the other taxa, the radius and tibia are underrepresented, and this has impacted all the subsequent taphonomic analyses.

We also observed that in "terrestrial taxa" (lizards, rodents), posterior elements are slightly more represented than anterior elements, while the opposite is true in "flying taxa" (bats, birds). This may be explained by several hypotheses: an anatomical specificity, the mode of consumption by the predator, a differential preservation or a sample effect. The first hypothesis is the most likely, since it is the limbs that ensure propulsion and are therefore the largest and most "robust" and the least subjected to differential conservation. Moreover, a reliable interpretation of these profiles remains difficult considering the sample size.

Our results are difficult to compare directly with other taphonomic analyses of Barn Owl pellet assemblages (*Tyto alba*) in varied geographical areas, which generally consider more element categories for the anatomical representation, and only for small mammals (e.g., [14,15,36,37]). However, it is interesting to note that most of them show a good preservation of "robust" bones, like mandibles, humeri, femora and tibiae, which are the elements we chose to consider and for which we observed a similar representation in small mammals.

4.2.3. Fragmentation

Despite some differences between observers (for squamates and bats), we generally observed that the fragmentation rate was very low, with a great majority of bones being complete, all taxa combined. The maximum percentage of fragmentation was obtained for bat radius (34%), the most elongated bone, which is more sensitive to breakage. These observations are in keeping with data on *Tyto alba*, which are known to present a low fragmentation of skulls and post-cranial elements and a high preservation rate of intact bones (e.g., [14,15,36,37]).

4.2.4. Digestion

Previous analyses of digestion traces on small vertebrate bones accumulated by barn owls have focused mainly on rodent incisors, molars and proximal femora, and are thus again difficult to compare directly with our results. Generally, *Tyto alba* produces low percentages and grades of digestion, with some moderate grades (e.g., [14,15,36,37]). In our study, we have shown that more or less half of the articular extremities are digested, all taxa combined (except *Thecadactylus rapicauda*), with a majority of light to moderate grades. In focusing on proximal femora of rodents, we see that on average 44% are digested, and 70% of them with a light intensity (Table S6), which is in keeping with the available data for *Tyto alba* (for example [15], 37.8% digested, among them 55% with light intensity). Moreover, [15] also noticed that 30.5% of bird remains were digested, of these 79.5% with light intensity, while in our study we obtained respectively 42% and 62%.

However, there can be important differences in the digestion pattern between taxa coming from the same pellets produced by the same predator. There may also exist a high variability between elements of the same species and between different parts of the same element. This variability in the evaluation of the digestion can be due either to an observer effect or to specific osteological characteristics of bones. For example, digestion more strongly affects bird mandibles and lizard dentaries than small mammal mandibles. This may be due to the very thin skin on lizard dentaries. For bird mandibles, this may result from the bone structure and/or the presence of rhamphotheca of keratin on their beak bone. In rodents, mandibles are protected by strong masticatory muscles, which may explain why digestion traces are weaker on these bones. As another example of inter-taxa variability, we observed that *Thecadactylus rapicauda* displays both a higher percentage of digestion and a higher proportion of moderate and strong grades for humerus and femur than all the other taxa. Currently, the role of peculiar bone structure specificities in this pattern remains unknown.

For lizards, the differences between observers may be due to the first step of the study —namely, the anatomical identification. For rodents, the main reason is that digestion is very difficult to evaluate on rodent mandibles and distal tibia. For bats, the main cause is probably the same as for the fragmentation—the absence of extremities is attributed to either fragmentation or extreme digestion, according to the observers. Moreover, as for rodent ulna, observation of digestion on bat tibia is difficult. More generally, the major reason may be that the two observers have evaluated differently the limit between digestion grades, notably between light and moderate grades. These differences do not follow a particular trend or pattern—that is, a specific observer will not necessarily over- or under-evaluate digestion compared to another researcher. Differences are randomly distributed; each researcher seems to evaluate digestion grades differently. Consequently, an error range may exist, and this must be taken into account when comparing reference models established by different people.

4.3. *Degree of Inter-Observer Differences and Potential Outcomes*

The differences identified between the two observers are not strong enough to bias the taphonomic description of the assemblage. The discrepancies observed in the quantification of the lizard ulna, which subsequently biased the quantification of the digestion traces on the same assemblage, encourage the consideration of criteria that are not too difficult to assess. Indeed, lizard ulnae are small and lack easily recognizable morphological criteria, which makes them difficult to identify, even by paleontologists who are very familiar with this group. In contrast, only minor and unsignificant differences were observed in the quantification of the other lizard bones. For bats (and small mammals in general), the absence of certain extremities may be due either to fragmentation or extreme digestion. In the case of extreme digestion there may be loss of a proximal or distal end, and the bone may then be considered "incomplete" from the point of view of the fragmentation pattern, instead of "complete" with extreme digestion. To avoid this type of "error" in the future, a clear characterization and definition of the indices allowing the observers to distinguish one or the other phenomenon will have to be clearly established.

However, when deciding whether an observation discrepancy is significant or not, we must consider the level of precision adhered to in the description of the taphonomic impact of a predator on a bone assemblage. In that respect, our results warn against the use of very minor taphonomic differences. Indeed, in addition to a fall in the variation between different observers, such variation might also indicate individual differences between specimens of the same species related to age/size/sex or environmental conditions.

It is clear that defining the category of predator (cf. [14,16]) responsible for the constitution of a bone accumulation is possible in spite of these different biases, however, distinguishing different predators of the same category might be more complex. To address this issue, the replication of our study protocol to compare different predators of similar size and lifestyle would be needed to assess whether the observer bias would make such a distinction meaningless or not using minor taphonomic differences.

4.4. Towards an "Inter-Taxa Calibration"

To help conduct multi-taxa analyses, we tried to establish an "inter-taxa calibration" on the basis of the digestion pattern.

Table 5 shows a proposition of classification based on the percentage of digestion. The same anatomical element is classified differently according to the considered species. For example, rodent mandibles are less digested (and more difficult to observe) than squamate dentaries or bird mandibles. For the same taxon, there are differences from one element to another, with a greater or lesser extension of values.

Table 5. Classification of the considered anatomical elements according to taxa and percentages of digestion.

	<30% Digested	30–50% Digested	50–80% Digested	>80% Digested
Aves		Hum, CMC, Fem, TMT	Md	
Squamata (*Thecadactylusrapicauda*)			Md, Hum, Ulna, Tib	Fem
Squamata (*Anolis* sp.)	Ulna	Tibia	Md, Hum, Fem	
Rodentia	Md	Hum	Ulna, Fem, Tib	
Chiroptera		Md, Fem, Tib	Hum, Radius	

Another proposition of classification based on the grades of digestion (Table 6) gives a result that is a little bit different from the previous one. Thus, the calibration will be different according to the considered parameter: percentage of digestion, grades of digestion, or other taphonomic impacts.

Table 6. Classification of the considered anatomical elements according to taxa and grade of digestion.

	Majority Not Digested	Majority Light Digestion	Majority Moderate Digestion	Majority Strong Digestion
Aves	Hum, CMC, Fem, TMT	Md		
Squamata (*Thecadactylusrapicauda*)		Md, ulna, Tib	Hum	Fem
Squamata (*Anolis* sp.)	Md, Hum, Ulna, Fem, Tib			
Rodentia	Md, Hum, Tib	Ulna, Fem		
Chiroptera	Md, Hum, Ulna, Fem, Tib			Radius

5. Conclusions

The present study focused on the analysis of pellet samples produced by *Tyto insularis* in Dominica. The prey spectrum reflects the diet of an opportunistic predator that exploits a variety of small vertebrates available in the environment. From a global point of view and for all the studied taxa, the taphonomic impact of *Tyto insularis* on prey bones seems very similar to that of *Tyto alba*: good preservation of mandibles and long bones and a high rate of complete bones, with more or less half of the material being digested with a majority of light to moderate grades.

From a methodological point of view, our work is an attempt to lay the foundations for a clear and simplified study protocol for the collaborative taphonomic analysis of multi-taxa assemblages. To do so, we defined a protocol that aims to limit the potential biases between observers and to allow for inter-taxa comparisons.

As a result, we observed high similarities in the taphonomic results obtained for the considered faunal groups, but some differences emerged at several levels: between observers, between taxa, between elements and even between different parts of the same element. There may be several reasons for these differences, including the way each person observes and evaluates digestion traces, and the way each part of the skeletal elements is affected by digestion.

However, the identification of these factors will remain out of reach until more comparison points using the same approach are available. Similar attempts considering other observers and taxa with the same protocol will open the way to understanding the taphonomic variability according to additional factors (predator age, season, environment and prey availability) and then to the application of our approach to multi-taxa microvertebrate fossil and sub-fossil assemblages.

From a regional point of view, this study constitutes the first neo-taphonomic comparison point available in the whole insular Caribbean for taphonomic studies of microvertebrate assemblages. Therefore, we hope that similar approaches will be further developed in the area to help the understanding of the accumulation process of the numerous fossil and sub-fossil assemblages investigated up until now and in the future. These studies will be of fundamental importance to better understand the biases that affect the vision of the paleobiodiversity provided by fossil and sub-fossil assemblages.

Supplementary Materials: The following are available online at https://www.mdpi.com/2571-550X/4/4/38/s1. Figure S1: Examples of digestion grades considered in the present study on bird ulnae; Table S1: Detailed data on the proportions of the prey species found in the three studied pellet samples from the island of Dominica, by site and by observer; Table S2: Detailed data on the size categories of the prey taxa found in the three studied pellet samples from the island of Dominica, by site and by observer; Table S3: Detailed data on the percentage of anatomical representation (PR) for each faunal group, by site and by observer; Table S4: Ratios calculated with the considered anatomical elements for each faunal group and each site, all observers combined; Table S5: Detailed data on the number of identified specimens (NISP) fragmented and percentage of fragmentation (PF) for each faunal group, by site and by observer; Table S6: Detailed data on the percentage of digestion (PD) for each faunal group, by anatomical element and by observer.

Author Contributions: E.S., S.B., C.B., D.C., M.G. and V.L. contributed to the choice and implementation of the taphonomic methodology, to the material analysis and to the result interpretations. E.S. collected, combined and compared the databases of all specialists. E.S. designed and wrote the original draft. All authors have read and agreed to the published version of the manuscript.

Funding: This study received financial support for collecting pellets in the field and for the organization of working meetings from the FEDER Program 2013–2015 "BIVAAG: Biodiversité Insulaire Vertébrée, floristique et malacologique Ancienne de l'Archipel de la Guadeloupe" (A. Lenoble dir.), with support from a European PO-FEDER grant 2007–2013 n_2/2.4/-33456, the Guadeloupe Regional Council, the DEAL of Guadeloupe, the DAC of Guadeloupe and the CNRS. This work also benefited from support by the CNRS-INEE GDR 3591 "TaphEnA: Taphonomie, environnement et archéologie" (J.P. Brugal dir.) and the CNRS-INEE GDR 3644 "BioArchoDat: Sociétés, biodiversité, environnement: données et resultats de l'archéozoologie et de l'archéobotanique sur le territoire de la France" (R.M. Arbogast dir.) for the organization of working meetings and participation in the ICAZ Taphonomy Working Group meeting in Paris in 2016. English reviewing fees were funded by the CNRS-INEE IRN 0871 "TaphEn: Taphonomy—a cross-disciplinary network of natural and social scientists dedicated to past remains accumulations, origins, formations and modifications, for the Quaternary period" (J.P. Brugal dir.).

Institutional Review Board Statement: Not applicable.

Informed Consent Statement: Not applicable.

Data Availability Statement: Data are contained within the article and the Supplementary Materials.

Acknowledgments: We thank Baptiste Angin (Ardops Environnement) for pellet collecting, as well as Arnaud Lenoble (PACEA UMR 5199), coordinator of the FEDER Program "BIVAAG", and Christiane Denys (ISYEB UMR 7205), coordinator of the micro-project "Référentiel(s) taphonomique(s) concernant les petits vertébrés: bilan et examen critique des méthodes d'étude" financed by the GDR TaphEnA (dir. J.P. Brugal). We also thank Barbara Wilson (Pensato Studios) for copy editing the English text, and the three anonymous reviewers for their constructive comments.

Conflicts of Interest: The authors declare no conflict of interest.

References

1. Ridgway, R. Birds of North and Middle America. *Bull. U.S. Nat. Mus.* **1914**, *50*, 594–825.
2. Wetmore, A.; Swales, B.H. The Birds of Haiti and the Dominican Republic. *Bull. U.S. Nat. Mus.* **1931**, *155*, 1–483. [CrossRef]
3. Weick, F. *Owls (Strigiformes): Annotated and Illustrated Checklist*; Springer: Berlin/Heidelberg, Germany, 2006.
4. Suarez, W.; Olson, S.L. Systematics and distribution of the living and fossil small barn owls of the West Indies (Aves: Strigiformes: Tytonidae). *Zootaxa* **2020**, *4830*, 544–564. [CrossRef] [PubMed]
5. Avibase. *The World Bird Database*; Lepage, D., Ed.; Available online: https://avibase.bsc-eoc.org/avibase.jsp (accessed on 15 July 2021).
6. Johnston, D.W. Food of the Barn Owl on Grand Cayman, BWI. *Q. J. Fla. Acad. Sci.* **1972**, *35*, 171–172. Available online: https://www.jstor.org/stable/i24313839 (accessed on 20 July 2021).
7. Buden, D.W. Prey remains of Barn Owls in the southern Bahama Islands. *Wils. Bull.* **1974**, *86*, 336–343.
8. McFarlane, D.A.; Garrett, K.L. The prey of common Barn Owls (*Tyto alba*) in dry limestone scrub forest of southern Jamaica. *Caribb. J. Sci.* **1989**, *25*, 21–23.
9. Arredondo, C.; Chirino, V.N. Consideraciones sobre la alimentación de *Tyto alba* furcata (aves: Strigiformes) con implicaciones ecológicas en Cuba. *J. Caribb. Ornithol.* **2002**, *15*, 16–24.
10. Hernández-Munoz, A.; Mancina, C.A. La dieta de la lechuza (*Tyto alba*) (Aves: Strigiformes) en hábitats naturales y antropogénicos de la región central de Cuba. *Rev. Mex. Biodiv.* **2011**, *82*, 217–226. [CrossRef]
11. López-Ricardo, Y. Alimentacion de la Lechuza (Tyto Alba Furcata) en Cuba Central: Presas Introducias y Autoctonas. Tesis de diploma, Universidad de La Habana, La Habana, Cuba, 2012.
12. Stoetzel, E.; Fraysse, A.; Bochaton, C.; Gala, M.; Grouard, S.; Lenoble, A.; Denys, C. Diet of the Lesser Antillean barn owl *Tyto insularis* (Aves: Strigiformes) in Dominica, Lesser Antilles. *Caribb. J. Sci.* **2016**, *49*, 91–100. [CrossRef]
13. Stoetzel, E.; Royer, A.; Cochard, D.; Lenoble, A. Late Quaternary changes in bat palaeobiodiversity and palaeobiogeography under climatic and anthropogenic pressure: New insights from Marie-Galante, Lesser Antilles. *Quat. Sci. Rev.* **2016**, *143*, 150–174. [CrossRef]
14. Andrews, P. *Owls, Caves and Fossils*; Natural History Museum Publications: London, UK, 1990.
15. Denys, C.; Stoetzel, E.; Andrews, P.; Bailon, S.; Rihane, A.; Huchet, J.B.; Fernandez-Jalvo, Y.; Laroulandie, V. Taphonomy of small predators accumulations: About palaeoecological implications. *Hist. Biol.* **2018**, *30*, 868–881. [CrossRef]
16. Fernández Jalvo, Y.; Andrews, P.; Denys, C.; Sese, C.; Stoetzel, E.; Marin-Monfort, D.; Pesquero, D. Taphonomy for taxonomists: Implications of predation in small mammal studies. *Quat. Sci. Rev.* **2016**, *139*, 138–157. [CrossRef]
17. Lorvelec, O.; Pascal, M.; Delloupe, X.; Chapuis, J.L. Les mammifères terrestres non volants des Antilles françaises et l'introduction récente d'un écureuil. *Rev. Écol.* **2007**, *62*, 295–314. Available online: https://www.google.com/url?sa=t&rct=j&q=&esrc=s&source=web&cd=&ved=2ahUKEwim0qDX7tvzAhXZ6OAKHWhZBJcQFnoECAIQAw&url=https%3A%2F%2Fecureuils.mnhn.fr%2Fsites%2Fdefault%2Ffiles%2Fdocuments%2Florvelec-et-al-rev-ecol-terr-vie.pdf&usg=AOvVaw20M02pZpFMS8O_t8aGb4gX (accessed on 20 July 2021).
18. Giovas, C.M. The beasts at large—Perennial questions and new paradigms for caribbean translocation research. Part II: Mammalian introductions in cultural context. *Environ. Archaeol.* **2018**, *24*, 294–305. [CrossRef]
19. Kemp, M.E.; Mychajliw, A.M.; Wadman, J.; Goldberg, A. 7000 years of turnover: Historical contingency and human niche construction shape the Caribbean's Anthropocene biota. *Proc. R. Soc. B* **2020**, *287*, 20200447. [CrossRef]
20. Rabinow, S.; Giovas, C. A systematic review of agouti (Dasyproctidae: *Dasyprocta*) records from the pre-1492 Lesser Antilles: New perspectives on an introduced commensal. *Int. J. Osteoarchaeol.* **2021**, *31*, 758–769. [CrossRef]
21. Mallye, J.B.; Cochard, D.; Laroulandie, V. Accumulation osseuse en périphérie de terriers de petits carnivores: Les stigmates de prédation et de fréquentation. *Ann. Paléontol.* **2008**, *3*, 187–208. [CrossRef]
22. Royer, A.; Montuire, S.; Gilg, O.; Laroulandie, V. A taphonomic investigation of small vertebrate accumulations produced by the snowy owl (*Bubo scandiacus*) and its implications for fossil studies. *Palaeogeogr. Palaeoclimatol. Palaeoecol.* **2019**, *514*, 189–205. [CrossRef]
23. Lebreton, L.; Bailon, S.; Guilllaud, E.; Testu, A.; Perrenoud, C. Multi-taxa referential of a modern Eurasian Eagle-Owl (*Bubo bubo*) aerie. *J. Archaeol. Sci. Rep.* **2020**, *32*, 102417. [CrossRef]
24. Erome, G.; Aulagnier, S. Contribution à l'identification des proies des Rapaces. *Bièvre* **1982**, *4*, 129–135.
25. Radulet, N. Comparative anatomy of the mandible in the mammal systematics (Mammalia: Insectvora, Chiroptera, Rodentia) from Romania (II). *Trav. Mus. Natl. Hist. Nat. Grigore Antipa* **2006**, *XLIX*, 365–374.

26. Radulet, N. Morphology of the inner side of the mandible in micromammals (Mammalia: Insectivora, Chiroptera, Rodentia) of Romania. *Trav. Mus. Natl. Hist. Nat. Grigore Antipa* **2007**, *L*, 371–393.
27. Fraysse, A. Contribution à L'identification Morphologique et Moléculaire des Chiroptères des Petites Antilles. Master's Thesis, Muséum National d'Histoire Naturelle, Paris, France, 13 September 2013.
28. Ibéné, B.; Leblanc, F.; Houllemare, C.; Tarlier, C.; Pentier, C. *Contribution à L'étude des Chiroptères de la Guadeloupe;* Diren Guadeloupe: Basse-Terre, France, 2007.
29. Arendt, W.J.; Faaborg, J.; Wallace, G.E.; Garrido, O.H. Biometrics of birds throughout the Greater Caribbean Basin. *Proc. West. Found. Vert. Zool.* **2004**, *8*, 1–33.
30. Bochaton, C.; Bailon, S.; Herrel, A.; Grouard, S.; Ineich, I.; Tresset, A.; Cornette, R. Human impacts reduce morphological diversity in an insular species of lizard. *Proc. R. Soc. B* **2017**, *284*, 20170921. [CrossRef] [PubMed]
31. Bochaton, C.; Daza, J.D.; Lenoble, A. Identifying gecko species from Lesser Antillean paleontological assemblages: Intraspecific osteological variation within and interspecific osteological differences between *Thecadactylus rapicauda* (Houttuyn, 1782) (Phyllodactylidae) and *Hemidactylus mabouia* (Moreau de Jonnès, 1818) (Gekkonidae). *J. Herpetol.* **2018**, *52*, 313–320. [CrossRef]
32. Lloveras, L.; Moreno-Garcia, M.; Nadal, J.; Thomas, R. Blind test evaluation of accuracy in the identification and quantification of digestion corrosion damage on leporid bones. *Quat. Int.* **2014**, *330*, 150–155. [CrossRef]
33. Goedert, J.; Cochard, D.; Lenoble, A.; Lorvelec, O.; Pisanu, B.; Royer, A. Seasonal demography of different black rat (*Rattus rattus*) populations under contrasting natural habitats in Guadeloupe (Lesser Antilles, Caribbean). *Mamm. Res.* **2020**, *65*, 793–804. [CrossRef]
34. Flikweert, M.; Prins, T.G.; de Freitas, J.A.; Nijman, V. Spatial variation in the diet of the Barn Owl *Tyto alba* in the Caribbean. *Ardea* **2007**, *95*, 75–82. [CrossRef]
35. Wiley, J.W. Food habits of the endemic ashy-faced owl (*Tyto glaucops*) and recently arrived barn owl (*T. alba*) in Hispaniola. *J. Raptor Res.* **2010**, *44*, 87–100. [CrossRef]
36. Lopez, J.M. Actualistic taphonomy of barn owl pellet-derived small mammal bone accumulations in arid environments of South America. *J. Quat. Sci.* **2020**, *35*, 1057–1069. [CrossRef]
37. Linchamps, P.; Stoetzel, E.; Hanon, R.; Denys, C. Neotaphonomic study of two *Tyto alba* assemblages from Botswana: Paleoecological implications. *J. Archaeol. Sci. Rep.* **2021**, *38*, 103085. [CrossRef]

quaternary

Article

The *Stegodon* Bonebed of the Middle Pleistocene Archaeological Site Mata Menge (Flores, Indonesia): Taphonomic Agents in Site Formation

Meagan J. Powley [1,2,*], Indra Sutisna [3], Katarina M. Mikac [2], Unggul Prasetyo Wibowo [3] and Gerrit D. van den Bergh [1,*]

1 Centre for Archaeological Science, School of Earth, Atmospheric and Life Sciences, University of Wollongong, Wollongong, NSW 2522, Australia
2 Centre for Sustainable Ecosystem Solutions, School of Earth, Atmospheric and Life Sciences, University of Wollongong, Wollongong, NSW 2522, Australia; kmikac@uow.edu.au
3 Geology Museum Bandung, Geological Agency, Jalan Diponegoro 57, Bandung 40122, Indonesia; indra11084@yahoo.com (I.S.); uungpw@yahoo.com (U.P.W.)
* Correspondence: mjp94@uowmail.edu.au (M.J.P.); gert@uow.edu.au (G.D.v.d.B.)

Abstract: The Middle Pleistocene fluvial channel site of the Upper Fossil-bearing Interval at Mata Menge in the So'a Basin, Flores, Indonesia, has yielded the earliest fossil evidence for *Homo floresiensis* in association with stone artefacts and fossils of highly endemic insular fauna, including *Stegodon*, giant rats, crocodiles, Komodo dragons, and various birds. A preliminary taphonomic review of the fossil material here aimed to provide additional context for the hominin remains in this bonebed. Analysis was performed on two subsets of material from the same fluvial sandstone layer. Subset 1 comprised all material from two adjacent one-metre square quadrants (n = 91), and Subset 2 all *Stegodon* long limb bones excavated from the same layer (n = 17). Key analytical parameters included species and skeletal element identification; fossil size measurements and fragmentation; weathering stages; bone fracture characteristics; and other biological and geological bone surface modifications. Analysis of Subset 1 material identified a highly fragmented assemblage with a significant bias towards *Stegodon*. A large portion of these bones were likely fractured by trampling prior to entering the fluvial channel and were transported away from the death-site, undergoing surface modification causing rounding. Subset 2 material was less likely to have been transported far based on its limited susceptibility to fluvial transport. There was no significant difference in weathering for the long limb bones and fragments, with the highest portion exhibiting Stage 2 weathering, indicating that prior to final burial, all material was exposed to prolonged periods of surface exposure. Approximately 10% of all material have characteristics of fracturing on fresh bone, contributing to the taphonomic context for this bonebed; however insufficient evidence was found for anthropogenic modification.

Keywords: taphonomy; *Stegodon*; Middle Pleistocene; hominins; Flores; fluvial; bonebed; Mata Menge; southeast Asia

1. Introduction

Extensive excavations from the early Middle Pleistocene archaeological site of Mata Menge, located near the northwestern margin of the So'a Basin, a ~400 km^2 depression on the island of Flores, Indonesia (Figure 1), have yielded in situ lithic artefacts and fossil vertebrate remains. The island is situated between the zoogeographical boundaries recognised as Wallace's Line in the west and Lydecker's Line in the east, in the region known as Wallacea. Flores has remained isolated from other landmasses even during periods of maximum low sea levels. The resultant low diversity fossil fauna was first examined in the 1950s when large fossilised bones, exposed by erosion, were excavated by Father Theodore Verhoeven, a priest affiliated with a nearby seminary, and identified as *Stegodon trigonocephalus florensis* [1,2]. The continued work for more than two decades uncovered a large assemblage of *Stegodon* remains, other vertebrate fauna, and stone artefacts at various localities, including Mata Menge (Figure 1), the site that is the focus of our paper [3–5]. Analysis of the abundant fossil material allowed the attribution of the Flores *Stegodon* to a distinct species, *Stegodon florensis*, which is smaller than *Stegodon trigonocephalus* from Java [6]. During the 1980s and 1990s, an Indonesian-Dutch team carried out investigations into the claims of Verhoeven that hominins had colonised Flores during the Pleistocene [5,7]. Since 1997, systematic excavations at Mata Menge and other sites have occurred, leading to chronometric age determinations and further validation of Verhoeven's findings, including details on the endemic insular fauna associated with stone artefacts embedded in stratigraphy representing over a million years [3,8–14]. Clear evidence for hominin occupation on Flores has been identified at various sites, including at two stratigraphic levels at Mata Menge, separated by 10 m of stratigraphy. A Zircon Fission-track age of 0.80 $\pm$ 0.07 Ma for the Lower Fossil-bearing Interval (LFI) has been reported [13]. A more recent palaeomagnetic study [15] revealed that the LFI has a normal polarity and corresponds with the base of the Brunhes Normal Epoch. The LFI is therefore thought to be slightly younger than the Matuyama-Brunhes boundary at 0.773 Ma, which is within the error range of the ZFT age. The Upper Fossil-bearing Interval (UFI) is bracketed by ages of 0.78 and 0.65 Ma [10,13,15]. At the nearby site Wolo Sege, stone artefacts occur directly below a widespread pumice tuff layer, dated at 1.0 Ma [9] and, combined with the evidence from Liang Bua [16], indicates hominin occupation on the island for at least one million years.

However, only limited detailed taphonomic analysis of all fossils located in the LFI and UFI has occurred. A taphonomic study on *Stegodon* fossils from the LFI suggested that the *Stegodon* bones had been affected by fluvial transport [4]. Two additional studies focusing on avian remains from the LFI found several bone surface modifications, including parallel grooves, spindle-like striations, and rows of pits, but no unambiguous evidence for modifications made by hominins [17,18]. Taphonomic observations and analysis of the fossilised material, as well as the study of sedimentological features, can improve the interpretation of the context and significance of the hominin activities.

As mentioned, to date, no proof has been found that the Mata Menge hominins were involved in scavenging or hunting activities of fauna despite the co-occurrence of stone artefacts [10,17]. In 2014 the UFI yielded a fragment of mandible and six isolated teeth from at least three small-bodied hominins [14], which consolidated previous evidence for the presence of hominins at Mata Menge. These fossils established Mata Menge as the site containing the oldest hominin remains on Flores that likely represent the ancestral population of *Homo florensiensis* [14], the small-bodied hominin from the Late Pleistocene of Liang Bua cave, which is located 70 km WNW of Mata Menge. Here we present a preliminary taphonomic analysis of the Mata Menge vertebrate fossils to clarify some of the taphonomic processes that have contributed to the formation of the bonebed and investigate potential evidence of hominin modification of the *Stegodon* bones.

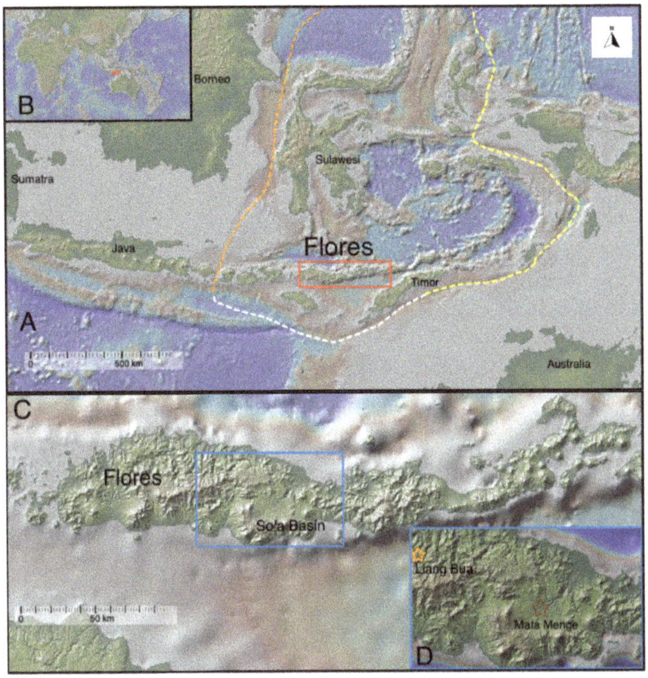

Figure 1. (**A**) Location of Flores (red rectangle) within the Indonesian archipelago. The orange line represents Wallace's Line; the yellow line represents Lydekker's Line; the white line connecting them represents the southern boundary of Wallacea [19]. (**B**) Location of the Indonesian archipelago. (**C**) Location of the So'a Basin (blue rectangle) on Flores, (**D**) Location of Mata Menge (red star) and Liang Bua (orange star) in the So'a Basin. (Map made with GeoMapApp (www.geomapapp.org)/CC BY (accessed on 15 February 2021)).

1.1. Geological Context

The stratigraphic sequence that fills the So'a Basin consists of a basement substrate known as the Ola Kile Formation (OKF) deposited by volcanic events with a total thickness of at least 100 m [4,9]. A fission track age of 1.86 ± 0.12 Ma for a tuff located at the top of OKF provides maximum age for the basin-fill sequence [13]. The Ola Bula Formation (OBF) sits unconformably on the slightly dipping OKF, with a maximum thickness of 120 m of volcaniclastic, fluviatile, and lacustrine deposits [10] (Figure 2).

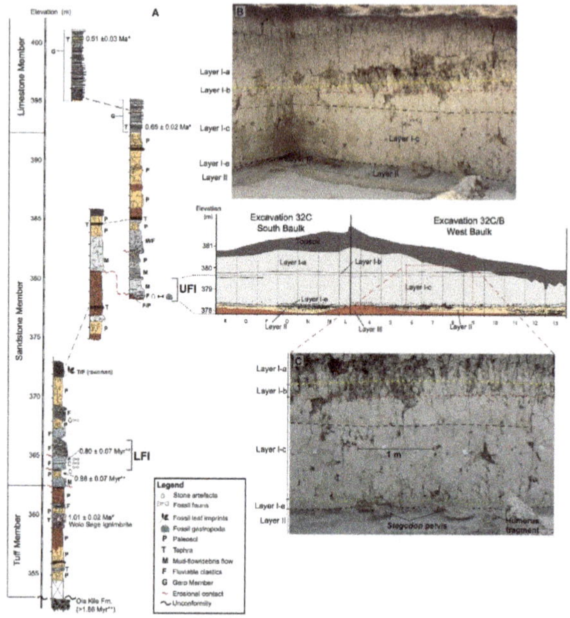

Figure 2. (**A**) Stratigraphy and chronology of the main fossil-bearing intervals at Mata Menge (Upper Fossil-bearing Interval, UFI, and Lower Fossil-bearing Interval LFI) as well as several basin-wide key marker tephra beds. (**B**) Photos and line drawings of the excavation baulks of Sector 32C (see also Materials and Methods for a plan view) in the UFI. The upper photo shows the view towards the SW corner of Sector 32C in 2015 when the up to ~30 cm thick sandstone lens that constitutes the main bonebed in the UFI (Layer II) was partly exposed. Layer II is overlaid by a sequence of massive, clay-rich mudflows (Layers Ia–e). Layer II has an erosional base and is deposited on top of a reddish paleosol (Layer III) with an irregular top surface, an elevated part of which is exposed in the corner. The line drawing in the middle shows the profiles of the south and west baulks in 2016 when Layer II was fully exposed in Sector 32C. (**C**) The bottom photo shows a detail of the west baulk. Two *Stegodon* pelvis bones and a humerus diaphysis are partly embedded in the sandstone of Layer II and are covered by the base of a 10 cm thick mudflow, Layer I-e, which contains abundant intraclasts eroded from the underlying strata and small fossils. Layer 1-d is not developed in Sector 32C, and Layer 1-e is here directly overlain by the massive mudflow unit Layer 1-c.

The OBF basal Tuff Member is characterised by an alternation of felsic pumice mass-flow layers and air-fall tephras, reworked tuffaceous silts, sandstones, and conglomerate lenses [13,20]. This part of the sequence was deposited by infrequent volcanic events and braided river systems, with minor lacustrine/palustrine conditions. However, deposition was interrupted by long periods of non-deposition, as evidenced by the development of pronounced palaeosols [4,10]. A widespread marker bed located within this basal Tuff Member named the Wolo Sege Ignimbrite has been $^{40}Ar/^{39}Ar$ dated at 1.01 ($\pm$0.02) Ma [9,10]. Only a single bonebed at the site Tangi Talo has been discovered within the Tuff Member, whereas stone artefacts have been shown to occur in fluvial conglomerates and associated overbank deposits just below the Wolo Sege Ignimbrite, providing a minimum age of ~1 Ma for the presence of hominins on the island [9]. The fauna from the bonebed at the locality Tangi Talo is characterised by few vertebrate taxa, with a giant tortoise, a 1 m tall pygmy proboscidean (*Stegodon sondaari*), and Komodo dragons (*Varanus komodoensis*), indicative of insular conditions [6,9,21].

The middle OBF Sandstone Member is characterised by sheet flow deposits, tuffaceous fluvial sandstones and siltstones, mudflow deposits, conglomerates, and a few mafic tuff layers [10,11,22]. This unit was deposited by a fluvial system and includes channel deposits, overbank deposits, and mudflows, and again frequently overprinted palaeosols. Compared with the basal Tuff Member, a decrease in volcanic events was noted during the deposition of this member [4]. More frequent fossil beds are found in the sandstone member, commonly associated with stone artefacts. Zircon fission-track samples collected in the Sandstone Member date the deposits between 0.88 ± 0.04 Ma and 0.70 ± 0.07 Ma, aligning closely with $^{40}Ar/^{39}Ar$ dated samples at 0.81 ± 0.04 [10,12,13]. Both fossil-bearing intervals near Mata Menge occur in this stratigraphic unit.

The OBF is capped by the Limestone Member featuring thin-bedded micritic freshwater limestones, thin-bedded silty clay layers, and numerous predominantly mafic air-fall tephra layers [10,20]. Deposition in this member was dominated by lacustrine conditions and volcanic ash influxes [4]. A tephra sample from the top of this member gave an $^{40}Ar/^{39}Ar$ age of 0.51 ± 0.03 Ma [10] and another sample from the base of this member, taken at 14 m above the UFI, provided an $^{40}Ar/^{39}Ar$ age of 0.65 ± 0.02 Ma [10], in accordance with two zircon fission-track samples from within the member that returned ages of 0.68 ± 0.07 Ma and 0.65 ± 0.07 Ma. Along the basin margins, the top of the OBF is overlaid by volcaniclastics from volcanoes surrounding the basin, some of which are still active [20].

1.2. Stratigraphic Context for Mata Menge

The Lower Fossil-bearing Interval (LFI), bracketed by a maximum age of 0.78 Ma based on palaeomagnetic evidence and a minimum age of 0.81 Ma [15,23], contains impoverished vertebrate fauna as well as in situ stone artefacts. The up to 2.2 m thick LFI represents a fining-upward sequence consisting of an amalgamated complex of sheets and lenses of sandstone alternating with tuffaceous siltstones and clay-rich mudflows. While no fossilised hominin remains, or evidence for hominin butchery practice have been identified from the LFI, simple stone technology excavated from this interval provided proxy evidence for the presence of hominins [4]. A taphonomic analysis on the *Stegodon* bone assemblage from the lower interval indicated that fluvial transport had removed smaller and more easily transportable bones from the assemblage, especially from the high-energy erosional basal sandy layer [4]. Additional excavations in the lower interval did not yield any hominin remains, and during 2013 fieldwork, excavations were expanded to a sandstone layer 10 m higher in the stratigraphic sequence, to what has been referred to as the c. 100,000 years younger Upper Fossil-bearing Interval (UFI) [10]. The fossil-bearing layer here comprises a less than 30 cm thick sandstone lens.

The sedimentology of the Upper Fossil-bearing Interval (UFI) has been described to some extent in a previous paper [10]. The fossils are concentrated in a sandstone layer (Layer II) with an erosive base, which fills in an irregular topography of a well-consolidated palaeosol (Layer III; Figure 2). Layer II has a maximum thickness of ~35 cm and wedges out towards the western margin of the site. Layer II consists of medium-coarse, poorly sorted sand with scattered rounded pebbles at the base. The average maximum diameter of the 15 largest pebbles is 91 mm (the largest pebble had a diameter of 115 mm). Using a Hjullström diagram, this would indicate a current transport velocity of over 100 cm/s to allow transport of such pebbles to the site. Layer II is fining upwards to medium to fine-grained sand at the top, suggesting decreasing flow velocity when the layer was deposited. Internal sedimentary structures are generally absent apart from vaguely preserved parallel laminations at the basal part of the layer. Layer II is covered by a 6.5 m thick series of massive clay-rich mudflow deposits (Layers I-a to I-f), which also contain few small fossil bone fragments and rounded mud intraclasts at the base. This sequence has been interpreted as the deposit of a small stream on a distal volcanic apron, cutting into consolidated palaeosol deposits, and at some stage rapidly covered by a series of mudflows, which most likely originated from a caldera structure with a lake directly north of Mata Menge [10]. It is notable that articulated fossils are very rare and that many *Stegodon* fossils show signs of rounding by fluvial transport. Some of the larger *Stegodon* fossils are half embedded in the fluvial sandstone and are covered by the mudflow (see Figure 2, lower photo). For the assemblage studied here, only fossil specimens from the sandstone Layer II were included in the analysis.

1.3. Location and Fauna

The isolation of Flores within the zoogeographical borders of Wallacea during maximum glacial events throughout the Pleistocene restricted access to the island to animals that could either fly, raft, or swim there (Figure 1) [7]. Endemic vertebrate fauna excavated from both levels at Mata Menge is restricted to *Stegodon florensis*, crocodile, *Varanus komodoensis*, a single giant rat species *Hooijeromys nusatenggara*, various types of birds, and frogs [4,10,17,24]. *S. florensis* is larger than *S. sondaari* from the underlying Tuff Member and represents a new immigrant to Flores, replacing *S. sondaari*. The Mata Menge *S. florensis* has been classified as a separate chrono-subspecies that was ancestral to *Stegodon florensis insularis* from the Late Pleistocene of Liang Bua [25].

From the thousands of fossils already excavated on Flores, no evidence for the presence of large mammalian predators has been found to date. Between 2010 and 2019, over 33,000 fossils were unearthed from excavation sites across the So'a Basin [26]. Of the fossil assemblage excavated from the UFI at Mata Menge until 2015, the largest portion comprised unidentified bone fragments (~44%), while the remainder of the fossils could be identified to the class taxonomic level. These relative proportions changed little after additionally incorporating the UFI fossils excavated during the period 2016–2019, totalling 15,535 specimens, which can be attributed as follows: unidentifiable fragments (42.9%); *Stegodon* (33.9%); Murine rodents (19.0%); Crocodilia (2.0%); Varanidae (0.68%); other Reptilia (0.7%); Aves (0.56%); Anura (0.28%); and Hominin 0.06% [26].

The bonebed at Mata Menge is important for interpreting formation processes of bonebeds that have no modifications made by mammalian carnivores. There are many continental examples of proboscidean bonebeds that have been modified by mammalian carnivores to various degrees, largely through disarticulation and dispersal of carcasses and bone surface modifications such as teeth punctures or gnawing on bone ends [27–30]. At Mata Menge, reptiles were the primary faunal predators. Reptilian predators can remove bones completely from the assemblage through digestion [31,32], or residual bones may retain bone surface modifications. Some researchers have suggested that the V-shaped linear grooves made by crocodile teeth cannot always be distinguished from those made by stone tools [33] unless advanced statistical tests are applied [34].

The question of whether hominins were involved with the accumulation of faunal remains here or what other agents led to their preservation remains unresolved at this time. Large numbers of *Stegodon* fossils, some with reported cutmarks, and evidence of anthropogenic bone accumulation of insular fauna attributed to *H. floresiensis* were identified at Liang Bua, revealing substantiation of hominin modification of *Stegodon* remains during the Late Pleistocene [35]. Whether hominins used stone artefacts with similar technology identified in both locations, which are separated in time by ~600,000 years [8,10], to hunt, scavenge/butcher, and consume *Stegodon* is one of the key motivators of this research. *Stegodon*, the largest fauna element on Flores during the Middle Pleistocene, could have provided substantial amounts of meat to hominins, which may have contributed to the decision to hunt this species. However, evidence for meat consumption by the Early to Middle Pleistocene hominins has so far not been uncovered at Mata Menge. In order to begin to interpret the material from the UFI at Mata Menge, analysis for this research was focused on identifying the presence of fractures on fresh bones, bone surface modifications and subsequently identifying the extent of surface exposure and possible fluvial transportation of the bones, processes that may have obliterated bone surface modifications.

1.4. Taphonomic Processes

Biological processes, including modifications made by hominins for food procurement and butchery [8,36–39], the use of bones as tools [40,41], and anthropogenic bone accumulation [25,42], have characteristics recognisable through the analysis of bone fracturing, bone surface modification, and location of bones in a bonebed. Similarly, faunal modifications are identified by characteristics on the bones that are associated with predation and feeding. Vertebrate tooth marks [31,34,43,44], disarticulation patterns such as those from large reptiles [34,43], and even the absence of bone material from the site when bones are completely removed through digestive corrosion [34], can provide additionalcontext and contribute to the identification of vertebrate modification of bones. Additionally, proboscidean assemblages may include taphonomic modifications through the process of trampling. Based on observations of modern elephants, *Stegodon* are likely to have repositioned bones of deceased animals with their feet and trunks. The fracturing of complete elements and increased fragmentation through continued trampling on the bones is highly likely to have occurred. Fully or partially submerging of bones below the ground surface due to trampling may have caused variable exposure to weathering for both the whole carcass as well as the individual skeletal elements. This is an additional taphonomic processing possibility [45].

Non-biological taphonomic processes likewise have characteristic features that are observable on fossilised material from bonebeds. Significantly for the Upper Interval at Mata Menge, evidence to support deposition by a small sinuous stream tributary identified by the occurrence of in situ water-worn volcanic pebbles, as well as locally faint parallel laminations in the lower part of Layer II, was previously reported [10]. Bones deposited in fluvial systems experience characteristic modifications, including bone rounding and abrasion of the bone surface, causing loss of anatomical features and other taphonomic surface modifications [46–48]. Bones entering a fluvial system are also susceptible to transport potential based on the type and size of element or body part [47–51], the intact nature of the bone [52], and the fluvial velocity [53]. Analysis of the extent of weathering on the bone while on the surface was performed to evaluate the consistency of weathering from across the site as an indicator of the amount of exposure the fossils experienced prior to either partial or final burial [49,54–56]. All the potential pre-burial taphonomic processes at Mata Menge are thought to have ceased abruptly after 6.5 m thick mudflows sealed off the sandstone layer and the bones that were present on the surface at the time.

2. Materials and Methods

2.1. Materials

Material from this preliminary study was excavated between 2013 and 2018 by a team of Indonesian and Australian researchers and is stored at the Geological Museum Bandung. Two representative subsets were selected for detailed analysis. Subset 1 included all vertebrate fossils excavated from two adjacent one-metre quadrants on the eastern margin of a 3-metre-wide stream gully infill (Layer II, quadrants O5-O6; see Figure 3). At this sampling spot, Layer II laps on against an elevated crest of the erosional subsurface palaeosol (Layer III), and three of the five hominin fossils that were not recovered during wet sieving, were retrieved from this stream gully infill, including a premolar from quadrant O6. Therefore, this area was chosen for the Subset 1 sample because the taphonomic analysis might provide clues as to why the hominin fossils were concentrated in this gully. The sample also contains fossils of all the most abundant taxa at Mata Menge and is representative of the fauna present in Layer II. Subset 2 included all *Stegodon* long limb bones and limb bone fragments excavated from Layer II of the Upper Interval at Mata Menge (n = 17). Although axial skeletal elements, ribs, and bones of the manus and pes are present in the assemblage, only the long limb bones (femur, tibia, fibula, humerus, radius, ulna) were chosen because certain types of butchery marks and indicators of hominin food procurement can be preserved on these representative proboscidean skeletal elements [41,57–61]. Accessing the medullar cavity for the consumption of highly nutritious marrow has been claimed at other proboscidean kill sites, and such activities could have left diagnostic fresh bone impact marks and fracture patterns on the long limb bones [60–62].

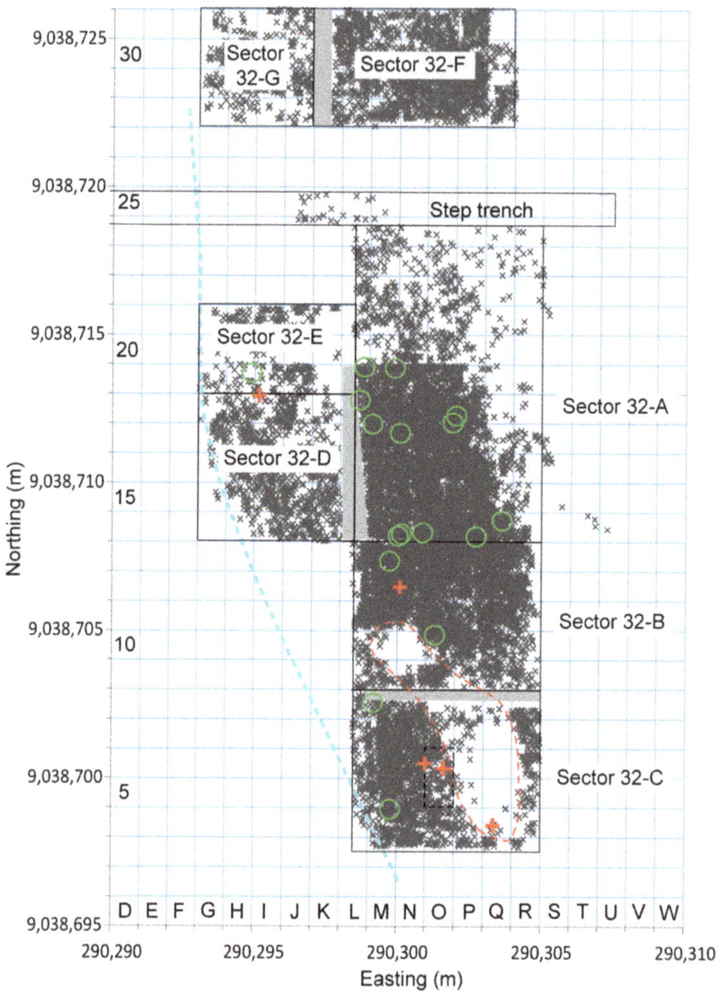

Figure 3. Plan view of the sectors of Excavation 32 in the Upper Fossil-bearing Interval at Mata Menge. Green circles indicate locations of Subset 2 long limb bones used in this study; red crosses indicate hominin fossil locations; grey crosses represent locations of all fossil specimens. The blue dotted line indicates the western termination of Layer II, where it pinches out against the elevated substratum of Layer III. This has been interpreted as the lateral bank of the fluvial stream channel. The black dotted rectangle indicates the location of the Subset 1 assemblage. Blue squares represent 1 m^2. Note that the northern half of Sector 32A and Sector 32B have been excavated to the base of Layer II, hence showing less dense spacing between finds. The area indicated with a red dashed line represents an elevated part of the eroded substratum of Layer III, where Layer II is thinner or absent, thus showing fewer fossils. Grey areas were not excavated.

2.2. Taphonomic Analysis

Identification of fossilised material was undertaken at the Geological Museum Bandung using *E. maximus* and *E. hysudrindicus* material for comparison. Fragments of bone with a cortical thickness greater than 4 mm were examined for characteristic crocodilian growth rings [63] and, if absent, the fragment was attributed to *Stegodon*. Enamel and tusk fragments were identified by their colour, texture, and shape, distinguishing them from fragmented bone. *Stegodon* enamel, usually thicker than 1 mm, has a translucent, double-layered, and wrinkled appearance [6] and is usually easy to distinguish from tooth enamel of other vertebrates, even tiny fragments. *Stegodon* tusk consists entirely of dentine, which is not covered with enamel but has a regular intersecting arcuate pattern in cross-section, referred to as Schreger Bands [64], which enabled macroscopic identification.

All fossils were examined macroscopically and, where required, with a handheld lens at 10× magnification. A caliper was used to take size measurements of all materials. The fossils of Subset 1 were measured along the longest length of the fossil to determine the Fragmentation Size Class [59]. The extent of fragmentation within a bonebed can be measured by comparing the number of Whole Bones and Part Bones with the remaining number of fragmented bones in the bonebed. Whole Bones are complete and unbroken (including complete teeth), whereas Part Bones represent fossils that constitute more than half of a recognisable bone element [58]. For example, a proximal portion of a long limb bone that was at least half of the original element was classed as a Part Bone. All material was assigned to one of ten size classes, based on the completeness of the bone element or its length (>10 mm, 10–20 mm, 20–30 mm, 30–40 mm, 40–50 mm, 50–60 mm, 60–70 mm, 70–80 mm, Part Bone, and Whole Bone).

Bone fractures were compared with descriptions defined by Outram [65] and were allocated a Fracture Freshness Index score of FFI 0–6. Although this analysis technique was not developed using proboscidean material, successful application has been observed on several large mammal species, including a preliminary review of mammoth material [66–71]. Criteria to define the FFI are Fracture Angle, Fracture Outline, and Fracture Texture, and were based on visual examination. For each of these three criteria, a score of 0, 1, or 2 was allocated for each specimen. A score of 0 was given if the fossil had only features indicative for fresh bone fracturing, a score of 1 if mixed fresh and unfresh fracture characteristics were present on a single specimen, and a score of 2 if only unfresh fractures could be determined. Consequently, a combined minimum FFI score of 0 would indicate all fractures occurred while the bone was fresh; alternatively, a maximum score of 6 indicates all fractures occurred while the bone was dry or fossilised.

Any fractures or surface modification attributed to sullegic or trephic processes were not included in FFI (or other palaeontological) analysis but were recorded separately. These modifications were clearly identifiable due to their characteristic fracturing patterns on fossilised material, usually following existing drying cracks and as a result of vibrations resulting from hammers and chisels used to break through the cemented sandstone in which the fossils were embedded. This process then exposed underlying material that was typically a different colour when compared with the outside of the fossil or old fractures.

All bones from both subsets were allocated a Stage of Weathering based on their physical characteristics described by Behrensmeyer [54], Stage 0–5:

Stage 0—No signs of cracking or flaking, bone may have tissue attached;
Stage 1—Cracking usually seen parallel to fibre structure, also mosaic cracking;
Stage 2—Outer bone show flaking, long thin to extensive flaking;
Stage 3—External bone patchy to mostly removed;
Stage 4—Coarsely fibrous rough bone surface, cracks well opened; or
Stage 5—Bone is fragile and easily broken, loss of original bone shape possible.

Other surface modifications, including biting and gnawing marks, dynamic impact marks, striation marks, corrosion, and abrasive rounding, were noted as absent or present. Residual sediment attached to Subset 1 fossils was macroscopically assessed for and allocated into very coarse to coarse, medium, or fine-grained to very-fine-grained sand categories according to the Wentworth Scale.

3. Results

3.1. Subset 1-Assemblage

The Subset 1 material attributed to *Stegodon* was 78 out of 91 (86%) fossils. The second-highest portion was Murinae (9%), followed by *Crocodilus* (4%) and hominin (1%). Approximately 54% of the highly fragmented *Stegodon* material was represented by unidentifiable bone fragments, where the provenance of the skeletal element was impossible to determine. The next highest portion of identifiable *Stegodon* fragments were very small fragments of tusk (21%) and tooth enamel (11%), but no complete or partially complete *Stegodon* tusks or teeth were identified from the subset (although scattered complete dental elements are present in Layer II). The remaining *Stegodon* fragments were small fragments of costa (5%), cranium (5%, consisting of cranial fragments characterised by thin-walled pneumatic bone tissue), and a single complete hyoid bone. Of the Murinae fossil material, 74% were either tooth or long bone fragments.

3.1.1. Fragmentation Analysis

Of the 91 fossils in Subset 1, extensive fragmentation was identified (94%). The 6% that was not fragmented included Part Bone classified teeth from *Crocodilus* and Murinae, as well as a single complete *Stegodon* hyoid bone. The majority of fragments were in the 10–20 mm class, followed by the 20–30 mm class and the <10 mm class, respectively. Combined, these classes represented 86% of all the fossils from Subset 1 and, additionally, 75% of fragments under 30 mm in length were attributed to the largest animal in the bonebed, *Stegodon*, based on the thickness of the cortical bone (Figure 4).

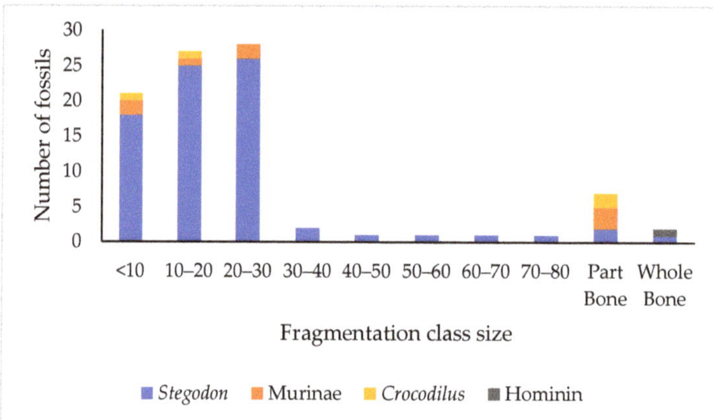

Figure 4. Results for analysis of the Fragmentation Classes of bone complete elements and bone fragments for Subset 1 (n = 91 NISP). Classes up to 80 mm are bone fragments. Part bone is more than half the original bone, and whole bone is entirely unbroken bone element. Portions of the column in various colours are representative of each identified species.

3.1.2. Weathering

None of the Subset 1 material was found to be in fresh or unweathered condition. The highest portion of fossils was Stage 2 weathering (56%), followed by Stage 1 (21%) and Stage 3 (16%); the remaining (3%) were each Stage 4 and Stage 5 (Figure 5). A significant difference was found ((χ^2 (4,91) = 13.135, p = 0.01, n = 91) when attributing fragmented material to a skeletal element for less weathered fragments, indicating that an increase in weathering on fossils reduces the likelihood of it being identified. No fragment from Subset 1 was identified to a skeletal element in Stage 4 or Stage 5 weathering.

Figure 5. Stages of Weathering from Subset 1 assemblage. (**A**) Fossil ID MM14-F392, Stage 1 weathering, the outer bone surface shows some small mosaic cracks (fracture edges a result of excavation damage). (**B**) Fossil ID MM14-F1389, Stage 2 weathering, moderate outer bone flaking. (**C**) Fossil ID MM16-4515, Stage 3 weathering, cortical bone patchy and mostly removed (**D**) Fossil ID MM14-F106, Stage 4 weathering, coarse fibrous bone exposed due to external cracks. (**E**) Fossil ID MM16-F4505, Stage 5 weathering, loss of original bone shape, and easily broken.

3.1.3. Bone Surface Modifications

No evidence of any biting or gnawing marks on the material was observed. No dynamic impact marks were observed on bone material that was likely to have been anthropogenic. The three incidents of crush style fractures had characteristics consistent with dry bone fracturing, and therefore are likely to have been caused by trampling. There were three *Stegodon* bone fragments with one or more linear surface marks. One costa fragment in weathering Stage 1 had multiple curved surface scraping style marks as well as two perpendicular parallel marks that contained in situ matrix within them (Figure 6). No evidence of any digestive corrosion was seen on any material. Moderate to heavy black staining penetrating the fossil bone surface up to 5 mm was seen on 55% of all material. These findings suggest that the bones have been exposed to dissolved manganese minerals in the water in which they were (intermittently) submerged [17]. There were incidents of dissolution circular pitting seen on the hominin premolar and some fossil surfaces (Figure 7). Some fragments presented with completely rounded edges, likely the result of abrasion during fluvial transportation, but further microscopic analysis would be required to confirm this (Figure 8). Macroscopic grain size assessment of the residual matrix on the Subset 1 fossils showed that 84% of the fossils had a matrix consisting of fine-grained sand, indicating that these fossils originated from the top of Layer II. A total of 14% was poorly sorted coarse-grained sand, and the remaining 2% was medium-grained sand.

Figure 6. Fossil ID MM14-394 *Stegodon* costa bone broken in two by excavation fracture with perpendicular parallel narrow striations (blue arrow), curved scrape type marks (orange arrows), some with in situ matrix are present.

(A) (B) (C)

Figure 7. Examples of dissolution circular pitting from Subset 1 assemblage. (**A**) Fossil ID MM14-F149 small fossilised *Stegodon* tusk fragment showing irregular dissolution pattern. (**B**) Fossil ID SOA-MM5, Hominin premolar with dissolution circular pitting (orange arrow) (Photo source: Kaifu 2016). (**C**) Fossil ID MM14-F1316, *Stegodon* bone fragment with multiple circular dissolution pits (orange arrows).

(A) (B) (C)

Figure 8. Examples of rounded fragments from Subset 1 assemblage likely as a result of abrasion in fluvial transportation. (**A**) Fossil ID MM14-F348, *Stegodon* bone fragment. (**B**) Fossil ID MM14-F487, *Stegodon* bone fragment, (**C**) Fossil ID MM14-F1413, *Stegodon* bone fragment.

3.2. Subset 2-Long Limb Bones

3.2.1. Identification

From the Subset 2 *Stegodon* long limb bone analysis (n = 17), 3 bones were complete elements, 10 were classified as Part Bones, consisting of at least half of the original element, and the remaining 4 were bone fragments. No elements from this subset were located in articulation. The most abundant element was the humerus, with three left and three right-side identified. There was a single complete humerus, and the remaining five were Part Bones. One left and one right complete femur was found; however, these were from different individuals as there was more than a 10 cm difference in length. Unpublished data from this site based on the identification of dental elements suggest an MNI of at least 35 individuals. The loss of postcranial material through transportation and fragmentation, following from the analysis of Subset 1 from this site, is therefore likely.

3.2.2. Weathering

From the weathering analysis of Subset 2, no fossil was found completely unweathered (Stage 0 weathering). The highest proportion was Stage 2 (44%), followed by Stage 3 (28%), both Stage 1 and Stage 4 (11%), and the remaining was Stage 5 (6%). In order to refine the analysis of weathering on the bone only material, enamel and tusk fragments from the Subset 1 data set were removed, and statistical analysis was performed. There was no statistical difference identified between Subset 1 and Subset 2 for weathering ($\chi^2_{(4,74)}$ = 4.832, p = 0.30, n = 74). There was no statistical difference identified for *Stegodon* only bone material between Subset 1 and Subset 2 for weathering (χ^2(4,68) = 5.053, p = 0.28, n = 68). Two incidences of differential weathering were observed on the proximal and distal ends of two left ulnae from Subset 2. Both bones were Stage 2 proximally and Stage 4 or Stage 5 distally (Figure 9).

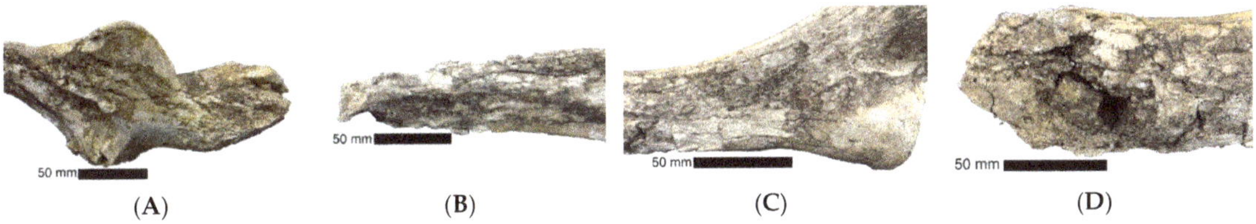

Figure 9. Differential weathering on two left ulna bones. Both proximal ends are less weathered when compared with the more pronounced weathering on the broken distal ends. Fossil ID MM13-F072 (**A**) Proximal. (**B**) Distal. Fossil ID MM13-F721 (**C**) Proximal. (**D**) Distal.

3.2.3. Fracturing

For both Subset 1 and Subset 2 fossils, the Fracture Freshness Index (FFI) score of 6 accounted for approximately 75% of the material. This indicates that all observed fractures occurred while the bone was either dry or fossilised, likely a result of non-anthropogenic modification. Of the remaining Subset 1 material, approximately 10% scored 3 or 4 on the FFI with both fresh and non-fresh fracture characteristics (Figure 10). No uniformity was detected in the fractures or their location on the Subset 1 fossils. For the FFI criterion Fracture Outline, a helically shaped fracture was observed on each of five humeri from Subset 2 (Figure 11). The humeri presented with the proximal portion of the bone missing and a helically shaped fracture differentially located along the shaft. In all cases, the anterior-medial aspect of the bone is the highest point of the fracture, and the lateral aspect is the lowest. Fractures were observed both proximal and distal to the deltoid tuberosity (Figure 11). However, the FFI characteristic evidence alone observed in either subset is not sufficient to confirm that the fractures occurred while the bone was fresh.

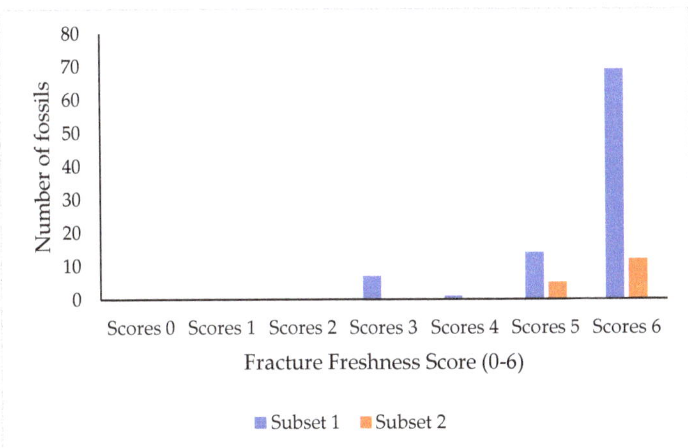

Figure 10. Numbers of fossils (NISP) in Subset 1 (n = 91) and Subset 2 (n = 17, long limb bones) represented by Fracture Freshness Index (FFI) Scores 0–6. The highest portion of material Score 6 indicating all fractures characteristic for unfresh bone fracture. Approximately 10% of material scoring 3 or 4 on the FFI.

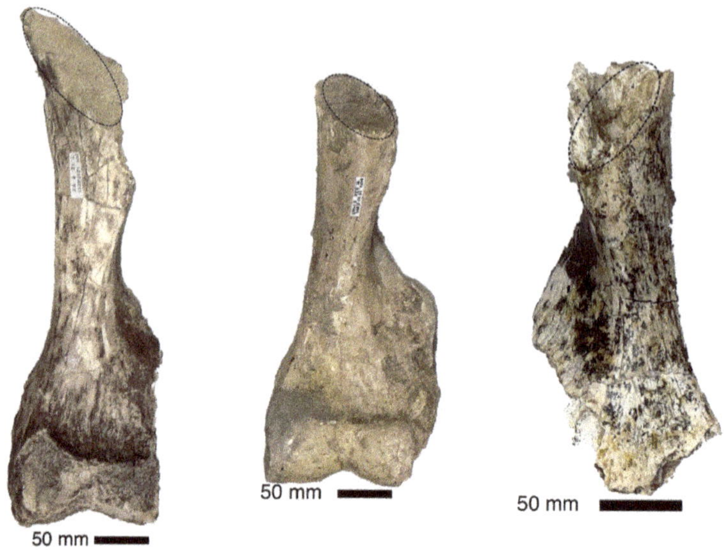

Figure 11. Examples of helical-shaped fractures on three humerus bones from Subset 2. The black outline represents the shape of the fracture. The highest point of fracture is medially, and the lowest point is lateral. Fracture presents both above and below deltoid tuberosity on different fossils. (**A**) Cranial view *Stegodon* Left humerus (MM13-179) (**B**) Cranial view *Stegodon* left humerus (MM15-1592). (**C**) Cranial view *Stegodon* right humerus (MM16-7610).

4. Discussion

The analysis of the Subset 1 fossils from Mata Menge Upper Interval (UFI) has revealed a significant bias towards bone fragments of the largest taxon present. This is not surprising, as the comparatively robust nature of the *Stegodon* bones has contributed to their preferred preservation, representing 86% of the assemblage, despite the fact that many of these *Stegodon* fossils represent small broken fragments of unidentifiable elements. Smaller vertebrates are mostly absent from the fossil record here, likely because of their susceptibility to rapid degradation caused by weathering after Stage 2. This commonly observed feature was recognised by Behrensmeyer (1978), who noted that animals under 100 kg, and especially those under 15 kg, after undergoing surface weathering are usually under-represented in fossil assemblages. During Mid Pleistocene in the So'a Basin, many birds, reptiles, and murids would be classified in the small vertebrate class size, therefore likely undergoing fast decay and rapid full bone degradation during surface exposure.

More than half of the *Stegodon* fragments are fragments of cortical bone, similar to observations made at other proboscidean assemblages where large numbers of fragmented cortical bone were located [40,72]. No other species in the So'a Basin during the Mid Pleistocene were large enough to have cortical bone of this size [6,73]. The highly fragmented nature of the *Stegodon* material in the bonebed contributed to difficulties in attributing bone fragments to a skeletal element. There was no increase in the successful identification of fragments to a skeletal element as their size increased. No fragment could be identified when showing Stage 4 or Stage 5 weathering. Fragments without any recognisable skeletal bone characteristics have undergone considerable surface modifications such as those seen in weathered material; however, abrasion and rounding caused during fluvial transportation certainly contributed considerably to modifications observed on these fossils.

Fractures on fresh bones have a close association with the biological dispersal of a carcass [74,75]. Biological agents in the Soa Basin are represented by either a reptilian carnivore, trampling, or other modification of dispersal by *Stegodon* or hominin. During Mid Pleistocene in the So'a Basin, the only faunal predators large enough to disarticulate and consume *Stegodon* included crocodiles and Komodo dragons [10]. However, these predators rarely fracture bones during feeding, preferring to either consume large disarticulated portions or complete smaller animals [31,33]. No evidence was found for any puncture style marks attributable to crocodile teeth. Ziphodont teeth of Komodo dragons are suitable for cutting flesh and not for puncturing bone. Based on the highly fragmented material in Subset 1 when compared to the selected complete element material from Subset 2, it is clear that the two subsets were exposed to different taphonomic processes by agents prior to entering the fluvial channel.

Five *Stegodon* humeri were observed with helical-shaped fractures located at various positions along the diaphysis of the bone. Observed fractures are located both proximally and distally to the deltoid tuberosity resulting in the differing presentations of these elements. While the helical nature of this fracture may not present sufficient evidence for the identification of a fracture occurring on fresh bone, the edges of these fractures were exposed to further taphonomic processes, including weathering and natural abrasion. These secondary processes have likely altered the fracture edge and reducing diagnostic interpretability. It is likely these fractures were the result of trampling, similarly observed on trampling fractures of modern elephants [45].

During recent actualistic experimentation and fieldwork observations in Africa with modern elephant bones (*Loxodonta africana*), the effects of trampling resulting in bone fractures were described [45]. Fractures due to trampling were observed in association with other bone surface marks, including scratches, grazes, and other marks distinguishable from cut marks. The fine-grained sand observed at the experiment location in Africa caused shallow, narrow, flat bottom marks that were multidirectional [45]. Acquisition of these marks was attributed to elephant trampling. While the characteristic trampling marks were not macroscopically observed on the fractured humeri from Mata Menge or on any of the long limb bones, conceivably, weathering and ablation may have obliterated scratches and grazes on the bone surface. The relatively well-preserved costa fragment from Subset 1 identified with multiple scrape marks were consistent with the description from experimental conditions and, therefore, likely the result of trampling; however, further microscopic analysis may provide additional context for this material.

Additionally, elephants frequently partially submerge bones in the sand surface, resulting in vertically reorientating the bones [45]. Some evidence supporting likely trampling-induced reorientation of bones was identified at Mata Menge with differential weathering observed on the proximal and distal ends of two *Stegodon* left ulna bones. Both ulnas had well-preserved Stage 2 weathering on the proximal end, while the distal ends of both bones were Stage 5 weathering. Intuitively, it is likely that the heavier and larger proximal end of the ulna bone could remain submerged after being pushed into the surface by force exerted by *Stegodon* feet, and therefore the proximal end remained protected from the exposure taphonomy of weathering.

Significant to the context for this bonebed is a further interpretation of the previously identified fluvial channel in which the fossils were preserved [10]. The sedimentology, identification of freshwater gastropods and water-worn volcanic pebbles led previous researchers to conclude that Layer II was deposited in a small stream [10]. The taphonomic analysis of the fossil assemblage is in accordance with this interpretation. The fragmented material in Subset 1 has undergone significant abrasion taphonomy and resulted in modified surface features caused by the significant removal of bone material in flowing water. The long limb bones underwent somewhat less abrasion in the fluvial channel, by comparison, resulting in some bone surface reduction, particularly on bone processus. Further interpretation of the two presentations of bone wear observed in this bonebed requires analysis of the mechanisms of the abrasion seen here. The mechanisms rely on the increasing hardness of the material and increasing velocity; however, whether the bone material was fresh or dry will influence the resultant abrasion [76]. Abrasion on bone material that is more ductile, such as fresh bone, is a result of sharp, angular cutting from the sediment and results in ablation-induced surface bone removal. Abrasion on dry bone is the result of deformation around existing cracks and impact points on the less elastic bone; subsequently, intersecting cracks lead to fragmentation and detachment of material [76]. These described conditions are exacerbated by the length of exposure time in the fluvial environment. The Mata Menge UFI bonebed is made up of material with multiple taphonomic histories, including: condition of bone entering the fluvial channel; dry or fresh; and interval of exposure to the fluvial conditions and post mortem breakage, much of which should be attributed to trampling.

Susceptibility to transportation in a fluvial system was also observed here. Maximum cortical bone thickness is considered to positively influence the transportability of fragments in a fluvial environment [52]. The over-representation of *Stegodon* material, as opposed to fossils of smaller taxa, would therefore be in accordance with a fluvial setting. It is probable that smaller bone material, including cancellous bone fragments and material from smaller vertebrates, was transported away from this site, further contributing to the bias identified here. Dry bones have a high transportation potential [50] as they are inclined to float when compared to fresh bones, whose high moisture content makes them more likely to become submerged. The high density of comparable bone fragments in both size and type in the assemblage reveals the likelihood of fragmented bone entering the fluvial channel following significant fracturing processes. The fine-grained sand adhering to the majority of the fossil fragments indicates that they originate from the upper part of Layer II. Fine-grained sand related to slower fluvial transportation [46] is associated with more bone edge rounding on fresh bone when compared with either dry or fossilised material in experimental research [77]. Based on the accumulation and deposition of the comparable sized and processed fragments, it is likely that following a significant fracturing event on fresh bone, considerable rounding by abrasion occurred in fluvial transportation by fine-grained sands. The fragments were also exposed to some surface exposure taphonomy resulting in typical Stage 2 bone weathering. The combination of both surface weathering and fluvial transportation has resulted in the loss of much of the original surface of the bone fragments.

The mostly Whole Bone and Part Bone *Stegodon* material of Subset 2 was likely transported over relatively short distances compared to the many small bone fragments from Subset 1, based on the susceptibility of bones to fluvial transport described by Voorhies [48]. Voorhies foundation experiments used disarticulated coyote and sheep skeletons to measure characteristic susceptibility to transport; however, further experimentation by Frison and Todd (1986) [47] using bones of a modern elephant augmented proboscidean fluvial transportation susceptibility. The resultant Fluvial Transport Index (FTI) grouped elephant bones based on their transport potential: a high FTI corresponds with high transport potential and a low FTI with low transport potential with a range of 1–100. Frison and Todd (1986) observed similar patterns to those of Voorhies and aligned an FTI of greater than 75 with the Voorhies Group 1 (sacrum, patella, astragalus, calcaneus, all vertebrae excluding atlas), FTI 50-74 with Voorhies Group 2 (ribs, scapula, humerus, tibia, metacarpals), and FTI less than 50 with Voorhies Group 3 (cranium, mandible, atlas, pelvis, femur, radius, ulna) [47].

Subset 2 long limb bones, selected as representing a potential indicator of hominin food procurement, corresponding with Voorhies Group 2/3 and FTI ranging between 50–100, have potential transport characteristics of bones gradually or intermittently moving in periods of higher velocity river discharge while staying in contact with the bottom. The long limb bones were less rounded when compared with the Subset 1 material; however, some loss of surface features was observed, such as rounding on protruding muscle attachments, tuberosities, and epiphyseal ends. Rounding was likely caused during water flow, caused by the bone abrading with particles on the bottom of the channel, and as particles moved above the bone as it remained settled on the base of the channel.

5. Conclusions

The present taphonomic analysis of the Upper Fossil-bearing Interval at Mata Menge supports previous interpretations that the site represents a stream channel. The analysis further suggests that fluvial transport has affected the assemblage, causing significant obliteration of bone surface characteristics and rounding during fluvial transport. This is particularly noted in the small fragments of bone for which the original skeletal element could not be identified due to their small size but that were attributed to *Stegodon* based on cortical bone thickness. While the possibility of identifying additional hominin material in this assemblage is high, it is very unlikely that any remains would be found in articulation. The high density of small fragments of bone that have undergone significant modification by weathering, fragmentation, and fluvial abrasion do align with the characteristics of the existing hominin fossils found at the UFI at Mata Menge, and additional material would likely present these same characteristics. The taphonomic processes identified from this study would likely have a significant effect on the remains of the small-bodied hominin, *Homo floresiensis*, identified from this location. The resultant disarticulation, fragmentation, transportation, and bone surface modifications certainly remove the likelihood of locating a fully or partially articulated hominin skeleton.

Overall the two examined subsets reveal an observed consistency in the extent of weathering for the examined material, with most material exhibiting Stage 2 weathering. Some evidence for *Stegodon* trampling bone modifications were identified with differential weathering observed on the proximal and distal ends of two ulna bones. It appears that the material in each subset is the result of different taphonomic processing after the death of the animal, with Subset 1 material exposed to further fragmentation processing, while Subset 2 remained mostly intact. Approximately 10% of all material scored 3 or 4 on the FFI, characteristic of fracturing on fresh and unfresh bone; however, extensive rounding, weathering, and further dry bone fracturing has possibly obscured any initial fresh bone fracturing. Subset 1 material was consistently highly fragmented, with more than half being unidentifiable bone fragments <30 mm in length with rounded fractured edges likely abraded in significant fluvial transportation. This preliminary taphonomic analysis at Mata Menge has identified differential presentation of material here, indicating multiple taphonomic processing events contributing to the formation of this bonebed.

Author Contributions: Conceptualisation, M.J.P. and G.D.v.d.B.; methodology, M.J.P.; validation, M.J.P., G.D.v.d.B., and K.M.M.; formal analysis, M.J.P.; investigation, M.J.P.; resources, I.S. and U.P.W.; data curation, M.J.P.; writing—original draft preparation, M.J.P.; writing—review and editing, M.J.P. and G.D.v.d.B.; visualisation, M.J.P.; supervision, G.D.v.d.B.; project administration, G.D.v.d.B.; funding acquisition, G.D.v.d.B. All authors have read and agreed to the published version of the manuscript.

Funding: This research was funded by an Australian Research Council grant to Gert van den Bergh, grant number DP190100164. Tickets to Bandung were financed by a student support grant from the Faculty of Science, Medicine and Health (SMAH) of the University of Wollongong.

Institutional Review Board Statement: Not applicable.

Informed Consent Statement: Not applicable.

Data Availability Statement: The data presented in this study are available on request from the corresponding author. The data are not publicly available due to the university policy on delaying access to thesis and related data until post-publication.

Acknowledgments: The authors would like to sincerely thank the three anonymous reviewers for their careful comments, which greatly improved the manuscript. We would also like to thank the academic editors for their thoughtful comments during this process too. Additional logistic and in-kind support from the Geological Survey Institute (GSI) and the Museum Geologi Bandung, Indonesia. Thank you to Iwan Kurniawan, Director of the Geology Museum Bandung, for access to the fossil collections. Scientific and technical personnel involved during the fieldwork campaigns, including: Ruly Setiawan, Dida Yurnaldi, Erick Setiyabudi, Ifan Yoga Pertama, Susan Hayes, supported in the field by 46 local people from the villages Mengeruda, Piga-I and Piga-II, and Fransiskus Bani and family for hospitality in Mengeruda.

Conflicts of Interest: The authors declare no conflict of interest.

References

1. Hooijer, D.A. A stegodon from flores. *Treubia* **1957**, *24*, 119–128.
2. Verhoeven, T. Pleistozäne Funde in Flores. *Anthropos* **1958**, *53*, 264–265.
3. Aziz, F.; Morwood, M.J. Introduction: geology, palaeontology and archaeology of the Soa Basin, Central Flores, Indonesia. In *Pleistocene Geology, Palaeontology and Archaeology of the Soa Basin, Central Flores, Indonesia*; Aziz, F., Morwood, M.J., van den Bergh, G.D., Eds.; Pusat Survei Geologi, Badan Geologi, Departemen Energi dan Sumber Daya Mineral: Bandung, Indonesia, 2009; pp. 1–18.
4. van den Bergh, G.D.; Kurniawan, I.; Morwood, M.J.; Lentfer, C.J.; Suyono; Setiawan, R.; Aziz, F. Environmental reconstruction of the Middle Pleistocene archaeological/palaeontological site Mata Menge, Flores, Indonesia. In *Pleistocene Geology, Palaeontology and Archaeology of the Soa Basin, Central Flores, Indonesia*; Aziz, F., Morwood, M.J., van den Bergh, G.D., Eds.; Pusat Survei Geologi, Badan Geologi, Departemen Energi dan Sumber Daya Mineral: Bandung, Indonesia, 2009; pp. 59–94.
5. van den Bergh, G.D.; Mubroto, B.; Aziz, F.; Sondaar, P.Y.; Vos, J. Did *Homo erectus* reach the Island of Flores? *Bull. Indo-Pac. Prehist. Assoc.* **1996**, *14*, 27–36. [CrossRef]
6. van den Bergh, G.D. The Late Neogene Elephantoid-Bearing Faunas of Indonesia and Their Palaeozoogeographic Implications: A Study of the Terrestrial Faunal Succession of Sulawesi, Flores and Java, Including Evidence for Early Hominid Dispersal East of Wallace's Line. Ph.D. Thesis, University of Utrecht, Utrecht, The Netherlands, 1999.
7. Sondaar, P.Y.; van den Bergh, G.D.; Mubroto, B.; Aziz, F.; de Vos, J.; Batu, U.L. Middle Pleistocene faunal turnover and colonization of Flores (Indonesia) by *Homo erectus*. *Comptes Rendus de l'Academie des Sciences de Paris* **1994**, *319*, 1255–1262.
8. Brumm, A.; Aziz, F.; van den Bergh, G.D.; Morwood, M.J.; Moore, M.W.; Kurniawan, I.; Hobbs, D.R.; Fullagar, R. Early stone technology on Flores and its implications for Homo floresiensis. *Nature* **2006**, *441*, 624–628. [CrossRef]
9. Brumm, A.; Jensen, G.M.; van den Bergh, G.D.; Morwood, M.J.; Kurniawan, I.; Aziz, F.; Storey, M. Hominins on Flores, Indonesia, by one million years ago. *Nature* **2010**, *464*, 748–752. [CrossRef]
10. Brumm, A.; van den Bergh, G.D.; Storey, M.; Kurniawan, I.; Alloway, B.V.; Setiawan, R.; Setiyabudi, E.; Gruen, R.; Moore, M.W.; Yurnaldi, D.; et al. Age and context of the oldest known hominin fossils from Flores. *Nature* **2016**, *534*, 249. [CrossRef]
11. Morwood, M.J.; Aziz, F.; van den Bergh, G.D.; Sondaar, P.Y.; De Vos, J. Stone Artefacts from the 1994 Excavation at Mata Menge, West Central Flores, Indonesia. *Aust. Archaeol.* **1997**, *44*, 26–34. [CrossRef]
12. Morwood, M.J.; O'Sullivan, P.B.; Aziz, F.; Raza, A. Fission-track ages of stone tools and fossils on the east Indonesian Island of Flores. *Nature* **1998**, *392*, 173–176. [CrossRef]
13. O'Sullivan, P.B.; Morwood, M.; Hobbs, D.; Aziz Suminto, F.; Situmorang, M.; Raza, A.; Maas, R. Archaeological implications of the geology and chronology of the Soa basin, Flores, Indonesia. *Geology* **2001**, *29*, 607. [CrossRef]
14. van den Bergh, G.D.; Kaifu, Y.; Kurniawan, I.; Kono, R.T.; Brumm, A.; Setiyabudi, E.; Aziz, F.; Morwood, M.J. Homo floresiensis-like fossils from the early middle Pleistocene of Flores. *Nature* **2016**, *534*, 245–248. [CrossRef]
15. Yurnaldi, D.; Setiawan, R.; Patriani, E. The Magnetostratigraphy and the Age of So'a Basin Fossil-Bearing Sequence, Flores, Indonesia. *Indones. J. Geosci.* **2018**, *5*. [CrossRef]
16. Sutikna, T.; Tocheri, M.W.; Morwood, M.J.; Saptomo, E.W.; Jatmiko; Awe, R.D.; Wasisto, S.; Westaway, K.E.; Aubert, M.; Li, B.; et al. Revised stratigraphy and chronology for Homo floresiensis at Liang Bua in Indonesia. *Nature* **2016**, *532*, 366–369. [CrossRef]
17. Meijer, H.J.M.; D'Errico, F.; Queffelec, A.; Kurniawan, I.; Setiabudi, E.; Sutisna, I.; Brumm, A.; van den Bergh, G.D. Characterization of bone surface modifications on an early to middle Pleistocene bird assemblage from Mata Menge (Flores, Indonesia) using multifocus and confocal microscopy. *Palaeogeogr. Palaeoclimatol. Palaeoecol.* **2019**, *529*, 1–11. [CrossRef]
18. Meijer, H.J.M.; Kurniawan, I.; Setiabudi, E.; Brumm, A.; Sutikna, T.; Setiawan, R.; van den Bergh, G.D. Avian remains from the early/middle Pleistocene of the So'a Basin, central Flores, Indonesia, and their palaeoenvironmental significance. *Palaeogeogr. Palaeoclimatol. Palaeoecol.* **2015**, *440*, 161–171. [CrossRef]
19. Ali, J.R.; Heaney, L.R. Wallace's line, Wallacea, and associated divides and areas: history of a tortuous tangle of ideas and labels. *Biol. Rev.* **2021**. [CrossRef]
20. Suminto; Morwood, M.J.; Kurniawan, I.; Aziz, F.; van den Bergh, G.D.; Hobbs, D.R. Geology and fossil sites of the Soa Basin, Flores, Indonesia. In *Pleistocene Geology, Palaeontology and Archaeology of the Soa Basin, Central Flores, Indonesia*; Aziz, F., Morwood, M.J., van de Bergh, G.D., Eds.; Pusat Survei Geologi, Badan Geologi, Departemen Energi dan Sumber Daya Mineral: Bandung, Indonesia, 2009; pp. 19–40.
21. Aziz, F.; van den Bergh, G.D.; Morwood, M.J.; Hobbs, D.R.; Kurniawan, I.; Collins, J.; Jatmiko. Excavations at Tangi Talo, central Flores, Indonesia. In *Pleistocene Geology, Palaeontology and Archaeology of the Soa Basin, Central Flores, Indonesia*; Aziz, F., Morwood, M.J., van den Bergh, G.D., Eds.; Pusat Survei Geologi, Badan Geologi, Departemen Energi dan Sumber Daya Mineral: Bandung, Indonesia, 2009; pp. 41–58.
22. Sondaar, P.Y. Faunal evolution and the mammalial biostratigraphy of Java. *Cour. Forsch. Inst. Senckenberg* **1984**, *69*, 219–235.
23. Morwood, M.J.; Aziz, F.; O'Sullivan, P.; Nasruddin; Hobbs, D.R.; Raza, A. Archaeological and palaeontological research in central Flores, east Indonesia: Results of fieldwork 1997-98. *Antiquity* **1999**, *73*, 273–286. [CrossRef]

24. van den Bergh, G.D.; de Vos, J.; Sondaar, P.Y. The late Quaternary palaeogeography of mammal evolution in the Indonesian Archipelago. *Palaeogeogr. Palaeoclimatol. Palaeoecol.* **2001**, *171*, 385–408. [CrossRef]
25. van den Bergh, G.D.; Awe, R.D.; Morwood, M.J.; Sutikna, T.; Jatmiko; Wahyu Saptomo, E. The youngest stegodon remains in Southeast Asia from the Late Pleistocene archaeological site Liang Bua, Flores, Indonesia. *Quat. Int.* **2008**, *182*, 16–48. [CrossRef]
26. van den Bergh, G.D. *Insular Fossil Fauna through Deep Geological Time of Wallacea*; Geological Survey Center Indonesia, University of Wollongong Australia: Wollongong, NSW, Australia, 2019.
27. Faith, J.T.; Behrensmeyer, A.K. Changing patterns of carnivore modification in a landscape bone assemblage, Amboseli Park, Kenya. *J. Archaeol. Sci.* **2006**, *33*, 1718–1733. [CrossRef]
28. Haynes, G. Longitudinal studies of African elephant death and bone deposits. *J. Archaeol. Sci.* **1988**, *15*, 131–157. [CrossRef]
29. Haynes, G. Mammoth landscapes: Good country for hunter-gatherers. *Quat. Int.* **2006**, *142*, 20–29. [CrossRef]
30. Haynes, G.; Krasinski, K. Taphonomic Fieldwork in Southern Africa and its Application in Studies of the Earliest Peopling of North America. *J. Taphon.* **2010**, *8*, 181–202.
31. D'Amore, D.C.; Blumenschine, R.J. Komodo monitor (*Varanus komodoensis*) feeding behaviour and dental function reflected through tooth marks on bone surfaces and the application to ziphodont paleobiology. *Paleobiology* **2009**, *35*, 525–552. [CrossRef]
32. Njau, J.K.; Blumenschine, R.J. A diagnosis of crocodile feeding traces on larger mammal bone, with fossil examples from the Plio-Pleistocene Olduvai Basin, Tanzania. *J. Hum. Evol.* **2006**, *50*, 142–162. [CrossRef]
33. Sahle, Y.; Zaatari Sireen, E.; White Tim, D. Hominid butchers and biting crocodiles in the African Plio–Pleistocene. *Proc. Natl. Acad. Sci. USA* **2017**, *114*, 13164. [CrossRef]
34. Domínguez-Rodrigo, M.; Baquedano, E. Distinguishing butchery cut marks from crocodile bite marks through machine learning methods. *Sci. Rep.* **2018**, *8*. [CrossRef]
35. van den Bergh, G.D.; Meijer, H.J.M.; Due Awe, R.; Morwood, M.J.; Szabo, K.A.; van den Hoek Ostenda, L.W.; Sutikna, T.; Saptomo, E.W.; Piper, P.J.; Dobney, K.M. The Liang Bua faunal remains: a 95 k.yr. sequence from Flores, East Indonesia. *J. Hum. Evol.* **2009**, *57*, 527–537. [CrossRef]
36. Badgley, C. Counting individuals in mammalian fossil assemblages from fluvial environments. *PALAIOS* **1986**, *1*, 328–338. [CrossRef]
37. Brumm, A.; Moore, M.W.; van den Bergh, G.D.; Kurniawan, I.; Morwood, M.J.; Aziz, F. Stone technology at the Middle Pleistocene site of Mata Menge, Flores, Indonesia. *J. Archaeol. Sci.* **2010**, *37*, 451–473. [CrossRef]
38. Dennell, R.W.; Coard, R.; Turner, A. Predators and scavengers in Early Pleistocene southern Asia. *Quat. Int.* **2008**, *192*, 78–88. [CrossRef]
39. Lyman, R.L. Archaeofaunas and Butchery Studies: A Taphonomic Perspective. *Adv. Archaeol. Method Theory* **1987**, *10*, 249–337.
40. Arroyo-Cabrales, J.; Johnson, E.; Morett, L. Mammoth bone technology in the Basin of Mexico. In *Bones for Tools—Tools for Bones: the Interplay Between Objects and Objectives*; Seetah, K., Gravina, B., Eds.; McDonald Institute for Archaeological Research: Cambridge, UK, 2012; pp. 97–113.
41. Morlan, R.E. Toward the definition of criteria for the recognition of artificial bone alterations. *Quat. Res.* **1984**, *22*, 160–171. [CrossRef]
42. Villa, P.; Mahieu, E. Breakage patterns of human long bones. *J. Hum. Evol.* **1991**, *21*, 27–48. [CrossRef]
43. Fernandez-Jalvo, Y.; Andrews, P.; Denys, C.; Sese, C.; Stoetzel, E.; Marin-Monfort, D.; Pesquero, D. Taphonomy for taxonomists; implications of predation in small mammal studies. *Quat. Sci. Rev.* **2016**, *139*, 138–157. [CrossRef]
44. Prasanth, T.; Saraswathi, T.R. Histopathological and radiographic evaluation of rat molar teeth after traumatic injury—A pilot study. *J. Oral Maxillofac. Pathol.* **2012**, *16*, 313–317. [CrossRef]
45. Haynes, G.; Krasinski, K.; Wojtal, P. Elephant bone breakage and surface marks made by trampling elephants: Implications for interpretations of marked and broken *Mammuthus* spp. bones. *J. Archaeol. Sci. Rep.* **2020**, *33*, 102491. [CrossRef]
46. Behrensmeyer, A.K. Vertebrate preservation in fluvial channels. *Palaeogr. Palaeoclimatol. Palaeoecol.* **1988**, *63*, 188–199. [CrossRef]
47. Frison, G.C.; Todd, L.C. *The Colby Mammoth Site: Taphonomy and Archeology of a Clovis Kill in Northern Wyoming*; University of New Mexica Press: Albuquerque, NM, USA, 1986.
48. Voorhies, M. *Taphonomy and Population Dynamics of an Early Pliocene Vertebrate Fauna, Knox County, Nebraska*; University of Wyoming: Laramie, WY, USA, 1969.
49. Behrensmeyer, A.K. The Taphonomy and Paleoecology of Plio-Pleistocene Vertebrate Assemblages East of Lake Rudolf, Kenya. *Bull. Museum Comp. Zool.* **1975**, *146*, 473–578.
50. Coard, R. One bone, two bones, wet bones, dry bones: Transport potential under experimental conditions. *J. Archaeol. Sci.* **1999**, *26*, 1369–1375. [CrossRef]
51. Coard, R.; Dennel, R.W. Taphonomy of some articulated skeletal remains: Transport potential in an artificial environment. *J. Archaeol. Sci.* **1995**, *22*, 441–448. [CrossRef]
52. Pante, M.C.; Blumenschine, R.J. Fluvial transport of bovid long bones fragmented by the feeding activities of hominin and carnivore. *J. Archaeol. Sci.* **2010**, *37*, 846–854. [CrossRef]
53. Hjulström, F. Studies of the morphological activity of rivers as illustrated by the River Fyris. *Bull. Geol. Inst. Univ. Uppsala* **1935**, *25*, 221–527.
54. Behrensmeyer, A.K. Taphonomic and Ecological Information from Bone Weathering. *Paleobiology* **1978**, *4*, 150–162. [CrossRef]

55. Behrensmeyer, A.K.; Miller, J.H. Building links between ecology and paleontology using taphonomic studies of recent vertebrate communities. In *Paleontology in Ecology and Conservation*; Louys, J., Ed.; Spring: Berlin/Heidelberg, Germany, 2012; pp. 69–91.
56. Lyman, R.L.; Fox, G.L. A critical evaluation of bone weathering as an indication of bone assemblage formation. *J. Archaeol. Sci.* **1989**, *16*, 293–317. [CrossRef]
57. Boschian, G.; Caramella, D.; Saccà, D.; Barkai, R. Are there marrow cavities in Pleistocene elephant limb bones, and was marrow available to early humans? New CT scan results from the site of Castel di Guido (Italy). *Quat. Sci. Rev.* **2019**, *215*, 86–97. [CrossRef]
58. Outram, A.K. The Identification and Palaeoeconomic Context of Prehistoric Bone Marrow and Grease Exploitation. Ph.D. Thesis, Thesis, University of Durham, Durham, UK, 1998.
59. Outram, A.K. A new approach to identifying bone marrow and grease exploitation: Why the "Indeterminate" fragments should not be ignored. *J. Archaeol. Sci.* **2001**, *28*, 401–410. [CrossRef]
60. Yravedra, J.; Panera, J.; Rubio-Jara, S.; Manzano, I.; Expósito, A.; Pérez-González, A.; Soto, E.; López-Recio, M. Neanderthal and *Mammuthus* interactions at EDAR Culebro 1 (Madrid, Spain). *J. Archaeol. Sci.* **2014**, *42*, 500–508. [CrossRef]
61. Yravedra, J.; Rubio-Jara, S.; Panera, J.; Uribelarrea, D.; Pérez-González, A. Elephants and subsistence. Evidence of the human exploitation of extremely large mammal bones from the Middle Palaeolithic site of PRERESA (Madrid, Spain). *J. Archaeol. Sci.* **2012**, *39*, 1063–1071. [CrossRef]
62. Yravedra, J.; Domínguez-Rodrigo, M.; Santonja, M.; Pérez-González, A.; Panera, J.; Rubio-Jara, S.; Baquedano, E. Cut marks on the Middle Pleistocene elephant carcass of Áridos 2 (Madrid, Spain). *J. Archaeol. Sci.* **2010**, *37*, 2469–2476. [CrossRef]
63. Company, J.; Pereda-Suberbiola, X. Long bone histology of a eusuchian crocodyliform from the Upper Cretaceous of Spain: Implications for growth strategy in extinct crocodiles. *Cretac.s Res.* **2017**, *72*, 1–7. [CrossRef]
64. Albéric, M.; Dean, M.N.; Gourrier, A.; Wagermaier, W.; Dunlop, J.W.C.; Staude, A.; Fratzl, P.; Reiche, I. Relation between the Macroscopic Pattern of Elephant Ivory and Its Three-Dimensional Micro-Tubular Network. *PLoS ONE* **2017**, *12*, e0166671. [CrossRef] [PubMed]
65. Outram, A.K. Bone Fracture and within-bone nutrients: an experimentally based method for investigating levels of marrow extraction. In *Consuming Passions and Patterns of Consumption*; Scarre, C., Ed.; McDonald Institute for Archaeology: Cambridge, UK, 2002; pp. 51–63.
66. Boyd, D.K.; Kenmotsu, N.A. *The Toyah Phase of Central Texas Late Prehistoric Economic and Social Processes*, 1st ed.; Texas A&M University Press: College Station, TX, USA, 2012.
67. Haynes, G.; Krasinski, K.; Wojtal, P. A Study of Fractured Proboscidean Bones in Recent and Fossil Assemblages. *J. Archaeol. Method Theory* **2020**. [CrossRef]
68. Johnson, E.V.; Parmenter, P.C.R.; Outram, A.K. A new approach to profiling taphonomic history through bone fracture analysis, with an example application to the Linearbandkeramik site of Ludwinowo 7. *J. Archaeol. Sci. Rep.* **2016**, *9*, 623–629. [CrossRef]
69. Karr, L.; Outram, A.; Hannus, L. A Chronology of Bone Marrow and Bone Grease Exploitation at the Mitchell Prehistoric Indian Village. *Plains Anthropol.* **2010**, *55*, 215–223. [CrossRef]
70. Karr, L.P. Brewster site zooarchaeology reinterpreted: understanding levels of animal exploitation and bone fat production at the Initial Middle Missouri type site. *STAR Sci. Technol. Archaeol. Res.* **2015**, *1*, 1–13. [CrossRef]
71. Outram, A. Distinguishing bone fat exploitation from other taphonomic processes: what caused the high level of bone fragmentation at the Middle Neolithic site of Ajvide, Gotland? In *The Zooarchaeology of Fats, Oils, Milk and Dairying. Proceedings of the 9th ICAZ Conference*; Oxbow Books: Oxford, UK, 2005.
72. Holen, S.R. Taphonomy of two last glacial maximum mammoth sites in the central Great Plains of North America: A preliminary report on La Sena and Lovewell. *Quat. Int.* **2006**, *142–143*, 30–43. [CrossRef]
73. van der Geer, A.A.E.; van den Bergh, G.D.; Lyras, G.A.; Prasetyo, U.W.; Due, R.A.; Setiyabudi, E.; Drinia, H. The effect of area and isolation on insular dwarf proboscideans. *J. Biogeogr.* **2016**, *43*, 1656–1666. [CrossRef]
74. Behrensmeyer, A.K. Patterns of natural bone distribution on recent land surfaces: implications for archaeological site formation. In *Animals and Archaeology: 1. Hunters and Their Prey*; Clutton, B., Grigson, C., Eds.; British Archaeological Reports International Series; British Archaeological: Oxford, UK, 1983; Volume 163.
75. Lyman, R.L. *Vertebrate Taphonomy*; Cambridge University Press: Cambridge, UK, 1994.
76. Thompson, C.E.L.; Ball, S.; Thompson, T.J.U.; Gowland, R. The abrasion of modern and archaeological bones by mobile sediments: the importance of transport modes. *J. Archaeol. Sci.* **2011**, *38*, 784–793. [CrossRef]
77. Fernandez-Jalvo, Y.; Andrews, P. Experimental effects of water abrasion on bone fragments. *J. Taphon.* **2003**, *1*, 147–163.

quaternary

Article

Preservation of Mammalian Teeth and Bones Influences Identification of Terminal Pleistocene to Middle Holocene Hunter-Gatherer Subsistence at Ban Rai Rockshelter, Northwest Thailand

Athiwat Wattanapituksakul [1,*], Rasmi Shoocongdej [1,2] and Cyler Conrad [3,4]

[1] Prehistoric Population and Cultural Dynamics in Highland, Pang Mapha Project, Princess Maha Chakri Sirindhorn Anthropology Centre, Bangkok 10170, Thailand
[2] Department of Archaeology, Faculty of Archaeology, Silpakorn University, Bangkok 10200, Thailand
[3] Department of Anthropology, University of New Mexico, Albuquerque, NM 87131, USA
[4] Environmental Stewardship, Los Alamos National Laboratory, Los Alamos, NM 87545, USA
* Correspondence: essathiwat@gmail.com

Abstract: Ban Rai Rockshelter in northwest Thailand, dating to the Terminal Pleistocene and Middle Holocene, includes evidence for hunter-gatherer exploitation of mammals, birds, reptiles, fish, and arthropods. Abundant faunal remains, identified throughout site deposits, include macaques (*Macaca* sp.) and Sambar deer (*Rusa unicolor*), but these identifications are influenced by an assemblage largely comprised of preserved tooth elements and fragmented bone. Area 3 at Ban Rai has the largest abundance and diversity of faunal remains recovered and identified in this study. Here, we examine the zooarchaeological assemblage from Ban Rai Rockshelter, to understand long-term hunter-gatherer subsistence change, influenced by site preservation, during and after the Pleistocene–Holocene transition. Our results support the presence of the exploitation of arboreal taxa during the Early and Middle Holocene in northwest Thailand.

Keywords: Pleistocene; Holocene; zooarchaeology; mainland Southeast Asia; Thailand; subsistence; hunter-gatherer; primate

1. Introduction

During the Terminal Pleistocene and Holocene, hunter-gatherer groups inhabited and foraged within the highland regions of northwest Thailand. Several sites (see Figure 1) support this record, including Tham Lod Rockshelter, Ban Rai Rockshelter, Spirit Cave, Steep Cliff Cave, and Banyan Valley Cave, suggesting that this region was a significant location for the development of human adaptations in the tropics of mainland Southeast Asia [1–10]. Zooarchaeological analyses in the 1960s suggested that during this prehistoric period, hunter-gatherers exploited a wide diversity of plant and animal taxa (see [11]), but recent studies have challenged this record. There was likely a more complex relationship between human population densities, subsistence, and paleoenvironmental change in this region [1,2,8,12–14]. For example, northwest Thailand experienced known faunal change. One of the major changes in ecological communities was the loss of the Sumatran serow (*Capricornis sumatraensis*)—a species which occurs in several archaeological sites within this region, and which was locally extirpated sometime during the Holocene [8,12,14]. Based on this evidence, we investigated the zooarchaeological assemblage from a Terminal Pleistocene–Middle Holocene archaeological site, Ban Rai Rockshelter, to understand shifts in hunter-gatherer subsistence over time, and the taphonomic processes that have influenced this assemblage. A record of broad-scale faunal and environmental change within Sundaland (an area encompassing portions of mainland and island Southeast Asia during the Last Glacial Maximum) during the Terminal Pleistocene–Middle Holocene [2,8] suggests that there was a shift from large mammal to small mammal exploitation during the period of human occupation at Ban Rai Rockshelter.

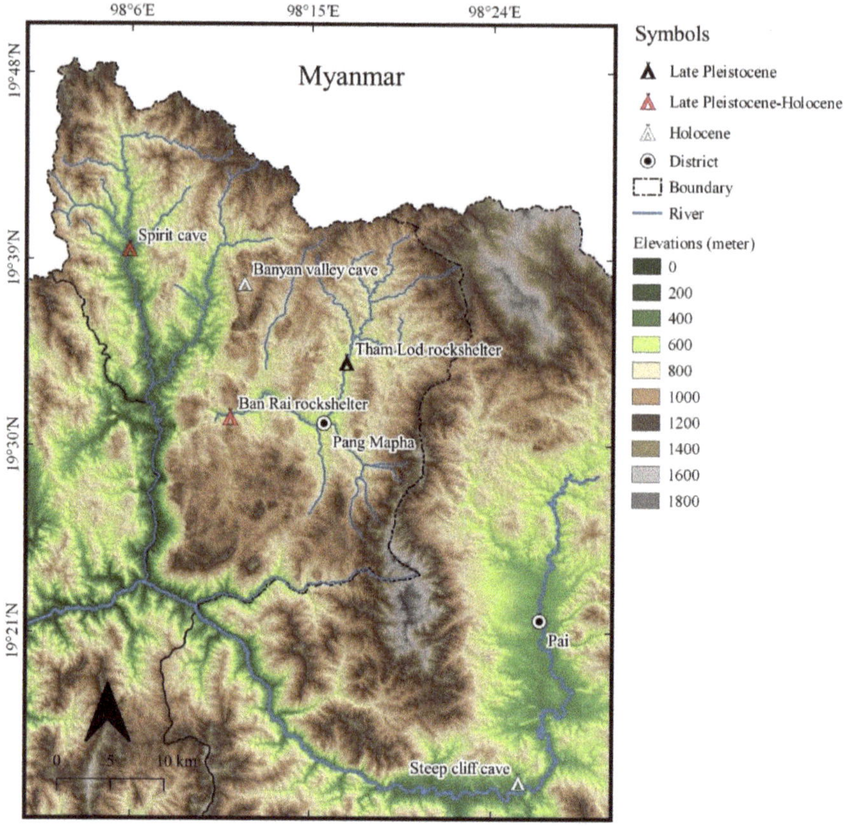

Figure 1. The location of archaeological sites within northwest Thailand discussed in the text, including Ban Rai Rockshelter in Pang Mapha District, Mae Hong Son Province.

In terms of paleoenvironmental change in Southeast Asia, evidence from hunter-gatherer archaeological sites and associated paleontological sites indicates that Sundaland (Figure 2) was likely a large, open grassland savannah corridor environment during the Last Glacial Maximum [12,13,15–19]. As deglaciation occurred during the Terminal Pleistocene, climates shifted to more 'modern' or recent conditions, and a corresponding shift occurred in Southeast Asian faunas and environments. Open grasslands transitioned towards tropical, monsoon-fed forests [19–22]. Holocene environments in mainland Southeast Asia became patchy and increasingly heterogeneous between regions that received different quantities of rainfall. These changes corresponded to changes in faunal assemblages.

For example, faunal change included a shift from the *Ailuropoda-Stegodon* complex (established by [23–25]), which recognized the co-occurrence of the giant panda (*Ailuropoda*) and the now-extinct Proboscideans (*Stegodon*), to 'modern' faunal communities [17,26,27]. There is evidence from this period for a larger biogeographic dispersal of certain faunas (e.g., *Naemorhedus goral, Rhinoceros unicornis, Sus barbatus*), strongly suggesting animal mobility within the Indochinese and Sundaic subregions bounded by the Kra Isthmus. Carnivore activity within this area suggests the co-importance of large herbivorous mammals and large carnivorous predators within this past habitat [28].

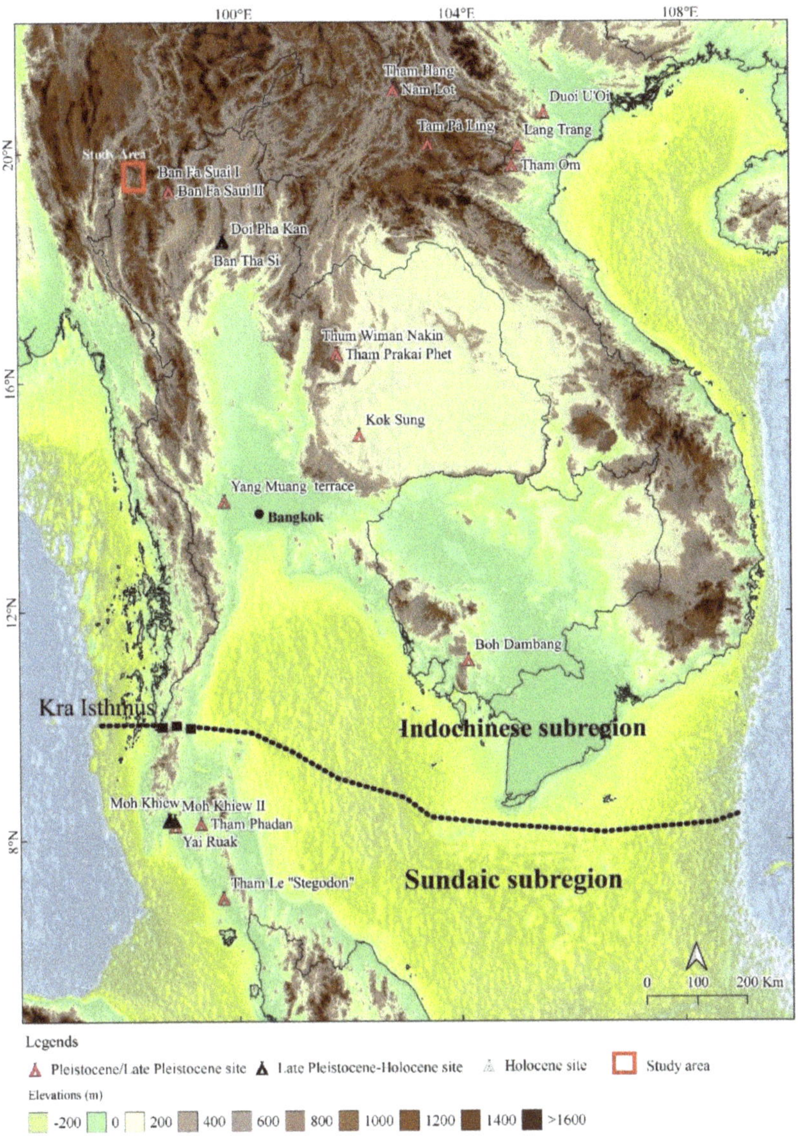

Figure 2. Prominent Pleistocene and Holocene sites in the zoogeographic subregions of mainland and island Southeast Asia (Sundaland) during the Last Glacial Maximum (~21,000 years ago). Sea level depicted at -116 m compared to modern levels [29].

As these changes occurred in past environments and within animal communities, hunter-gatherers continued to forage and adapt to shifting conditions throughout this region. Ban Rai Rockshelter, situated in a highland location of mainland Southeast Asia, and spanning the critical period of the Terminal Pleistocene to Middle Holocene, thus provides an important opportunity to investigate the dynamics of this process, as it relates to human subsistence strategies, palaeoecologies, and taphonomic processes.

2. Site Excavation, Chronology, and Stratigraphy

Ban Rai Rockshelter (Figure 3) is located on the southern edge of the Lang River Valley, approximately 2–3 km from the nearby village of Ban Rai in Pang Mapha District, Mae Hong Son Province, northwest Thailand. The rockshelter is positioned on the top of a steeply sided valley, approximately 740–760 m above sea level, or approximately 200 m above the Lang River, which flows westward into a sinkhole [7,9]. The lithology of the valley is comprised of Lower Carboniferous limestone, and is thus karstic [30]. Six forest types occur within the vicinity of the site (mixed deciduous, dry dipterocarp forest, hill evergreen forest, dry evergreen forest, bamboo forest, and limestone forest [6]). Except for the roof of the rockshelter (top of the karst) where dry dipterocarp forest occurs, mixed deciduous forest is present throughout the valley [7].

Figure 3. Ban Rai Rockshelter: view from the northeast wing of the rockshelter.

The rockshelter is semicircular in shape, measuring 105 × 142 m and 30 m high from the surface to the hanging wall edge at the center of the shelter. Large limestone debris on the rockshelter floor, especially on the western side of the site, is a component of rock fall from prior geologic events. There are large wooden posts and log coffins throughout the rockshelter, representing the local Log Coffin Culture, dating to the Late Holocene [31]. The Log Coffin Culture occurred throughout northwest Thailand during the past several millennia, and likely represented mixed forager–farmer groups exploiting the upland regions throughout Pang Mapha, using this specific type of interment (see below). A series of prehistoric, realistic, and idealistic rock paintings on the eastern hanging wall of the site currently remains undated [7,9].

The Highland Archaeology Project in Pang Mapha District surveyed and selected the rockshelter for excavation in 2001. Excavation occurred in three areas beneath the log coffins unearthed immediately below the surface, after clearing the rockshelter floor. Each area included two types of units—trenches (4 × 4 m) and baulks (1 × 4 m)—oriented north–south (Figure 4; [7,9]). The excavations included:

Area 1: a ~28 m^2 trench at the center of the rockshelter, with five pits (including S1E4, S2E12, Baulk S2E12, S3E19, and Baulk S3E19) and eight primary layers, totaling ~0.7–2 m in depth.

Area 2: a ~24 m^2 trench about 5 m east of Area 1, with three pits (including S1E6, S2E14, and Baulk S2E14) and five primary layers, totaling ~1.5–1.8 m in depth.

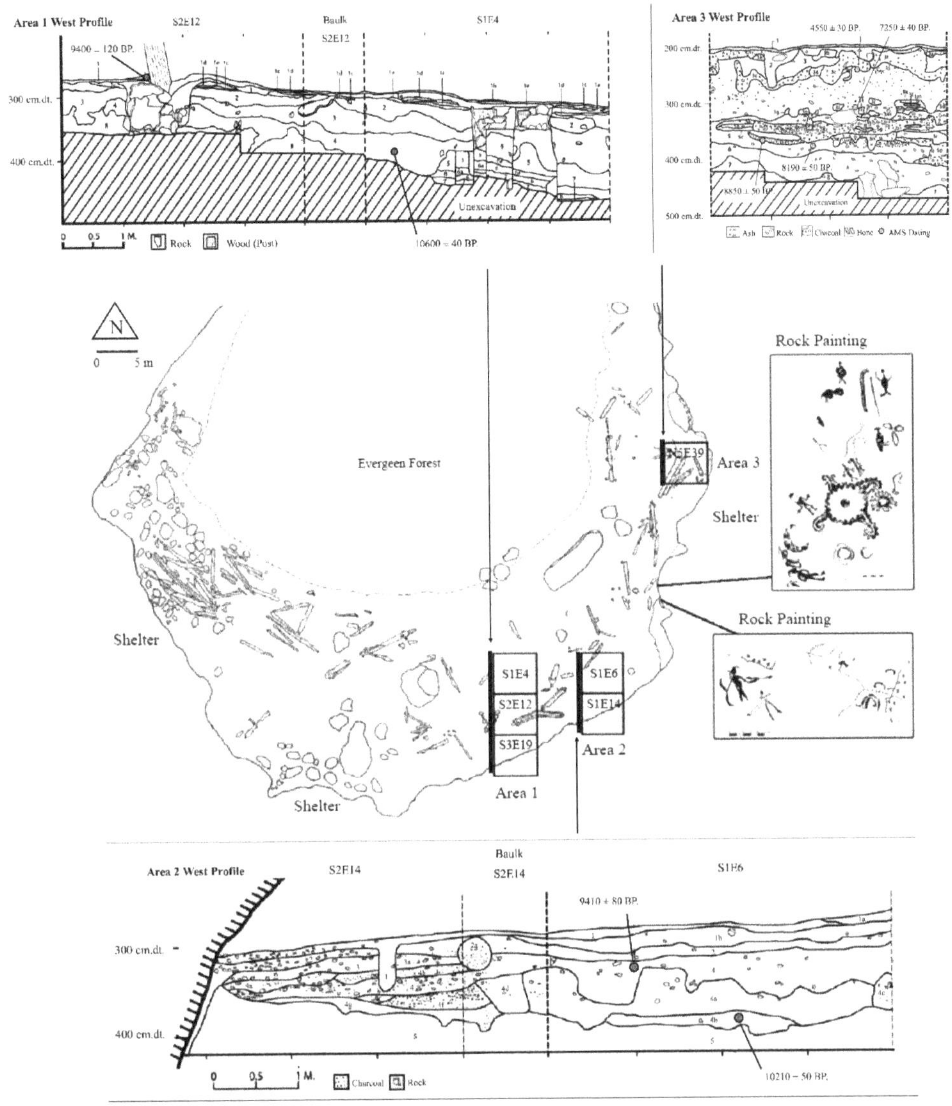

Figure 4. Ban Rai Rockshelter, showing the location of excavation units, radiocarbon determinations, and the west wall profiles from each area (Area 1, 2, and 3) (drawn by Pipad Krajaejun).

Area 3: a ~16 m^2 trench at the eastern edge of the rockshelter, with one pit (N5E39) and seven primary layers, totaling 3 m in depth.

Excavations recovered 11,600 lithic items (including 801 items from Area 1, 999 items from Area 2, and 9800 items from Area 3). These stone artifacts included sumatraliths, short axes, wasted cores and flakes, and utilized cores and flakes: they represented a stone tool assemblage, with hunter-gatherers exploiting material far distant from Ban Rai, within patchy and diverse ecological resource zones [7,9,32].

Two standard radiocarbon and seven Accelerator Mass Spectrometry (AMS) radiocarbon conventional dates established a chronological sequence in successive layers, from 10,600 ± 40 radiocarbon (^{14}C) years BP (Beta-168216) in Area 1-S1E4 to 4550 ± 30 ^{14}C years BP (Beta-538749) in Area 3-N5E39 (Table 1). The profile of each area straddled sequential cultural phases, resulting in a two-phased contextual chronological framework from the Terminal Pleistocene to the Middle Holocene [7,9,33].

Table 1. Radiocarbon determinations from Ban Rai Rockshelter [7,9,33].

Laboratory Code	Area	Layer	Material	$\delta^{13}C$	Radiocarbon Years BP	±SD	Calibrated Year BP	Context
Beta-168215	1	2	Charcoal	−25.0	9400	120	11,110–10,250	Hunter-Gatherer
Beta-168216	1	4	Charcoal	−28.8	10,600	40	12,910–12,610, 12,490–12,350	Hunter-Gatherer
Beta-168218	2	4	Charcoal	−25.0	9410	80	11,060–10,950, 10,780–10,420	Hunter-Gatherer
Beta-168217	2	4	Charcoal	−30.8	9720	50	11,210–11,090, 10,920–10,890	Hunter-Gatherer
Beta-168219	2	4	Charcoal	−28.4	10,210	50	12,340–11,670	Hunter-Gatherer
Beta-534121	3	-	Wood	−24.6	1770	30	1744–1605	Log Coffin
Beta-534122	3	-	Wood	−24.4	1840	30	1864–1708	Log Coffin
Beta-534118	3	-	Wood	−24.3	1880	30	1884–1728	Log Coffin
Beta-534119	3	-	Wood	−25.9	1920	30	1947–1812	Log Coffin
Beta-534120	3	-	Wood	−25.5	2040	30	2067–1924	Log Coffin
Beta-538749	3	-	Sediment	-	4550	30	5189–5053, 5318–5257	Hunter-Gatherer
Beta-168220	3	3	Charcoal	−28.0	7250	40	8340–8140	Hunter-Gatherer
Beta-168222	3	5	Charcoal	−29.7	8190	50	9290–9120	Hunter-Gatherer
Beta-168221	3	5	Charcoal	−27.7	8850	50	10,170–9720	Hunter-Gatherer

2.1. The First Cultural Phase

The first cultural phase was located in the lowest portion of each unit in all excavation areas, specifically spanning Layers 5–8 in Area 1, Layer 5 and 4b in Area 2, and Layer 7 in Area 3. These contexts included the oldest directly dated materials, and suggested a Terminal Pleistocene human occupation of nearly 10,600 ^{14}C years BP [7,9]. Based on lenses of charcoal and ash identified throughout each layer and area, hunter-gatherers likely first settled at Ban Rai Rockshelter during this time; however, there were no faunal remains or stone tools from these excavated contexts, only charcoal and ash stains [7].

2.2. The Second Cultural Phase

The second cultural phase consisted of Layers 2–4 in Area 1, Layer 4 and 4a in Area 2, and Layers 3–6 in Area 3. The successive dating of these layers suggested an occupation between 9400 ^{14}C years BP and 4550 ^{14}C years BP (the Terminal Pleistocene to Late-Middle Holocene). Evidence for human occupation included abundant faunal remains and lithic artifacts associated with charcoal and ash. There was also a human skeleton in a primary flexed burial position within a roughly circular pit (64 cm in diameter), buried approximately 90–110 cm deep in Area 2, dating to this period: this was the fragile skeleton of an adult male.

Considering the chronological dating of this burial—from charcoal in sediments near the skeleton—the individual died around 9800 ^{14}C years BP (Beta-168217). According to the sequence of occupation correlated to this absolute chronology, hunter-gatherers likely occupied the center of the rockshelter (Areas 1 and 2) during this initial period. The latest and terminal occupation of the rockshelter shifted to areas near the left wall (in Area 3), based on radiocarbon dating (see Table 1). The second cultural phase consisted of spatially diverse forager activities [7,9], including Late Holocene activity. Regardless, there was a clear Terminal Pleistocene and Early Holocene occupation in Area 1–2, while Area 3 was extensively used from the Early to Middle Holocene. As a result, all areas included evidence of ancient occupations. Simultaneously, the consistency between the dates assigned to the earliest cultural levels identified in Area 3, as well as the latest dates retrieved from Areas 1 and 2, may indicate a spatial movement across the site, with the main habitation area shifting from the site's center to its eastern side (see [9]).

The uppermost layers of this area (and Areas 1 and 2) suggest the presence of the Log Coffin Culture in the Late Holocene (see [31]). In Highland Pang Mapha, the Log Coffin Culture is an ancient mortuary practice dating to around 2120–1250 B.P. (see [33]).

In dry limestone caves and rockshelters, coffins were purposefully placed on wooden posts and beams. The majority of the coffins were made from hollowed-out teak wood trunks; the exterior surface included carvings of bodies and heads, of various shapes and sizes. Human remains are often found in log coffins used for funerals. Log Coffin burials also include faunal remains, earthenware, glass beads, bronze ornaments, iron implements, wooden objects, and textiles (see [33]).

3. Materials and Methods

3.1. Study Materials

During excavations in 2001, archaeologists collected a large assemblage of faunal remains from Ban Rai Rockshelter [7,9,34]. Between 2003 and 2006, an initial analysis of this assemblage identified 55,929 specimens, weighing 71,334.15 g [35]. Subsequent analysis suggested that only 5816 specimens (2409.9 g) were taxonomically unidentifiable—these were extremely fragmented and degraded specimens. A preliminary analysis of the Area 3 specimens identified 21 taxa present in this context at the site [34] (see also [8]).

A recent publication that included analysis of the Ban Rai faunal assemblage [8] has a slightly different taxonomic composition (including counts) than reported here: this is due to the presence of updated identifications and counts of the assemblage for the purposes of this paper (see below). Our results reflect the most recent and up-to-date analysis of the assemblage [33,34].

All the specimens identified and analyzed here (by A.W.) derived from six pits throughout the site: Pit S1E4 and S2E12 in Area 1; Pit S1E6, S2E14, and Baulk S2E14 in Area 2; and Pit N5E39 in Area 3. Compared to pits, baulk samples from the site typically included a smaller number of faunal specimens: for example, 1364, 140, and 83 specimens derived from Baulks S2E12, S3E19, and S3E19, respectively, [34].

A.W. identified specimens, using a combination of comparative skeletal material and reference guides [27,36–43]. Comparative skeletons provided the final identification criteria, whenever possible, due to known inconsistencies within published reference guides. Comparative specimens included skeletons from the Chulalongkorn University Museum of Natural History, the Natural History Museum, National Science Museum Thailand in Bangkok, and the Palaeontological Research and Education Centre in Mahasarakham University, Mahasarakham Province.

3.2. Zooarchaeological Methods

Measuring taxonomic abundance included quantification of both the number of identified specimens (NISP) and the minimum number of individuals (MNI) [44,45]. NISP is the counting of the abundance of all skeletal elements, including fragments of specimens which are identifiable to the smallest possible taxonomic levels (species and subspecies). MNI quantification involved identifying individual taxonomic abundance, based on the number of diagnostic skeletal elements of particular taxa, using overlapping features (e.g., element side and age) and other contextual data from the archaeological units, areas, and cultural layers.

Given zooarchaeological evidence from nearby hunter-gatherer sites in northwest Thailand suggesting the presence of various taphonomic influences affecting faunal assemblages [1,2], we examined a linear model of Ban Rai's faunal data between taxonomic abundance and the number of taxa identified or taxonomic richness (NTAXA) (see [44,46–48]): this provided the examination of the relationship between NISP and NTAXA or NISP–NTAXA, as well as between MNI and NTAXA or MNI–NTAXA, that enabled us to understand how sample size biases impacted the assemblage. We did not log-transform zooarchaeological count data for this analysis, due to known issues with this technique (see [49]). We also explored the relationship between identified tooth and bone skeletal elements in Area 3, to evaluate the role and influence of tooth enamel preservation on our identifications and analysis. Finally, we examined the diachronic relationships between abundances of specific faunas, to evaluate hunter-gatherer subsistence patterns at Ban

Rai Rockshelter during the Pleistocene and Holocene, given these taphonomic constraints. These analyses allowed for an interpretation of long-term hunting strategies during periods of environmental change.

In addition, A.W. also recorded evidence of weathering, butchery marks, root etching, burning, and rodent/carnivore gnawing on the surface of all bone and tooth specimens [35]. We analyzed all data, and created visualizations in R (v.4.0.4) and RStudio (v.1.4.1717; see [50] for source code) (see: https://github.com/cylerc/BanRai)(accessed on 1 November 2022).

4. Results

4.1. Taxonomic Composition

Zooarchaeological analysis of the Ban Rai Rockshelter assemblage resulted in the identification of over 1000 mammalian, avian, reptilian, fish, and arthropod specimens throughout Areas 1, 2, and 3 (Table 2, Tables S1–S3). Our analysis does not currently include molluscan taxa. In total, A.W. identified 1353 specimens (NISP), representing 292 individuals (MNI). Bone specimens were highly fragmented at the site and, thus, identification often occurred through the presence of morphologically distinguishing features on mammalian tooth fragments (see [8]). Examination of faunal remains from Ban Rai Rockshelter included identification of surface modifications (e.g., cut marks, bone breaks, burning), but are not further reported here.

Table 2. Summary of identified taxa in the Ban Rai Rockshelter zooarchaeological assemblage. Area 1 includes Layers 2–4 of both S1E4 and S2E12. Area 2 includes only Layer 4 of pit S1E6, S2E14, and Baulk S2E14. Finally, Area 3 includes Layers 3, 4, 5, and 6 of pit N5E39. Taxonomic abundance listed as the number of identified specimens (NISP) followed by the minimum number of individuals (MNI); additional data in Supplementary Tables S1–S3.

Taxon	Common Name	Area 1 S1E4		Area 1 S2E12		Area 2 S1E6		Area 2 S2E14		Area 2 Baulk S2E14		Area 3 N5E39	
		NISP	MNI	NISP	MNI	NISP	MNI	NISP	MNI	NISP	MNI	NISP	MNI
Mammalia	Mammal	51	4	29	6	21	4	17	3	51	4	88	5
Manis sp.	Pangolin											1	1
Nycticebus sp.	Loris											1	1
Macaca mulatta	Rhesus macaque											13	6
Macaca arctoides	Stump-tailed macaque											11	7
Macaca sp.	Macaque	1	1	2	1			5	1	2	1	243	19
Cercopithecidae	Old World monkeys	4	2			1	1	8	3			1	1
Trachypithecus cf. *phayrei*	Phayre's langur											2	2
Colobinae	Langur					1	1	2	1	1	1	73	9
Hylobatidae	Gibbon											10	2
Ursus thibetanus	Asia black bear											1	1
Ursus spp.	Bear	1	1									4	2
Arctonyx collaris	Hog badger											1	1
Arctictis binturong	Binturong											1	1
Canis sp.?	Dog											8	2
Panthera sp.	Tiger or leopard											2	1
Felidae	Felids											6	3
Carnivora	Carnivors	1	1	1	1	1	1	3	2			26	5
Elephas sp.	Elephant											4	1
Rhinocerotidae	Rhinoceros											10	2
Sus scrofa	Eurasian wild pig			1	1	3	3					8	4
Muntiacus sp.	Muntjac	8	3	2	2							13	4
Muntiacus feae?	Fea's muntjac							4	2				
Panolia eldii	Eld's deer	10	3	2	2	3	1						
Rusa unicolor	Sambar deer	30	7	20	5	30	4	22	4			33	7
Panolia/Rusa sp.	Sambar/Eld's deer											18	4
Axis sp.	Hog deer	9	5	4	2	8	3	12	1			19	6
Cervidae	Cervids	3	2			1	1					19	4
Bubalus arnee	Wild water buffalo	1	1	1	1							5	3
Bos gaurus	Gaur							1	1			2	2
Bos javanicus	Banteng					1	1						
Bos sp.	Wild cattle	2	2			4	2					6	3
Bovinae	Bovid					1	1	1	1				
Capricornis sumatraensis	Sumatran serow	2	2			5	3			1	1	8	2
Naemorhedus griseus	Chinese goral	1	1										
Artiodactyla	Even-toed ungulates			1	1								
Callosciurus finlaysonii	Variable squirrel											1	1
Petauristinae	Flying squirrel											1	1

Table 2. *Cont.*

Taxon	Common Name	Area 1 S1E4		Area 1 S2E12		Area 2 S1E6		Area 2 S2E14		Area 2 Baulk S2E14		Area 3 N5E39	
		NISP	MNI	NISP	MNI	NISP	MNI	NISP	MNI	NISP	MNI	NISP	MNI
Sciuridae	Squirrel											5	2
Leopoldamys sabanus	Long-tailed giant rat											1	1
Bandicota sp.	Bandicoot rat											1	1
Rhizomyidae	Bamboo rat					3	1	1	1	2	2	6	2
Hystricidae	Old World porcupine											19	3
Rodentia	Rodent											29	6
Aves	Bird											1	1
Lacertilia	Lizards											3	2
Testudinata	Turtle	22	5	30	6	2	2	12	3	15	4	53	4
Indotestudo elongata	Elongated tortoise					28	2						
Cyprinidae	Carp or minnow			2	2	1	1			2	1	61	15
Eumalacostraca	Crab									1	1		
	Total	155	43	96	29	119	33	90	24	74	15	818	150

Considering taxa identified to the genus and species level, Ban Rai was comprised of a large number of primates and artiodactyls. Macaques (*Macaca* sp.) and Sambar deer (*Rusa unicolor*) were abundant throughout site contexts, particularly in Area 3. Unidentified turtles, tortoises, and Cyprinidae fishes (carps and minnows) also occurred relatively abundantly throughout each area.

Taxonomic diversity and abundance within each excavation area at Ban Rai differed, but Area 3 was especially distinct in terms of its zooarchaeological assemblage (Figure 5). Areas 1 and 2 at Ban Rai had relatively small abundances of taxa (Table 2 and Supplemental Tables S1–S3). Primate diversity and abundance—with three taxa in only two Families and one genus identified (*Macaca* sp., Cercopithecidae, Colobinae)—were almost identical between these two areas: Area 1, NISP = 7 and MNI = 3; Area 2, NISP = 11 and MNI = 5. Only NISP = 3 or MNI = 2 of Macaque (*Macaca* sp.) occurred in Area 1, and NISP = 7 in Area 2; likewise, there were NISP = 4 (or MNI = 2) Cercopithecidae specimens in Area 1, and only NISP = 2 (or MNI = 2) in Area 2. Additionally, NISP = 2 (or MNI = 1) Colobinae primate specimens were present in Area 2. In contrast, Area 3—represented by a single unit, N5E39—included specimens from four families of primates, two genera, and three species (*Nycticebus* sp., *Macaca mulatta*, *Macaca arctoides*, *Macaca* sp., Cercopithecidae, *Trachypithecus* cf. *phayrei*, Colobinae, and Hylobatidae) with a total of NISP = 354 or MNI = 47. The proportion of primate specimens varied from 2.8% of the total NISP (NISP = 7/251) or 5.5% of the total MNI (MNI = 4/72) in Area 1 to 6.7% of the total NISP (NIS*p* = 19/283) or 12.5% of the total MNI (MNI = 9/72) in Area 2; however, Area 3 had a higher proportion of primates than the other Areas: 43.3% of the total NISP (NISP = 354/818) or 31.3% of the total MNI (MNI = 47/150).

While Areas 1 and 2 included carnivores, they occurred in relatively minor abundance (Area 1: NISP = 3, MNI = 3; Area 2: NISP = 4, MNI = 3) compared to Area 3 (NISP = 49, MNI = 16): for example, identification included one Asian black bear (*Ursus thibetanus*) within Area 1, and only one unidentified Carnivora specimen in Area 2; yet, Asian black bear, unidentified bears (*Ursus* sp.), hog badger (*Arctonyx collaris*), binturong (*Arctictis binturong*), possibly dog (*Canis* sp.?), tiger or leopard (*Panthera* sp.), and unidentified Felids all occurred within Area 3. Area 3's analysis identified several elephant (*Elephas* sp.: NISP = 4, MNI = 1) and rhinoceros (Rhinocerotidae) specimens (NISP = 10, MNI = 2)—the only such elements within the site.

Although there were turtles and tortoises found throughout each area (Area 1: NISP = 52, MNI = 9; Area 2: NISP = 61, MNI = 11; Area 3: NISP = 53, MNI = 4), the unidentified bird (NISP = 1, MNI = 1) and lizard (NISP = 3, MNI = 2) elements appeared only within Area 3. This was largely true for all identified rodent taxa as well, including the variable squirrel (*Callosciurus finlaysonii*; NISP = 1, MNI = 1), the flying squirrel (Petauristinae; NISP = 1, MNI = 1), the squirrel (Sciuridae; NISP = 5, MNI = 2), the long-tailed giant rat (*Leopoldamys sabanus*; NISP = 1, MNI = 1), the bandicoot rat (*Bandicota* sp.; NISP = 1; MNI = 1), the bamboo rat (Rhizomyidae; NISP = 6, MNI = 2), the Old World porcupine (Hystricidae;

NISP = 19, MNI = 3), and unidentified rodents (NISP = 29, MNI = 6). Only unidentified rodent specimens occurred within Area 1 (NISP = 29, MNI = 6), as well as bamboo rats (Rhizomyidae) in Area 2 (NISP = 6, MNI = 4). In similar form to primates, there was a large abundance of Cyprinidae within Area 3 (NISP = 61, MNI = 15), with few specimens from Areas 1 (NISP = 2, MNI = 2) and 2 (NISP = 3, MNI = 2). A single crab (NISP = 2, MNI = 1) occurred in Area 2.

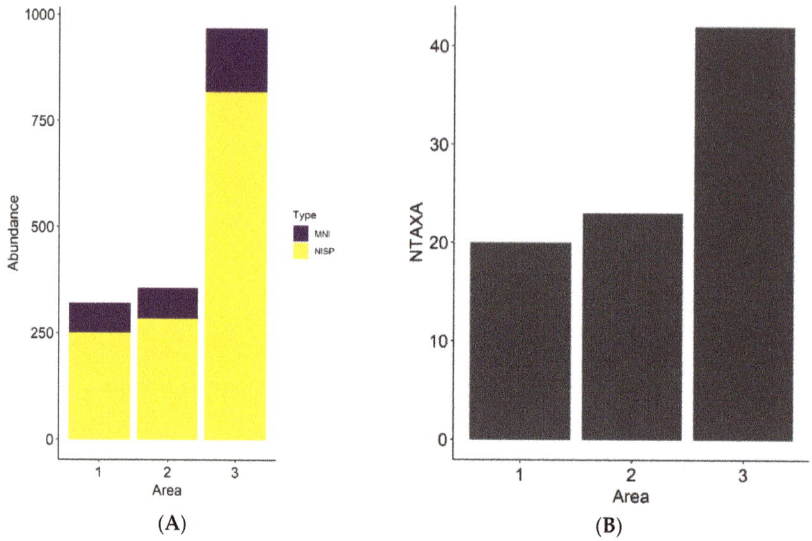

Figure 5. Abundance of fauna and number of identified taxa identified in each area at Ban Rai Rockshelter: (**A**) abundance (NISP, number of identified specimens and MNI, minimum number of individuals) of faunas; (**B**) number of identified taxa (NTAXA).

Identification and analysis of artiodactyls suggested a more complex pattern within the site. In general, the largest NISP of all identified artiodactyls occurred within Area 3 (NISP = 131, MNI = 39), compared to the largest MNI in Area 1 (NISP = 99, MNI = 42) and Area 2 (NISP = 100, MNI = 31). There were large abundances of Sambar deer (*Rusa unicolor*) in each area (Area 1: NISP = 50, MNI = 12; Area 2: NISP = 52, MNI = 8; Area 3: NISP = 33, MNI = 7), as well as the smaller hog deer (*Axis* sp.) (Area 1: NISP = 13, MNI = 7; Area 2: NISP = 20, MNI = 4; Area 3: NISP = 19, MNI = 6). The ecologically significant—due to their extirpated and threatened status [13,15]—Caprinae from Ban Rai included the Sumatran serow (*Capricornis sumataensis*) and the Chinese goral (*Naemorhedus griseus*). The latter taxon only occurred in Area 1 (NISP = 1, MNI = 1), while the serow was present in all areas (Area 1: NISP = 2, MNI = 2; Area 2: NISP = 6, MNI = 4; Area 3: NISP = 8, MNI = 2).

Within each area, there was also a diachronic change in taxonomic presence and abundance. Area 1, with relatively large abundances of artiodactyls and turtles/tortoises, experienced accumulation of large numbers of artiodactyls in Pit S1E4, from the deeper layers to the surface (Figure 6). In Area 1 Pit S2E12, accumulation of artiodactyls and turtles/tortoises increased from the deeper layers up to Layer 3, where abundance peaked, then decreased towards the surface layers.

In contrast, Area 2, with lower taxonomic abundances overall, had experienced minimal change over time in the accumulation of faunas (Figure 7). Pit S2E14 and Pit S2E14 Baulk both had small faunal abundances, with a decrease in the relative abundance of artiodactyls from the deeper layers to the surface levels. This trend was also present in Pit S1E6, where artiodactyls decreased over time, while turtle and tortoise abundance tended to increase over time.

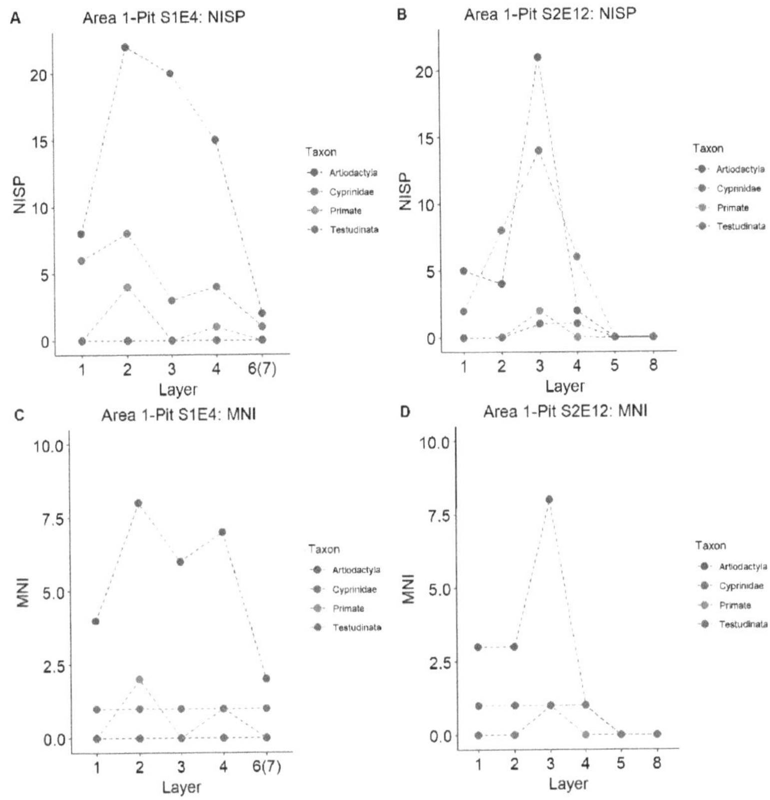

Figure 6. Diachronic change in taxonomic abundance (NISP, number of identified specimens and MNI, minimum number of individuals) within Area 1: (**A**,**C**) S1E4; and (**B**,**D**) S2E12.

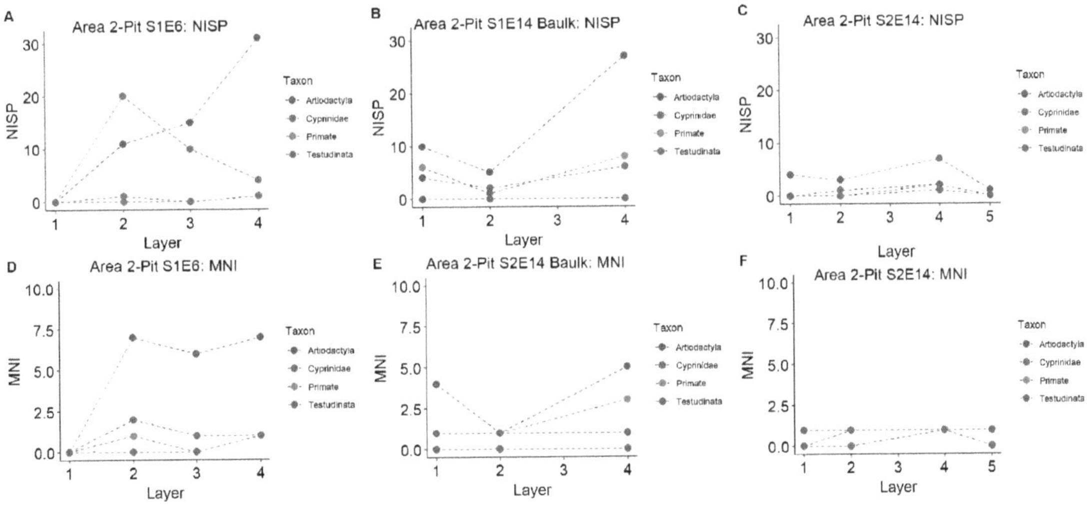

Figure 7. Diachronic change in taxonomic abundance (NISP, number of identified specimens and MNI, minimum number of individuals) within Area 2: (**A**,**D**) S1E6; (**B**,**E**) S2E14 Baulk; and (**C**,**F**) S2E14.

Area 3, with the largest taxonomic richness and abundance of fauna, experienced a relatively consistent pattern of change over time, with artiodactyls, primates, turtles/tortoises, and fishes increasing, up to surface Layers 1–2 (Figures 8 and 9). Primates dominated the taxonomic abundance of this unit in Area 3, suggesting significant exploitation and/or accumulation of these taxa within Middle Holocene contexts at Ban Rai Rockshelter.

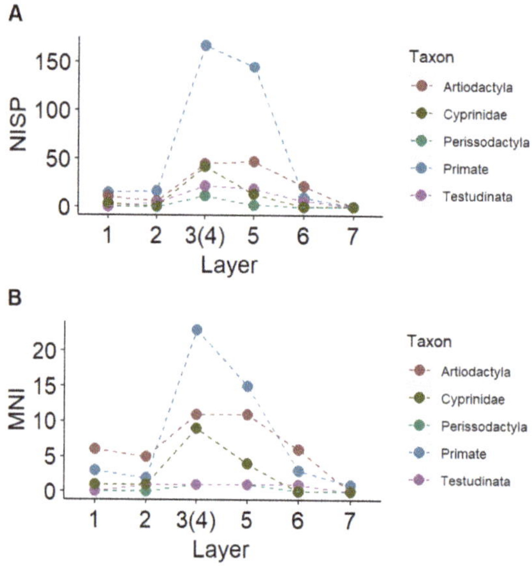

Figure 8. Diachronic change in taxonomic abundance within Area 3 N5E39: (**A**) NISP, number of identified specimens; and (**B**) MNI, minimum number of individuals.

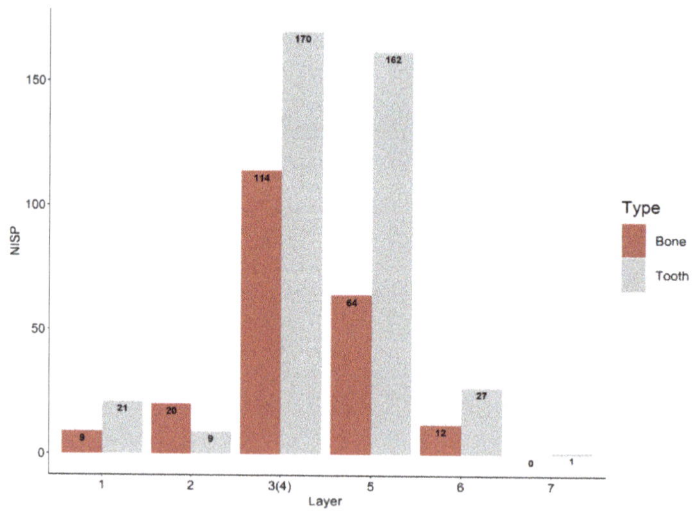

Figure 9. Relationship between identified bone and tooth for Artiodactyla, Perissodactyla, Testudinata, Primates, and Cyprinidae within Area 3 at Ban Rai Rockshelter, listed as the number of identified specimens (NISP).

4.2. Taphonomic Insights

Preservation of tooth enamel clearly influenced the composition and identification of faunal remains within Area 3, and was representative of the faunal assemblage as a whole from Ban Rai (Figure 5). In each area and unit, there tended to be larger assemblages of identified teeth in comparison to bone, and in some layers within Area 3 this difference was considerable. For example, in Layer 5, ~72% of the identified zooarchaeological assemblage derived from tooth fragments.

There was also clear sample size influence occurring within the Ban Rai Rockshelter faunal assemblage—largely driven by the abundance and diversity of the Area 3 assemblage, compared to Areas 1 and 2. When including Area 3 (Figure 10), a positive correlation existed between NISP–NTAXA ($r^2 = 0.94$, $p \leq 0.01$) and MNI–NTAXA ($r^2 = 0.97$, $p \leq 0.01$). The inclusion of Area 3 contributed significant leverage to this correlation: when Area 3 was removed from this model, its influence became clear for both NISP ($r^2 = 0.46$, $p = 0.12$) and MNI ($r^2 = 0.63$, $p = 0.07$).

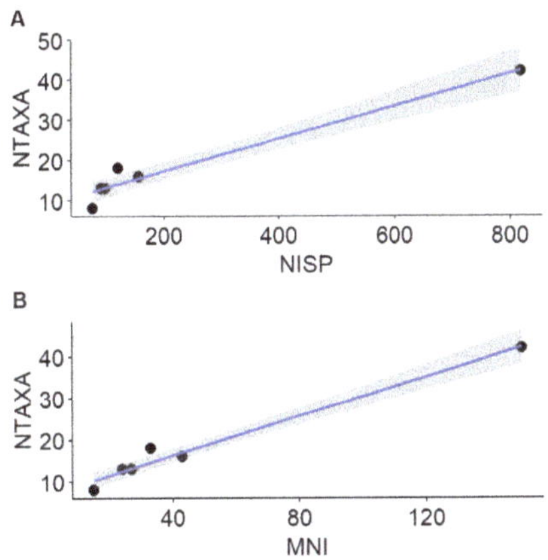

Figure 10. The linear model between taxonomic abundance (NISP and MNI) and taxonomic richness (NTAXA) at Ban Rai Rockshelter (the points represent values per unit in each Area): (**A**) NISP–NTAXA and (**B**) MNI–NTAXA. NISP: the number of identified specimens, MNI: the minimum number of individuals, and NTAXA: the number of taxa identified.

Previous zooarchaeological research at Ban Rai suggests that differential preservation influenced the assemblage [35]. A subset of the assemblage was examined here, and it was discovered that the specimens from Ban Rai contained evidence for bone and tooth weathering, root etching, and rodent/carnivore gnawing [35]. Chemical processes also occurred in specimens, such as calcium carbonate concretions covering the surface of the bones. The nearby and related site of Tham Lod Rockshelter also showed evidence of calcium carbonate formation on bones [7,35] (see Tables 3–5).

Table 3. Taphonomic evidence in Area 1 of the Ban Rai Rockshelter, in numbers of specimens.

Layer	Butchery Mark [1]	Bone Tool [1]	Burning	Weathering [1]	Chemical Process [1]	Gnawing Damage [1]	Root Etching [1]	Total
1	1	3	15	0	1	5	0	**25**
2	6	7	25	0	0	5	1	**44**
3	8	14	21	1	3	6	0	**53**
4	7	8	10	0	2	4	0	**31**
5	0	0	0	0	0	0	0	**0**
6	0	0	0	0	0	0	0	**0**
7	0	0	0	0	0	0	0	**0**
8	0	0	1	0	0	0	0	**1**
Total	**22**	**32**	**72**	**1**	**6**	**20**	**1**	**154**

[1] Previous data [35].

Table 4. Taphonomic evidence in Area 2 of the Ban Rai Rockshelter, in numbers of specimens.

Layer	Butchery Mark [1]	Bone Tool [1]	Burning	Weathering [1]	Chemical Process [1]	Gnawing Damage [1]	Root Etching [1]	Total
1	3	12	11	0	0	6	0	**32**
2	0	9	21	0	0	4	1	**35**
3	0	4	13	0	1	3	0	**21**
4	5	27	37	0	0	7	0	**76**
5	0	0	3	0	0	0	0	**3**
Total	**8**	**52**	**85**	**0**	**1**	**20**	**1**	**167**

[1] Previous data [35].

Table 5. Taphonomic evidence in Area 3 of the Ban Rai Rockshelter, in numbers of specimens.

Layer	Butchery Mark [1]	Bone Tool [1]	Burning	Weathering [1]	Chemical Process [1]	Gnawing Damage [1]	Root Etching [1]	Total
1	2	2	14	0	1	4	0	**23**
2	1	2	21	0	1	5	0	**30**
3,4	7	12	72	0	4	11	0	**106**
5	12	24	38	0	0	3	0	**77**
6	1	2	13	0	1	1	0	**18**
Total	**23**	**42**	**158**	**0**	**7**	**24**	**0**	**254**

[1] Previous data [35].

For aspects of taphonomic information in Area 1 (Table 3), there was a lack of clear evidence in the lower layers (Layers 5–8), with only one burned specimen observed in Layer 8. The upper layers of Area 1 (Layers 1–4) contained the bulk of the evidence for specimens exhibiting taphonomic processes. In Layer 4, the first evidence of ancient human activity on faunal elements (including butchering, tool marks, and burning) occurred, but the greatest abundance appeared in Layer 3. The upper layers contained several bones with butchering marks, and bones worked into tools. This layer also included a moderate amount of evidence for bone burning. There was a general lack of evidence for chemical alteration in bones from Layers 1–4, or for plant root etching. Little evidence existed from these layers for carnivore and rodent scavenging.

Area 2 exhibited a taphonomic pattern which was similar to Area 1 (see Table 4). Only burned elements occurred consistently in Area 2. From Layer 4 to Layer 1, there were small pieces of possibly worked bone tools, with a small number in each layer. Layer 4 contained the oldest evidence from Ban Rai for human modification of bone, including five butchering marks, 27 bone tools, and a burned element. Butchering evidence only occurred in Layers 1 and 4 with a small number of elements, and there was no evidence for bone weathering. Calcium carbonate concretions occurred on a specimen in Layer 3, and there was root etching evidence in Layer 2. Layers 1–4 also showed moderate evidence of carnivore/rodent gnawing.

Table 5 shows that the trend in Area 3 differed from the trends in Areas 1 and 2. Layer 5 in Area 3 had the most butchered specimens (12 specimens) and the most bone tool specimens (24 specimens) combined. Layers 3 and 4 had the most burnt specimens (72 specimens). There was no weathering and root etching evidence for non-ancient human activity in Area 3. There was some evidence of calcium carbonate concretions in almost all layers, except Layer 5. Consistent evidence for carnivore/rodent gnawing also occurred in all layers.

Consequently, according to taphonomic evidence [35], ancient human activity at Ban Rai Rockshelter was clearly different in Area 3, which corresponded to the area with the largest abundance of faunal specimens and the highest density of charcoal and ash. Root etching and calcium carbonate concretions were generally rare at the site, but carnivore/rodent gnawing and disturbance likely occurred over time. It is important to note that during excavations, porcupines would visit Ban Rai Rockshelter, and there was evidence of areas being disrupted by recent scavengers.

5. Discussion

Zooarchaeological analysis of specimens recovered from excavations in Areas 1–3 at Ban Rai Rockshelter provides insights into the patterns of human subsistence within highland, mainland Southeast Asia during the Terminal Pleistocene to Middle Holocene. Our discussion of the results focuses on: (1) the composition of the assemblage, (2) diachronic patterns in faunal exploitation, and (3) overall changes in paleoenvironments and faunas in mainland Southeast Asia.

5.1. Sample Size Influence, Tooth Preservation, and Area 3

Our examination of sample size in the Ban Rai Rockshelter zooarchaeological assemblage clearly indicated that Area 3 was distinct from Areas 1 and 2. When Area 3 was included in the regression model, there was a significant relationship between increasing numbers of identified specimens and increasing numbers of taxa, and this was also true for minimum numbers of individuals; however, the removal of Area 3 from the model also removed the significant statistical relationship between these same variables.

An examination of the taxa of interest (Primate, Artiodactyla, Perissodactyla, Testudinata, and Cyprinidae) indicated that tooth preservation likely influenced the underlying presence or absence of faunal remains. Within Area 3, tooth preservation drove the identification of the majority of taxa, especially taxa identified at species level. Turtles and tortoises were all identified based on bone elements, and thus were identified broadly. When species were identified to genus and/or species level (e.g., *Bos gaurus*), this was typically due to the presence of tooth elements.

While our data did not allow for a further examination of additional taphonomic and fragmentation metrics (e.g., [44,51]; see other examples at Spirit Cave [2]), they did indicate that understanding the spatial context and preservation of the faunal remains at Ban Rai was significant. Areas 1 and 2 included smaller total abundances of specimens and species richness. In contrast, Area 3 was abundant in both species richness and specimens, but the presence of a large number of preserved tooth specimens influenced these counts. The influence of Area 3 on these regression models thus suggested that the unique characteristics of this area's faunal assemblage was meaningful for understanding human subsistence activities at the site.

For example, there was little evidence from the faunal assemblage at Ban Rai Rockshelter to suggest that a non-anthropogenic agent accumulated animal bones and teeth within this site—aside from carnivores and rodents gnawing at bones deposited by past forager occupants. The clear human association with these specimens, and the distinct nature of lithic assemblages—including sumatraliths, core tools, and flakes—in Area 3, suggested that humans occupied and inhabited this portion of the site more frequently than other areas. Area 3 was likely a butchering and habitation area, based on used stone tool debris, faunal remains, and thin charcoal layers. This area of the site also received the most sunlight. According to chronological analyses, the main habitat area at Ban Rai shifted from the center of the rockshelter (Areas 1 and 2) to the east wing (Area 3) [7]. This type of selective hunter-gatherer use of space matches known patterns from elsewhere in Southeast Asia: for example, at Niah Cave, where faunal debris in the inside cave area moves away from the entrance [52,53].

Other contemporaneous prehistoric archaeological sites in Thailand also exhibit evidence for this type of spatially distinct occupation within individual sites. Lang Kamnan in west-central Thailand is a useful comparative example of this type of context [54,55]. At this site, hunter-gatherers occupied certain areas, and then discarded lithic materials and faunal remains against a specific portion of the rockshelter wall. It is likely that a similar situation existed at Ban Rai Rockshelter in Area 3; however, further geoarchaeological research is required to understand if a specific taphonomic process facilitated the preferential preservation of tooth enamel within Area 3, in comparison to Areas 1–2.

5.2. Long-Term Subsistence Change

Recent evidence from several sites suggests that at some time during the Holocene, a shift towards the exploitation of primates and other arboreal taxa occurred throughout mainland and island Southeast Asia [2,8,56,57]. Paleoenvironmental research indicates the replacement of open grasslands by forested areas after the Last Glacial Maximum period [12,58,59], and this likely drove a transition away from higher abundances of Artiodactyla in the environment. During this period, arboreal and terrestrial forest taxa appear to have increased in the environment, and hunter-gatherers appear to have adapted their exploitation strategies to hunt these newly abundant preys. As with other nearby and contemporary sites (e.g., Ban Tha Si and Doi Pha Kan in Lam Pang province [60–62], and Spirit Cave, Steep Cliff Cave, and Banyan Valley Cave in Mae Hong Son province [1–3]), the fauna of Ban Rai Rockshelter is dominated by arboreal taxa, especially primates in Area 3. This differs slightly from sites dominated by artiodactyls during this period (e.g., Doi Pha Kan and Steep Cliff Cave), and suggests that the transition towards arboreal fauna was relatively consistent throughout northwest Thailand during the Early–Middle Holocene.

At Ban Rai Rockshelter Areas 1 and 2, both dating to the Terminal Pleistocene, assemblage data included evidence for the exploitation of artiodactyls and turtles/tortoises; however, there was possible change over time in the abundance of artiodactyls in each area—or, in this case, change in hunter-gatherer exploitation strategies. Sambar deer abundance likely influenced this change, as they were not present in the deepest layers of either area, but increased in abundance through time, before decreasing again in the upper layer of each context. A decrease in artiodactyls during this Terminal Pleistocene period is relevant, as it relates to the faunal assemblage from Area 3.

In Area 3 deposits dating to the Early and Middle Holocene, there was a substantial increase in the abundance of primates, beginning in Layer 5. Artiodactyls, turtles/tortoises and fishes also increased over time, but in lower comparative abundance than primates. In these contexts, within Area 3, primates dominated the total NISP and MNI for the unit: they only decreased in abundance, along with all other taxa, in the upper layers of this area.

In light of the above considerations, we diagnosed the relationship between artiodactyl exploitation in Areas 1–2 and primate exploitation in Area 3, based on the differences in the chronological age of each context. The findings suggest that hunter-gatherers occupying Ban Rai Rockshelter shifted their exploitation of arboreal prey to include primate taxa during the Pleistocene–Holocene transition or shortly thereafter.

Evidence from Late Pleistocene sites, such as Tham Lod Rockshelter [7,35], Ban Tha Si [60,61], and the Late Pleistocene Layers of Niah Cave [52,53] suggest that the subsistence patterns present during the Late Pleistocene focused on ground-dwelling prey, which shifted during the Early Holocene. Increased primate and arboreal exploitation identified throughout Southeast Asia at this transitional period—including Holocene Layers at sites such as Moh Khiew II [63], Niah Cave [52,53], and Song Gupuh [64]—helps support this record.

A relevant example to the pattern of shifting faunas and human exploitation identified at Ban Rai Rockshelter was found at Braholo Cave and Song Terus [56,57], in eastern Java. Braholo Cave and Song Terus, also dating to the Terminal Pleistocene/Early–Middle Holocene, are dominated by non-human primate specimens. Javan langur (*Trachypithecus auratus*) is the most abundant Cercopithecidae present, accompanied by long-tailed macaque (*Macaca fascicularis*) and Javan surili (*Presbytis comata*)—all of which were identified through analysis of dental morphology and metrics. In contrast to Ban Rai Rockshelter, the identified skeletal assemblages from Braholo Cave and Song Terus contained a large number of post-cranial (non-tooth) elements: these specimens were specifically collected by past hunter-gatherers focused on the manufacture of bone tools—especially at Song

Terus, where there was dedicated production of bone tools. This suggests that, while our determination of animal exploitation and change identified within the Ban Rai Rockshelter deposits is similar to patterns found throughout Southeast Asia, the exact preservation and taphonomic processes occurring within each site's zooarchaeological assemblage significantly impacts the way in which those determinations are supported. At Ban Rai, preserved tooth elements support a shift towards arboreal taxa during the Holocene.

5.3. Changes in Paleoenvironments and Paleofaunas

During the Terminal Pleistocene to Middle Holocene, the overall paleoenvironment of northern Thailand was a closed-canopy semi-tropical rain forest with more stable humidity and temperature [65], and limited montane forest at higher elevation, separated by a grassland and open-canopy woodland in the inland [12,13,20,21]. Meanwhile, sea level increased and transgressed in Thailand's central plain and south and east coastlines [65–67], as supported by marine environmental proxy data from continental sites far distant from the modern sea level [68,69]: coastal mud, mangrove pollen, marine shell, and corresponding high precipitation, and increased wetland and water resource areas, occurred in Thailand's interior [68,69].

Thus, an increased arboreal faunal diet for hunter-gatherers at Ban Rai Rockshelter is a useful proxy for understanding the highly stratified closed-canopy rainforest change that occurred during the end of the Pleistocene to the Middle Holocene. The remains of ground dwelling fauna—such as gaur, serow, hog deer, sambar deer, elephant, long-tailed giant rat, hog badger, and more (following previous studies [1,8] and an earlier isotope analysis on faunal teeth [12,13])—suggest the presence of a denser rainforest. A few of these ground-dwelling animals appear at Ban Rai Rockshelter, such as Asian black bear, Eld's deer, bamboo rat, goral, and wild water buffalo, which matches the evidence we expect for a transition from a partly open environment (such as an open-canopy woodland/grassland) to a montane forest and swamps (or muddy swamps) around the valley river [1,8,70,71].

Zooarchaeological evidence dating from the Terminal Pleistocene to the Middle Holocene at Ban Rai Rockshelter highlights the importance of arboreal taxa in prehistoric human diets, and supports a pattern that matches broader scale paleoenvironmental and paleoclimate changes in this region.

6. Conclusions

Three excavated areas at Ban Rai Rockshelter in Pang Mapha District, Mae Hong Son Province, northwest Thailand, establishes a zooarchaeological sequence suggesting specific exploitation of artiodactyls, primates, turtles, tortoises, and fishes throughout the Terminal Pleistocene and Holocene. Area 3, with the largest and most diverse faunal assemblage, suggests that hunter-gatherers focused their occupation of the site in this area; however, poor preservation of bone influenced the identification of taxa within this site assemblage. The shifting abundance of primate exploitation over time indicates that arboreal fauna became increasingly important for human subsistence during the Holocene era. These results add to a growing body of research suggesting that paleoenvironmental change, tied with faunal shifts and changing human population densities, led to differing human adaptations throughout the Southeast Asian tropics after the Last Glacial Maximum.

Supplementary Materials: The following supporting information can be downloaded at: https://www.mdpi.com/article/10.3390/quat5040051/s1: Table S1—Taxonomy and profile of Pit S1E4 and S2E12 of Area 1; Table S2—Taxonomy and profile of Pit S2E14, Baulk S1E14 and S1E6 of Area 2; Table S3—Taxonomy and profile of Pit N5E39 of Area 3.

Author Contributions: Conceptualization, A.W. and R.S.; methodology, A.W. and C.C.; software, C.C.; validation, All; formal analysis, A.W. and C.C.; investigation, R.S.; resources, R.S. and A.W.; data curation, A.W.; writing—original draft preparation, A.W.; writing—review and editing, R.S. and C.C.; visualization, C.C. and A.W.; supervision, R.S.; project administration, R.S. and A.W.; funding acquisition, R.S. All authors have read and agreed to the published version of the manuscript.

Funding: This research was funded by Thailand Science Research and Innovation (TSRI), grant number RTA6080001.

Institutional Review Board Statement: Not applicable.

Informed Consent Statement: Not applicable.

Data Availability Statement: All data and source code is available here: https://github.com/cylerc/BanRai (accessed on 10 October 2022).

Acknowledgments: This research is part of the Prehistoric Population and Cultural Dynamics in Highland Pang Mapha Project. We thank Juan Rofes, Janine Ochoa, and Emmanuelle Stoetzel for inviting us to contribute to this special issue. Thank you to three anonymous reviewers for their valuable feedback on the text.

Conflicts of Interest: The authors declare no conflict of interest.

References

1. Conrad, C. Mainland Southeast Asia in the Longue Durée: A Zooarchaeological Test of the "Broad Spectrum Revolution" in Northern Thailand. Ph.D. Thesis, The University of New Mexico, Albuquerque, New Mexico, 2018.
2. Conrad, C.; Higham, C.; Eda, M.; Marwick, B. Palaeoecology and forager subsistence strategies during the Pleistocene-Holocene transition: A reinvestigation of the zooarchaeological assemblage from Spirit Cave, Mae Hong Son Province, Thailand. *Asian Perspect.* **2016**, *55*, 2–27. [CrossRef]
3. Gorman, C.F. Excavations at Spirit Cave, North Thailand: Some Interim Interpretations. *Asian Perspect.* **1970**, *13*, 79–107.
4. Reynolds, T.E.G. Excavations at Banyan Valley Cave, Northeast Thailand: A Report on the 1972 Season. *Asian Perspect.* **1992**, *31*, 77–97.
5. Shoocongdej, R. Paleoenvironment during Late Pleistocene to Late Holocene in highland Pang Mapha District, Mae Hong Son Province. In *The Proceeding of the Conference on People, Culture, and Paleoenvironment in Highland Pang Mapha, Mae Hong Son Province, Silpakorn University (Tha Phra Palace), 20'21 February 2003*; Silpakorn University: Bangkok, Thailand, 2003; pp. 292–317. (In Thai)
6. Shoocongdej, R. Late Pleistocene activities at the Tham Lod rockshelter in highland Pang Mapha, Mae Hong Son province, northwestern Thailand. In *Uncovering Southeast Asia's Past: Selected Papers from the 10th International Conference of the European Association of Southeast Asian Archaeologists, the British Museum, London, 14'17 September 2004*; Bacus, E.A., Glover, I.C., Pigott, V., Eds.; NUS Press: Singapore, 2006; pp. 22–37.
7. Shoocongdej, R.; Phumijumnong, N.; Chintakanon, K.; Pureepatpong, N.; Hoontrakul, U.; Treerayapiwat, C. *Final Report of Highland Archaeology in Pang Mapha Distric, Mae Hong Son Province Phase 2*, 2007; Volume 2. Archaeology, Submitted to Thailand Research Fund (TRF) (in Thai)
8. Shoocongdej, R.; Wattanapituksakul, A. Faunal assemblages and demography during the Late Pleistocene (MIS 2-1) to Early Holocene in Highland Pang Mapha, Northwest Thailand. *Quat. Int.* **2020**, *563*, 51–63. [CrossRef]
9. Treerayapiwat, C. Patterns of habitation and burial activity in the Ban Rai rockshelter, northwestern Thailand. *Asian Perspect.* **2005**, *44*, 231–245. [CrossRef]
10. Conrad, C.; Shoocongdej, R.; Marwick, B.; White, J.C.; Thongcharoenchaikit, C.; Higham, C.; Feathers, J.K.; Tumpeesuwan, S.; Castillo, C.C.; Fuller, D.Q.; et al. Re-evaluating Pleistocene–Holocene occupation of cave sites in north-west Thailand: New radiocarbon and luminescence dating. *Antiquity* **2021**, *96*, 280–297. [CrossRef]
11. Gorman, C. The Hoabinhian and After: Subsistence Patterns in Southeast Asia during the Late Pleistocene and Early Recent Periods. *World Archaeol.* **1971**, *2*, 300–320. [CrossRef]
12. Suraprasit, K.; Jaeger, J.-J.; Shoocongdej, R.; Chaimanee, Y.; Wattanapituksakul, A.; Bocherens, H. Long-Term Isotope Evidence on the Diet and Habitat Breadth of Pleistocene to Holocene Caprines in Thailand: Implications for the Extirpation and Conservation of Himalayan Gorals. *Front. Ecol. Evol.* **2020**, *8*, 67. [CrossRef]
13. Suraprasit, K.; Shoocongdej, R.; Chintakanon, K.; Bocherens, H. Late Pleistocene human paleoecology in the highland savanna ecosystem of mainland Southeast Asia. *Sci. Rep.* **2021**, *11*, 16756. [CrossRef]
14. Wattanapituksakul, A.; Filoux, A.; Amphansri, A.; Tumpeesuwan, S. Late Pleistocene Caprinae assemblages of Tham Lod rockshelter (Mae Hong Son province, northwest Thailand). *Quat. Int.* **2018**, *493*, 212–226. [CrossRef]
15. Bacon, A.-M.; Duringer, P.; Westaway, K.; Joannes-Boyau, R.; Zhao, J.; Bourgon, N.; Dufour, E.; Pheng, S.; Tep, S.; Ponche, J.-L.; et al. Testing the savannah corridor hypothesis during MIS2: The Boh Dambang hyena site in southern Cambodia. *Quat. Int.* **2018**, *464*, 417–439. [CrossRef]
16. Heaney, L.R. A synopsis of climatic and vegetational change in Southeast Asia. *Clim. Change* **1991**, *19*, 53–61. [CrossRef]
17. Louys, J.; Roberts, P. Environmental drivers of megafauna and hominin extinction in Southeast Asia. *Nature* **2020**, *586*, 402–406. [CrossRef]
18. Suraprasit, K.; Jongautchariyakul, S.; Yamee, C.; Pothichaiya, C.; Bocherens, H. New fossil and isotope evidence for the Pleistocene zoogeographic transition and hypothesized savanna corridor in peninsular Thailand. *Quat. Sci. Rev.* **2019**, *221*, 105861. [CrossRef]

19. Wurster, C.M.; Bird, M.I.; Bull, I.D.; Creed, F.; Bryant, C.; Dungait, J.A.J.; Paz, V. Forest contraction in north equatorial Southeast Asia during the Last Glacial Period. *Proc. Natl. Acad. Sci.* **2010**, *107*, 15508–15511. [CrossRef] [PubMed]
20. Kealhofer, L.; Penny, D. A combined pollen and phytolith record for fourteen thousand years of vegetation change in northeastern Thailand. *Rev. Palaeobot. Palynol.* **1998**, *103*, 83–93. [CrossRef]
21. Penny, D. A 40,000 year palynological record from north-east Thailandl; implications for biogeography and palaeo-environmental reconstruction. *Palaeogeogr. Palaeoclimatol. Palaeoecol.* **2001**, *171*, 97–128. [CrossRef]
22. Sun, X.; Li, X.; Luo, Y.; Chen, X. The vegetation and climate at the last glaciation on the emerged continental shelf of the South China Sea. *Palaeogeogr. Palaeoclimatol. Palaeoecol.* **2000**, *160*, 301–316. [CrossRef]
23. Bien, M.N.; Chia, L.P. Cave and rock-shelter deposits in Yunnan. *Bull. Geol. Soc. China* **1938**, *18*, 325–347. [CrossRef]
24. Von Koenigswald, G. The relationship between the fossil mammalian faunae of Java and China, with special reference to early man. *Peking Nat. Hist. Bul.* **1939**, *13*, 293–298.
25. Kahlke, H.D. On the complex of the *Stegodon-Ailuropoda*-fauna of southern China and the chronological position of *Gigantopithecus blacki* V. Koenigswald. *Vert. PalAs.* **1961**, *6*, 83–108.
26. Louys, J. Quaternary extinctions in Southeast Asia. In *Mass Extinction*; Springer: Berlin, Germany, 2007; pp. 159–189.
27. Suraprasit, K.; Jaeger, J.-J.; Chaimanee, Y.; Chavasseau, O.; Yamee, C.; Tian, P.; Panda, S. The Middle Pleistocene vertebrate fauna from Khok Sung (Nakhon Ratchasima, Thailand): Biochronological and paleobiogeograpgical implications. *Zookeys* **2016**, *613*, 1–157.
28. Bacon, A.-M.; Westaway, K.; Antoine, P.O.; Duringer, P.; Blin, A.; Demeter, F.; Ponche, J.-L.; Zhao, J.-X.; Barnes, L.; Sayavongkhamdy, T.; et al. Late Pleistocene mammalian assemblages of Southeast Asia: New dating, mortality profiles and evolution of the predator-prey relationships in an environmental context. *Palaeogeogr. Palaeoclimatol. Palaeoecol.* **2015**, *422*, 101–127. [CrossRef]
29. Sathiamurthy, E.; Voris, H.K. Maps of Holocene sea level transgression and submerged lakes on the Sunda Shelf. *Nat. Hist. J. Chulalongkorn Univ.* **2006**, *2*, 1–43.
30. Fontaine, H.; Hoang, T.T.; Jaunngam, S.; Kavinate, S.; Salyapongse, S.; Sutheethorn, V.; Vachard, D. *Paleontology and Stratigraphy of the Northwest Thailand: Paleogeographical Implications*; Department of Mineral Resources: Bangkok, Thailand, 2009.
31. Grave, P. Beyond the Mandala: Buddhist Landscapes and Upland-Lowland Interaction in North-West Thailand AD 1200–1650. *World Archaeol.* **1995**, *27*, 243–265.
32. Marwick, B. Human behavioural ecology and stone artefacts in northwest Thailand during the terminal Pleistocene and Holocene. In Proceedings of the Southeast Asian Prehistory, the 11th EurASEAA Conference, Musée de Bougon, Paris, 25–30 September 2006; pp. 37–49.
33. Shoocongdej, R.; Venunan, P.; Chintakanon, K.; Kutanan, W.; Wonnapinij, P.; Surat, W.; Rodcharoen, P.; Wattanapituksakul, W.; Na Nongkai, W. *Final Report of Prehistoric Population and Cultural Dynamics in Highland Pang Mapha, Mae Hong Son Province*, 2021; Volume 2. Archaeology, Submitted to Thailand Science Research and Innovation (TSRI) (in Thai)
34. Siraboot, D.; Pureepatpong, N. *Result of animal bone analysis. In Final Report of Highland Archaeology in Pang Mapha District, Mae Hong Son Province*, 2003; Volume 2, 238–248. Submitted to Thailand Research Fund (TRF) (in Thai)
35. Amphansri, A. Animal bone. *Final Report of Highland Archaeology in Pang Mapha District, Mae Hong Son Province Phase 2*, 2007; Volume 2. Submitted to Thailand Research Fund (TRF) (in Thai)
36. Coolidge, H.J., Jr. The Indo-Chinese Forest Ox or Kouprey. *Mem. Mus. Comp. Zoölogy Harv. Coll.* **1940**, *54*, 421–531.
37. Hooijer, D.A. *Fossil Bovidae from the Malay Archipelago and the Punjab*; Rijksmuseum van Natuurlijke: Leiden, Netherlands, 1958; Volume 38.
38. Guérin, C. *Gallogoral* (nov. gen.) *meneghini* (Rütimeyer, 1878): Un Rupicaprine du Villafranchien d'Europe Occidentale. Ph.D. Thesis, University de Lyon, Lylon, France, 1966.
39. Thein, T. La Fauna Néolithique du Phnom Loang (Cambodge). Ph.D. Thesis, University Paris VI, Paris, France, 1974.
40. De Vos, J.; Long, V.T. Systematic Discussion of the Lang Trang Fauna, Vietnam. 1993; Unpublished Report.
41. Bacon, A.M.; Demeter, F.; Duringer, P.; Helm, C.; Bano, M.; Long, V.T.; Thuy, N.T.K.; Antoine, P.O.; Mai, B.T.; Huong, N.T.M.; et al. The Late Pleistocene Duoi U'Oi cave in northern Vietnam: Palaeontology, sedimentology, taphonomy and palaeoenvironments. *Quat. Sci. Rev.* **2008**, *27*, 1627–1654. [CrossRef]
42. Bacon, A.M.; Duringer, P.; Antoine, P.-O.; Demeter, F.; Shackelford, L.; Sayavongkhamdy, T.; Sichanthongtip, P.; Khamdalavong, P.; Nokhamaomphu, S.; Sysuphanh, V.; et al. The Middle Pleistocene mammalian fauna from Tam Hang karstic deposit, northern Laos: New data and evolutionary hypothesis. *Quat. Int.* **2011**, *245*, 315–332. [CrossRef]
43. Filoux, A.; Wattanapituksakul, A.; Lespes, C.; Thongcharoenchaikit, C. A Pleistocene mammal assemblage containing Ailuropoda and Pongo from Tham Prakai Phet cave, Chaiyaphum province, Thailand. *Geobios* **2015**, *48*, 341–349. [CrossRef]
44. Lyman, R.L. *Quantitative Paleozoology*; Cambridge University Press: Cambridge, UK, 2008.
45. Reitz, E.J.; Wing, E.S. *Zooarchaeology*; Cambridge University Press: Cambridge, UK, 2008.
46. Cannon, M.D. A Mathematical Model of the Effects of Screen Size on Zooarchaeological Relative Abundance Measures. *J. Archaeol. Sci.* **1999**, *26*, 205–214. [CrossRef]
47. Cannon, M.D. Archaeofaunal Relative Abundance, Sample Size, and Statistical Methods. *J. Archaeol. Sci.* **2001**, *28*, 185–195. [CrossRef]

48. Lyman, R.L. On the variable relationship between NISP and NTAXA in bird remains and in mammal remains. *J. Archaeol. Sci.* **2015**, *53*, 291–296. [CrossRef]
49. O'Hara, R.; Kotze, J. Do not log-transform count data. *Nat. Preced.* **2010**. [CrossRef]
50. Conrad, C.; Wattanapituksakul, A.; Shoocongdej, R. Compendium for the paper Preservation of Mammalian Teeth and Bones Influences Terminal Pleistocene to Middle Holocene Hunter-Gatherer Subsistence at Ban Rai Rockshelter, Northwest Thailand. 2022. Available online: https://github.com/cylerc/BanRai (accessed on 1 November 2022).
51. Lyman, R.L. *Vertebrate Taphonomy*; Cambridge University Press: Cambridge, UK, 1994.
52. Barton, H.; Piper, P.J.; Rabett, R.; Reeds, I. Composite hunting technologies from the Terminal Pleistocene and Early Holocene, Niah cave, Borneo. *J. Archaeol. Sci.* **2009**, *36*, 1708–1714. [CrossRef]
53. Piper, P.J. Human cultural, technological and adaptive changes from the end of the Pleistocene to the mid-Holocene in Southeast Asia. In *The Routledge Handbook of Bioarchaeology in Southeast Asia and the Pacific Islands*; Routledge: London, UK, 2015; pp. 52–72.
54. Shoocongdej, R. Forager Mobility Organization in Seasonal Tropical Environments: A View from Lang Kamnan Cave, Western Thailand. Ph.D. Dissertation, Department of Anthropology, University of Michigan, Ann Arbor, MI, USA, 1996.
55. Shoocongdej, R. Forager mobility organization in seasonal tropical environments of western Thailand. *World Archaeol.* **2000**, *32*, 14–40. [CrossRef]
56. Amano, N.; Moigne, A.-M.; Ingicco, T.; Sémah, F.; Awe, R.D.; Simanjuntak, T. Subsistence strategies and environment in Late Pleistocene–Early Holocene Eastern Java: Evidence from Braholo Cave. *Quat. Int.* **2016**, *416*, 46–63. [CrossRef]
57. Ingicco, T.; Amano, N.; Setiagama, K.; Moigne, A.M.; Budiman Sémah, A.M.; Simanjuntak, T.; Sémah, F. From Food to Grave Good: Nonhuman Primate Exploitation in Early to Mid-Holocene Eastern Java (Indonesia). *Curr. Anthropol.* **2020**, *61*, 264–277. [CrossRef]
58. Marwick, B.; Gagan, M.K. Late Pleistocene monsoon variability in northwest Thailand: An oxygen isotope sequence from the bivalve *Margaritanopsis laosensis* excavated in Mae Hong Son province. *Quat. Sci. Rev.* **2011**, *30*, 3088–3098. [CrossRef]
59. White, J.C.; Penny, D.; Kealhofer, L.; Maloney, B. Vegetation changes from the late Pleistocene through the Holocene from three areas of archaeological significance in Thailand. *Quat. Int.* **2004**, *113*, 111–132. [CrossRef]
60. Frère, S.; Auetrakulvit, P.; Zeitoun, V.; Sophady, H.; Forestier, H.; Tan, N.H. Doi Pha Kan (Thailand), Ban Tha Si (Thailand) and Laang Spean (Cambodia) Animal Bone Assemblages: A New Perception of Meat Supply Strategies for Early olocene Mainland Southeast Asia. In Proceedings of the Advancing Southeast Asian Archaeology 2019: Selected Papers from the Second SEAMEO SPAFA International Conference on Southeast Asian Archaeology, Noel Hidalgo Tan, Bangkok, Thailand, May 2016; pp. 100–108.
61. Zeitoun, V.; Auetrakulvit, P.; Forestier, H.; Zazzo, A.; Davtian, G.; Nakbunlung, S.; Tiamtinkrit, C. Discovery of a Mesolithic burial near the painted rock-shelter of Ban Tha Si (Lampang province, Northern Thailand): Implications for regional mortuary practices. *CR Palevol* **2013**, *12*, 127–136. [CrossRef]
62. Zeitoun, V.; Auetrakulvit, P.; Zazzo, A.; Pierret, A.; Frère, S.; Forestier, H. Discovery of an outstanding Hoabinhian site from the Late Pleistocene at Doi Pha Kan (Lampang province, northern Thailand). *Archaeol. Res. Asia* **2019**, *18*, 1–16. [CrossRef]
63. Auetrakulvit, P. Faunes du Pléistocène final à l'Holocène de Thaïlande: Approche archéozoologique. Ph.D. Dissertation, Université Aix-Marseille I, Marseille, France, 2004.
64. Morwood, M.J.; Sutikna, T.; Saptomo, E.W.; Westaway, K.E.; Jatmiko Due, R.A.; Moore, M.W.; Yuniawati, D.Y.; Hadi, P.; Zhao, J.; Turney, C.S.M.; et al. Climate, people and faunal succession on Java, Indonesia: Evidence from Song Gupuh. *J. Archaeol. Sci.* **2008**, *35*, 1776–1789. [CrossRef]
65. Higham, C. *Early Mainland Southeast Asia from first Humans to Angor*; River Books Co., Ltd.: Bangkok, Thailand, 2014.
66. Anderson, D.D. *Lang Rongrien rockshelter: A Pleistocene-Early Holocene Archaeological Site from Krabi, Southwestern Thailand*; The University Museum: Philadelphia, PA, USA, 1990.
67. Mudar, K.; Anderson, D. New evidence for Southeast Asian Pleistocene foraging economies: Faunal remains from the early levels of Lang Rongrien Rockshelter, Krabi, Thailand. *Asian Perspect.* **2007**, *46*, 298–334. [CrossRef]
68. Wohlfarth, B.; Klubseang, W.; Inthongkaew, S.; Fritz, S.C.; Blaauw, M.; Reimer, P.J.; Chabangborn, A.; Löwemark, L.; Chawchai, S. Holocene environmental changes in northeast Thailand as reconstructed from a tropical wetland. *Glob. Planet. Change* **2012**, *92*, 148–161. [CrossRef]
69. Chawchai, S.; Chabangborn, A.; Kylander, M.; Löwemark, L.; Mörth, C.-M.; Blaauw, M.; Klubseang, W.; Reimer, P.J.; Fritz, S.C.; Wohlfarth, B. Lake Kumphawapi-An archive of Holocene paleoenvironmental and paleoclimatic changes in northeast Thailand. *Quat. Sci. Rev.* **2013**, *68*, 59–75. [CrossRef]
70. Lekagul, B.; McNeely, J.A. *Mammals of Thailand*; Kurusapa Ladprao Press: Bangkok, Thailand, 1988.
71. Francis, C.M. *A Guide to the Mammals of Southeast Asia*; Princeton University Press: Princeton, NJ, USA, 2008.

quaternary

MDPI

Article

Neanderthal Fossils, Mobile Toolkit and a Hyena Den: The Archaeological Assemblage of Lateral Gallery 1 in Cova Del Gegant (NE Iberian Peninsula)

Joan Daura [1,*], Montserrat Sanz [1,*], Manuel Vaquero [2,3], Juan Manuel López-García [2,3], Hugues-Alexandre Blain [2,3] and Antonio Sánchez Marco [4]

1 Grup de Recerca del Quaternari (GRQ-SERP), Departament d'Història i Arqueologia, Universitat de Barcelona, Carrer Montalegre 6, 08001 Barcelona, Spain

2 Departament d'Història i Història de l'Art, Universitat Rovira i Virgili (URV), Avinguda de Catalunya 35, 43002 Tarragona, Spain; manuel.vaquero@urv.cat (M.V.); jmlopez@iphes.cat (J.M.L.-G.); hablain@iphes.cat (H.-A.B.)

3 Institut Català de Palaeoecologia Humana i Evolució Social (IPHES-CERCA), Campus Sescelades URV (Edifici W3), Zona Educacional 4, 43007 Tarragona, Spain

4 Institut Català de Palaeontologia Miquel Crusafont, Campus de la UAB, 08193 Bellaterra, Spain; antonio.sanchez@icp.cat

* Correspondence: jdaura_lujan@ub.edu (J.D.); montsesanzborras@ub.edu (M.S.)

Abstract: Lateral Gallery 1 (GL1) in Cova del Gegant is a Middle Palaeolithic assemblage yielding diagnostic Neanderthal remains, together with Mousterian tools and faunal remains. It is a good archive for evaluating the environmental conditions of the coastal areas during MIS 4 and MIS 3 in the NE of the Iberian Peninsula, and also the Neanderthals' behaviour and mobility. Here we provide a comprehensive assessment of all of the data available from GL1, such as lithics, human remains, fauna and chronostratigraphic details. The biotic ecofacts studied point to the development of a coastal plain in front of the cave and indicate that local conditions likely favoured a large variety of ecosystems characterised by open environments and woodland-edge taxa, and favoured repeated visits by humans during the Middle Palaeolithic. The evidence suggests that the gallery was mainly used by carnivores, such as hyenas, and also by Neanderthals as a brief stopping place, in view of the presence of transported and abandoned ergonomic lithic artifacts and/or the placement of bodies (or parts of bodies). The regional context suggests high human mobility and emphasises the variability of Neanderthal behaviour.

Keywords: late Pleistocene; Neanderthals; palaeoenvironment; NE Iberian Peninsula; hyena

1. Introduction

Archaeological sites are generally palimpsests, mixing remains from the successive activities of carnivores and humans carried out during their temporary stays. Thus, sites have often lost their individuality (i.e., their stratigraphic resolution or diachrony) and integrity (i.e., their internal organisation or synchrony). Archaeology tries to dissect these sites by applying diverse methods, technical means and interdisciplinary lines of research. For the most part, karst deposits with long archaeological occupations have been found to be low-resolution palimpsests that hinder the identification of human activities and occupational patterns [1–4].

Lateral Gallery 1 (GL1) in Cova del Gegant has been described in several publications, mostly concerning the chronological framework [5,6] and human remains. The first descriptions of two Neanderthal specimens from GL1 (namely Gegant-1 and Gegant-2) [7–10] led to renewed interest in the site. A chronostratigraphic approach was proposed with the goal of placing both remains in sequence [5,11]. Faunal remains were also analysed to explore the role of carnivores in the accumulation [10,12,13]. The first results concerning

GL1 were published subsequently [12,13], and showed that the cave was mainly used as a carnivore den; there were carnivore remains, coprolites and evidence of damage inflicted to ungulate bones, mainly by hyenas.

Neanderthals inhabited an extensive geographical area with a wide range of environmental conditions, extending from the westernmost part of Europe to Central Asia between Middle and Upper Pleistocene, 350–37.5 ka. They are known to be descendants of Middle Pleistocene regional populations [14,15]. The fate of the Neanderthal groups remains an important focus of research into human evolution, and much of this debate centres on south-western Europe [16]. After decades of investigation, many topics in Neanderthal research remain unclear, including, among others, the influence of palaeoenvironmental variability on the economy [17], the ways of life of the hunter-gatherer groups, their technology, and their symbolic behaviour.

To further examine the role of agents in the accumulation and to refine the Neanderthal record of the cave, this study aims to reconstruct the context in which the Neanderthal remains (Gegant-1 and Gegant-2) were found. Cova del Gegant is one of the palaeolithic sites in the NE of the Iberian Peninsula that has yielded Neanderthal specimens in association with Middle Palaeolithic artifacts and Pleistocene faunal remains [7,8,18]. Some of the recovered artifacts have been presented previously [5] in a study that provided a preliminary assessment of the Neanderthal context from GL1. This paper expands on that previous study, providing (i) a more detailed description of the chronostratigraphic sequence of GL1, (ii) an improved context for the Neanderthal remains Gegant-1 and Gegant-2, and (iii) a contextualisation of human occupancies inside the Middle Palaeolithic record. To achieve these goals, we present a synoptic overview of the chronostratigraphy and data on the large mammal, small vertebrate, bird, lithic and human remains for GL1. The multi-proxy analysis conducted reveals a dynamic of increasing occupation during the Upper Pleistocene in the Iberian Peninsula.

2. Study Area and Site Description

Cova del Gegant is located in the Garraf Massif, a horst composed mainly of Mesozoic rocks [19]. The massif is a low-relief mountain range that rises to a height of almost 600 m, and the central part is dominated by karst landform [20]. The morphological elements that make up this sector are defined by the structural history, lithology, and sea level fluctuations. The cliffy coast is mainly composed of calcareous Mesozoic outcrops, and sandy beaches are created by the longshore dynamics [21]. The most important landform in the area is the Riera de Ribes delta where the present-day shore is eroded by transgression. The detailed regional context, including geology and bathymetry, is shown in Figure 1. The area is located in the meso- and thermo-Mediterranean zone, which is composed primarily of evergreen oak forest *(Quercion ilicis)* with a few deciduous oak communities. The presence of carbonate outcrops has favoured the development of thermo-Mediterranean vegetation (Oleo-Ceratonion) characterised by the presence of *Quercus coccifera, Olea europaea, Pistacia lentiscus, Pinus halepensis* and *Chamaerops humilis* [22].

Cova del Gegant (Punta de les Coves, Sitges, Barcelona) is a complex karstic system located in the south-west edge of the massif in front of the Mediterranean sea (1°46′27.33″ E, 41°13′24.75″ N) (Figure 2(1,2)). A small joint system in the limestone rock, sea level fluctuations and water table oscillations favoured the development of several caves in the area, which include the adjacent Cova Llarga or de la Trompeta (connected by a narrow passage, GL-T), Cova del Musclo, Cova de la Masia de les Coves, Cova Verda and Cova de l'Aina. The caves are accessible via a horizontal entrance located at sea level and partially eroded by sea waves (Figure 2(4)). Cova del Gegant has several galleries (Figure 2(5)) located a few metres above sea level, where archaeological sediments have been preserved. Cova del Gegant is mainly formed by three different chambers, one large gallery that is 22 m long (GP1 + GP2) and two lateral galleries (GL) that are almost parallel to each other; one closer to the sea (GL1) (Figure 2(6)) and the other deeper inside the cave (GL2) (Figure 2(5)). Two small remnants of cemented sediments are preserved from sea erosion, one between Cova del Gegant and Cova Llarga and the other close to the GL1 entrance (BPII). The former yielded a Levallois point (Figure 2(3)).

Previous excavations at Cova del Gegant were mainly carried out in Lateral Gallery 1 (GL1). The site was probably discovered in 1880 during the construction of a rail trench [23], photographed at the beginning of the twentieth century and first excavated in 1952 by the Agrupació Muntanyenca de Sitges (AMUNT), under the supervision of the palaeontologist Santiago Casanova [24]. Following this, excavations took place during the 1960s and 1970s [25–27], and additional campaigns were undertaken in 1985 and 1989 to save material threatened by coastal erosion [28–31]. The current fieldwork were started in 2007 by the Grup de Recerca del Quaternari at the University of Barcelona. This group has focused on a section of preserved sediments at the back of the main gallery (GP2).

The publication of the Gegant-1 and Gegant-2 specimens [7–9] recovered from a lateral gallery (GL1) has renewed interest in the cave. The chronostratigraphic proposal was constructed using previous stratigraphic data and sediments preserved in the cave with the goal of contextualizing the human remains [5,11]. Subsequently, two new Neanderthal remains (namely Gegant-4 and Gegant-5) were recovered at the rear of the main gallery (GP2) [18]. Faunal remains, a few stone tools and a large number of coprolites from within the archaeological layers point to hominin and carnivore occupation of the site in the past [32].

The stratigraphic sequence at Cova del Gegant was grouped [6–8] into eight site formation episodes from the Late Pleistocene (Episodes 1–3), ca. 145–30 ka, to the Holocene (Episodes 4–7), alternating between periods of continental and marine sediment deposition and periods of coastal erosion. Episode 0 occurs at the base of the sequence and is represented by layer XV (GL1 and GP2), which consists of a thin layer of endokarstic red clay deposit. This is followed by Episode 1, which corresponds to Middle Palaeolithic occupancies and is represented by several layers (XXX, V and probably also layer XVII, XVIII and XVa/b). Above this deposit, Episodes 2 and 3 contain the Upper Palaeolithic record, the former represented by layers IV, (GP2), XIIb (GL2) and VIII (GP1) and the latter composed of layers III, II and IIb (GP, GL2), and part of layer IX (GP1). The chronological framework of the episodes is mainly constructed by the existing profile preserved at the rear of the GP2, the Middle Palaeolithic sequence covering ~94–52 ka (layer V 94–59 ka and XVII/III ca. 60–52 ka), a Châtelperronian/Aurignacian section spanning ~43–39 ka (layer IV), and a Late Aurignacian/Gravettian section spanning ~34–32 ka (layer III–II).

Episode 4 (Late Bell Beaker) is the oldest of the Holocene layers and is represented by layer VI and Ic_2 (GP and GL2) and layer XXV (GP2). Two additional storage pits (Silo 1 and Silo 2) are probably related to the same episode formation. Episode 5, represented by layers VII, X and XI in the GP1 sector and Ib_{2dsup} and Ib_{d2base} (in GP2), corresponds to an erosive transgression that emptied part of the deposit and accumulated beach sand between the Iron Age and the Middle Ages. Episode 6 is the most recent depositional event and is characterised by layers Ia, Ia_3, Ib_{2b}, Ib2c, Ic, Ie, and If (Middle Ages to Early Modern Period). Episode 7 corresponds to current sea erosion [5].

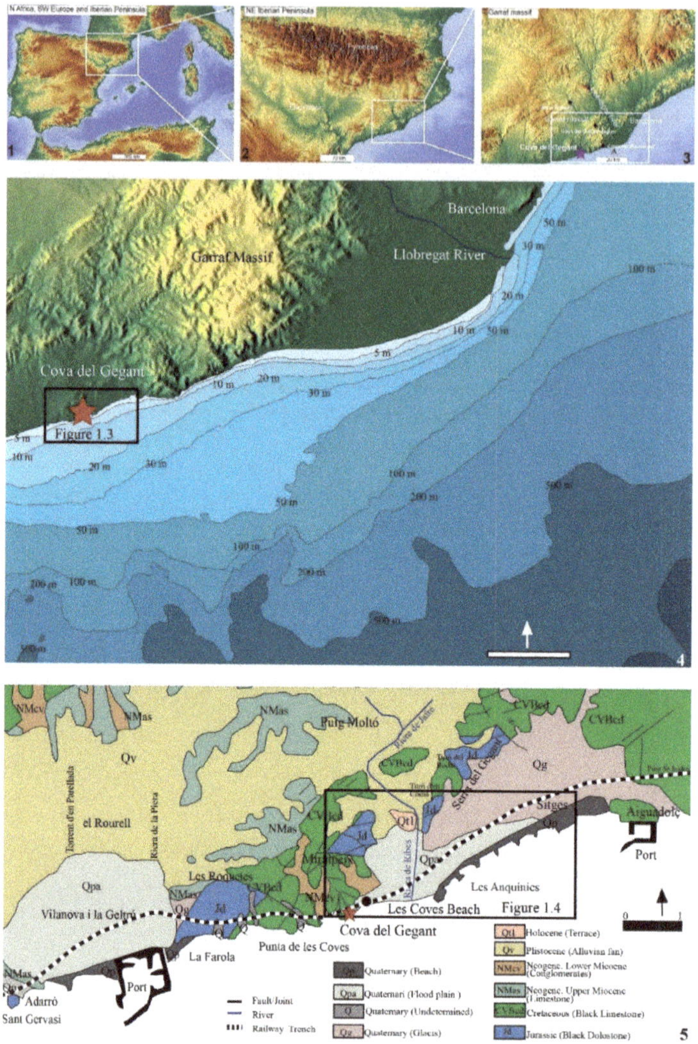

Figure 1. Regional context of Cova del Gegant. (**1–3**): Location of the archaeological site. The map was downloaded from http://maps-for-free.com/ (© OpenStreetMap contributors, accessed on 21 December 2021). The cartography in the OpenStreetMap map tiles is licensed under CC BY-SA (www.openstreetmap.org/copyright, accessed on 21 December 2021). The licence terms can be found on the following link: http://creativecommons.org/licenses/by-sa/2.0/ (accessed on 21 December 2021). (**4**): Topography and bathymetry of the Barcelona area, showing major isobaths. (**5**): Regional geology based on ICGC data base.

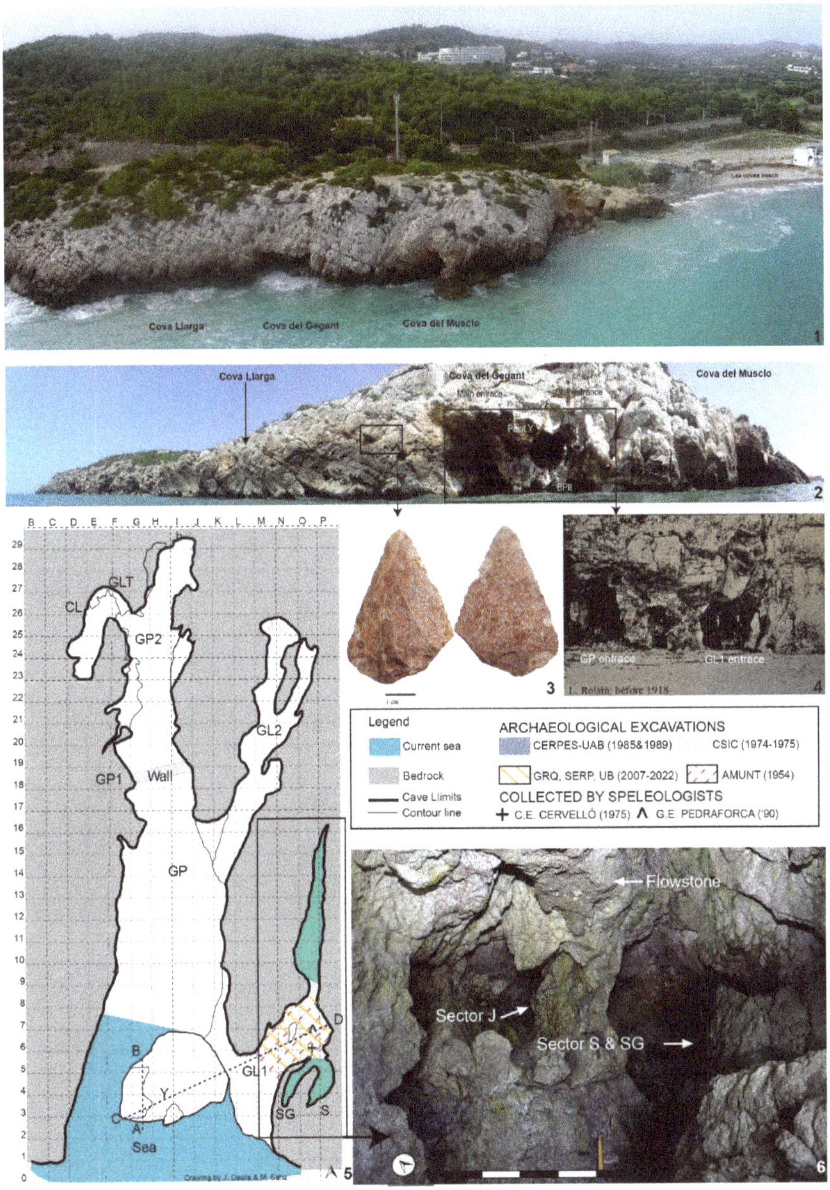

Figure 2. Cova del Gegant. (**1**): Drone view in 2017. (**2**): Panoramic view in 2012. (**3**): Levallois point recovered from the external breccia. (**4**): Cova del Gegant and Cova del Musclo before 1918 (L. Roisin, Arxiu Històric de Sitges, J. Mates Collection). (**5**): Plan of the site indicating fieldwork carried out in GL1. (**6**): Detailed view of GL1 in 2005.

3. Materials and Methods

The present study includes the revision of archaeological materials recovered during several field seasons conducted between 1956 and 2022. Table 1 summarises the archaeological excavations in GL1 and the museums where the materials are stored. The materials analysed here include small vertebrates, large mammals, birds, lithic artifacts and human remains. All of the archaeological and palaeontological collections from previous campaigns have been analysed with the exception of the materials from 1985 and 1989 because permission has not been obtained. However, the accounts have been included and are summarised here according to the published data.

3.1. Dating Methods

Two different dating methods were applied in the GL1 gallery to construct the chronological framework: U-Th dating of speleothems and bones, and luminescence dating of sediments. A detailed methodological description for each technique and the sample provenience has been explained in detail in Daura et al. [5,6]. In summary U-Th dating of the flowstone sealing the GL1 (sample #2) was performed using alpha-spectrometers BR-024-450-100 ORTEC OCTETE PLUS at the Institute of Earth Sciences Jaume Almera (ICTA-CSIC) [5]. The chemical separation and purification followed the procedure described by Bischoff et al. [33] and the age was calculated using the UDATE program devised by Rosenbauer [34]. The same speleothem was re-sampled (samples #5 and #6) using updated methods, and isotopes were measured on a ThermoFinnigan Neptune MC-ICPMS with a Cetac Aridus II and a Savillex PFA 50 μL/min microconcentric nebuliser. The pre-treatment processes used for the U–Th samples are described in Hoffmann et al. [35], while the measurement methods and protocols follow Hoffmann [36] and Hoffmann et al. [37,38]. The Neanderthal mandible (Gegant-1) (samples #3 and #4) was directly dated using the Difusion-Adsorption (D-A) model [39] to account for uranium uptake. U-Th isotopes were measured by the laser ablation multi collector ICP-MS hosted at the University of Bristol using the methodology described by Pike et al. [40].

The luminescence dating (IRSL) of polymineral fine grains (sample #1) was carried out at the Laboratorio de Datación y Radioquímica, Universidad Autónoma de Madrid following Valero-Garcés et al. [41]. The chemical and physical treatment followed procedures described by Aitken [42], and the conversion factors of Nambi and Aitken [43] were used to derive alpha, beta and gamma dose rate estimates from measured elemental concentrations and specific activities. Measurements were carried out on a Risø TL-DA-10 reader equipped with IR LEDs and a calibrated 90Sr/90Y beta source.

3.2. Faunal Remains

The faunal remains described here include all of the collections from GL1 that are housed at the institutions listed in Table 1 [13]. The collection of the Universitat Autonoma de Barcelona (UAB) was not available for this study, and for this reason the data from this collection included here are from a previous archaeological report [29]. The information on the rhinoceros is from previous studies [44] and the quantification was updated using available data [45].

The analysis of large mammal bones (not including leporids) involved the classification of the skeletal specimens identifiable at genus or species level following conventional zooarchaeological criteria [3,46]. The unidentifiable bones were catalogued to bone tissue types (i.e., long, flat, or spongy bones). The number of remains (NR), the minimum number of individuals (MNI) and the number of identified specimens (NISP) were estimated and calculated for a given taxon/genus [47–49]. Five age categories, such as foetal/neonatal, infant, juvenile, adult and senile, are used in the classification of age groups.

Bones were analysed following standard taphonomic methods [50,51]. Bone surfaces were examined under a binocular device (MOTIC SFC-121 GG). Human modifications were identified by butchery damage and breakage [52,53]. Carnivore damage was diagnosed by gnawing modifications and breakage [54,55]. Burning categories were described according to surface colour, fractures, in relation to angle, outline and edge and weathering based on the criteria proposed by Behrensmeyer [56].

Table 1. Information summarising excavations conducted at GL1.

Year	Direction	Institution	Materials Stored	Acronyms
1954	S. Casanova	AMUNT	Arxiu Històric Municipal de Sites, Museu Mar i Cel, Museu Arqueològic de Catalunya	AHSI MMC MAC
1972, 1974–75	R. Viñas-J. Villalta	Secció Ecologia Quaternari- Institut Jaume Almera	Museu Geològic de Barcelona Museu Arqueològic de Catalunya	MGB MAC
'70	Collected by speleologists (A. Asensio)	Centre Espeleològic Cervelló	Universitat de Barcelona	UB-1
1985	J. Martínez- R. Mora- I. Muro- J. Miret	Centre de Recerques Palaeoecosocials de Girona i de la Universitat Autònoma de Barcelona	Universitat Autònoma de Barcelona	UAB
1989	J. Martínez- R. Mora- J. Parcerisas- G. Roca	Centre de Recerques Palaeoecosocials de Girona i de la Universitat Autònoma de Barcelona	Universitat Autònoma de Barcelona	UAB
'90	Collected by speleologists (M. Nebot)	Grup Espeleològic Pedraforca	Institut Català de Paleontologia	ICP
2007–2022	J. Daura-M. Sanz	Grup Recerca del Quaternari-SERP (Universitat Barcelona)	Universitat de Barcelona	UB-2

Avian remains and the small vertebrates studied here come from the excavations undertaken by Viñas and Villalta in 1974–1975. Sediments were water screened using 2 mm and 5 mm meshes. These remains were hosted at the Museu de Geologia de Barcelona and small vertebrates were previously listed in two catalogues of this museum's collections [57,58] and then analysed [59]. Birds were loaned for a comprehensive study of the birds of the Iberian Quaternary [60].

3.3. Lithic Assemblage

The lithic assemblage from Cova del Gegant is the result of various field seasons carried out at the cave. Most artifacts were found during the 1974–1975 (n = 30) and 1985 (n = 40) excavations. The main part of this collection was located in gallery GL1, although some lithics were also found in the principal chamber (GP) and gallery GL2. According to Mir [61], the 1974–1975 assemblage was made up of four sidescrapers, three denticulates, three splintered pieces, eighteen flakes and two cores. Besides the high number of retouched implements, this industry was characterised by the presence of Levallois flakes, and a resemblance to the Charentian Mousterian of Ferrassie type was suggested. The 1985 excavation yielded seven sidescrapers, eight denticulates, one point, one endscraper, twenty flakes, one core and two fragments [28]. Additional information on the 1985 materials can be found in other unpublished documents [31,62]. At first sight, the 1974–1975 and 1985 collections appeared similar due to the high proportion of retouched artifacts, the balance between sidescrapers and denticulates, and the predominant use of flint as the raw material. Only four quartz artifacts were found in the 1985 excavation. Some more were recovered in the 1989 excavation, in GL1 and GL2, but they remain unpublished [29]. Finally, four additional artifacts were found in the 2007 field season: three in GL1 and one in GP.

Here, we have analysed 33 artifacts, including the collection from the 1974–1975 excavation housed at the Museu d'Arqueologia de Catalunya (Barcelona), and three artifacts from GL1 found in the 2007–2022 excavation. Although this is only a part of the Cova del Gegant lithic assemblage, it allows us to make some qualitative assessments of their technological features.

The method used in the study applied an attribute analysis. The attributes related to dimensions (length, width, and thickness) were recorded for all the artifact classes (cores, flakes and retouched tools). Lithics were classified according to five size categories established by multiplying length by width: very small <500 mm^2, small 500–1000 mm^2, medium 1000–1500 mm^2, large 1500–2000 mm^2 and very large >2000 mm^2.

The attributes recorded for the cores were related to their structure, including the number and characteristics of striking platforms and flaking surfaces. We also considered the hierarchisation of the flaking surfaces and reduction degree. The attributes recorded for flakes were the following: presence of cortex in the striking platform (cortical or uncortical), striking platform type (flat, linear or punctiform), striking platform faceting (unprepared, dihedral, multifaceted), amount of cortex in the dorsal face, bulb of percussion type (marked or diffuse), and curvature of the ventral face (concave, convex, straight or winding). For the retouched artifacts, we used the attribute analysis and the typological system of Laplace [63].

3.4. Human Remains

Two Neanderthal remains were identified in GL1. The mandible Gegant-1 was recovered during the 1952 excavation conducted by Santiago Casanova and was not recognised until 2001 [7]. The mandible was stored between 1998 and 2008 at the Arxiu Historic de Sitges and is now exhibited at the Museu d'Arqueologia de Catalunya.

A Neanderthal tooth (incisor) was recovered during the 1974 and 1975 excavations and stored at the Museu de Geologia de Barcelona as part of the Villalta collection [64,65]. It was kept with other dental remains and was not identified as Neanderthal until the study of the palaeontological collection of faunal remains [10], and was subsequently published [8].

The tooth was labelled twice. One label refers to the museum reference (V-2828) and the second to the excavation record showing grid provenance (c-5) and depth (0–30 cm below the 1975 datum).

4. Results

4.1. GL1 Stratigraphy and Chronology

GL1 is the closest gallery to the sea and today is exposed to sea erosion. Remnant sediments are still visible, adhering in the breccia at point Y and to the rear of GL1. Figure 2(5) shows the location of the profiles described here and Figure 3(3,4,6) shows the sedimentary architecture of this cave section as preserved in 2007. Profile C–D is a 10-m-long section, which allows us to correlate the layers where the Cova del Gegant's human mandible (Gegant-1) and incisor (Gegant-2) were found with the rest of the cavity. Figure 3(1,2) correlates the previous stratigraphic sequences of the 1954 campaign, Masriera's sedimentological study [66], Viñas and Villalta [25] and the 1985 excavation [62] with the current stratigraphic log. A stalagmitic flowstone (Figure 3(2,5)) located at the top of the sequence seals part of the sediment fill and has been dated by means of alpha-spectrometry and TIMS methodology (samples #2 and #5/6 respectively). Two different, distinct units have been identified in the preserved sediments. These units are related to episodes described in Daura et al. [5]. Episode 1 is preserved at the top of the gallery entrance (breccia at point Y) (Figure 3(3,4)) and at the rear of GL1 (Figure 3(5)). In GL1, this episode is represented by layer XVa, preserved under the stalagmitic flower and at the vertex of the triangle formed by the gallery morphology (XVb). It is a hard-cemented breccia composed of angular pebbles and pink (Munsell 7.5YR 7/4) lutitic matrix. At the breccia at point Y, concretion has helped preservation from maritime activity and the identified layers present similarities with XVa/b. However, this correlation is tentative because there is a large horizontal discontinuity between the two remnants. Layers XVII (a/b) and XVIII present boulder and angular pebbles in a brown lutitic matrix. Bones and coprolites are visible. Between the cave entrance and the sea, another remnant of cemented breccia (BPII) is stuck to the bedrock, is very strongly cemented and is difficult to study. Its existence indicates that most of Cova del Gegant's external part was filled with sediment, which is no longer present (Figure 3(3)). Episode 0 is only located at the gallery base and is represented by layer XVI. It consists of red endokarstic clay with no archaeological remains.

The speleothem located at the top of GL1 and capping layer XVa has been dated twice. On the first occasion, the alpha spectrometry (Table 2, sample #2) method was used to provide an approach to the age range. The nominal age obtained of 49.4 $\pm$ 1.8 ka may be slightly too high as a consequence of 232 Th contamination. MC-ICPMS (Table 2, sample #5 and 6) were used to re-sample this speleothem and 232Th was used as a proxy for detrital contribution, which in general is not high. The uncorrected nominal age for the basal speleothem layer shows similarities with the alpha methodology results. However, after correction, it gives an age of 47.1 $\pm$ 1.2 ka (2 σ uncertainty) for the basal layer and 45.5 $\pm$ 0.4 ka for the uppermost layer. Assuming that the Neanderthal mandible was located at the top of layer XVa and that the uranium adsorption was rapid, its direct dating result of 52.3 $\pm$ 2.3 ka (2 σ uncertainty) (Table 2, samples #5 and #6) is consistent with the capping flowstone. The IRSL date for the medial and basal deposit (XVb) from the gallery at 60.1 $\pm$ 3.9 ka (1 σ) provides the maximum age for the entire deposit preserved at GL1 (Table 2, sample #1). Figure 3(2) summarises and correlates the previous excavation and the layer identified in a single log related to the current site episodes (Figure 3(1)).

Table 2. Final ages of the GL1 gallery from the Cova del Gegant site. Luminescence mean ± total uncertainty (1 σ or 68% confidence interval), calculated as the quadratic sum of the random and systematic uncertainties. U-series dating results for speleothems from Cova del Gegant (2 σ confidence interval). All ratios are activity ratios calculated from isotope concentration ratios using decay constants according to Jaffey et al. [67] (λ238), Cheng et al. [68] (λ234 and λ230) and Holden [69] (λ232).

			Luminescence (IRSL)					
#	**Sample Name**	**Layer**	**Mineral**	**Grain Size (μm)**	**Total Dose Rate (Gy/ka)**	**D_e (Gy)**		**Age ka (1σ)**
1	MAD-5642	XVb	Polymineral	2–10	1.59 ± 0.10	95.5 ± 2.2		60.1 ± 3.9
				U-series				
	Lab #	**Layer**	**^{238}U**	**^{232}Th**	**$^{234}U/^{238}U$**	**$^{230}Th/^{238}U$**	**$^{230}Th/^{232}Th$**	**Age ka (2 σ)**
2	6006	XVa	0.22	0.03	1.16 ± 0.03	-	10.48	49.4 ± 1.8
3	JD52a	XVa/b	n/a	n/a	1.262 ± 0.017	0.495 ± 0.023	>100	53.2 ± 3.3
4	JD52b	XVa/b	n/a	n/a	1.262 ± 0.017	0.495 ± 0.023	>100	53.2 ± 3.3
5	UEVA 1028	XVa	367.64 ± 3.58	37.63 ± 0.39	1.1615 ± 0.0024	0.4281 ± 0.0023	12.78 ± 0.06	47.11 ± 1.21
6	UEVA 1029	XVa	37.63 ± 0.70	0.191 ± 0.004	1.1621 ± 0.0030	0.3933 ± 0.0028	237.01 ± 2.40	44.51 ± 0.42

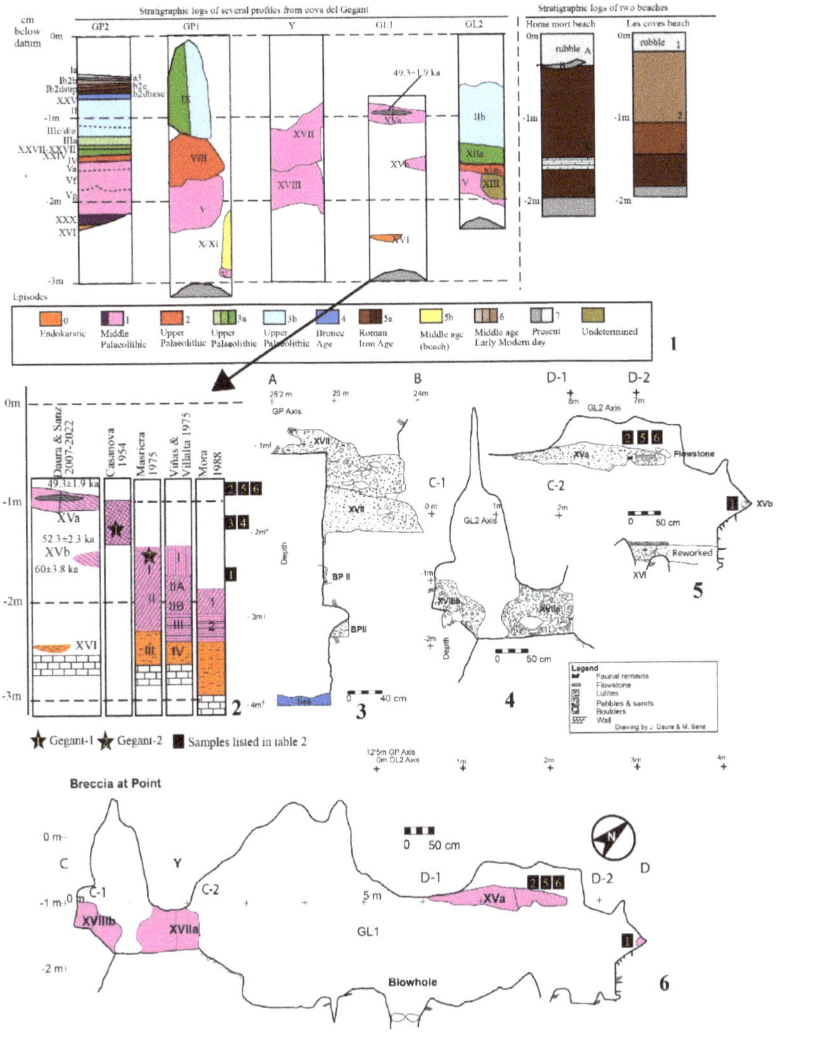

Figure 3. (**1**): Stratigraphy and correlation of the Cova del Gegant's logs. (**2**): GL1 logs integrating previous excavations. (**3**): Transversal profile in the point Y breccia. (**4**): Transversal profile in the point Y breccia. (**5**): GL-1 section. (**6**): GL-1 cross section.

4.2. Neanderthal Remains

Two Neanderthal remains have been recovered at GL1 (Figure 4). The mandible Gegant-1 (Figure 4(1) and Supplementary Materials) is represented by three fragments that comprise part of the mandibular corpus from the right M1 tooth socket to the mesial margin of the left M3 alveolus. No teeth are preserved along the specimen, but the distal root of the left M2 is present in its root socket. The alveolar margin of the specimen is abraded. The absence of teeth make it hard to determine the age at death, but the presence and development of the preserved root and root sockets indicate a minimum age at death of around 15 years. The mandible presents clear archaic features which correspond to Neanderthals: (i) the lack of a bony chin, (ii) the posterior placement of the anatomical structures and (iii) the low position of the mental foramen.

Gegant-2 (Figure 4(2)) is a lower left lateral permanent incisor (I2) which preserves the entire crown but approximately a third of the apical root is missing. The size of the anterior teeth helps to assess the taxonomic affinities of the Cova del Gegant tooth. The buccolingual and mesiodistal dimensions and discriminant analysis of these two variables characterise it as a Neanderthal specimen. The root tip was broken postmortem, making it difficult to determine the time of death. The degree of tooth wear suggests that the tooth belonged to a second individual from the site, who was perhaps around 10 years old.

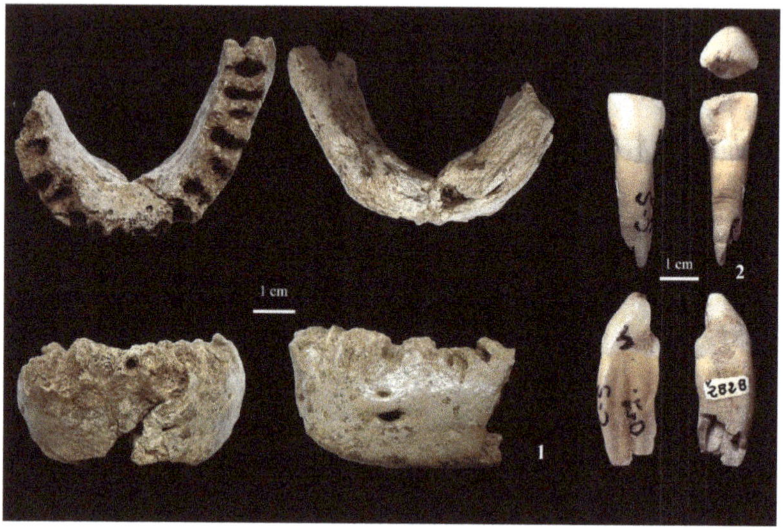

Figure 4. Neanderthal remains identified at GL1 from Cova del Gegant. (**1**): Different views of the adult mandible Gegant-1. (**2**): The lower incisor, Gegant-2, from an individual of 10 years old.

4.3. Large Mammals

Table 3 shows the results of the identification of large mammals in GL1. Equids are the most abundant taxa (34%NISP, NISP = 894), followed by red deer (17%NISP, NISP = 448) and large bovids (7%NISP, NISP = 186). Carnivores are also represented in the assemblage. The most abundant is the hyena (7%NISP, NISP = 179). Other taxa are less frequent, and include rhinoceros (1%NISP, NISP = 36), ibex (≤1%NISP, NISP = 18), proboscideans (≤1%NISP, NISP = 7) and wild boar (≤1%NISP, NISP = 10). Other less frequent carnivores are the wild cat (1%NISP, NISP = 40), Iberian lynx (1%NISP, NISP = 30), leopard (≤1% del NISP, NISP = 13), wolf (1%NISP, NISP = 28), brown bear (≤1%NISP, NISP = 19), fox (≤1%NISP, NISP = 11), dhole (≤1% del NISP, NISP = 6) and badger (≤1% del NISP, NISP = 3). Following Discamps [70] and based on MNI values, we estimated the ungulate biomass index for the GL1 assemblage, which equals 0.7. This value is closer to the index for modern-day grasslands (index = 1), with a biomass higher than 3 ton/km^2, than to the value for forest and tundra biomes (index = 0), with a biomass lower than 1 ton/km^2. This implies the presence of vegetation that can support and feed large herbivores and indicates an open landscape environment rather than forested areas close to the study site.

Adult individuals (ungulates and carnivores) are the most abundant (56%) in the assemblage (Figure 5). Hyena individuals are represented by several ontogenetic ages. Infants and juveniles are most abundant among the carnivores. Equids are dominated by adults and the same can be said for deer and large bovids. A single foetal remain has been identified, probably from a perissodactyl.

Table 3. Number of Identified Specimens (NISP) and Minimum Number of Individuals (MNI) of the large mammals identified at GL1.

	NISP	MNI
Carnivora	329	55
Canis lupus	28	5
Vulpes vulpes	11	1
Ursus sp./*U. arctos*	19	2
Meles meles	3	1
Crocuta crocuta	179	24
Felis silvestris	40	8
Lynx pardinus	30	7
Panthera pardus	13	5
Cuon alpinus	6	2
Artiodactyla	662	47
Sus scrofa	10	3
Cervus elaphus	448	22
Bos/Bison	103	19
Bison priscus	18	
Bos primigenius	65	
Capra pyrenaica	18	3
Perissodactyla	930	49
Equus ferus	823	40
Equus hydruntinus	71	7
Stephanorhinus hemitoechus	36	2
Proboscidea	7	1
Elephas sp.	3	
Proboscidea indet.	4	1
Undetermined		
Macrofauna indet.	674	
Total	2602	152

The skeletal frequencies are similar for ungulates and carnivores. The cranial elements are the most abundant (NR = 592 for ungulates and NR = 161 for carnivores), and isolated dentition is the most represented. Autopodium elements are also relevant (NR = 305 for ungulates and NR = 55 for carnivores). The other skeletal elements are less frequent.

The main taphonomic processes that had affected the bones were concretion (34%) and manganese coatings (23%), but they show only a low degree of alteration. Abrasion is observed in 15% of the remains and presents different degrees of alteration, in some cases affecting only a small area and in others a large part of the surface. Exfoliation (5%) and desquamation (7%) are also present. Less frequent are rodent activity (2%), trampling (2%), root etching, sediment pressure and dissolution (≤1%).

The abundant biological activity observed in bones is due to carnivore ravaging. Furrows are frequent (3%), epiphyses may have been gnawed away and hollowed out (2%), and pitting is present (1%). Digested bones are scarce (≤1%). No direct anthropogenic activity on the bones is documented (e.g., cut-marks, diagnostic elements of bone breakage, etc.), and burned bones are scarce (≤1%) (Figure 6).

Two fracture morphologies are observed: the first by oblique angles (34%) with a curved outline (59%) and a smooth surface (39%), which suggests fresh-bone breakage, and the second with right (28%) and mixed angles (38%), transversal (34%) and an irregular fracture surface (61%), which suggests dry-bone breakage.

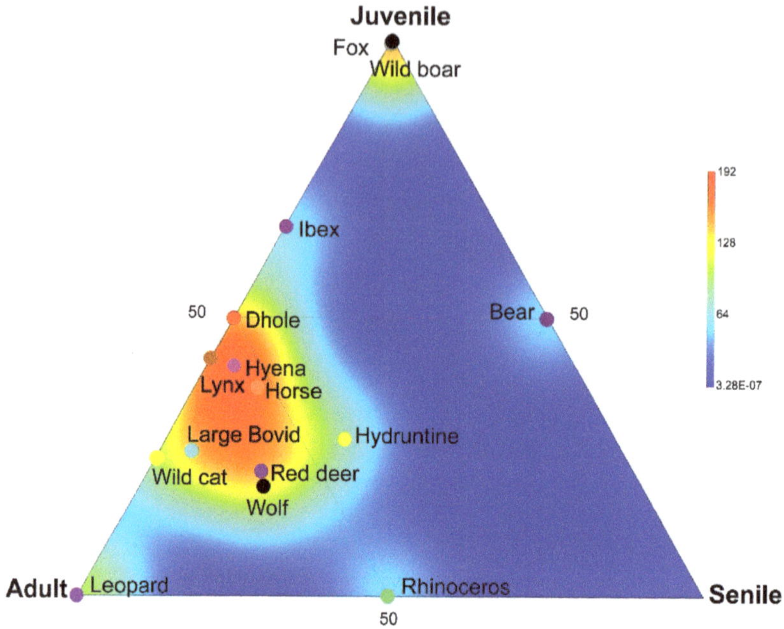

Figure 5. Ternary plot distribution of large mammals from Cova del Gegant-GL1. Dots indicate the percentage median of individuals for juveniles, adults, and senile individuals. Analysis was performed using PAST software [71].

Equid and large bovid metapodials are almost complete, and damage inflicted by carnivores is restricted to distal condyles. The consumption of stylopodium epiphyses reduced them to bone cylinders. Medium-sized and small ungulates are totally different, with a predominance of broken shafts.

Complete specimens are scarce in the assemblage (≤7%). Specimens with some preserved epiphyses mainly have a shaft length of less than a half (70%). A total of 95% of the shafts analysed preserve less than half of their original circumference and 97% less than half of the bone length, while complete sections and lengths are very few in number.

4.4. Birds

The skeletal avian elements recovered in GL1 are grouped into four distribution layers by Viñas and Villalta (1975) and compose quite a wide, varied taxonomic assemblage (Table 4). These remains have been studied previously by Sánchez-Marco [60,72]. Findings under the heading "unknown layer of provenance" (the rightmost column in Table 4) have not been considered in the present study as they appeared to be mixed with fossils from more modern periods [60]. The approximately 25 recognised species indicate a complex landscape, made up of diverse habitats. The avian remains attributed to the red-legged partridge (*Alectoris rufa*) are among the earliest evidence of this species. The earliest dates recorded for this species at eastern and southern Iberian sites are around 50 to 40 ka, which can be understood as evidence of its appearance around that time in the Iberian Mediterranean region [73].

Species that today exhibit a resident phenological pattern form the majority of the entire avian assemblage, in contrast to the low representation of wintering and breeding species. However, according to current phenological patterns, all species may hold resident status. The occurrence of the Manx shearwater, *Puffinus puffinus*, is likely explained by its behaviour of digging burrows in soft coastal terrains. *Anthus spinoletta* is a montane passeriform, linked to bodies of water. An environment with open spaces is denoted by many taxa, such as all corvids (*P. pica*, both *Pyrrhocorax* and both *Corvus*), *Sturnus*, *E. calandra*, *C. coturnix*, *Columba*, *A. noctua* and *F. tinnunculus*, species that are very common in the Iberian and European Pleistocene [74].

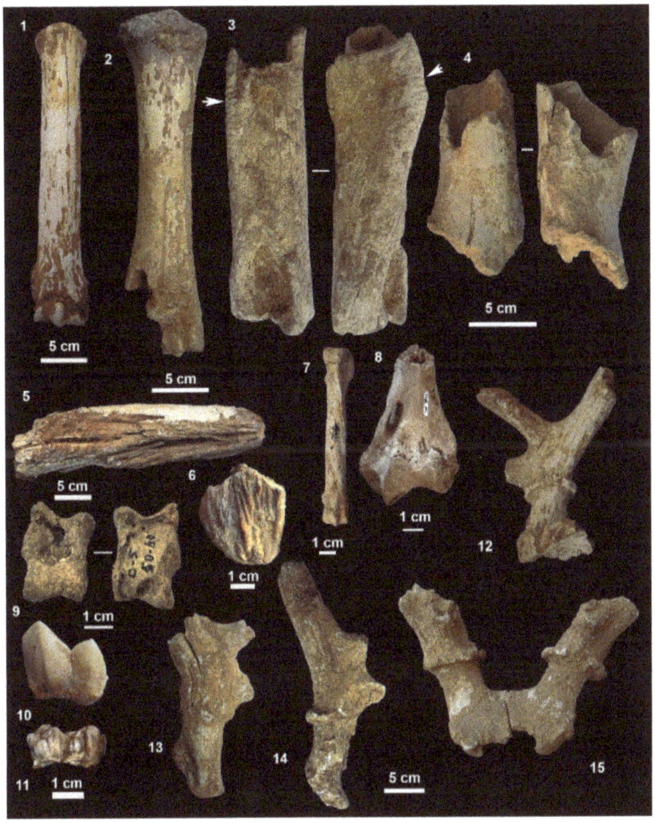

Figure 6. Faunal remains from GL1 stored at Museu de Geologia de Barcelona (MGB) and Arxiu Històric de Sitges (AHSI). (**1**): Equid metatarsal (AHSI). (**2**): Large bovid metacarpal (AHSI) showing a distal condyle gnawed away. (**3**,**4**): Shaft cylinders of equid humeri showing carnivore damage ((**3**): AHSI; (**4**): UB-1). (**5**): Proboscidean tusk fragment (MGB). (**6**): Dental fragment of a proboscidean (UB-1). (**7**): Hyena metacarpal (UB-1). (**8**): Leopard humerus (ICP). (**9**): Digested astragalus (MGB). (**10**): Hyena tooth (AHSI). (**11**): Wild boar tooth (MGB). (**12–15**): Red deer frontal bones with antlers ((**12**): UB-1; (**13–15**): AHSI).

More unusual in the Iberian avian record is the high representation of species linked to forest habitats documented here. *Fringilla coelebs* and *Carduelis chloris* live in woodland edges with thickets. *Strix aluco*, *Loxia pytyopsittacus*, *Pyrrhula pyrrhula*, *Coccothraustes coccothraustes* and *Pinicola enucleator* are characteristic inhabitants of more or less open forests [75,76]. Among these birds, *L. pytyopsittacus* and *P. enucleator* are northern irruptive species. Their presence could be explained by the steady Mediterranean refugia hypothesis, an interpretation of the recording of species with currently northern distributions within associations characterised by Mediterranean species [72]. According to this theory, species, such as *Alectoris*, have an important role in palaeoclimatic interpretations. These species are typical of the Mediterranean region, have sedentary behaviour and perform short movements throughout the year. Their occurrence is incompatible with cold weather conditions. The occurrence of northern irruptive birds in the warmer areas of the European peninsula indicate colder conditions at high continental latitudes, probably due to one of the cold pulses of the late Pleistocene.

The bird assemblage may have been conditioned by taphonomy. It includes species with different behavioural patterns, such as cave-dwelling species, species that used caves sporadically, species that used indoor rocky shelters for night roosting, and others that did not enter caves. Birds linked to rocky habitats may have died in the cavity itself. The large presence of choughs (*Pyrrhocorax pyrrhocorax* and *Pyrrhocorax graculus*) in avian assemblages has been commonly attributed to chough nesting and roosting behaviour in rocky cavities. Very recently it has been observed that this pattern of overabundance does not occur with other rock dweller species, and is only observed in caves with evidence of human presence, and as such it has been proposed that these birds may have been hunted since Neanderthal times [77]. Most avian bones recorded correspond to diurnal species, but there are also some nocturnal raptors. All nocturnal birds documented at Cova del Gegant could have been preyed upon by the eagle owl because it eats everything that flies at night; however, this assumption is speculative because there is no evidence of digestion or other signs of predation. Small diurnal birds could have been caught by falcons because they hunt all small birds flying during the day, but this is also speculative. It seems clear that there was no single cause for the accumulation of the birds inside the cave. But regardless of the vectors that introduced the birds into the cave, there is no doubt that the climatic conditions were temperate, as indicated by the presence of partridge (resident status) and quail, and that there were extensive forest masses, totally or partially constituted by conifers, as indicated by a large number of strict indicator species of forest habitats.

4.5. Small Vertebrates

A total of 225 small vertebrate remains have been identified, including amphibians, squamate reptiles, insectivores, bats and rodents, corresponding to 96 individuals representing at least 22 species. The small vertebrate (Table 5; Figure 7) association identified in Cova del Gegant is mainly represented by taxa related to Mediterranean forest environments, as in the case of *Apodemus sylvaticus*, *Eliomys quercinus*, and *Hystrix* (*A.*) cf. *brachyura vinogradovi*. However, species that prefer open environments are also represented in the association, such as *Microtus* (*Terricola*) *duodecimcostatus*, *Microtus arvalis*, *Erinaceus europaeus*, *Crocidura russula*, and *Pelobates cultripes*, together with squamates in general. Then, the presence of certain taxa, not currently represented in the Garraf massif, such as *Microtus agrestis*, *Sorex gr. coronatus-araneus* and *Talpa europaea* (Table 6), indicates wetter conditions around the cave than in the present. Nevertheless, the herpetofaunal species *Pelobates cultripes* and *Zamenis scalaris* are indicators of drier environments.

To evaluate palaeoclimatic parameters based on the small vertebrate assemblages described in López-García et al. [59], we used the Mutual Ecogeographic Range method [78,79]. This method involves defining the climatic conditions of the area where the fauna of the site currently lives (Table 6). Species like porcupine that do not have extant representatives in the Iberian Peninsula, are not included, but are considered for the interpretation. This method places the small vertebrate assemblage of Cova del Gegant in three 10 × 10 km UTM squares of the external Sierras of the Pyrenees in Huesca. These results suggest lower mean annual temperatures ($MAT_{Gegant} = 10 \pm 2.6$ °C) than at present, where $MAT_{Current} = 15.3$ °C according to [80] at Barcelona airport meteorological station, nearly 30 km to the north of the cave. The mean annual precipitations ($MAP_{Gegant} = 850 \pm 150$ mm) are higher than present ($MAP_{Current} = 659$ mm) and the warmest monthly temperature ($MTW_{Gegant} = 20.1 \pm 1$ °C) is lower than today ($MTC_{Current} = 8.8$ °C; $MTW_{Current} = 23.0$ °C).

Table 4. Avian remains documented at GL1.

Taxa	Layers												Total NISP		Total MNI	
	I		I–II		II		III		IV		Unknown					
	NISP	MNI	NISP	MNI	NISP	MNI	NISP	MNI	NISP	MNI	NISP	MNI	NISP	%NISP	MNI	%MNI
Puffinus puffinus					1	1							1	0.3	1	0.7
Falco tinnunculus	2	2									3	1	5	1.4	3	2.1
Accipiter nisus					1	1							1	0.3	1	0.7
Accipitriformes indet.	1	1									1	1	2	0.5	2	1.4
Alectoris rufa	5	2			6	4	1	1	1	1	2	1	15	4.1	9	6.2
Coturnix coturnix	5	2			1	1	1	1			4	2	11	3.0	6	4.1
Columba livia s. *oenas*	9	3			5	3	2	2			5	2	21	5.7	10	6.8
Tyto alba					1	1					1	1	2	0.5	2	1.4
Bubo bubo											2	1	2	0.5	1	0.7
Athene noctua	2	1			3	2					1	1	6	1.6	4	2.7
Strix aluco					1	1							1	0.3	1	0.7
Apus apus			1	1									1	0.3	1	0.7
Delichon urbica											1	1	1	0.3	1	0.7
Anthus spinoletta	1	1			2	1							3	0.8	2	1.4
Turdus viscivorus					1	1							1	0.3	1	0.7
Emberiza calandra	1	1			2	2							3	0.8	3	2.1
Fringilla coelebs					1	1							1	0.3	1	0.7
Carduelis chloris					1	1	1	1			1	1	3	0.8	3	2.1
Loxia pytyopsittacus	1	1											1	0.3	1	0.7
Pinicola enucleator			1	1									1	0.3	1	0.7
Pyrrhula pyrrhula	1	1											1	0.3	1	0.7
Coccothraustes coccothraustes	1	1			1	1					1	1	3	0.8	3	2.1
Sturnus sp.											2	1	2	0.5	1	0.7
Pica pica			1	1	1	1					3	1	5	1.4	3	2.1
Pyrrhocorax pyrrhocorax	76	16	7	3	48	15	13	5			62	9	206	55.8	48	32.9
Pyrrhocorax graculus	10	6	14	6	10	8					14	2	48	13.0	22	15.1
Corvus monedula	8	4	2	2	2	2					8	4	20	5.4	12	8.2
Corvus corone					1	1					1	1	2	0.5	2	1.4
Total													369	100	146	100

All of these data suggest that the Neanderthal populations that inhabited the surroundings of the Cova del Gegant lived in temperate summers (−1.7 °C) and harsher winters (−5.3 °C), but in a Mediterranean climate. In addition, the precipitations recorded support the presence of the porcupine, suggesting a slightly wetter climate in the area around the cave than nowadays. The small-vertebrate assemblage of Cova del Gegant may have been conditioned by taphonomy. According to Andrews [81], owls in particular may have been responsible for the accumulation of micromammals. Because this predator is opportunistic, their prey must reflect the local environmental conditions [82]. In support of this idea, it appears likely that the owl species responsible for the assemblage was a category 1 nocturnal bird of prey, based on a taphonomic bias that is not visible (primarily digesting evidence). In contrast, bats and porcupines are cave dwellers and their presence in the Cova del Gegant may be related to in-situ mortality rather than predator accumulation. This is a preliminary taphonomic study and the environmental inferences using small-mammals as a proxy should be interpreted with caution.

Table 5. Number of Identified Specimens (NISP) and Minimum Number of Individual (MNI) of the small vertebrates identified at GL1.

	Taxon	NISP	MNI
Amphibians	*Pelobates cultripes*	4	1
	Bufo gr. *B. bufo*	4	2
	Epidalea calamita	11	3
	Bufonidae indet.	15	0
	Ranidae indet.	2	1
	Anura indet.	1	0
Testudines	*Testudo hermanni*	20	2
Squamates	*Timon* cf. *lepidus*	2	1
	Malpolon monspessulanus	10	1
	Zamenis scalaris	14	1
	Colubrinae indet.	2	0
	Vipera sp.	2	1
	Ophidia indet.	7	0
Insectivores	*Erinaceus europaeus*	1	1
	Sorex gr. *coronatus-araneus*	6	4
	Crocidura russula	10	9
	Talpa europaea	2	1
Bats	*Rhinolophus ferrumequinum*	2	1
	Myotis myotis	3	3
	Myotis sp.	1	1
	Miniopterus schreibersii	3	2
Rodents	*Microtus arvalis*	5	4
	Microtus agrestis	5	4
	Microtus (*Iberomys*) *cabrerae*	11	7
	Microtus (*Terricola*) *duodecimcostatus*	12	9
	Microtus (*Terricola*) *pyrenaicus*	2	2
	Apodemus sylvaticus	69	29
	Eliomys quercinus	15	6
	Hystrix (*Acanthion*) cf. *brachyura-vinogradovi*	4	2
	Total	245	98

Table 6. Small mammal association of Cova del Gegant (left) in comparison with current small mammals that inhabit the Garraf Massif (right), according to Palomo et al. [83].

Cova del Gegant	Currently at Garraf Massif
Order Eulipotyphla	
Erinaceus europaeus	
Talpa europaea	
Crocidura russula	*Crocidura russula*
Sorex gr. *coronatus -araneus*	
Order Rodentia	
Hystrix (A.) cf. *brachyura-vinogradovi*	
	Sciurus vulgaris
	Arvicola sapidus
Microtus arvalis	
Microtus agrestis	
Microtus (Iberomys) cabrerae	
Microtus (Terricola) duodecimcostatus	*Microtus (Terricola) duodecimcostatus*
Microtus (Terricola) pyrenaicus	
Apodemus sylvaticus	*Apodemus sylvaticus*
	Rattus rattus
	Rattus norvegicus
	Mus spretus
	Mus musculus
Eliomys quercinus	*Eliomys quercinus*

4.6. Lithic Assemblage

The lithic artifacts analysed present an intense white patina covering most or all of their surface; only two elements are unaltered. Moreover, three of them show evidence of edge damage in the form of microfractures, suggesting a post-depositional mechanical alteration. The less patinated artifacts correspond to four items wrapped in paper labelled as "from the Bronze Age", although all the lithics included in the 1974–1975 collection were previously considered Middle Palaeolithic. According to Mir [61], these artifacts were found in different stratigraphic units—one in unit I and three in unit IIb—and we do not know why they were labelled separately. These four pieces—one core-on-flake (Figure 8(10)), one denticulate (Figure 8(12)), one flake (Figure 8(11)) and one fragmented flake (Figure 9(2))—are not considered as a part of the present study. The distribution of the remaining artifacts by size and the main lithic categories is presented in Table 7.

Among the 26 remaining lithics, two cores have been identified. Both exhibit a well-defined reduction strategy. The first is a Levallois core showing a preferential flaking surface with mainly unidirectional detachments. Although some small removals were struck from the opposite direction, they seem designed to prepare the distal convexity. The opposite surface remains largely cortical, with the exception of the small proximal detachments used to prepare the striking platforms. In addition, the left edge of the preferential surface is retouched, which suggests that this core was secondarily used to manufacture a sidescraper. The second is a core-on-flake with bifacial removals and a typical discoid structure (Figure 9(4)). Both cores are characterised by a relatively large size ($65 \times 58 \times 15$ mm and $56 \times 44 \times 20$ mm, respectively), which indicates that they were far from being exhausted.

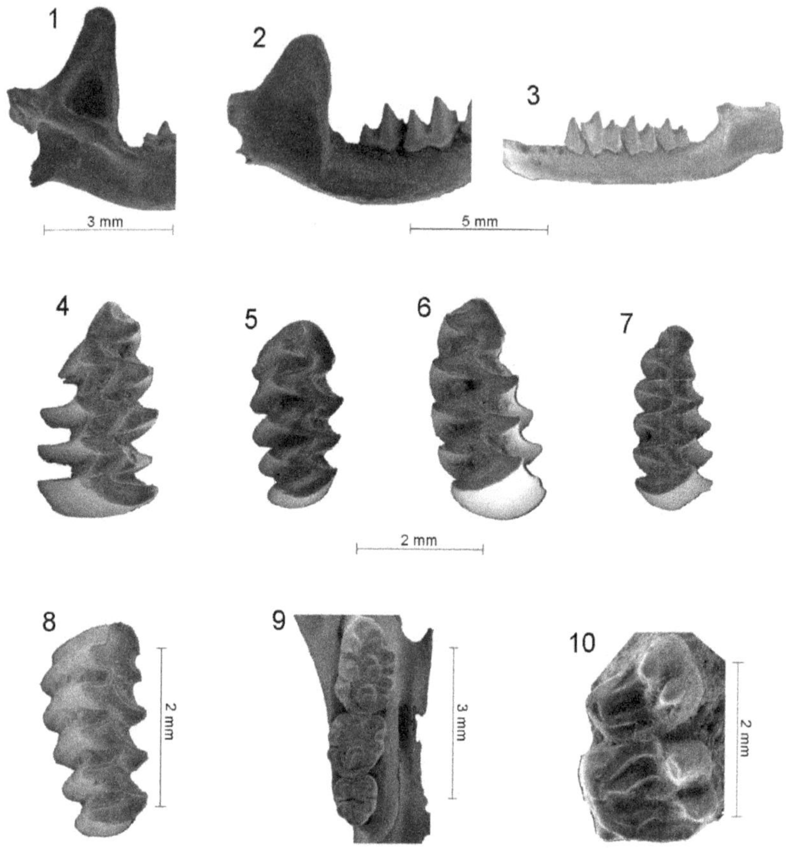

Figure 7. Some small mammal remains identified from Cova del Gegant. (**1**): Right mandible of *Crocidura russula* (lingual view). (**2**): Right mandible of *Myotis myotis* (buccal view). (**3**): Left mandible of *Miniopterus schreibersii* (buccal view). (**4–6**): One left, two right m1 of *Microtus (Iberomys) cabrerae* (occlusal view). (**7**): Right m1 of *Microtus agrestis* (occlusal view). (**8**): m1 right *Microtus (Terricola) duodecimcostatus* (occlusal view). (**9**): Right M1-M3 of *Apodemus sylvaticus* (occlusal view). (**10**): Left P4–M1 of *Eliomys quercinus* (occlusal view). ESEM (Environmental Scanning Electron Microscope) images of mandibles 1 to 3 showing diagnostic criteria for taxonomic identification.

Among the unretouched products, we identified eighteen complete flakes and three fragmented flakes. Among the complete products, very small and small flakes (n = 14) are predominant, although large and very large ones are also represented (n = 4). There are no medium-sized flakes. In spite of the presence of a unidirectional Levallois core, no blades or elongated products have been documented. Only three flakes preserve cortical residues on their dorsal surfaces. The most striking platforms are flat and unprepared, but three dihedral and three facetted butts have been found. The presence of one Levallois flake (Figure 8(6)) and one pseudo-Levallois point (Figure 8(7)) should be noted. No naturally backed flakes have been identified. In general, asymmetrical products opposing an abrupt side to a cutting edge are scarce (only the pseudo-Levallois flake and the denticulate manufactured on a debordant flake can be included in this category).

Six retouched implements were identified: three lateral sidescrapers, one point and two denticulates. Different kinds of blanks were used to make these artifacts. One sidescraper (Figure 8(8)) suggesting a post-depositional mechanical alteration and the point (Figure 8(4)) were manufactured on Levallois blanks, one of them showing a faceted striking platform. Another sidescraper (Figure 8(5)) and the denticulates (Figures 8 and 9(15)) were on ordinary flakes and the third sidescraper was on a cortical flake (Figure 8(2)). The retouched artifacts were manufactured on medium-sized, large and very large flakes. There are two artifacts—one sidescraper (Figure 8(5)) and one denticulate (Figure 9(15))—which, in addition to the retouched edges, show large removals on the ventral surface. In one case, the ventral removals postdated the formation of the white patina, indicating that this artifact was modified after a phase of abandonment (Figure 9(15)). These ventral removals have often been considered as actions aimed at removing the bulb and thinning the blanks. However, these artifacts can also be interpreted as representing an expedient exploitation of certain blanks to obtain a small number of flakes.

Figure 8. Lithic artifacts from Cova del Gegant. (**1,3,11**): Flake. (**2,5,8**): Sidescraper. (**4**): Point. (**6**): Levallois flake. (**7**): Pseudo-Levallois point. (**9,12**): Denticulate. (**10**): Core-on-flake.

Figure 9. Lithic artifacts from Cova del Gegant. (**1**): Levallois core. (**2**,**6**,**10**,**12**): Flake fragment. (**3**,**5**,**7–9**,**11**,**13**,**14**,**16–18**): Flake. (**4**): Discoid core. (**15**). Denticulate.

Table 7. Distribution of the artifacts according to technical and size categories.

	Very Small	Small	Medium	Large	Very Large	Total
Cores					2	2
Flakes	6	8		3	1	18
Flake fragments	2	1				3
Retouched artifacts			3	1	2	6
Total	8	9	3	4	5	29

5. Discussion

5.1. GL1 History and Correlation

Cova del Gegant is one of the largest horizontal caves located on the Mediterranean sea board of the NE of the Iberian Peninsula. This cave presents a complex system of galleries and narrow passages containing lithics, faunal bones and anthropological remains. Previous studies have attempted to summarise the cave's chronological framework, based on the sedimentological profile preserved at the rear of the GP2 [6] and cemented remnants adhered to the cave bedrock [5]. These previous studies identified at least three main anthropogenic occupational episodes at the GP2, grouped between MIS 5 and MIS 3: (a) the earliest presence occurred between ~93 and ~59 ka and corresponds to short Middle Palaeolithic visits alternating with a hyena den, (b) the second presence occurred at around 43–39 ka in the form of sporadic fireplaces placed during the Châtelperronian/Aurignacian time period, and (c) the later presence occurred between ~34 and ~31 ka during the Later Aurignacian/Gravettian. In this context, the Neanderthal human remains documented in the GP2 section are from layer Vf and could be dated between 72 and 67 ka. The second section with human presence is in the gallery described here (GL1).

However, the correlation between the two galleries is difficult because there is a large discontinuity without sediments s between GL1 and GP2 and the layers identified in both areas show significant variation. These variations are related with the distance from the source, i.e., the cave entrance, and the slope processes, which condition the amount of detrital elements present in the areas of the cave, thus implying significant intra-layer variation in grain size. In addition, the hard cementation of the sediments close to the sea, i.e., GL1, point Y and GP1, contrast with the soft sediments of the rear of GP2 introducing colour alterations. Even so, the chronological range obtained from GL1 (49–65 ka) is consistent with that of layer V documented at GP2 and we propose to correlate the infilling preserved in this area with the sedimentation accumulated at the top of layer V. Assuming this correlation, the Middle Palaeolithic human remains, which are documented in layer Vf of the GP2 and XVb of GL1 of the cave, were accumulated over two periods during MIS 4 and MIS 3: the earliest fossils (Gegant-4 and Gegant-5) accumulated ~72–67 ka (GP2) and additional remains (Gegant-1 and Gegant-2) were deposited ~52 ka (GL1).

Cova del Gegant is now located in front of the sea. However, during the lowered sea levels of the Last Glacial Period (MIS 4-3-2) an extended continental platform emerged, permitting connectivity across the Mediterranean coast of the Iberian Peninsula. The entrance to Cova del Gegant, oriented towards the south, and the dimensions of the GP made the cave an attractive space for human and carnivore habitation. At the time of the Neanderthals, the main human occupation must have been located at the escarpment-foot and at the cave entrance. The presence of a Levallois point in brecciated sediments adhered to the escarpment bedrock (Figure 2(5)) and located at the same altitude as GL1 reinforces the idea of external occupation eroded by sea wave action. Now, the original entrance is partially flooded and eroded by the sea, and a non-archaeological record has been observed in this area: only small-brecciated remnants adhered to the cave wall have been preserved. Being located ~1.5 above the main chamber; GL1 was protected from sea erosion. However, the low ceiling and narrow dimensions would have meant that it was uninhabitable for hunter-gatherer groups. Infilling by a gravitational process may have been the main mechanism for the accumulation of layer XVa/b, which is located deep inside, and sediments accumulated at the cliff. Few lithics (NR = 26) and human remains (NR = 2) have been recovered in this gallery, probably due to the pattern of very short-term occupations. The absence of hearths, scarce thermoaltered bones ($\leq$1%) and the absence of cut-marks in the faunal assemblage reinforces the idea of very short occupation, slightly displaced along the sedimentation talus from the primary locus of deposition in the main gallery and/or in the entrance of the cave.

5.2. GL1 Neanderthal Presence and Palaeoenvironmental Record

No clear evidence of human action on the bones has been reported. Carnivore damage to ungulate bones observed in GL1, such as reduction of the epiphyses, furrows and other gnawing patterns, are in keeping with the pattern described for hyenas [84,85]. Differences in consumption patterns according to prey size are also observed, i.e., complete bones, such as metapodials, and bone cylinders are frequent for equids and large bovids; in contrast, broken shafts predominate in the medium-sized and small ungulate bones. However, the activity of other carnivores documented in GL1 cannot be totally ruled out.

This type of archaeological context—carnivore dens or trapped animals in caves with a few lithic remains—is common in the Iberian Middle Palaeolithic record [86–90] and identifying the nature of the anthropogenic processes carried out at the sites is a particularly challenging task. Some of the lithic assemblages recovered at these sites have been interpreted as the result of short visits by humans in the context of resource provisioning. However, it has also been suggested that, in some instances and especially in carnivore dens, lithic and faunal remains might be accidentally associated, since artifacts may have been moved into the caves by natural gravitational processes [91] or washed in [88]. In addition, we should bear in mind that archaeological assemblages tend to be comprised of large time-averaged palimpsests resulting from the outcome of an indeterminate, superimposed number of events occurring over a long time span [1,2]. The nature of these events may be diverse and there is no reason to expect that all of the remains were accumulated by the same agent and/or at the same time. Accordingly, several taphonomic signatures have been observed on faunal remains, such as well-preserved bones, bones in anatomical connection, or bones showing evidence of water erosion [13,45]. These are indicative of the agents and processes that occurred in the assemblage.

The origin of the human remains is difficult to discern. Neanderthal bones found in these contexts are usually isolated elements without articulation, which makes it difficult to relate them to intentional deposition or burial. During the Middle Palaeolithic, caves and rock shelters were occupied by both humans and carnivores and encounters between them are plausible. Several Neanderthal human remains show marks inflicted by carnivores, such as punctures, notches or gnawing marks [92]. The nature of these marks is difficult to distinguish; many of these fossil remains may have originated from carnivore predation of hominins, rather than intentional practices of corpse deposition, but the most common pattern of the taphonomic analyses are indicative of scavenging scenarios of Neanderthal corpses [93,94].

More controversial discussions surround the mortuary behaviour of the Middle Pleistocene European populations that evolved into Neanderthals. At least two major practices have been inferred: the treatment of corpses (cannibalism) [95,96] and the accumulation of bodies in specifically designated places [97]. The presence of cranial fragments at GL1 and in the rear of GP2 may indicate that the skull was present, and probably the most parsimonious explanation for the accumulation of human remains could be related to disturbed burials, as proposed by other scholars [98]. However, the low number of skeletal elements and the lack of sediments in the entrance and in the main chamber of Cova del Gegant means that this interpretation should be considered with caution.

Concerning the lithic remains, the assemblage is another issue of the Neanderthal presence to be discussed. The small component of lithic remains suggests a marginal role of humans in the accumulation. This is consistent with the information provided by previous faunal remains studies [13,45], which indicates that carnivores were the main agents responsible for the accumulation of bone assemblage. The absence of precise intra-site spatial distribution makes it difficult to assess this issue in more depth.

Our assessments are therefore based on post-depositional alterations and the distribution of lithics according to size and technical category. First, the intense white patina and the edge damage observed in some artifacts indicate that the lithics were affected by post-depositional processes. Although none of these alterations necessarily mean that the lithics were moved from elsewhere by natural agents, this scenario cannot be ruled out. Second, no size sorting is evident, as would be expected if natural dynamics were the main agent responsible for the accumulation. Both large and small artifacts are represented at Cova del Gegant, and the latter are dominant (Table 7), although they are less common than in records derived from knapping activities carried out on the spot. In this regard, we should consider that the excavation carried out in 1974–1975 applied a different methodology to the one used today and perhaps involved the occasional recovery of the smallest items.

Distribution by technical categories can provide additional clues on the formation dynamics of the lithic assemblage. The high proportion of retouched artifacts cannot be attributed to natural processes. This feature is not a bias of the recovery methods used during the 1974–1975 excavation because it was also observed in the 1985 campaign and during the current excavations developed at the GP2 section. Whatever the role played by postdepositional processes, they acted on an assemblage that had already been selected by humans, who mainly transported retouched implements into the site. Moreover, we should bear in mind that the artifacts were found in a lateral gallery (GL1), but the deposits from the main gallery (GP) were eroded by sea waves. The GP and, particularly, the zone close to the cave entrance was probably the area that was most suitable for human occupation, as previously mentioned. For this reason, we cannot rule out the idea that artifacts found in GL1 were related to hypothetical knapping activities carried out in these living areas and that we are therefore dealing with a marginal sample of a larger assemblage. As highlighted in other Palaeolithic contexts [99], the lithic assemblages found in the outer and more illuminated areas of caves can be very different from those found in the inner and darker areas. In fact, the human occupation of the cave has been confirmed by a hearth found in the innermost area of the main gallery of Cova del Gegant (GP2) [32], which strengthens the probability of more substantial human inputs in the outer zone.

Even so, considering the assemblage found in GL1, it seems clear that knapping activities were very restricted in this cave area. However, the presence of cores and small/very small flakes suggest that some limited knapping events were carried out. The distribution of flakes by size shows a marked discontinuity between the small/very small and the large/very large items due to the absence of medium-sized products. This feature indicates that these products correspond to two different reduction sequences and were associated with two different inputs. The large and very large flakes, together with the retouched artifacts, tend to be part of mobile toolkits and were therefore introduced into the sites as single items [86,100–102]. These two components—the transported vs. the locally produced artifacts—can be recognised in most cases and the balance between them determines many characteristics of the lithic assemblages. In Cova del Gegant, the transported artifacts are highly visible due to the limited role played by on-site production.

The archaeological record from the GL1 gallery helps to reconstruct the environmental conditions of Neanderthal populations and the occupational patterns. The faunal assemblage is composed of a rich collection of large mammals, birds and small vertebrates. It reflects diverse habitats and thus suggests either the presence of a complex environment or a large catchment area. This is relevant for the bird species and for the small vertebrates living in Mediterranean woodland edges and other taxa more typical of open environments. The large mammal record suggests an open landscape and, together with the absence of marine taxa, indicates that a littoral platform emerged in front of the cave. Thus, the cave mouth would have opened onto a large plain that could sustain and feed large herbivores, such as horses, proboscideans and large bovids. The presence of *Equus hydruntinus* in GL1 supports this idea of the landscape and also indicates an arid and open terrain [103–105]. Red deer, which are also abundant in the assemblage, are intermediate feeders, but can feed on grasses or in light forests. Forest dwellers are also present, such as wild boar, ibex, small vertebrates and some birds, such as greenfinch, parrot crossbill, pine grosbeak, bullfinch and hawfinch. The presence of porcupine has been discussed elsewhere [106]; those authors suggest that these remains located at the top of GL1 may have been re-elaborated from other layers. Based on the small vertebrate proxies, the environmental conditions of the GL1 assemblage suggest lower temperatures (−2.7 °C) than at present in the same area, with temperate summers (−1.7 °C) and harsher winters (−5.3 °C). However, there is nothing in the composition of the faunal assemblage to indicate the presence of a marine component. The only taxon related to this environment corresponds to a single bone of Manx shearwater, possibly from a specimen breeding in the soft sediments of the cave floor. The avian record is dominated by species adapted to open landscapes. In terms of MNI, the red-billed and yellow-billed choughs represent together 48% of the total. This is a species adapted to rocky environments.

5.3. Cova Del Gegant in the Regional Context

The regional context of Cova del Gegant comprises a rich set of archaeo-palaeontological record characterised by small lithics with a huge amount of faunal remains accumulated by carnivores or natural agents. In this local context, Cova del Rinoceront and Cova del Coll Verdaguer, 15 and 20 km respectively from Cova del Gegant, exhibit rather different lithic assemblages, reflecting the variability that can be discerned in what Brugal and Jaubert [107] defined as "palaeontological sites with scarce human presence". Cova del Rinoceront was used as a carnivore den (layers I and II) and was a natural trap (III to VII) [108], but no evidence of human activity was documented on bones [83]. Only 22 artifacts were recovered, and very small items were clearly dominant. This evidence of size sorting, together with the edge damage shown in most artifacts, suggested that lithics may have been dragged in by post-depositional processes.

Cova del Coll Verdaguer yielded a rich faunal assemblage, mainly accumulated by hyenas that used the cave as a den, although the role of other carnivores is present [89,109]. Fourteen lithics were documented and very small artifacts are represented only by two flake fragments. The rest of the assemblage was characterised by high volumetric variability, including small (n = 5), medium (n = 4), large (n = 2) and very large (n = 1) flakes. Six of these flakes exhibited asymmetrical cross-sectional profiles, opposing a cutting edge to an abrupt back. This feature may suggest that the items in the assemblage were selected, and that the accumulation was related to the human transport of single artifacts. In marked contrast to Cova del Gegant, neither cores nor retouched artifacts were found in Cova del Coll Verdaguer. Some knapping was identified through refitting.

The comparison between Cova del Gegant and Cova del Coll Verdaguer is also significant in terms of the criteria used to select the mobile toolkit, since both assemblages are largely made up of transported single items. In spite of the constraints on mobility and transport common to both sites, the artifacts from Cova del Gegant and Cova del Coll Verdaguer exhibit clear differences. The first is the high frequency of retouched tools in Cova del Gegant and their absence in Coll Verdaguer. Second, artifacts with an asymmetrical cross-sectional profile (naturally backed flakes, debordant flakes, pseudo-Levallois points) are common in Cova del Coll Verdaguer, but scarce in Gegant. This suggests that the occupants of Cova del Coll Verdaguer were more concerned with selecting this kind of blanks, whose ergonomic properties have been emphasised [110]. There are also some technological differences, even though the assemblages are too small to carry out a thorough assessment of the reduction strategies. The use of the Levallois method is clear in Cova del Gegant, where a typical Levallois core and at least three Levallois blanks have been documented, but less evident in Cova del Coll Verdaguer. These differences emphasise the behavioural variability associated with artifact transport and the range of options available to Neandertal groups facing similar challenges in similar environmental settings.

6. Conclusions

The GL1 of Cova del Gegant provides a good archive for evaluating the environmental conditions of coastal areas during MIS 4 and MIS 3 in the NE of Iberian Peninsula. The vertebrate record, comprising ~70 taxa of faunal remains, is indicative of the richness and diversity of species and providing a clear picture of a period between 49 and 65 ka in the Mediterranean coast. The palaeoenvironmental reconstruction presented here suggests that an open landscape dominated the vicinity of Cova del Gegant close to woodland areas, in agreement with previous studies and palaeoenvironmental reconstructions in the NE of the Iberian Peninsula. The absence of marine elements at GL1 is notable and suggests the development of a large coastal platform during the cold climatic periods due to lower sea levels and plain bathymetric curves. Thus, the cave entrance was roughly 7–9 km further from the coastline than today.

The low level of human activity observed at GL1 may indicate that the site was a brief stopping-place used by highly mobile Neanderthal hunter-gatherer groups. The age-at-death of Neanderthal individuals from Cova del Gegant, not restricted to adults, may suggest that the cave was used as a short stop during residential mobility involving all the group (i.e., all ages). Indeed, the lithics from GL1 seem to be part of a mobile toolkit rather than instruments used for knapping and manufacturing activities at the cave. The absence of butchering or other domestic activities supports this idea. Thus, it seems plausible that the Neanderthal remains from GL1 could be the result of repeated visits when groups halted in their camp-to-camp displacements along this coastal platform, rather than the result of hunting parties or groups procuring raw materials. Indeed, the regional context presents evidence of a wide variability in habitat types, site occupation, environmental conditions, intensity and geographic settings. This knowledge of the local resources supports the idea that subsistence strategies of the Neanderthal groups involved a certain amount of forward planning.

Supplementary Materials: The following supporting information can be downloaded at: https://www.mdpi.com/article/10.3390/quat5010012/s1, Figure S1: The 3d model of Cova del Gegant mandible (Gegant-1).

Author Contributions: J.D. and M.S. conceived the manuscript, designed the research, and conducted current excavations. J.D. and M.S. wrote the manuscript with contributions from other authors (J.D. stratigraphy and chronology, M.V. lithics; M.S. large mammals; J.M.L.-G. and H.-A.B. small vertebrates, A.S.M. birds). All authors have read and agreed to the published version of the manuscript.

Funding: This research received no external funding.

Institutional Review Board Statement: Not applicable.

Informed Consent Statement: Not applicable.

Acknowledgments: This study is part of the output of a research project entitled: "Els canvis climàtics durant el plistocè superior a la costa central catalana i l'impacte en les poblacions neandertals i humans anatòmicament moderns" (CLT009/18/00022-Generalitat de Catalunya), supported by projects 2017SGR-00011 (Generalitat de Catalunya) and PID2020-113960GB-100 (Ministerio de Ciencia e Innovación). J. Daura holds a Ramon y Cajal contract (RYC-2015-17667). M. Sanz was supported by a Juan de la Cierva postdoctoral grant (IJCI-2017-33908). Thanks to the Castelldefels City Council (La Guixera Laboratory) where the archaeological analyses were carried out. We extend our thanks to I. Bergamnn for photographic documentation of Neanderthal mandible images and 3d model. We are grateful to the staff of the Museu d'Arqueologia de Catalunya (MAC), Museu de Geologia de Barcelona (MGB) and Arxiu Històric de Sitges (AHS) for their assistance during the study of the materials analysed here. This research was supported by the 2017-SGR-836 and 2017-SGR-859 projects of the Catalonian Government and the French International Research Network (IRN 0871 CNRS-INEE): Taphonomy European Network (TaphEN). J.M.L.-G. was supported by a Ramon y Cajal contract (RYC-2016-19386) from the Spanish Ministry of Science and Innovation. ASM has been supported by the project CGL2016-76431-P and CGL2017-82654-P (AEI/FEDER, EU).

Conflicts of Interest: The authors declare no conflict of interest.

References

1. Bailey, G. Time Perspectives, Palimpsests and the Archaeology of Time. *J. Anthropol. Archaeol.* **2007**, *26*, 198–223. [CrossRef]
2. Bailey, G.; Galanidou, N. Caves, Palimpsests and Dwelling Spaces: Examples from the Upper Palaeolithic of South-East Europe. *World Archaeol.* **2009**, *41*, 215–241. [CrossRef]
3. Stiner, M.C. *The Faunas of Hayonim Cave, Israel: A 200,000-Year Record of Paleolithic Diet, Demography, and Society*; Harvard University Press: Cambrige, MA, USA, 2005.
4. Daura, J.; Sanz, M.; Oms, F.X.; Pedro, M.; Martínez, P.; Mendiela, S.; Oliva Poveda, M.; Gibaja, J.F.; Mozota, M.; Alonso-Eguíluz, M.; et al. Deciphering Neolithic Activities from a Cardial Burial Site (Cova Bonica) on the Western Mediterranean Coast. *J. Archaeol. Sci. Rep.* **2019**, *23*, 324–347. [CrossRef]
5. Daura, J.; Sanz, M.; Pike, A.W.G.; Subirà, M.E.; Fornós, J.J.; Fullola, J.M.; Julià, R.; Zilhão, J. Stratigraphic Context and Direct Dating of the Neandertal Mandible from Cova Del Gegant (Sitges, Barcelona). *J. Hum. Evol.* **2010**, *59*, 109–122. [CrossRef] [PubMed]
6. Daura, J.; Sanz, M.; Demuro, M.; Arnold, L.J.; Costa, A.M.; Moreno, J.; Freitas, M.C.; Lopes, V.; Égüez, N.; Hoffman, D.L.; et al. A New Chronological Framework and Site Formation History for Cova Del Gegant (Barcelona): Implications for Neanderthal and Anatomically Modern Human Occupation of NE Iberian Peninsula. *Quat. Sci. Rev.* **2021**, *270*, 107–141. [CrossRef]
7. Daura, J.; Sanz, M.; Subirá, M.E.; Quam, R.; Fullola, J.M.; Arsuaga, J.L. A Neandertal Mandible from the Cova Del Gegant (Sitges, Barcelona, Spain). *J. Hum. Evol.* **2005**, *49*, 56–70. [CrossRef] [PubMed]
8. Rodríguez, L.; García-González, R.; Sanz, M.; Daura, J.; Quam, R.; Fullola, J.M.; Arsuaga, J.L. A Neanderthal Lower Incisor from Cova Del Gegant (Sitges, Barcelona, Spain). *Bol. R. Soc. Esp. Hist. Nat. Sec. Geol.* **2011**, *105*, 25–30.
9. Arsuaga, J.L.; Quam, R.; Daura, J.; Sanz, M.; Subira, M.E.; Dalén, L.; Götherström, A. Neandertal MtDNA from a Late Pleistocene Human Mandible from the Cova Del Gegant (Spain). In *Vertebrate Paleobiology and Paleoanthropology*; Springer: Dordrecht, The Netherlands, 2011; pp. 213–217. [CrossRef]
10. Sanz, M. Patrons d'acumulació de Restes de Fauna Del Plistocè Superior Al Nord-Est Peninsular (Àrea Del Massís Del Garraf-Ordal). Ph.D. Thesis, Universitat de Barcelona, Barcelona, Spain, 2013.
11. Daura, J.; Sanz, M. Procedencia Estratigráfica de Los Restos Humanos Neandertales de La Cova Del Gegant (Sitges, Barcelona). *Mainake* **2011**, *33*, 215–232.
12. Samper, S.C. *Who Let the Hyenas Out? Taphonomic Analysis of the Faunal Assemblage from Upper Pleistocene Site of Cova Del Gegant (Sitges, Spain)*; The University of Edinburg: Edinburg, Scotland, 2011.
13. Sanz, M.; Daura, J. La Fauna Del Pleistoceno Superior Asociada a Los Restos Humanos Neandertales de La Galería Lateral 1 de La Cova Del Gegant (Sitges, Barcelona). *SAGVNTVM* **2020**, *20*, 81–98.
14. Arsuaga, J.L.; Martinez, I.; Arnold, L.J.; Aranburu, A.; Gracia-Tellez, A.; Sharp, W.D.; Quam, R.M.; Falgueres, C.; Pantoja-Perez, A.; Bischoff, J.; et al. Neandertal Roots: Cranial and Chronological Evidence from Sima de Los Huesos. *Science* **2014**, *344*, 1358–1363. [CrossRef]
15. Meyer, M.; Arsuaga, J.-L.; de Filippo, C.; Nagel, S.; Aximu-Petri, A.; Nickel, B.; Martínez, I.; Gracia, A.; de Castro, J.M.B.; Carbonell, E.; et al. Nuclear DNA Sequences from the Middle Pleistocene Sima de Los Huesos Hominins. *Nature* **2016**, *531*, 504–507. [CrossRef]
16. Zilhão, J. The Late Persistence of the Middle Palaeolithic and Neandertals in Iberia: A Review of the Evidence for and against the "Ebro Frontier" Model. *Quat. Sci. Rev.* **2021**, *270*, 107098. [CrossRef]
17. Zilhão, J.; Angelucci, D.E.; Igreja, M.A.; Arnold, L.J.; Badal, E.; Callapez, P.; Cardoso, J.L.; D'Errico, F.; Daura, J.; Demuro, M.; et al. Last Interglacial Iberian Neandertals as Fisher-Hunter-Gatherers. *Science* **2020**, *367*, eaaz7943. [CrossRef]
18. Quam, R.; Sanz, M.; Daura, J.; Robson Brown, K.; García-González, R.; Rodríguez, L.; Dawson, H.; Rodríguez, R.F.; Gómez, S.; Villaescusa, L.; et al. The Neandertals of Northeastern Iberia: New Remains from the Cova Del Gegant (Sitges, Barcelona). *J. Hum. Evol.* **2015**, *81*, 13–28. [CrossRef]
19. Salas, R. El Malm i El Cretaci Inferior Entre El Massís Del Garraf i La Serra dEspadà. PhD Thesis, Universitat de Barcelona, Barcelona, Spain, 1987.
20. Daura, J.; Sanz, M.; Fornós, J.; Asensio, A.; Julià, R. Karst Evolution of the Garraf Massif (Barcelona, Spain): Doline Formation, Chronology and Archaeo-Palaeontological Archives. *J. Cave Karst Stud.* **2014**, *76*, 69–87. [CrossRef]
21. Marqués, M.A.; Julià, R. Geomorphological Mapping of Mediterranean Coastal Features, Northeast Spain. *J. Coast. Res.* **1987**, *3*, 29–36.
22. Daura, J.; Sanz, M.; Ramos, J.; Riera, S.; Miras, Y.; Allué, E.; Picornell-Gelabert, L.; López-Reyes, D.; Albert, R.M.; Macià, L.; et al. Palaeoenvironmental Record of the Cal Maurici Wetland Sediment Archive in Barcelona (NE Iberian Peninsula) between c. 6000 and 4000 Cal. Yr BP. *Holocene* **2016**, *26*, 1020–1039. [CrossRef]
23. Virella, A. *Vilanova i La Geltrú. Imatges de La Ciutat i de La Comarca*; Joan Rius Vila: Vilanova i la Geltrú, Spain, 1949.
24. Daura, J.; Sanz, M. Nota Necrológica: Santiago Casanova i Giner (Forcall 1922–Barcelona 2011). *Cypsela* **2017**, *20*, 7–8.
25. Viñas, R.; Villalta, J. *El Depósito Cuaternario de La "Cova Del Gegant." In Speleon V Symposium de Espeleología*; Centre Excursionista Catalunya: Barcelona, Spain, 1975; pp. 19–33.
26. Viñas, R. Observaciones Sobre Los Depósitos Cuaternarios de La Cova Del Gegant. Sitges (Barcelona). *Speleon* **1972**, *19*, 115–126.

27. Bellmunt, J. Crónica de La Sección Arqueológica. *Boletín Bibl. -Mus. Balaguer* **1958**, *5*, 132–134.
28. Martínez-Moreno, J.; Miret, J.; Mora, R.; Muro, I. Excavacions a La Cova Del Gegant. *Butlletí d'estudis Sitgetans* **1985**, *32–33*, 1.
29. Martínez-Moreno, J. *Informe Técnico de Los Restos Óseos de La Cova Del Gegant (Sitges, Garraf). Servei d'Arqueologia i Paleontologia*; Centre d'Informació i Documentació del Patrimoni Cultural de la Generalitat de Catalunya: Barcelona, Spain, 1990.
30. Martínez-Moreno, J.; Mora, R.; Roca, G.; Parcerisas, J. *Memoria d'Excavació a La Cova Del Gegant 1989. Servei d'Arqueologia i Paleontologia*; Centre d'Informació i Documentació del Patrimoni Cultural de la Generalitat de Catalunya: Barcelona, Spain, 1990.
31. Martínez-Moreno, J.; Mora Torcal, R.; Muro Morales, I.; Miret i Mestre, J. *Memòria de l'excavació d'urgència de La Cova Del Gegant (Sitges, Garraf) 1985*; Centre d'Informació i Documentació del Patrimoni Cultural de la Generalitat de Catalunya: Barcelona, Spain, 1985.
32. Sanz, M.; Daura, J.; Égüez, N.; Cabanes, D. On the Track of Anthropogenic Activity in Carnivore Dens: Altered Combustion Structures in Cova Del Gegant (NE Iberian Peninsula). *Quat. Int.* **2017**, *437*, 102–114. [CrossRef]
33. Bischoff, J.L.; Julia, R.; Mora, R. Uranium-Series Dating of the Mousterian Occupation at Abric Romani, Spain. *Nature* **1988**, *332*, 68–70. [CrossRef]
34. Rosenbauer, R.J. UDATE1: A Computer Program for the Calculation of Uranium-Series Isotopic Ages. *Comput. Geosci.* **1991**, *17*, 45–75. [CrossRef]
35. Hoffmann, D.L.; Pike, A.W.G.; García-Diez, M.; Pettitt, P.B.; Zilhão, J. Methods for U-Series Dating of CaCO3 Crusts Associated with Palaeolithic Cave Art and Application to Iberian Sites. *Quat. Geochronol.* **2016**, *36*, 104–119. [CrossRef]
36. Hoffmann, D.L. 230Th Isotope Measurements of Femtogram Quantities for U-Series Dating Using Multi Ion Counting (MIC) MC-ICPMS. *Int. J. Mass Spectrom.* **2008**, *275*, 75–79. [CrossRef]
37. Hoffmann, D.L.; Prytulak, J.; Richards, D.A.; Elliott, T.; Coath, C.D.; Smart, P.L.; Scholz, D. Procedures for Accurate U and Th Isotope Measurements by High Precision MC-ICPMS. *Int. J. Mass Spectrom.* **2007**, *264*, 97–109. [CrossRef]
38. Hoffmann, D.L.; Standish, C.D.; García-Diez, M.; Pettitt, P.B.; Milton, J.A.; Zilhão, J.; Alcolea-González, J.J.; Cantalejo-Duarte, P.; Collado, H.; de Balbín, R.; et al. U-Th Dating of Carbonate Crusts Reveals Neandertal Origin of Iberian Cave Art. *Science* **2018**, *359*, 912–915. [CrossRef]
39. Millard, A.R.; Hedges, R.E.M. A Diffusion-Adsorption Model of Uranium Uptake by Archaeological Bone. *Geochim. Et Cosmochim. Acta* **1996**, *60*, 2139–2152. [CrossRef]
40. Pike, A.W.G.; Eggins, S.; Grün, R.; Hedges, R.E.M.; Jacobi, R.M. U-Series Dating of the Late Pleistocene Mammalian Fauna from Wood Quarry (Steetley), Nottinghamshire, UK. *J. Quat. Sci.* **2005**, *20*, 59–65. [CrossRef]
41. Valero-Garcés, B.L.; González-Sampériz, P.; Gil-Romera, G.; Benito, B.M.; Moreno, A.; Oliva-Urcia, B.; Aranbarri, J.; García-Prieto, E.; Frugone, M.; Morellón, M.; et al. A Multi-Dating Approach to Age-Modelling Long Continental Records: The 135 Ka El Cañizar de Villarquemado Sequence (NE Spain). *Quat. Geochronol.* **2019**, *54*, 101006. [CrossRef]
42. Aitken, M.J. *Thermoluminescence Dating*; Academic Press: London, UK, 1985.
43. Nambi, K.S.V.; Aitken, M.J. Annual Dose Conversion Factors for TL and ESR Dating. *Archaeometry* **1986**, *28*, 202–205. [CrossRef]
44. Santafé, J.V.; Casanovas, M.L. Dicherorhinus hemitoechus (Falconer, 1868) (Mammalia, Perissodactyla) Del Yacimiento Pleistocénico de La Cueva Del Gegant (Garraf, Barcelona). *Empúries Rev. Món Clàssic Antig. Tardana* **1989**, *48–50*, 310–322.
45. Samper Carro, S.C.; Martínez-Moreno, J. Who Let the Hyenas out? Taphonomic Analysis of the Faunal Assemblage from GL-1 of Cova Del Gegant (Sitges, Spain). *Quat. Int.* **2014**, *330*, 19–35. [CrossRef]
46. Reitz, E.J.; Wing, E.S. *Zooarchaeology*; Cambridge University Press: Cambridge, UK, 2008.
47. Poplin, F. Un Problème d'ostéologie Quantitative: Calcul d'effectif Initial d'après Appariements. Généralisation Aux Autres Types de Remontages et à d'autres Matériels Archéologiques. *Rev. D'archéométrie* **1981**, *5*, 159–165. [CrossRef]
48. Lyman, R.L. *Quantitative Paleozoology*; Cambridge University Press: Cambridge, UK, 2008. [CrossRef]
49. Brain, C.K. *The Hunters or the Hunted?: An Introduction to African Cave Taphonomy*; University of Chicago Press: Chicago, CA, USA, 1981.
50. Lyman, R.L. *Vertebrate Taphonomy*; Cambridge University Press: Cambridge, UK, 1994. [CrossRef]
51. Fernández-Jalvo, Y.; Andrews, P. *Atlas of Taphonomic Identifications*; Vertebrate Paleobiology and Paleoanthropology; Springer: Dordrecht, The Netherlands, 2016. [CrossRef]
52. Johnson, E. Current Developments in Bone Technology. In *Advances in Archaeological Method and Theory*; Elsevier: Amsterdam, The Netherlands, 1985; pp. 157–235. [CrossRef]
53. Blumenschine, R.J.; Selvaggio, M.M. Percussion Marks on Bone Surfaces as a New Diagnostic of Hominid Behaviour. *Nature* **1988**, *333*, 763–765. [CrossRef]
54. Haynes, G. A Guide for Differentiating Mammalian Carnivore Taxa Responsible for Gnaw Damage to Herbivore Limb Bones. *Paleobiology* **1983**, *9*, 164–172. [CrossRef]
55. Fosse, P.; Avery, G.; Selva, N.; Smietana, W.; Okarma, H.; Wajrak, A.; Fourvel, J.B.; Madelaine, S. Taphonomie Comparée Des Os Longs d'ongulés Dévorés Par Les Grands Prédateurs Modernes d'Europe et d'Afrique (C. Lupus, P. Brunnea). In *Prédateurs dans tous leurs états. Evolution, Biodiversité, Interactions, Mythes, Symboles. XXXIe Rencontres Internationales d'Archéologie et d'Histoire d'Antibes*; Brugal, J.-P., Gardeisen, A., Zucker, A., Eds.; APDCA: Antibes, France, 2011; pp. 127–156.
56. Behrensmeyer, A.K. Taphonomic and Ecologic Information from Bone Weathering. *Paleobiology* **1978**, *4*, 150–162. [CrossRef]
57. Blain, H.-A.; Bailon, S. Catalogue of Spanish Plio-Pleistocene Amphibians and Squamate Reptiles from the Museu de Geologia de Barcelona. *Treb. Mus. Geol. Barc.* **2006**, *14*, 61–80.

58. López-García, J.M.; Agustí, J.; Cuenca-Bescós, G. Catalogue of the Late Quaternary Iberian Micromammals (Rodentia, Eulipothypla, Chiroptera) from the Museu de Geologia de Barcelona. *Treb. Mus. Geol. Barc.* **2007**, *15*, 5–23.
59. López-García, J.M.; Blain, H.-A.; Cuenca-Bescós, G.; Arsuaga, J.L. Chronological, Environmental, and Climatic Precisions on the Neanderthal Site of the Cova Del Gegant (Sitges, Barcelona, Spain). *J. Hum. Evol.* **2008**, *55*, 1151–1155. [CrossRef]
60. Sánchez Marco, A. Avifaunas Cuaternarias de La Península Ibérica: Sistemática, Paleocologia, Paleozoografia. Ph.D. Thesis, Universidad Autónoma de Madrid, Madrid, Spain, 2005.
61. Mir, A. La Industria Lítica de La Cova Del Gegant. Sitges (Barcelona). *Speleon. Monogr. I. V Symp. Espeleol.* **1975**, 39–48.
62. Mora, R. El Paleolítico Medio En Catalunya: Yacimientos En Cueva y Al Aire Libre. Ph.D. Thesis, Unversitat de Barcelona, Barcelona, Spain, 1988.
63. Laplace, G. La Typologie Analytique et Structurale. Base Rationalle d'étude Des Industries Lithiques et Osseuses. In *Branques des Données Archéologiques*; Centre National de la Recherche Scientifique: Paris, France, 1974; pp. 91–143.
64. Gómez-Alba, J. Catálogo Razonado de Los Vertebrados Fósiles de España Del Museo de Geología de Barcelona (1882-1982). *Treb. Mus. Geol. Barc.* **1997**, *6*, 1–296.
65. Daura, J.; Sanz, M. Historiografia Dels Jaciments Plistocens Al Massís Del Garraf i Curs Baix Del Riu Llobregat. *Treb. Mus. Geol. Barc.* **2009**, *16*, 5–38.
66. Masriera, A. *Observaciones Sedimentológicas Sobre El Depósito Cuaternario de La Cova Del Gegant (Sitges, Barcelona). Speleon V Symposium de Espeleología*; Centre Excursionista de Catalunya: Barcelona, Spain, 1975; pp. 35–38.
67. Jaffey, A.H.; Flynn, K.F.; Glendenin, L.E.; Bentley, W.C.; Essling, A.M. Precision Measurement of Half-Lives and Specific Activities of 235U and 238U. *Phys. Rev. C* **1971**, *4*, 1889. [CrossRef]
68. Cheng, H.; Edwards, R.L.; Hoff, J.; Gallup, C.D.; Richards, D.A.; Asmerom, Y. The Half-Lives of Uranium-234 and Thorium-230. *Chem. Geol.* **2000**, *169*, 17–33. [CrossRef]
69. Holden, N.E. Total Half-Lives for Selected Nuclides. *Pure Appl. Chem.* **1990**, *62*, 941–958. [CrossRef]
70. Discamps, E. Ungulate Biomass Fluctuations Endured by Middle and Early Upper Paleolithic Societies (SW France, MIS 5-3): The Contributions of Modern Analogs and Cave Hyena Paleodemography. *Quat. Int.* **2014**, *337*, 64–79. [CrossRef]
71. Hammer, Ø.; Harper, D.A.T.; Ryan, P.D. Past: Paleontological Statistics Software Package for Education and Data Analysis. *Palaeontol. Electron.* **2001**, *4*, 1–9.
72. Sánchez Marco, A. Avian Zoogeographical Patterns during the Quaternary in the Mediterranean Region and Paleoclimatic Interpretation. *Ardeola* **2004**, *51*, 91–132.
73. Sánchez Marco, A. Avifauna de La Peña de Estebanvela (Segovia). In *La Peña de Estebanvela: Grupos Magdalenienses en el Sur del Duero*; Cacho, C., Ripoll, S., Muñoz, F.J., Eds.; Arqueología en Castilla y León 17; Arqueología en Castilla y León; Consejería de Cultura y Turismo de la Junta de Castilla y León: Valladolid, Spain, 2007; pp. 141–144.
74. Tyrberg, T. *Pleistocene Birds of the Palearctic: A Catalogue*; Mass: Cambridge, MA, USA, 1998; Volume 27.
75. Díaz, M.; Asensio, B.; Tellería, J.L. *Aves Ibéricas: I. No Paseriformes.*; Reyero, J.M., Ed.; Reyero, J.M.: Madrid, Spain, 1996.
76. Tellería, J.L.; Asensio, B.; Díaz, M. *Aves Ibéricas: II. Paseriformes*; Reyero, J.M., Ed.; Reyero, J.M.: Madrid, Spain, 1999.
77. Blanco, G.; Sánchez-Marco, A.; Negro, J.J. Night Capture of Roosting Cave Birds by Neanderthals: An Actualistic Approach. *Front. Ecol. Evol.* **2021**, *9*, 9. [CrossRef]
78. Blain, H.-A.; Bailon, S.; Cuenca-Bescós, G.; Arsuaga, J.L.; Bermúdez de Castro, J.M.; Carbonell, E. Long-Term Climate Record Inferred from Early-Middle Pleistocene Amphibian and Squamate Reptile Assemblages at the Gran Dolina Cave, Atapuerca, Spain. *J. Hum. Evol.* **2009**, *56*, 55–65. [CrossRef] [PubMed]
79. Blain, H.-A.; Lozano-Fernández, I.; Agustí, J.; Bailon, S.; Menéndez Granda, L.; Espígares Ortiz, M.P.; Ros-Montoya, S.; Jiménez Arenas, J.M.; Toro-Moyano, I.; Martínez-Navarro, B.; et al. Refining upon the Climatic Background of the Early Pleistocene Hominid Settlement in Western Europe: Barranco León and Fuente Nueva-3 (Guadix-Baza Basin, SE Spain). *Quat. Sci. Rev.* **2016**, *144*, 132–144. [CrossRef]
80. Font-Tullot, I. *Climatología de España y Portugal*; Ediciones Universidad de Salamanca: Salamanca, Spain, 2000.
81. Andrews, P. *Owls, Caves and Fossils. Predation, Preservation and Accumulation of Small Mammañ Bones in Caves, with an Analysis of the Pleistocene Cave Faunas from Westbury-Sub-Mendip, Somerset, United Kingdom*; Oxford University: Oxford, UK, 1990.
82. Pokines, J.T.; Kerbis Peterhans, J.C. Barn Owl (Tyto Alba) Taphonomy in the Negev Desert, Israel. *Isr. J. Ecol. Evol.* **1998**, *44*, 19–27. [CrossRef]
83. Palomo, L.J.; Gisbert, J.; Blanco, J.C. *Atlas y Libro Rojo de Los Mamíferos Terrestres de España*; Dirección General para la Biodiversidad-SECEM-SECEMU: Madrid, Spain, 2007.
84. Cruz-Uribe, K. Distinguishing Hyena from Hominid Bone Accumulations. *J. Field Archaeol.* **1991**, *18*, 467–486.
85. Fosse, P. Variabilité Des Assemblages Osseux Créés Par l'hyène Des Cavernes. *Paléo* **1997**, *9*, 15–54. [CrossRef]
86. Alcaraz-Castaño, M.; Alcolea-González, J.; Kehl, M.; Albert, R.-M.; Baena-Preysler, J.; de Balbín-Behrmann, R.; Cuartero, F.; Cuenca-Bescós, G.; Jiménez-Barredo, F.; López-Sáez, J.-A.; et al. A Context for the Last Neandertals of Interior Iberia: Los Casares Cave Revisited. *PLoS ONE* **2017**, *12*, e0180823. [CrossRef] [PubMed]
87. Arsuaga, J.L.; Baquedano, E.; Pérez-González, A.; Sala, N.; Quam, R.M.; Rodríguez, L.; García, R.; García, N.; Álvarez-Lao, D.J.; Laplana, C.; et al. Understanding the Ancient Habitats of the Last-Interglacial (Late MIS 5) Neanderthals of Central Iberia: Paleoenvironmental and Taphonomic Evidence from the Cueva Del Camino (Spain) Site. *Quat. Int.* **2012**, *275*, 55–75. [CrossRef]

88. Daura, J.; Sanz, M.; Julià, R.; García-Fernández, D.; Fornós, J.J.; Vaquero, M.; Allué, E.; López-García, J.M.; Blain, H.A.; Ortiz, J.E.; et al. Cova Del Rinoceront (Castelldefels, Barcelona): A Terrestrial Record for the Last Interglacial Period (MIS 5) in the Mediterranean Coast of the Iberian Peninsula. *Quat. Sci. Rev.* **2015**, *114*, 203–227. [CrossRef]
89. Daura, J.; Sanz, M.; Allué, E.; Vaquero, M.; López-García, J.M.; Sánchez-Marco, A.; Domènech, R.; Martinell, J.; Carrión, J.S.; Ortiz, J.E.; et al. Palaeoenvironments of the Last Neanderthals in SW Europe (MIS 3): Cova Del Coll Verdaguer (Barcelona, NE of Iberian Peninsula). *Quat. Sci. Rev.* **2017**, *177*, 34–56. [CrossRef]
90. Vaquero, M.; van der Made, J.; Blain, H.-A.; Ibáñez, N.; López-García, J.M.; Rivals, F.; Alonso, S.; Ameijenda, A.; Bennàsar, M.; Fernández-García, M.; et al. Fauna, Environment and Human Presence during MIS5 in the North of Spain: The New Site of Valdavara 3. *Comptes Rendus Palevol* **2018**, *17*, 557–593. [CrossRef]
91. Villa, P.; Soressi, M. Stone Tools in Carnivore Sites: The Case of Bois Roche. *J. Anthropol. Res.* **2000**, *56*, 187–215. [CrossRef]
92. Trinkaus, E.; Maki, J.; Zilhão, J. Middle Paleolithic Human Remains from the Gruta Da Oliveira (Torres Novas), Portugal. *Am. J. Phys. Anthropol.* **2007**, *134*, 263–273. [CrossRef]
93. Camarós, E.; Cueto, M.; Lorenzo, C.; Villaverde, V.; Rivals, F. Large Carnivore Attacks on Hominins during the Pleistocene: A Forensic Approach with a Neanderthal Example. *Archaeol. Anthropol. Sci.* **2016**, *8*, 635–646. [CrossRef]
94. Camarós, E.; Cueto, M.; Rosell, J.; Díez, J.C.; Blasco, R.; Duhig, C.; Darlas, A.; Harvati, K.; Jordá, J.; Montes, L.; et al. Hunted or Scavenged Neanderthals? Taphonomic Approach to Hominin Fossils with Carnivore Damage. *Int. J. Osteoarchaeol.* **2017**, *27*, 606–620. [CrossRef]
95. Garralda, M.D.; Giacobini, G.; Vandermeersch, B. Neanderthal Cutmarks: Combe-Grenal and Marillac (France). A SEM Analysis. *Anthropologie* **2005**, *43*, 189–198. [CrossRef]
96. Orschiedt, J. Der Fall Krapina—Neue Ergebnisse Zur Frage von Kannibalismus Beim Neandertaler. *Quartär* **2008**, *55*, 63–81.
97. Pettitt, P. The Neanderthal Dead. *Before Farming* **2002**, *2002*, 1–26. [CrossRef]
98. Arsuaga, J.L.; Villaverde, V.; Quam, R.; Martínez, I.; Carretero, J.M.; Lorenzo, C.; Gracia, A. New Neandertal Remains from Cova Negra (Valencia, Spain). *J. Hum. Evol.* **2007**, *52*, 31–58. [CrossRef]
99. Zilio, L.; Hammond, H.; Karampaglidis, T.; Sánchez-Romero, L.; Blasco, R.; Rivals, F.; Rufà, A.; Picin, A.; Chacón, M.G.; Demuro, M.; et al. Examining Neanderthal and Carnivore Occupations of Teixoneres Cave (Moià, Barcelona, Spain) Using Archaeostratigraphic and Intra-Site Spatial Analysis. *Sci. Rep.* **2021**, *11*, 4339. [CrossRef]
100. Neruda, P.; Kaminská, L. Neanderthals at the Open-Air Site of Bojnice III: The Issue of "Missing" Artifacts. In *Settlement Dynamics of the Middle Paleolithic and Middle Stone Age IV.*; Conard, N.J., Delagnes, A., Eds.; Diane Marie Kerns Kerns Verlag: Tübingen, Germany, 2015; pp. 205–226.
101. Picin, A.; Chacón, M.G.; Gómez de Soler, B.; Blasco, R.; Rivals, F.; Rosell, J. Neanderthal Mobile Toolkit in Short-Term Occupations at Teixoneres Cave (Moia, Spain). *J. Archaeol. Sci. Rep.* **2020**, *29*, 102165. [CrossRef]
102. Vaquero, M.; Chacon, M.G.; Garcia-Anton, M.D.; de Soler, B.G.; Martinez, K.; Cuartero, F. Time and Space in the Formation of Lithic Assemblages: The Example of Abric Romani Level. *J. Quat. Int.* **2012**, *247*, 162–181. [CrossRef]
103. Sanz-Royo, A.; Sanz, M.; Daura, J. Upper Pleistocene Equids from Terrasses de La Riera Dels Canyars (NE Iberian Peninsula): The Presence of Equus Ferus and Equus Hydruntinus Based on Dental Criteria and Their Implications for Palaeontological Identification and Palaeoenvironmental Reconstr. *Quat. Int.* **2020**, *566–567*, 78–90. [CrossRef]
104. Burke, A.; Eisenmann, V.; Ambler, G.K. The Systematic Position of Equus Hydruntinus, an Extinct Species of Pleistocene Equid. *Quat. Res.* **2003**, *59*, 459–469. [CrossRef]
105. Daura, J.; Sanz, M.; García, N.; Allué, E.; Vaquero, M.; Fierro, E.; Carrión, J.S.; López-García, J.M.; Blain, H.A.; Sánchez-Marco, A.; et al. Terrasses de La Riera Dels Canyars (Gavà, Barcelona): The Landscape of Heinrich Stadial 4 North of the "Ebro Frontier" and Implications for Modern Human Dispersal into Iberia. *Quat. Sci. Rev.* **2013**, *60*, 26–48. [CrossRef]
106. López-García, J.M.; Blain, H.-A.; Sanz, M.; Daura, J. A Coastal Reservoir of Terrestrial Resources for Neanderthal Populations in North-Eastern Iberia: Palaeoenvironmental Data Inferred from the Small-Vertebrate Assemblage of Cova Del Gegant, Sitges, Barcelona. *J. Quat. Sci.* **2012**, *27*, 105–113. [CrossRef]
107. Brugal, J.-P.; Jaubert, J. Les Gisements Paléontologiques Pléistocènes à Indices de Fréquentation Humaine: Un Nouveau Type de Comportement de Prédation? *Paléo* **1991**, *3*, 15–41. [CrossRef]
108. Sanz, M.; Daura, J. Taphonomic Analysis of an Ungulate-Dominated Accumulation at the Pleistocene Cova Del Rinoceront Site near Barcelona, Spain (Northeastern Iberian Peninsula). *Palaeogeogr. Palaeoclimatol. Palaeoecol.* **2018**, *498*, 24–38. [CrossRef]
109. Sanz, M.; Daura, J. Carnivore Involvement in Bone Assemblages Based on Taphonomic and Zooarchaeological Analyses of Cova Del Coll Verdaguer Site (Barcelona, Iberian Peninsula). *Hist. Biol.* **2018**, *30*, 807–820. [CrossRef]
110. Beyries, S.; Boëda, E. Etude Technologique et Traces d'utilisation Des "Éclats Débordants" de Corbehem (Pas-de-Calais). *Bull. Société Préhistorique Français* **1983**, *80*, 275–279. [CrossRef]

quaternary

Article

Contextual, Taphonomic, and Paleoecological Insights from Anurans on Tiwanaku Sites in Southern Peru

Juliana Rubinatto Serrano [1], Maria Camila Vallejo-Pareja [2], Susan D. deFrance [1,*], Sarah I. Baitzel [3] and Paul S. Goldstein [4]

1 Department of Anthropology, University of Florida, Gainesville, FL 32611, USA; julianarubinatto@ufl.edu
2 Department of Biology and Florida Museum of Natural History, University of Florida, Gainesville, FL 32611, USA; maria.vallejo@ufl.edu
3 Department of Anthropology, Washington University in St. Louis, St. Louis, MO 63130, USA; sbaitzel@wustl.edu
4 Department of Anthropology, University of California San Diego, La Jolla, CA 92093, USA; psgoldstein@ucsd.edu
* Correspondence: sdef@ufl.edu; Tel.: +1-352-281-5873

Abstract: We examine the processes that resulted in the deposition of bones of at least three anuran genera on four archaeological sites associated with the Tiwanaku culture occupied between 700–1100 CE in the Moquegua Valley of far southern Peru. We review archaeological data and ethnographic accounts of Andean peoples using frogs and toads for food and in rain-quest rituals. Anuran bones are common in prehispanic cemeteries, but far less common in habitational and ceremonial sites. The quantity of anuran remains in certain cemeteries is anomalous in comparison to other archaeological sites in the valley and to Tiwanaku sites in other geographic settings. We conclude that anurans are most common where abandoned below-ground rock-covered tombs are likely to have been reused by nesting owls, and propose that most anuran remains in archaeological contexts in Moquegua are the result of predation. We consider environmental, cultural and taphonomic explanations and posit that the abundance of anuran remains relates to the 14th-century Miraflores ENSO event. This event generated increased rainfall in the desert, creating conditions favorable for frogs and toads, and predation by owls. We also advocate for the use of fine-screening to recover small-sized animal remains, such as anurans, that can be used to understand taphonomic processes and paleoenvironmental conditions.

Keywords: Burrowing Owl (*Athena cunicularia*); ENSO; herpetofauna; Middle Horizon; Miraflores event; Moquegua

1. Introduction

Archaeofaunal remains can provide a critical window into ancient human diet, behavior and practices. However, archaeological sites also often contain animal remains that are not the direct result of human behavior, but that nonetheless provide important insights into the past. Animals that people did not consume and that did not pass through a cultural selection process can serve as indicators of past habitats and palaeoecological conditions. In the south-central Andes, anurans (frogs and toads) are primarily a non-food animal whose presence in archaeological deposits has cultural implications or may be used as a paleoenvironmental proxy. Here, we examine deposits of anuran remains on sites in the hyper-arid, mid-elevation setting (900–1400 masl) of the Moquegua Valley of far southern Peru. We report on archaeological evidence of frogs and toads from four sites associated with the Tiwanaku culture occupation of the region (~700–1100 CE). The original goal of the zooarchaeological analysis was focused on the identification of economic and ritual uses of animals by the Tiwanaku. However, the unexpected recovery of large numbers of frog and toad remains in certain contexts of the faunal assemblages warranted investigation.

Some archaeologists working in the Central Andes have interpreted the scant remains of frogs and toads as food remains [1]. Other possible explanations for anuran remains are as non-food refuse or ritual animals depending on the archaeological context. In the recent past and among modern populations in the Lake Titicaca region and the city of La Paz in Bolivia, people are known to consume frogs, particularly the very large-sized Titicaca water frog, *Telmatobious culeus* [2,3]. It is possible that people in the past ate these large-sized frogs as well as smaller frogs that naturally inhabit the south-Central Andes. Ethnographic accounts of human–anuran interactions primarily focus on Andean Indigenous peoples collecting frogs for a variety of therapeutic purposes and using them in ceremonies to bring on the rainy season [3–6]. Frogs also appear in Andean ceramic iconography attesting to the animals' cultural significance [7,8].

Today, numerous species of frogs and toads are common in the highlands, particularly near and in Lake Titicaca, the homeland of the Tiwanaku civilization, as well as in the eastern Andean tropical region [9]. Conversely, frogs and toads are uncommon in the western Andean valleys and Sierra due to the extreme aridity that characterizes this region. Therefore, the discovery and identification of abundant anuran remains in excavations at Tiwanaku sites in Moquegua warranted further analysis.

Here, we examine the possible explanations for the deposition of frog and toad bones on sites that had been occupied by Tiwanaku communities in the second half of the first millennium CE. Based on the contexts and patterns of deposition, and on comparisons with other archaeological deposits in the region and south-Central Andes, we consider natural death, human deposition as food or ritual deposits, animal predation, and other natural processes as explanations. Our analysis of archaeological contexts and different processes leads us to hypothesize that the anuran remains resulted primarily from animal predation. Our hypotheses regarding the taphonomic processes responsible for the accumulation of anuran remains are preliminary and await more detailed taphonomic analysis. In addition to our initial suggestions of the agents responsible for the deposition of anuran remains, we then use the frog and toad remains to hypothesize past paleoclimatic conditions in the Moquegua region during the first half of the second millennium CE. By combining the study of herpetofauna, site abandonment processes and paleoclimate in the central Andes, this paper contributes new data on anuran deposition at open-air archaeological sites and desert environments in line with a growing knowledge of anurans and taphonomy in the ancient Americas [10].

Anurans and Taphonomy

Taphonomic studies have proposed different agents responsible for accumulations of animal remains in archaeological sites. These studies tend to rely on the analysis of scats and pellets of birds of prey and carnivore mammals to identify taphonomic patterns caused by animal predation, such as the accumulation of small mammal remains in caves [11–16]. Regarding anuran remains, Pinto Llona and Andrews [17] and Stoetzel et al. [18] have compiled markers of animal predation for several predatory species and applied them to contexts with high anuran accumulations in Spanish and Moroccan archaeological sites, respectively. Besides physical marks of predations in specimens (gnawing, perforations), other markers encompass taxa-level digestion, breakage, and element representation patterns that, together, can indicate species-specific predation. Additional studies have expanded on these foundational studies while considering multiple lines of evidence, from archaeological and cultural contexts, to site taphonomy and paleoenvironmental conditions [19–23]. Biton et al. [22] and Bisbal-Chinesta et al. [23] hypothesized that the accumulation of anuran remains in an open-air site in Israel and a Spanish cave resulted from diverse processes such as natural death, human exploitation for consumption or ritual/symbolic use, accidental human encounter, and animal predation. Blain and Villa [19], Blain [20], and Martin et al. [21] also explored the taphonomic processes responsible for anuran accumulation in archaeological deposits, but complement their interpretations with discussions on paleoclimatic and paleoenvironmental conditions. This previous research has improved our ability to more accurately distinguish between natural and anthropogenic accumulation patterns, and it guides our use here of anuran remains to reconstruct paleoenvironmental conditions.

2. Anurans in the Andes

2.1. Cultural Entanglements

Andean peoples have had diverse cultural associations with frogs and toads, in particular through ritual practice and artistic expression. Despite modern consumption of Lake Titicaca frogs, there is no definitive archaeological evidence for the use of frogs and toads as food. Warwick [5] (p. 51) summarizes previous anthropological work on the different uses of amphibians in Aymara and Quechua communities of the Lake Titicaca basin to treat maladies and diseases. For example, LaBarre [6] describes the use of claws for toothache and the whole animal for cataplasm. He also concludes that Aymara communities did not use frogs as a food resource. Tschopik [4] (pp. 277–278) describes the Rain Making Rite (or *haju luraña*) that was practiced by Aymara communities of Peru and Bolivia. This ceremony was performed only in periods of extreme drought. He describes one ceremony sponsored by the Chucuito Village in 1939. This community paid for a "magician" and an orchestra to go to Lake Titicaca to collect frogs (presumably, the large lake frog *Telmatobius culeus*) for the rite. While playing and singing the Frog Song, the orchestra and local Chucuito inhabitants accompanied the "magician" in collecting frogs that were then placed on a shrine in the village, where they all prayed for rain [4] (p. 278). Similar ceremonies have been reported elsewhere in the region, indicating a possible long history of the cultural significance of frogs.

Frogs and toads also feature in prehispanic Central Andean iconography. In particular, the Formative period Chiripa culture (600–100 BCE) of the southern Lake Titicaca basin depicted frogs, tadpoles, and/or toads along with other natural and supernatural animals in stone sculpture and other media as part of the Yaya-Mama Religious Tradition [7]. At the Tiwanaku (500–1100 CE) site of Pariti on the Lake Titicaca island, archaeologists excavated fragments of at least 435 intentionally "killed" ceramic vessels, several of which have painted polychrome images of frogs or toads and possible anuran eggs [8] (pp. 109–111; Plates 37 and 38). In the contemporaneous highland Huari culture further north, frogs or toads are painted on ceramics [24]. Additionally, both the coastal Moche and Chimu cultures of northern Peru produced zoomorphic effigy vessels in the form of frogs or toads [25].

2.2. Anurans Species and Ecology in the Western Andes

Peru today has a rich anuran diversity with at least 550 identified species [26]. A large number of the country's endemic anuran species are found in the Andean cordillera at or above 1000 meters above sea level [27]. This is expected because the Tropical Andes are considered a hotspot of biodiversity [28] resulting from the altitudinal gradient, topographic complexity and the long climatic and geologic history that have helped shaped the biota of the Central Andes. In addition, the thermal gradient and dispersal limitation likely have contributed to the high number of endemism [29,30].

The greatest diversity of anurans in Peru is concentrated in the northern Andes and on the eastern slopes of the cordillera. These areas feature many ecoregions with environmental characteristics—such as greater humidity and vegetation coverage found in cloud and montane humid forest—that offer suitable microhabitats along the altitudinal gradient to sustain a variety of anuran species [27,31].

The western or Pacific slopes of central and southern Peru are characterized by drier and open environments with environmental conditions less favorable for the establishment of anuran populations; therefore, anuran diversity is significantly reduced along the western Andes and the Pacific coast [31] (pp. 15,41, Anexo 1). There are four anuran species—*Rhinella spinulosa* (Bufonidae); *Pleurodema marmoratum* (Leptodactylidae); *Telmatobius arequipensis* and *Telmatobius peruvianus* (Telmatobiidae)—for the departments of

Arequipa, Moquegua and Tacna, all with altitudinal ranges from 1000 to 4000 m above sea level [27]. None of these species are burrowing frogs [26,27]. All of these species are listed with some degree of threat in the IUCN Red List.

3. Moquegua Valley

3.1. The Environmental Setting

The Moquegua Valley forms part of the Osmore drainage, one of the southernmost river valleys of Peru. Between the highlands and the coastal plain, the mid-section of the valley located 40 to 70 km inland from the Pacific coast ranges in elevation from 900–1400 masl. In contrast to the steeply incised highland valleys upriver, the mid-valley is a broad, flat expanse consisting of a fertile strip of floodplain roughly 28 km north–south by. 75 km east–west along the banks of the Moquegua River. Along the margins of the valley are barren bluffs, dry river gorges, and mountains. At the lower portion of the middle valley at ~900 masl, the river channel descends into a deep gorge and emerges in a more open main channel 17 km from the coast.

The coastal plain and the mid-valley of the Moquegua drainage form the northern reaches of the Atacama Desert, one of the driest places on earth. Prior to a late 1990s large-scale hydrological engineering project that altered seasonal water pattern to generate river flow throughout the year, the river and its upper-valley tributaries primarily carried seasonal runoff from the western cordillera during the austral rainy season (December to April) when precipitation falls in the highlands above ~4000 masl [32]. Because mid-valley precipitation is less than <100 mm/year [33] and seasonal water was scarce, agriculture and food production during the archaeological past were dependent on the construction of canals and channels to irrigate the arid landscape [34].

In this desert region, the pattern of predictably minimal seasonal rainfall is occasionally disturbed by El Niño Southern Oscillation (ENSO) climatic events. During ENSO events warm waters originating from the tropical Pacific Ocean travel west along the equator and then south along the Pacific coast of South America resulting in the suppression of the cold upwelling currents that naturally dominate the region. The climatic consequences of ENSO often reverse the normal highland and coastal patterns of precipitation. Highland regions that generally have seasonal rainfall experience extreme drought while hyper-arid coastal and low elevation areas experience above-average to heavy precipitation [35].

In southern Peru and elsewhere, ENSO rainfall on the desert coast and the western foothills can trigger catastrophic flooding and debris flows that commonly travel down dry ravines or quebradas [33,36]. In mid-elevation settings, rainfall can result in floods that overwhelm the main channels of rivers as well as their tributaries. Overbanking and flood waters can destroy settlements as well as agricultural and irrigation systems [37]. Although ENSO events resulted in catastrophic disasters, ancient peoples also employed cultural mechanisms to mitigate ancient ENSO occurrences; in some instances, they even took advantage of ENSO-recharged aquifers to increase agricultural production [38,39].

In far southern Peru, and the Moquegua Valley in particular, researchers have documented the history of ENSO events [36], as well as their impact on archaeological sites and irrigation systems [33,34,40]. One well-studied ENSO episode, known as the Miraflores event, occurred during the first half of the 14th century [38,41]. This ENSO generated catastrophic flooding along the southern coast that destroyed villages and agricultural systems [42–44]. We will explore the potential of studying anuran remain accumulations in providing insights into past ENSO events.

3.2. The Cultural Setting

The Moquegua Valley has a long history of settlement dating to the end of the Pleistocene. Following several millennia of hunter-gatherer-forager lifestyles, populations increased in size and settled into villages where they practiced agriculture along with the rearing of domesticated animals (camelids and guinea pigs) in a local Formative cultural phase known as Huaracane (ca cal 385 BCE–CE 700) [45]. The fertile valley and agropastoral productivity of local cultures were attractive to expansive empires who sought to extend their political economies. In the second half of the first millennium CE, two empires, the Huari and the Tiwanaku, colonized the Moquegua Valley.

Originating from a homeland in the Bolivian altiplano at over 3800 masl near Lake Titicaca, Tiwanaku colonists traveled over 250 km and settled in the mid-valley desert oasis of the Osmore drainage beginning in the 7th century AD. In contrast to the Huari, who settled the high sierra of the valley, the Tiwanaku occupied the highly productive middle Moquegua valley where their occupation was concentrated in four large cosmopolitan townsites at Omo, Chen Chen/Los Cerrillos, Rio Muerto, and Cerro Echenique (Figure 1). The Tiwanaku colonies thrived for over four centuries as evidenced by their construction of large towns, a sunken court temple, massive field and canal systems, geoglyphs, and extensive cemeteries [45,46]. At the height of settlement, the colony supported populations between 8000 and 20,000 people. Residents practiced intensive agriculture and produced crops amenable to the mid-valley setting, such as maize and legumes [45]. People transported locally produced foodstuffs to their highland homeland via llama caravans [47]. The demise of the Tiwanaku colony began around 1000 CE. At that time, the population dispersed into smaller settlements and none of the subsequent cultural groups reoccupied the original Tiwanaku townsites in the valley.

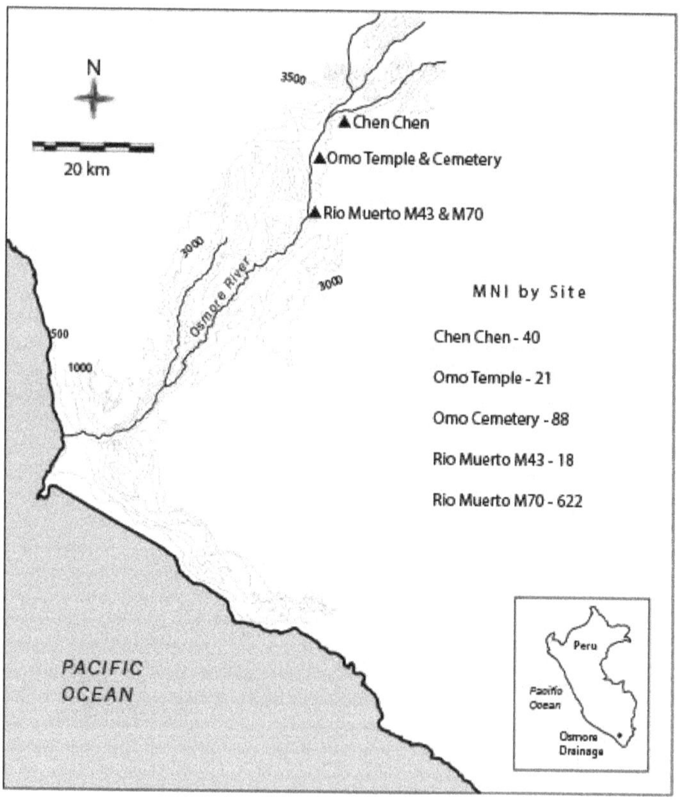

Figure 1. Map of analyzed Tiwanaku sites with anuran MNI by site.

4. Materials and Methods

4.1. Archaeological Sites

This study presents the analysis of anuran remains and associated taxa from 803 contexts excavated from habitational, ceremonial, and cemetery sectors at three of the four Tiwanaku site groups in the Moquegua valley. The sites are:

4.1.1. Chen Chen (M1)

The Chen Chen M1 site is located northeast of the city of Moquegua at 1490 masl, representing the largest known Tiwanaku settlement and mortuary complex in the valley. Contexts analyzed for anurans include excavations in three large habitational sectors (8.6 ha), and a smaller habitational and ceremonial sector (2.5 ha) located near irrigation canals and a 90 hectare agricultural field system [45] (pp. 136,148). Analyzed tomb contexts from Chen Chen are from a large-scale rescue excavation conducted prior to the construction of a modern irrigation system and urban expansion [48]. As in all but one of the other Tiwanaku cemeteries from which fauna was examined, tombs at Chen Chen consisted of stone-lined cylindrical cists or unlined pits in which a single individual was interred.

4.1.2. Omo *Temple and Cemeteries (M10), and Domestic Sectors (M10, M12)*

The Omo M10 site is located at 1200 masl on a desert bluff 20 m above the agricultural canal line, and adjacent to the most productive natural spring of this hyper-arid region. The M10 temple and cemeteries make up 1.5 ha of the M10 townsite (9.5 ha within the 35.5 ha Omo site group). The contexts that we analyzed for anurans are from the cemeteries and temple, which is the only sunken temple in the valley and the only Tiwanaku sunken temple outside of the highland capital in Bolivia, probably serving as ritual and administrative center [45] (p. 160). According to excavation field notes field crews recovered the anuran remains from below or around M10 cemetery capstones, high above the intentionally deposited burial contents. The anuran remains from the M10 temple are also from deposits closer to the surface, not from deeply buried contexts that would suggest they were offerings or ritual deposits.

4.1.3. Rio Muerto (M43 and M70)

The Rio Muerto complex located at 1025 masl, is the third-largest Tiwanaku settlement in Moquegua [45] (p. 149). Unlike Omo, Rio Muerto is not on a blufftop, but is located in a dry gulch tributary just outside of, but only slightly higher than, the canal-irrigated valley. Rio Muerto includes two sites, M43 (6.7 ha), which is closest to the modern canal, and M70 (2.8 ha), which is further removed from the valley bottom. At M70, the small cemetery M70B, located northwest of the domestic sector M70A, is of greatest interest to this study. The M70B cemetery is unique for our sample and unusual for the Moquegua Tiwanaku culture because the approximately 90 tombs are located under or within a massive rockpile up to 1m in height of loosely stacked stones (Figure 2) [45]. Excavators recovered anuran remains in deposits above the intact tombs. Anuran remains were not in direct association with the burials themselves, indicating that they are not offerings.

4.2. Methods

The zooarchaeological assemblage analyzed in this study was collected from excavations and surface materials from the four archaeological sites described above during field projects directed by various excavators (Berta Vargas at Chen Chen in 1987, Goldstein at Omo and Rio Muerto since 1986 and Baitzel in 2010 at Omo). Sediment from (near) surface levels was screened through 1/4″ (6.35 mm) screens; all tomb fill and domestic and temple floors and features were screened through 1/16″ (1.8 mm) mesh. We recognize that some differences in anuran remain abundance across the contexts we examine may be attributed to different excavation and recovery methods over multiple projects, notably in salvage work at Chen Chen; however, other recovery techniques were largely consistent. In all cases, faunal material was bagged separately and analyzed in the archaeological laboratory of the Contisuyo Museum in Moquegua.

Figure 2. Rockpile cemetery feature at Rio Muerto, M70B prior to excavation (photo by Paul S. Goldstein).

Using the laboratory facilities at the Contisuyo Museum (Moquegua, Peru), deFrance completed identification of the anuran specimens. At the time of taxonomic identifications in Peru, deFrance identified the majority of elements only to the order Anura. Additionally, deFrance photographed selected specimens from Rio Muerto sites, mainly elements from the large assemblage of anuran humeri (and a single urostyle) from M70B. Although the ilium is highly diagnostic for anuran identification [49] and ilia are present in a modern owl pellet (see discussion below), this element was not well-preserved in the studied archaeological materials (Figure 3). Therefore, we used the humerus, which is the most abundant element in our assemblages, for quantification and more precise taxonomic identification of genera.

Using the photographs of humeri, Vallejo-Pareja used dry specimens and X-ray computed microtomography (CT-scans) of alcohol-preserved specimens from the comparative collection at the Herpetology Division of the Florida Museum of Natural History to identify a subsample of anuran bones to make more specific taxonomic identifications (SI Table S1). The CT scans were used to digitally segment and create three-dimensional (3D) models of the humerus of specimens used for comparisons using VG StudioMax v. 3.4 (Volume Graphics, Heidelberg, Germany). The 3D models of the humerus are available through

Morphosource.org. Taxonomic identification was hindered by not having access to the assemblage, which is currently curated in Peru, and therefore, relied on the photographs of selected specimens from sites M43 and M70.

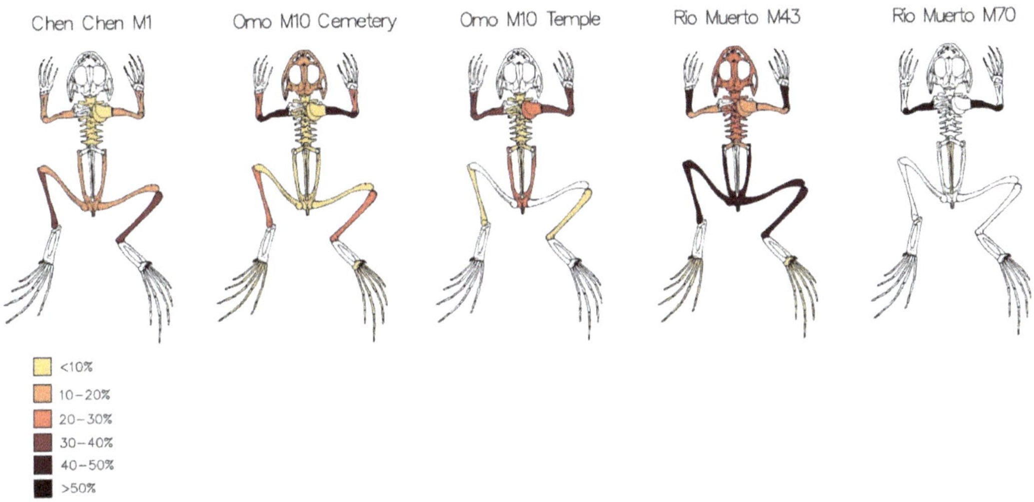

Figure 3. Diagram of anuran element relative abundance by site.

Sexual dimorphism in some groups of South American anurans may present as marked differences in size with females being larger than males [50] or as morphological differences in humeri—with additional crests or spines in males that help during the amplexus or during combats for female section [51,52]. However, because the photographic material examined in this research does not display these dimorphic traits, we did not attempt to sex the specimens.

For the preliminary taphonomic analysis, deFrance recorded evidence of human con-sumption, such as burning and butchering traces. Digestion and other evidence of non-an-thropogenic factors were not considered for the present study. Bone breakage was most commonly classified using whole vs shaft/fragment categories (though some were classified through whole, 3/4, 1/2, 1/4 or shaft/fragment). We tabulated the Number of Identified Specimens (NISP) and calculated the Minimum Number of Individuals (MNI) by context for both anurans and other associated taxa according to Reitz and Wing's recommendations [53]. The relative abundance of anuran elements was also calculated by site according to Andrews [11] (p. 145).

5. Results

The sites in this study yield a combined total of 5647 anuran specimens (NISP) and a minimum of 789 individuals (MNI) (Table 1; for data, consult SI Table S2). Anuran remains are most abundant at Rio Muerto cemetery M70B, totaling 4808 specimens, followed by the Omo M10 Cemeteries with 541, Rio Muerto M43 with 158, Omo M10 Temple with 80, and Chen Chen with at least 60 specimens (NISP was not calculated from the rescue project tombs, which were not screened). The order of sites changes slightly when considering MNI. Anurans are still most abundant at M70, with a minimum of 622 individuals, followed by Omo cemeteries with 88, Chen Chen with 40, Omo Temple with 21, and M43 with 18.

Table 1. Anura NISP and MNI by site and context type.

Site	NISP	Total MNI	MNI from Tombs	Percentage MNI from Tombs	MNI from Domestic Contexts	Percentage MNI from Domestic Contexts	MNI from Ceremonial Contexts	Percentage MNI from Ceremonial Contexts	MNI from Midden	Percentage MNI from Midden
Chen Chen M1	60	40	30	75.0	8	20.0	2	5.0	n/a	n/a
Omo Cemetery	541	88	85	96.6	n/a	n/a	n/a	n/a	3	3.4
Omo Temple	80	21	n/a	n/a	n/a	n/a	21	100.0	n/a	n/a
Rio Muerto M43	158	18	11	61.1	7	38.9	0	0.0	n/a	n/a
Rio Muerto M70	4808	622	352	56.6	1	0.2	n/a	n/a	269	43.2
TOTAL	5647	789	478	60.6	16	2	23	2.9	272	34.5

5.1. *Taxonomic Identifications of Anurans and Non-Anuran Fauna*

Vallejo-Pareja identified at least three Anura genera based on the photographs of 149 humeri, representing a minimum of 87 individuals (MNI), from tomb M70B-66 at the Rio Muerto M70B cemetery (M70-2961; Table 2). *Rhinella* sp. is the most common taxon, represented by 58 humeri (38.9% of the total humeri in this context). The humerus of *Rhinella* can be identified and differentiated from other taxa found in the region by having a distal condyle (humeral head) that is wider than long, and more oval than rounded. The ulnar and radial epicondyle are reduced and there are no lateral or medial crests along the shaft. The ventral crest is well developed (Figure 4). *Telmatobius* sp. is represented by 17 humeri (11.4%). This genus can be diagnosed and differentiated because of its well-developed medial crest that is wide and extends proximally past the middle of the humerus shaft. The lateral crest is well developed, wide and short. The radial epicondyle is not developed, the ulnar epicondyle is well developed, and the humeral head is rounded (Figure 4). The third taxon identified from the sample is *Pleurodema* sp. with 10 humeri (6.7%). It can be diagnosed and differentiated by its robust humerus with a rounded humeral head, a poorly developed radial epicondyle, and a well-developed ulnar epicondyle. There are no lateral or medial crests but the humeral shaft is laterally expanded towards the humeral head (Figure 4). The remaining 42 humeri (28.2%) could not be identified to genus, and 22 (14.9%) are classified as possibly belonging to one of these three genera. Although not all humeri from the cemetery context could be identified to genus, the three identified taxa account for 73% of the humeri present. Additionally, a preserved owl pellet from Rio Muerto, M70-4075 (22 NISP, MNI = 2) contains remains of at least one *Rhinella* sp. individual as well as a smaller anuran, probably another juvenile *Rhinella* sp. (see discussion below). We do not know if other mortuary and non-mortuary contexts have similar ratios of anuran taxa or if they contain other unidentified anuran species.

Table 2. Number of Anura humeri from M70-2961 identified to the genus level by NISP and MNI (photos reproduced with permission: *Rhinella* sp. [54], *Telmatobius* sp. [55], and *Pleurodema* sp. [56]).

Image	Taxon	Probable species	NISP	% of NISP	MNI	% of MNI
	*Rhinella* sp.	*Rhinella limensis* or *spinulosa*	58	38.9	32	36.8
	Rhinella sp. (possible)	https://calphotos.berkeley.edu/cgi/img_query?seq_num=904525&one=T	10	6.7	9	10.3
	*Telmatobius* sp.	*Telmatobius arequipensis, marmoratus* or *peruvianus*	17	11.4	11	12.6
	Telmatobius sp. (possible)	https://calphotos.berkeley.edu/cgi/img_query?enlarge=0000+0000+0810+0233	8	5.4	8	9.2
	*Pleurodema* sp.	*Pleurodema cinereum* or *marmoratum*	10	6.7	5	5.7
	Pleurodema sp. (possible)	https://calphotos.berkeley.edu/cgi/img_query?seq_num=364994&one=T	4	2.7	2	2.3
	unidentified		42	28.2	20	23
	All		149	100	87	100

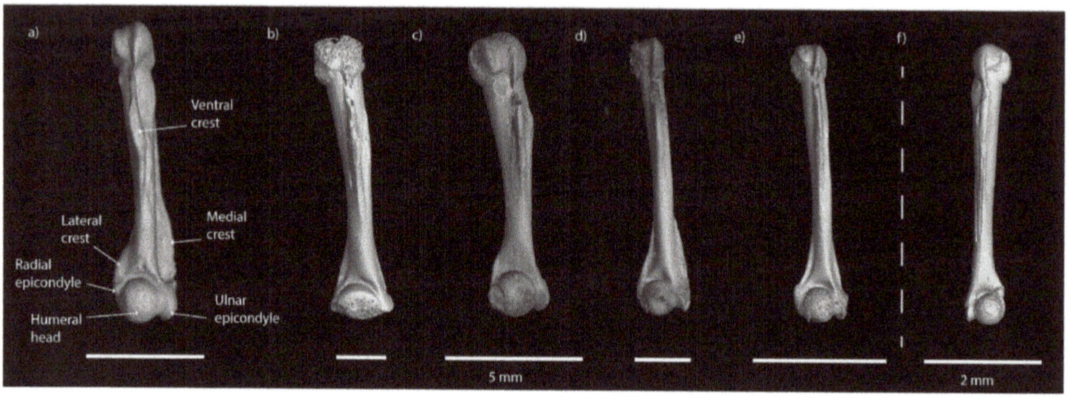

Figure 4. Humeri of anuran species used for comparison with fossil material. (**a**) *Telmatobius carrillae;* (**b**) *Rhinella marina;* (**c**) *Pleurodema nebulosa;* (**d**) *Hyla boans;* (**e**) *Gastrotheca nebulanastes;* (**f**) *Psychrophrynella bagrecito.* (**a–e**) Right humerus, scale bar = 5 mm; (**f**) left humerus, scale bar = 2 mm. [51] (Anatomical nomenclature). Composite image by Vallejo-Pareja from photographs of digitally segmented specimens (ct-scans) at the Herpetology Division of the Florida Museum of Natural History.

During excavations field crews collected the remains of a variety of other taxa from contexts with anuran bones, including camelids, rodents, carnivores, bats, cervids, fish, crustacean, mollusca, and coral (Table 3). Notably, small-sized rodents exhibit the strongest co-occurrence with anurans, perhaps either because they occupy similar site environments or because both have the same origin of accumulation. Rodent, bat, and squamate remains were found in 69 out of the 106 contexts with anurans (65.1%) and represent a minimum of 225 individuals across all sites (in contrast to the anuran MNI of 789) (Table 4; SI Table S3).

5.2. Abundance and Ubiquity of Anuran Remains

The relative abundance of elements is highest for limb bones across all sites (Figure 3). Regarding breakage, the vast majority of remains are classified as shaft/fragment (89.2%). Few specimens are whole (4.5%) or mostly whole (3/4, 3.7%). No specimen from any sites exhibits butchering, burning, or other human modifications. The only exceptions to the fragmentary remains are two complete mummified frogs (species unidentified) wrapped in wool recovered from M10A-1, an apparent offering context in the Omo temple (Figure 5).

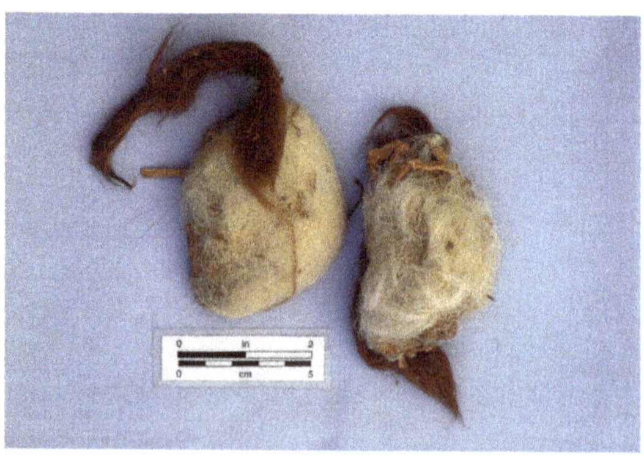

Figure 5. Mummified frogs wrapped in camelid fiber from Omo M10 Temple (photo by Susan D. deFrance).

Table 3. Non-anuran taxa present in analyzed site contexts.

Site\Taxa	Vertebrata	Mammalia	Camelidae	Rodentia	Carnivora	Chiroptera	Artiodactyla and Cervidae	Aves	Squamata	Osteichthyes	Crustacean	Mollusca	Coral
Chen Chen M1	X	X	X	X	X						X	X	
Omo Cemetery	X	X	X	X	X	X	X	X		X		X	X
Omo Temple	X	X	X	X		X			X	X	X	X	
Rio Muerto M43	X	X	X	X	X		X	X	X	X	X	X	X
Rio Muerto M70	X	X		X		X		X	X	X	X		

Table 4. Anurans and other possible owl prey taxa MNI by site.

Site	Anura MNI	Rodentia MNI	Chiroptera MNI	Squamata MNI
Chen Chen	40	15	0	0
Omo Cemetery	88	43	2	0
Omo Temple	19	13	1	1
RM M43	17	10	0	1
RM M70	622	144	11	1
TOTAL	**786**	**225**	**14**	**3**

The ubiquity index of anuran remains by site and by context type is given in Table 5. At Chen Chen, frog and toad remains are present in eight habitation, two ceremonial, and 12 tomb contexts (15.3% of the 144 contexts analyzed for this site). At the Omo Temple (specifically M10A), anurans are present in only 13 contexts (4.3% of 304 analyzed contexts from this sector). Among the 142 contexts analyzed from Omo Cemeteries, anurans occur in 26 tombs and three non-tomb contexts from the cemeteries (20.4% of analyzed tombs). Interestingly, tomb M10H-24 contained 133 anuran remains representing 21 individuals. This deposit alone accounts for 23.8% of the Omo M10 site NISP. The excavators noted an abundance of small bones under the capstones prior to opening this undisturbed burial. The Rio Muerto site M43 samples are from eight tombs and six domestic contexts (9.2% of the 153 analyzed contexts from this site). At Rio Muerto site M70, the anuran bones are from 23 tombs and four non-tomb contexts from the M70B rockpile cemetery, and from only one domestic context from M70C. Thus, 43.8% of the 64 analyzed contexts from M70 contain anurans.

Frog and toad remains are most common in cemeteries and tombs. At the Rio Muerto M70B cemetery, 23 of the 27 analyzed tomb contexts yielded 352 anuran individuals (see Tables 1 and 5) resulting in an anuran ubiquity index of 85.2%. At Omo M10, 26 of the 65 studied tomb contexts produced an MNI of 85 anuran individuals (40.0% ubiquity index). The assemblage from Rio Muerto M43 included eight individuals collected from eight tombs, while the other 34 tombs excavated did not yield any anuran bones indicating a ubiquity index of 23.5%. Similarly, the assemblage from Chen Chen cemeteries yields 30 individuals from 12 tombs out of a total of 36 analyzed tombs (ubiquity index of 33.3%). The contrast between M43 and M70B is notable, considering that both cemeteries are located at a similar elevation but M43 is slightly closer to water sources where frogs would naturally occur.

Anuran remains were considerably more commonly associated with Tiwanaku cist tombs than in other contexts, and anurans were remarkably ubiquitous in the unique M70B "rockpile" cemetery. The quantity of frog/toad individuals in and around tombs contrasts with the small number of anuran individuals in habitational and ceremonial areas of Chen Chen (MNI = 10), in the ceremonial site of the Omo Temple (MNI = 21), in the habitation areas of Rio Muerto M43 (MNI = 7), and in habitation areas of Omo M10 (178 contexts) and M12 (123 contexts) where no anurans were found. Even though anurans are present in some Moquegua Tiwanaku monumental and domestic contexts, they are rare in comparison to tombs and cemeteries.

Further contextual evidence is found from specific anuran contexts in cemeteries. All anuran bones in the Omo cemeteries were recovered from pit or cist tombs whose capstones were found in various states of intactness, collapse, or anthropogenic disturbance, allowing for at least gradual infilling of sediment and accumulation of intrusive fauna (Figure 6).

Table 5. Contexts and Ubiquity Index by site and context type.

Site	Total Contexts Analyzed	Total Contexts with Anurans	Ubiquity Index (%)	Total Tombs Analyzed	Total Tombs with Anurans	Tomb Ubiquity Index (%)	Total Domestic Contexts Analyzed	Total Domestic with Anurans	Domestic Ubiquity Index (%)	Total Ceremonial Contexts Analyzed	Total Ceremonial with Anurans	Ceremonial Ubiquity Index (%)	Total Midden Contexts Analyzed	Total Midden with Anurans	Ceremonial Ubiquity Index (%)
Chen Chen	144	22	15.3	36	12	33.3	90	8	8.9	18	2	11.1	n/a	n/a	n/a
Omo Cem.	142	29	20.4	65	26	40.0	n/a	n/a	n/a	n/a	n/a	n/a	80	3	
Omo Tem.	304	13	4.3	n/a	n/a	n/a	n/a	n/a	n/a	304	13	4.3	n/a	n/a	n/a
Rio M43	153	14	9.2	34	8	23.5	100	6	6.0	4	0	0	n/a	n/a	n/a
Rio M70	64	28	43.8	27	23	85.2	31	1	3.2	n/a	n/a	n/a	6	4	n/a
TOTAL	807	106	13.1	162	69	42.6	221	15	6.8	326	15	4.6	86	7	8.1

Figure 6. Top left, diagram of cist tomb view from above and cross section; modified from [45]; **top right**, Omo cemetery tomb with intact capstone; **bottom left**, collapsed rock capstone of an Omo cist tomb; **bottom right**, Omo cemetery collared tomb on the left (U10) and radiocarbon dated pit (U8) next to tomb. Arrow is pointing to volcanic tephra from the February 1600 eruption of the Huaynaputina volcano. Note the openings allowing infill sediment and creating a burrow for animals. (photos by Sarah I. Baitzel).

6. Discussion

6.1. Anurans and Agents of Accumulation

We consider the multiple possibilities for the accumulation and distribution of anuran remains. One possibility is that the anuran assemblages resulted from natural death. As mentioned previously, frog and toad species common to the Moquegua Valley are not burrowing anurans [26,27]; therefore, they would not have naturally sought shelter in the archaeological deposits described here. Today farmers occasionally capture anurans in fields, but anurans are not found in the hyper-arid settings outside of irrigated fields. Given the low frequencies of frogs or toads today around these sites and the low frequencies of their remains at highland Tiwanaku sites, where frogs are naturally abundant, the quantity of anuran remains at the hyper-arid Moquegua Tiwanaku sites is anomalous [57–59]. Thus, we do not think that natural death is a sufficient explanation to account for the high number of anurans accumulated.

Secondly, we consider the possibility of frog or toad consumption as a cultural practice among Tiwanaku colonists in Moquegua. Because anuran remains are mostly absent in other (highland) Tiwanaku sites, we argue that Tiwanaku peoples did not consider anurans as regular food resource. Thus far, archaeological evidence indicates that the use of toads and frogs in Tiwanaku culture was primarily ritual. Specimens identified as Lake Titicaca frogs (*Telmatobius culeus*) and unidentified anurans are included along with other animals in a Lake Titicaca underwater ritual site, Khoa Reef near the Island of the Sun [60]. This site is interpreted as a ritual locale where religious specialists made repeated state offerings in the lake. Although the authors suggest that the frog remains from this site could be natural inclusions, it is also possible that people intentionally included frogs as offerings along with camelids and birds. The only unequivocally ritual use of anurans at Tiwanaku sites in Moquegua is the deposition of the two wool-wrapped complete anurans in an offering context at the Omo M10A temple. An X-ray revealed both frogs were mummified, possibly desiccated, and fully articulated. There are no other known examples of mummified frogs wrapped in wool from Tiwanaku sites; therefore, we hypothesize that these objects may have been ritual talismans used as a dedication or other offering.

In contrast, the quantity, uneven distributions, and contexts of disarticulated frog and toad bones at the Moquegua sites suggests that the majority of anuran remains accumulated as the result of non-cultural processes rather than anthropogenic factors (e.g., food discard, which consists primarily of butchered and fragmentary camelid remains in domestic contexts, or other intentional human action). Although we cannot discount the possibility that cultural practices (i.e., ceremonial or ritual behavior, or human consumption) resulted in the deposition of some of the anuran bones, the most parsimonious explanation is that frogs and toads are non-cultural deposits.

As taphonomic studies of anuran remains the world over show, one of the principal taphonomic agent for introducing these bones into archaeological assemblages are the natural predators of frogs/toads [17–23,61,62]. In Moquegua, identifying one of the probable taphonomic agents that produced the accumulations of frog and toad skeletons described here was largely serendipitous. During a field visit to the post-Tiwanaku site of Estuquiña (~1250–1470 CE), located north of the city of Moquegua, one of us (deFrance) observed several pairs of adult Burrowing Owls and owlets of the species *Athene cunicularia* near ancient below-ground tombs that had been disturbed and partially opened, probably by recent and past looting activity (Figure 7). In subsequent years, deFrance and others observed Burrowing Owls also nesting in tombs at this site. Significantly, no one observed predatory birds other than Burrowing Owls occupying the cemeteries.

The Burrowing Owl (*Athene cunicularia*) is a common species across western and southern Peru (Figure 8). These owls favor dry, open habitats, feeding on insects, small mammals, reptiles, and amphibians (Figure 9) [63–65]. They are known to nest in abandoned burrows excavated by small mammals. The owl parents at Estuquiña became highly agitated when researchers approached them, suggesting nests were nearby. The abandoned Estuquiña tombs with their cavities protected by surface rocks resemble the burrows typically used by Burrowing Owls and serve as a suitable space for nidification. Inspection of the open tombs revealed that roughly 20 cm below the ground surface several tombs contained abundant remains of owl pellets and nesting material (e.g., feathers, animal fur, small-sized twigs and dry vegetation). The pellets consisted of the remains of small rodents, scorpions, and frogs and toads among other owl food refuse. These observations align with studies of the Burrowing Owl pellets across South America [66–68].

Figure 7. Burrowing Owl (*Athene cunicularia*) surrounded by seven (7) tombs at the site of Estuquiña (photo by Susan D. deFrance).

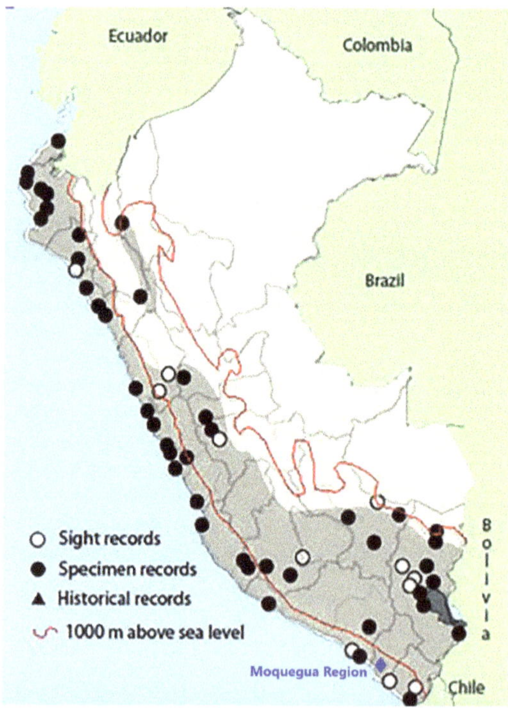

Figure 8. Map of Burrowing Owl distribution in Peru [69], and study region location in southern Peru.

Figure 9. Burrowing Owl eating a frog at Fort Lauderdale, FL [70].

Similar to the modern Burrowing Owls using Estuquiña tombs for their nests and deposits of prey, we hypothesize the abundance of anuran remains at Tiwanaku sites is the result of Burrowing Owl predation (see Figures 7–9). The shelter offered by the intact and collapsed capstones of Tiwanaku tombs at Chen Chen, Omo M10 and Rio Muerto M43 mimic the spaces Burrowing Owls tend to inhabit. In particular, the crags and crannies of the M70B rockpile tombs seems to have made this cemetery highly suitable to owl needs (see Figures 2 and 6). The abandoned tombs in the Moquegua valley would have provided a favorable habitat for the Burrowing Owls by offering nesting spaces where they would bring back prey to their offspring. Modern owl pellets collected from Estuquiña corroborate our hypothesis that some tomb fill resulted from owl predation because the pellets contain large numbers of frog elements together with remains of other taxa; the faunal composition of modern pellets analyzed resembles that found around capstones and in infill sediment of tombs from several Tiwanaku sites (Figure 10). The higher relative abundance of limb bones, and the relative scarcity of cranial fragments and extremities may be in keeping with patterns observed for owl predation in the literature [17–23,61,62]. The remains of Chiroptera (bat), *Phyllotis* sp. (leaf-eared mouse) and Squamata (lizards) might also have been introduced as owl prey, or they could have taken shelter in the tombs.

Non-Anura taxa such as bats, mice, and lizards seem unlikely to have been included as grave goods or offerings. Only fauna with known value as food or offerings in Tiwanaku culture (e.g., camelids, guinea pigs, fish, river crayfish, mollusks) are interpreted as possibly intentional inclusions in the tombs or cemetery surface as offerings.

Several other birds of prey inhabit the Moquegua valley, such as hawks (*Geranoaetus* sp. and *Parabuteo unicinctus*), other owls (*Tyto alba*, *Glaucidium peruanum* and *Bubo virginianus*), falcons (*Falco* sp. and *Phalcoboenus megalopterus*), and vultures (*Vultur gryphus* and *Cathartes aura*) [71]. However, these birds are either rare in the region or have vastly different nesting habits (high terrain or near cliffs) than the Burrowing Owl, and their hunting behavior would be an unlikely explanation for anuran accumulation in the tombs [72,73].

However, we caution that a full taphonomic study would be warranted and would like to consider the limitations of our analysis. As previously discussed, the anuran remains are currently curated in Peru and additional analysis of digestion, fragmentation, and other modifications (such as beak marks and gnawing) is not possible at this time. The lack of access to materials also hindered a more careful quantified analysis of the contents of modern owl pellets. Future deductive studies of these aspects, together with broader anuran taxonomic identification, will allow for a more comprehensive examination of the hypotheses considered here. That would include comparisons with results to many studies on Burrowing Owl pellets in similar settings across South America, some of which include the identification of markers as the ones described in the section on anurans and taphonomic studies [61,62,66–68,74–76].

Figure 10. Owl pellet with *Rhinella* sp. remains from Rio Muerto tomb (M70B-4075) (photo by Susan D. deFrance).

The quantity and context of frog/toad remains suggest that paleoenvironmental conditions in the mid-valley were highly favorable for anurans and owl predators, particularly after the abandonment of Tiwanaku settlements. None of the sites were reoccupied during subsequent periods and there was little subsequent human presence to discourage animals from using the remnants of these townsites and cemeteries. This suggests that sometime after the abandonment of the Tiwanaku settlements around 1100 CE, there was greater moisture and accumulated water in the valley leading to an increase in the frog/toad population. In the desert setting of Moquegua, changes in the normal seasonal pattern of rainfall signal a larger-scale climatic anomaly.

6.2. Anurans and Palaeoecology

The proliferation of anurans that resulted in their deposition in the hyper-arid setting of the Moquegua Valley requires consideration of past climatic conditions, especially given that the archaeological sites examined here are located away from/above the canal/fields. The anuran genera identified from these sites, *Telmatobius*, *Pleurodema* and *Rhinella*, are

Andean taxa found at altitudes between 1000 to 4000 masl, in open areas, grasslands, dry forest or arid environments [27,77]. The *Telmatobius* and *Pleurodema* species of the Peruvian Andes are found near or in permanent water bodies and their dispersion and distribution is limited by the presence and connectivity of these water bodies such as rivers and ponds [78]; few species of *Telmatobius* inhabit humid *paramos* and forest [79]. At least three species of *Telmatobius* are found in the rivers and ponds of the central Peruvian coastal valleys but none have overlapping distribution since their dispersal is limited by the hyper-arid environmental conditions [77]. *Pleuroderma* species are well adapted for environments with strong seasonality and are explosive breeders, an adaptation for living in dry environments, in which the presence of water bodies is crucial for reproduction, which is characterized by laid foam-nests during rainy seasons [78,80,81]. Because southwestern Peru is a hyper-arid region, we hypothesize that the region would not have provided an adequate environment for the survival of the large number of anurans identified from these sites under predictable minimal seasonal rainfall. However, observations from other Andean location have shown a close relationship between these anuran genera and changes in environmental conditions and water availability. This is the case for *Telmatobius marmoratus, Rhinella spinulosa* and *Pleurodema marmoratum* in the Cordillera Vilcanota, southeast Peru, where new ponds created after ice recession expanded the altitudinal range of this species [82]. This evidence suggests that periods of intense rainfall and water accumulation in ponds or other settings created a more favorable habitat for anuran populations to thrive. Significantly, increased rainfall in the Moquegua Valley is historically associated with the onset of ENSO [33,40].

In the mid-valley of Moquegua, researchers documented an early14th-century flood event occurring in the lower valley tributary that passes within 30m of Rio Muerto (M43 and M70) [33]. This flood deposit almost certainly represents the Miraflores event. Here, rainfall triggered a high-velocity flood deposit adjacent to the abandoned Rio Muerto Tiwanaku site. Because major ENSO events may correspond to extended periods of anomalous increased rainfall, the mid-valley probably experienced significant precipitation prior to and after the flooding [83]. Over the longer term, additional effects likely included recharging of subterranean aquifers, replenishing and enhancing local springs, and creating wetter local microenvironments.

We hypothesize that the onset of this 14th-century Miraflores ENSO altered the local habitat making the otherwise arid environment suitable for anurans. Under non-ENSO conditions, local frogs and toads would be restricted to active watercourses, irrigated fields and local springs/wetlands. Under ENSO, small-scale episodes of rainfall or residual pools from quebrada floods generated habitats of standing water suitable for the reproduction of anuran populations. Larger frog/toad populations in turn would have inhabited the area, potentially dying if stuck in flooded stone-lined tombs at Chen Chen, Omo, and Rio Muerto, a process more closely resembling cases of catastrophic death. The infill stratigraphy of some Tiwanaku tombs provides further evidence for tombs as places of localized water accumulation that might have attracted or trapped anurans. Tombs at Rio Muerto and Omo M10 had silty infill sediments that were sometimes laminated and compacted from periods of heavy rain. If anuran populations occupied these habitats, they would have attracted predation by Burrowing Owls that also nested in the abandoned tombs. However, these sites and the analyzed contexts are above (higher elevation) than nearby field systems, streams, and probable water habitats that would have attracted anurans. Because of the location of the cemeteries and analyzed contexts these areas would not have been habitats suitable for anurans in the past.

Pools of standing rainwater, springs, and canals could also fill and regenerate vegetation growth in relic and abandoned canals and field systems; such seasonal habitats increased insects and other anuran prey that attract frogs and other animals to the areas near sites. In north-central Chile, increased plant productivity during ENSO supported periods of abundance of small mammals, including *Phyllotis* sp. (leaf-eared mouse), a small rodent present in our assemblage [84]. On the one hand, increased rainfall in other environments, including arid ones, has been linked to higher anuran egg masses and call production [85–87]. At Rio Muerto and Omo, even a small amount of rainfall could have potentially resulted in standing pools suitable for frogs to colonize and to deposit eggs. On the other hand, even after ENSOs and with lower availability of anurans, the established populations of Burrowing Owls could rely more heavily on insects, *Phyllotis* sp., and Squamata (lizards) as food sources, and continue to occupy the tombs [66,74,76]. We therefore suggest that ENSO in Moquegua may have created conditions favorable for frog and toad growth, in turn sustaining predator populations.

The 14th-century Miraflores ENSO was among the first such climate alterations that made the otherwise arid environment suitable for anurans. That the Miraflores ENSO event created favorable conditions for anurans is supported by the recovery of frog/toad remains from a radiocarbon-dated non-mortuary context at Omo M10 that corresponds to the Miraflores ENSO time frame. In 2010, Baitzel uncovered a small carbon-filled pit (Feature M10U-8) next to an intact tomb rock collar (Burial Feature M10U-10) that was initially interpreted as an offering or burning event associated with the tomb (see Figure 6). However, a radiocarbon date obtained from charcoal recovered in the pit produced a calibrated (SHCal20) date of 1328-1455 cal CE (2-sigma) (^{14}C age BP: 547 $\pm$ 38; AA101909) indicating the deposit postdates the Tiwanaku occupation but overlaps with the Miraflores ENSO event. This pit may have been used by nesting owls feeding on anurans. Future direct radiocarbon dating of anuran specimens could help identify the period of accumulation and examine if it corresponds to the Miraflores ENSO, or possibly to a yet unidentified ENSO event. Other future work would consist in monitoring anuran populations in the region during the next ENSO to determine if these events would be linked to anuran population explosions.

Although ENSO may create conditions that are highly favorable for frog and toad proliferation in arid regions such as Moquegua, other research on the impact of El Niño events on anuran populations shows that increased rainfall can lead to higher embryonic mortality, proliferation of infectious diseases, and decreased species diversity and range [85,88]. These studies, however, have been conducted in areas where frogs and toads tend to be naturally abundant; in those settings, higher-than-normal rainfall might exceed the range to which local anurans are adapted. Our study indicates that the effects of ENSO on herpetofauna need to consider the role of geographic variability on population outcomes.

Today, the impact of climate change is altering the rainfall patterns associated with ENSO. Therefore, it is probable that more frequent and irregular El Niño events, together with increasing temperatures and habitat fragmentation, could lead to permanent declines in frog and toad populations [85,89–91]. Our work points to the potential of zooarchaeological research in supplementing contemporary studies of anuran populations, refining our understanding of the impacts of climate change in amphibians and their related ecosystems, and improving the development of conservation strategies.

7. Conclusions

Zooarchaeological assemblages from four Tiwanaku-culture sites in the Moquegua Valley of far southern Peru contain abundant skeletal remains of frogs and/or toads. Anuran remains are most common in tombs, with far lesser quantities in habitation and ceremonial sites. Our analysis indicates that at least three genera of anurans are present on these sites. Andean people have a long history of ritual interaction with frogs, particularly to bring about rainfall during drought conditions, but other than two complete mummified frogs from a temple context at Omo, we have no evidence that the inhabitants of these sites intentionally deposited anurans. On the basis of modern Burrowing Owl behavior and their preferential use of abandoned archaeological tombs for nests, we interpret the frog and toad remains on these sites as accumulations primarily from the hunting activities of nesting Burrowing Owls, and secondarily from natural death more likely post-ENSO event. At one of the sites in particular, Rio Muerto's M70B "rockpile cemetery," the large quantity of anuran remains may be related to a 14th-century El Niño event that generated rainfall and created conditions favorable for frog populations near an unusual Tiwanaku cemetery that was uniquely suited to owl nidification. Future research on the assemblages of frog and toad remains studied here can contribute to a more comprehensive examination of the hypotheses we have considered. Future research on similar assemblages in other areas of the Central Andes may serve as a proxy for past El Niño events and complement existing biological data for the study of species distribution and environmental change. Our results also point out considerations on best practices to reduce collection bias with the consistent use of 1/16″ screens to improve chances of recovering important small animal remains; many researchers have highlighted the importance of smaller screens for zooarchaeological and paleoenvironmental research [53,92–94]. Furthermore, we caution against automatically interpreting the remains of non-food animals as possible ritual inclusions, particularly in burial contexts, and advocate for archaeologists to explore taphonomic explanations.

Supplementary Materials: The following supporting information can be downloaded at: https://www.mdpi.com/article/10.3390/quat5010016/s1, Table S1: Comparative specimens used in this analysis. Table S2: Data from anuran specimens included in this analysis. Table S3: Context and MNI of Rodentia, Chiroptera, and Squamata compared to Anuran MNI.

Author Contributions: Conceptualization, S.D.d.; methodology, S.D.d.; validation, S.D.d., J.R.S., P.S.G., S.I.B. and M.C.V.-P.; formal analysis, S.D.d., J.R.S. and M.C.V.-P.; investigation, S.D.d., J.R.S., P.S.G. and S.I.B.; data curation, S.D.d. and J.R.S.; writing—original draft preparation, S.D.d., J.R.S., P.S.G., S.I.B. and M.C.V.-P.; writing—review and editing, S.D.d., J.R.S., P.S.G., S.I.B. and M.C.V.-P.; visualization, S.D.d., J.R.S. and M.C.V.-P.; funding acquisition, S.D.d., P.S.G. and S.I.B. All authors have read and agreed to the published version of the manuscript.

Funding: The zooarchaeological research was funded by National Science Foundation Grants No. BCS-1152520 (deFrance), Field work was funded by Fulbright-Hays, Wenner Gren, HJ Heinz III foundation and NSF Award No 1067986 (Goldstein) and NSF BCS-1240079 and Fulbright-Hays DDRA (Baitzel).

Data Availability Statement: Data are contained in the article and the Supplementary Materials. CT-scans and 3D reconstruction of the humeri of comparative species are open access and available on Morphosource.org.

Acknowledgments: The Contisuyo Museum provided research space for the original analysis of the curated zooarchaeological materials. Comments by Prudence Rice and Nicola Sharratt improved this manuscript. We also thank anonymous reviewers and the volume editors whose comments and suggestions sig-nificantly improved the manuscript. All omission and errors are our own.

Conflicts of Interest: The authors declare no conflict of interest.

References

1. Moore, K.M.; Steadman, D.; Defrance, S. Herds, Fish, and Fowl in the Domestic and Ritual Economy of Formative Chiripa. In *Early Settlement at Chiripa*; Hastorf, C.A., Ed.; Research of the Taraco Archaeological Project: La Paz, Bolivia, 1999; pp. 105–116, ISBN 1-882744-01-1.
2. Aguayo, R. Anfibios. In *Libro rojo de la fauna de vertebrados de Bolivia*; Aguirre, L., Balderrama, J., Aguayo, R., Cortez, C., Tarifa, T., Damme, P., Arteaga, L., Eds.; Ministerio del Medio Ambiente y Agua: La Paz, Bolivia, 2009; ISBN 978-99954-0-699-8.
3. Vellard, J. Associated Animal Communities: The Amphibia. In *Lake Titicaca: A Synthesis of Limnological Knowledge*; Dejoux, C., Iltis, A., Eds.; Monographiae Biologicae; Springer: Dordrecht, The Netherlands, 1992; pp. 449–457, ISBN 978-0-7923-1663-3.
4. Tschopik, H. *The Aymara of Chucuito, Peru; Anthropological Papers of the American Museum of Natural History*; AMNH Division of Anthropology: New York, NY, USA, 1951; Volume 44, Pt. 2.
5. Warwick, M.C. *In the Shadow of the Peñon: A Zooarchaeological Study of Formative Diet, Economy, and Sociopolitics in the Río Pukara Valley, Peru*; The University of Wisconsin-Milwaukee: Milwaukee, WI, USA, 2012.
6. La Barre, W. *The Aymara Indians of the Lake Titicaca Plateau, Bolivia. Memoirs of the American Anthropological Association*; Pan American Institute of Geography and History: Bloomington, IN, USA, 1948.
7. Chávez, K.L.M. The Significance of Chiripa in Lake Titicaca Basin Developments. *Exped. Mag. Univ. Pa.* **1988**, *30*, 17–26.
8. Korpisaari, A.; Pärssinen, M. *Pariti: The Ceremonial Tiwanaku Pottery of An Island in Lake Titicaca*; Finnish Academy of Science and Letters: Helsinki, Finland, 2011; ISBN 978-951-41-1081-8.
9. Catenazzi, A.; Von May, R. Conservation Status of Amphibians in Peru1. *Herpetol. Monogr.* **2014**, *28*, 1–23. [CrossRef]

10. Whyte, T.R.; Compton, J.M. Explaining Toad Bones in Southern Appalachian Archaeological Deposits. *Am. Antiq.* **2020**, *85*, 305–330. [CrossRef]
11. Andrews, P. *Owls, Caves, and Fossils: Predation, Preservation, and Accumulation of Small Mammal Bones in Caves, with An Analysis of the Pleistocene Cave Faunas from Westbury-Sub-Mendip, Somerset, UK*; University of Chicago Press: Chicago, IL, USA, 1990; ISBN 978-0-226-02037-2.
12. Hockett, B.; Haws, J. Taphonomic and Methodological Perspectives of Leporid Hunting During the Upper Paleolithic of the Western Mediterranean Basin. *J. Archaeol. Method Theory* **2002**, *9*, 269–302. [CrossRef]
13. De Cupere, B.; Thys, S.; Van Neer, W.; Ervynck, A.; Corremans, M.; Waelkens, M. Eagle owl (Bubo bubo) pellets from Roman Sagalassos (SW Turkey): Distinguishing the prey remains from nest and roost sites. *Int. J. Osteoarchaeol.* **2008**, *19*, 1–22. [CrossRef]
14. Lloveras, L.; Moreno-García, M.; Nadal, J. The eagle owl (Bubo bubo) as a leporid remains accumulator: Taphonomic analysis of modern rabbit remains recovered from nests of this predator. *Int. J. Osteoarchaeol.* **2008**, *19*, 573–592. [CrossRef]
15. Serra, A.S.; Margalef, C.R.; Pérez, J.V.M.; Ripoll, M.P.; Cuñat, C.T.; Marco, Y.C.; Jordá, G.P.; Gómez, A.R.; Marqués, J.B.; Bonilla, V.V. Towards the identification of a new taphonomic agent: An analysis of bone accumulations obtained from modern Egyptian vulture (*Neophron percnopterus*) nests. *Quat. Int.* **2014**, *330*, 136–149. [CrossRef]
16. Hockett, B. Taphonomic comparison of bones in mountain lion scats, coyote scats, golden eagle pellets, and great-horned owl pellets. *Quat. Int.* **2018**, *466*, 141–144. [CrossRef]
17. Llona, A.C.P.; Andrews, P.J. Amphibian taphonomy and its application to the fossil record of Dolina (middle Pleistocene, Atapuerca, Spain). *Palaeogeogr. Palaeoclim. Palaeoecol.* **1999**, *149*, 411–429. [CrossRef]
18. Stoetzel, E.; Denys, C.; Bailon, S.; El Hajraoui, M.A.; Nespoulet, R. Taphonomic Analysis of Amphibian and Squamate Remains from El Harhoura 2 (Rabat-Témara, Morocco): Contributions to Palaeoecological and Archaeological Interpretations. *Int. J. Osteoarchaeol.* **2011**, *22*, 616–635. [CrossRef]
19. Blain, H.-A.; Villa, P. Amphibians and squamate reptiles from the early Upper Pleistocene of Bois Roche Cave (Charente, southwestern France). *Acta Zoöl. Cracoviensia–Ser. A Vertebr.* **2006**, *49*, 1–32. [CrossRef]
20. Blain, H.-A. Contribution de La Paléoherpétofaune (Amphibia & Squamata) à La Connaissance de l'Evolution Du Climat et Du Paysage Du Pliocène Supérieur Au Pléistocène Moyen d'Espagne. *Treb. Del Mus. De Geol. De Barc.* **2009**, *16*, 39–170.
21. Martin, C.; Szyndlar, Z.; Sanchiz, B. Herpetofauna Del Pleistoceno Superior de La Cueva de Cobrante Upper Pleistocene Herpetofauna from Cueva de Cobrante. *Sautuola* **2009**, *XV*, 107–114.
22. Biton, R.; Bailon, S.; Birkenfeld, M.; Bridault, A.; Khalaily, H.; Valla, F.R.; Rabinovich, R. The anurans and squamates assemblage from Final Natufian Eynan (Ain Mallaha, Israel) with an emphasis on snake-human interactions. *PLoS ONE* **2021**, *16*, e0247283. [CrossRef]
23. Bisbal-Chinesta, J.F.; Bañuls-Cardona, S.; Fernández-García, M.; Cáceres, I.; Blain, H.-A.; Vergès, J.M. Elucidating anuran accumulations: Massive taphocenosis of tree frog Hyla from the Chalcolithic of El Mirador cave (Sierra de Atapuerca, Spain). *J. Archaeol. Sci. Rep.* **2020**, *30*, 102277. [CrossRef]
24. Cooke, A.G. (American University, Washington, DC, USA). Personal Communication, 2020.
25. Kidder, A., II. *Two Peruvian Frogs*; Expedition Magazine–Penn Museum: Philadelphia, PA, USA, 1968.
26. Jarvis, L.E.; Angulo, A.; Catenazzi, A.; von May, R.; Brown, J.L.; Lehr, E.; Lewis, J. A Re-Assessment of Priority Amphibian Species of Peru. *Trop. Conserv. Sci.* **2015**, *8*, 623–645. [CrossRef]
27. Aguilar, C.; Ramírez, C.; Rivera, D.; Siu-Ting, K.; Suárez, J.; Torres, C. Anfibios andinos del Perú fuera de Áreas Naturales Protegidas: Amenazas y estado de conservación. *Rev. Peru. Biol.* **2011**, *17*, 005–028. [CrossRef]
28. Myers, N.; Mittermeier, R.A.; Mittermeier, C.G.; da Fonseca, G.A.B.; Kent, J. Biodiversity hotspots for conservation priorities. *Nature* **2000**, *403*, 853–858. [CrossRef]
29. Janzen, D.H. Why Mountain Passes are Higher in the Tropics. *Am. Nat.* **1967**, *101*, 233–249. [CrossRef]
30. Polato, N.R.; Gill, B.A.; Shah, A.A.; Gray, M.M.; Casner, K.L.; Barthelet, A.; Messer, P.W.; Simmons, M.P.; Guayasamin, J.M.; Encalada, A.C.; et al. Narrow thermal tolerance and low dispersal drive higher speciation in tropical mountains. *Proc. Natl. Acad. Sci. USA* **2018**, *115*, 12471–12476. [CrossRef]
31. MINAM Situación Actual de Las Especies de Anfibios y Reptiles Del Perú. 2018. Available online: https://cdn.www.gob.pe/uploads/document/file/355102/Publicacion_Anfibios_y_Reptiles_final_15.08.19.pdf (accessed on 1 July 2021).
32. Oficina Nacional de Evaluación de Recursos Naturales (ONERN) Mapa ecológico del Perú: Guía explicativa. Autoridad Nacional del Agua 1976. Available online: https://sinia.minam.gob.pe/mapas/mapa-perfil-ambiental-peru (accessed on 1 July 2021).
33. Magilligan, F.J.; Goldstein, P.S. El Niño Floods and Culture Change: A Late Holocene Flood History for the Rio Moquegua, Southern Peru. *Geology* **2001**, *29*, 431–434. [CrossRef]
34. Williams, P.R. Disaster in the Development of Agriculture and the Evolution of Social Complexity in the South-Central Andes. Ph.D. Thesis, University of Florida, Gainesville, FL, USA, 1997.
35. Caviedes, C.N.; Waylen, P.R. El Niño y Crecidas Anuales En Los Ríos Del Norte Del Perú. *Bull. Inst. Fr. Études Andines* **1987**, *16*, 1–19.
36. Keefer, D.K.; Defrance, S.D.; Moseley, M.E.; Richardson, J.B.; Satterlee, D.R.; Day-Lewis, A. Early Maritime Economy and El Niño Events at Quebrada Tacahuay, Peru. *Science* **1998**, *281*, 1833–1835. [CrossRef] [PubMed]
37. Manners, R.B.; Magilligan, F.J.; Goldstein, P.S. Floodplain Development, El Niño, and Cultural Consequences in a Hyperarid Andean Environment. *Ann. Assoc. Am. Geogr.* **2007**, *97*, 229–249. [CrossRef]

38. Caramanica, A.; Mesia, L.H.; Morales, C.R.; Huckleberry, G.; Quilter, J. El Niño resilience farming on the north coast of Peru. *Proc. Natl. Acad. Sci. USA* **2020**, *117*, 24127–24137. [CrossRef]
39. Nesbitt, J. Ancient agriculture and climate change on the north coast of Peru. *Proc. Natl. Acad. Sci. USA* **2020**, *117*, 24617–24619. [CrossRef]
40. Goldstein, P.S.; Magilligan, F.J. Hazard, risk and agrarian adaptations in a hyperarid watershed: El Niño floods, streambank erosion, and the cultural bounds of vulnerability in the Andean Middle Horizon. *CATENA* **2011**, *85*, 155–167. [CrossRef]
41. Satterlee, D.R.; Moseley, M.E.; Keefer, D.K.; Tapia A., J.E. The Miraflores El Niño Disaster: Convergent Catastrophes and Prehistoric Agrarian Change in Southern Peru. *Andean Past* **2000**, *6*, 8.
42. Satterlee, D.R. Impact of a Fourteenth Century El Niño Flood on an Indigenous Population near Ilo, Peru. Ph.D. Thesis, University of Florida, Gainesville, FL, USA, 1993.
43. Moseley, M.E.; Defrance, S.D.; Vining, B.R. Droughts, floods, and farming at Quebrada Tacahuay from late prehispanic to colonial times. *Ñawpa Pacha* **2017**, *37*, 25–37. [CrossRef]
44. Reycraft, R.M. The Terminal Chiribaya Project: The Archaeology of Human Response to Natural Disaster in South Coastal Peru. Ph.D. Thesis, The University of New Mexico, Albuquerque, NM, USA, 1998.
45. Goldstein, P.S. *Andean Diaspora: The Tiwanaku Colonies and the Origins of South American Empire*, 1st ed.; University Press of Florida: Gainesville, FL, USA, 2005; ISBN 978-0-8130-2774-6.
46. Goldstein, P.S. Multiethnicity, pluralism, and migration in the south central Andes: An alternate path to state expansion. *Proc. Natl. Acad. Sci. USA* **2015**, *112*, 9202–9209. [CrossRef]
47. Stanish, C.; de la Vega, E.; Moseley, M.; Williams, P.R.; Chávez J., C.; Vining, B.; LaFavre, K. Tiwanaku trade patterns in southern Peru. *J. Anthr. Archaeol.* **2010**, *29*, 524–532. [CrossRef]
48. Vargas, V.B. *Informe Final Del Proyecto: "Rescate Arqueológico Del Cementerio de Chen Chen"*; Instituto Nacional de Cultura, Sucursal Departamental Moquegua: Moquegua, Peru, 1988.
49. Sanchiz, B. *Salientia; Handbuch der Paläoherpetologie/begr. von Oskar Kuhn. Hrsg. von Peter Wellnhofer. Unter Mitarb. von R. M. Appleby*; Pfeil: München, Germany, 1998; ISBN 978-3-931516-27-7.
50. Shine, R. Sexual Selection and Sexual Dimorphism in the Amphibia. *Copeia* **1979**, *1979*, 297–306. [CrossRef]
51. Keeffe, R.; Blackburn, D.C. Comparative morphology of the humerus in forward-burrowing frogs. *Biol. J. Linn. Soc.* **2020**, *131*, 291–303. [CrossRef]
52. Lynch, J.D. Evolutionary Relationships, Osteology, and Zoogeography of Leptodactyloid Frogs. *Misc. Publ.–Univ. Kans. Mus. Nat. Hist.* **1971**, *53*, 1–238.
53. Reitz, E.J.; Wing, E.S. *Zooarchaeology*; Cambridge manuals in archaeology; Cambridge University Press: Cambridge, NY, USA, 1999; ISBN 978-0-521-48069-7.
54. Miranda, P.M. Rhinella Spinulosa. Available online: https://calphotos.berkeley.edu/cgi/img_query?seq_num=904525&one=T (accessed on 4 June 2021).
55. Kosch, T. Telmatobius Marmoratus. Available online: https://calphotos.berkeley.edu/cgi/img_query?seq_num=328239&one=T (accessed on 4 June 2021).
56. Kosch, T. Pleurodema Marmoratum. Available online: https://calphotos.berkeley.edu/cgi/img_query?seq_num=364994&one=T (accessed on 4 June 2021).
57. Park, J.E. Food from the Heartland: The Iwawi Site and Tiwanaku Political Economy from A Faunal Perspective. Master's Thesis, Simon Fraser University, Vancouver, BC, Canada, 2001.
58. Vallières, C. A Taste of Tiwanaku: Daily Life in an Ancient Andean Urban Center as Seen through Cuisine. Ph.D. Thesis, McGill University, Toronto, ON, Canada, 2012.
59. Webster, A.D. The Role of Camelid in the Development of the Tiwanaku State. Unpublished. Ph.D. Thesis, University of Chicago, Chicago, IL, USA, 1993.
60. Delaere, C.; Capriles, J.M.; Stanish, C. Underwater ritual offerings in the Island of the Sun and the formation of the Tiwanaku state. *Proc. Natl. Acad. Sci. USA* **2019**, *116*, 8233–8238. [CrossRef]
61. Montalvo, C.I.; Fernández, F.J.; Tomassini, R.L.; Mignino, J.; Kin, M.S.; Santillán, M.A. Spatial and temporal taphonomic study of bone accumulations of the burrowing owl (Athene cunicularia) in central Argentina. *J. Archaeol. Sci. Rep.* **2020**, *30*. [CrossRef]
62. Montalvo, C.I.; Tejerina, P. Análisis tafonómico de los huesos de anfibios y roedores depredados por Athene cunicularia (Strigiformes, Strigidae) en La Pampa, Argentina. *Mamül Mapu Pasado Y Presente Desde La Arqueol. Pampeana* **2009**, *1*, 323–334.
63. Burrowing Owl Conservation Network Burrowing Owl Facts. Burrowing Owl Conservation Network. Available online: http://burrowingowlconservation.org/ (accessed on 4 June 2021).
64. Arana, M.; Ruíz-Luna, M.L.; María, S.S.; Ramirez, O. Population fluctuations of the house mouse in a Peruvian loma and the functional response of burrowing owls. *Austral Ecol.* **2006**, *31*, 956–963. [CrossRef]
65. Cavalli, M.; Isacch, J.P.; Baladrón, A.V.; Martínez, G.; Bó, M.S. Prey selection and food habits of breeding Burrowing Owls (Athene cunicularia) in natural and modified habitats of Argentine pampas. *Emu–Austral Ornithol.* **2013**, *114*, 184–188. [CrossRef]
66. Andrade, A.; Nabte, M.J.; Kun, M.E. Diet of the Burrowing Owl (Athene cunicularia) and its seasonal variation in Patagonian steppes: Implications for biodiversity assessments in the Somuncurá Plateau Protected Area, Argentina. *Stud. Neotropical Fauna Environ.* **2010**, *45*, 101–110. [CrossRef]

67. Cadena-Ortiz, H.; Garzón, C.; Villamarín-Cortéz, S.; Pozo-Zamora, G.M.; Echeverría-Vaca, G.; Yánez, J.; Brito, J. Diet of the Burrowing Owl Athene cunicularia, in two locations of the inter-Andean valley Ecuador. *Rev. Bras. Ornitol.* **2016**, *24*, 122–128. [CrossRef]
68. Zilio, F. Diet of Falco Sparverius (Aves: Falconidae) and Athene Cunicularia (Aves: Strigidae) in a Dune Region on North Coast of Rio Grande Do Sul, Brazil. *Rev. Bras. Ornitol.* **2006**, *14*, 379–392.
69. Peru Aves Burrowing Owl (Athene Cunicularia). Available online: https://www.peruaves.org/strigidae/burrowing-owl-athene-cunicularia/ (accessed on 30 May 2021).
70. Johnston, D. Breakfast with the Burrowing Owls. Available online: https://dinascitywildlife.com/2012/04/21/breakfast-with-the-burrowing-owls/ (accessed on 4 June 2021).
71. Schulenberg, T.S.; Stotz, D.F.; Rico, L. Distribution Maps of the Birds of Peru, Version 1.0. Environment, Culture & Conservation (ECCo). Available online: http://fm2.fieldmuseum.org/uw_test/birdsofperu/ (accessed on 4 June 2021).
72. The IUCN Red List of Threatened Species. Available online: https://www.iucnredlist.org/en (accessed on 4 June 2021).
73. eBird EBird Homepage. Available online: https://ebird.org/home (accessed on 9 November 2021).
74. Nabte, M.J.; Pardiñas, U.J.F.; Saba, S.L. The Diet of the Burrowing Owl, Athene Cunicularia, in the Arid Lands of Northeastern Patagonia, Argentina. *J. Arid Environ.* **2008**, *72*, 1526–1530. [CrossRef]
75. Schlatter, R.P.; Yáñez, J.L.; Núñez, H.; Jaksić, F.M. The Diet of the Burrowing Owl in Central Chile and Its Relation to Prey Size. *Am. Ornithol. Soc.* **1980**, *97*, 616–619.
76. Carevic, F.; Carmona, E.; Muñoz-Pedreros, A. Seasonal diet of the burrowing owl Athene cunicularia Molina, 1782 (Strigidae) in a hyperarid ecosystem of the Atacama desert in northern Chile. *J. Arid. Environ.* **2013**, *97*, 237–241. [CrossRef]
77. Catenazzi, A.; Vargas, V.; Lehr, E. A new species of Telmatobius (Amphibia, Anura, Telmatobiidae) from the Pacific slopes of the Andes, Peru. *ZooKeys* **2015**, *480*, 81–95. [CrossRef]
78. Faivovich, J.; Ferraro, D.P.; Basso, N.G.; Haddad, C.F.; Rodrigues, M.T.; Wheeler, W.C.; Lavilla, E.O. A phylogenetic analysis of Pleurodema (Anura: Leptodactylidae: Leiuperinae) based on mitochondrial and nuclear gene sequences, with comments on the evolution of anuran foam nests. *Cladistics* **2012**, *28*, 460–482. [CrossRef]
79. De La Riva, I.; Aparicio, J.; Ríos, J.N. New Species of Telmatobius (Anura: Leptodactylidae) from Humid Paramo of Peru and Bolivia. *South Am. J. Herpetol.* **2005**, *39*, 409–416. [CrossRef]
80. Sousa, J.G.; Ávila, R. Body size, reproduction and feeding ecology of Pleurodema diplolister (Amphibia: Anura: Leiuperidae) from Caatinga, Pernambuco state, Northeastern Brazil. *Acta Herpetol.* **2015**, *10*, 129–134. [CrossRef]
81. Duellman, W.E.; Trueb, L. *Biology of Amphibians*; Johns Hopkins University Press: Baltimore, MD, USA, 1994; ISBN 978-0-8018-4780-6.
82. Seimon, T.A.; Seimon, A.; Yager, K.; Reider, K.; Delgado, A.; Sowell, P.; Tupayachi, A.; Konecky, B.; McAloose, D.; Halloy, S. Long-term monitoring of tropical alpine habitat change, Andean anurans, and chytrid fungus in the Cordillera Vilcanota, Peru: Results from a decade of study. *Ecol. Evol.* **2017**, *7*, 1527–1540. [CrossRef]
83. Mendoza, R. Moquegua: Puente Montalvo Colapsó Por Intensas Lluvias. Available online: https://diariocorreo.pe/peru/moquegua-puente-montalvo-colapso-por-intensas-lluvias-869564/ (accessed on 30 May 2021).
84. Holmgren, M.; Stapp, P.; Dickman, C.; Gracia, C.; Graham, S.; Gutiérrez, J.R.; Hice, C.; Jaksic, F.; Kelt, D.A.; Letnic, M.; et al. Extreme climatic events shape arid and semiarid ecosystems. *Front. Ecol. Environ.* **2006**, *4*, 87–95. [CrossRef]
85. Carey, C.; Alexander, M.A. Climate Change and Amphibian Declines: Is There a Link? *Divers. Distrib* **2003**, *9*, 111–121. [CrossRef]
86. Jensen, J.B.; Bailey, M.A.; Blankenship, E.L.; Camp, C.D. The Relationship between Breeding by the Gopher Frog, Rana capito (Amphibia: Ranidae) and Rainfall. *Am. Midl. Nat.* **2003**, *150*, 185–190. [CrossRef]
87. Pérez-Granados, C.; Schuchmann, K.-L.; Ramoni-Perazzi, P.; Marques, M.I. Calling behaviour of Elachistocleis matogrosso (Anura, Microhylidae) is associated with habitat temperature and rainfall. *Bioacoustics* **2019**, *29*, 670–683. [CrossRef]
88. Alexander, M.A.; Eischeid, J.K. Climate Variability in Regions of Amphibian Declines. *Conserv. Biol.* **2001**, *15*, 930–942. [CrossRef]
89. Carey, C.; Heyer, W.R.; Wilkinson, J.; Alford, R.A.; Arntzen, J.W.; Halliday, T.; Hungerford, L.; Lips, K.R.; Middleton, E.M.; Orchard, S.A.; et al. Amphibian Declines and Environmental Change: Use of Remote-Sensing Data to Identify Environmental Correlates. *Conserv. Biol.* **2001**, *15*, 903–913. [CrossRef]
90. Kiesecker, J.M.; Blaustein, A.R.; Belden, L. Complex causes of amphibian population declines. *Nature* **2001**, *410*, 681–684. [CrossRef]
91. Blaustein, A.R.; Walls, S.C.; Bancroft, B.; Lawler, J.J.; Searle, C.; Gervasi, S.S. Direct and Indirect Effects of Climate Change on Amphibian Populations. *Diversity* **2010**, *2*, 281. [CrossRef]
92. Cannon, M.D. A Mathematical Model of the Effects of Screen Size on Zooarchaeological Relative Abundance Measures. *J. Archaeol. Sci.* **1999**, *26*, 205–214. [CrossRef]
93. Lyman, R.L. *Quantitative Paleozoology*; Cambridge University Press: Cambridge, UK, 2008; ISBN 978-0-511-38463-9.
94. Lyman, R.L. The influence of screen mesh size, and size and shape of rodent teeth on recovery. *J. Archaeol. Sci.* **2012**, *39*, 1854–1861. [CrossRef]

quaternary

Article

Who Killed the Small Mammals of Ittenheim (Northeastern France)? An Integrative Approach and New Taphonomic Data for Investigating Bone Assemblages Accumulated by Small Carnivores

Aurélien Royer [1,*], Jean-Baptiste Mallye [2], Maxime Pelletier [3] and Sylvain Griselin [4]

1 Biogéosciences, UMR 6282 CNRS, EPHE, Université Bourgogne Franche-Comté, 6 Boulevard Gabriel, 21000 Dijon, France
2 PACEA, UMR 5199, CNRS, 33615 Pessac, France; jean-baptiste.mallye@u-bordeaux.fr
3 Archaeology, History, Culture and Communication Studies, Faculty of Humanities, University of Oulu, P.O. Box 8000, FI-90014 Oulu, Finland; Maxime.Pelletier@oulu.fi
4 INRAP, UMR 7041, Centre Archéologique de Strasbourg, 10 Rue d'Altkirch, 67100 Strasbourg, France; sylvain.griselin@inrap.fr
* Correspondence: aurelien.royer@u-bourgogne.fr

Abstract: Small carnivores are susceptible to regularly accumulating small- to medium-sized mammal remains in both natural and archaeological sites. However, compared to nocturnal birds of prey, these accumulations are still poorly documented and are generally based on a limited number of samples, including those of relatively small size. Here, we present an analysis of European hamster remains from a rescue excavation at Ittenheim (Bas-Rhin, Grand-Est, France), which were recovered from an infilled burrow, three meters below the current surface. The remains are well preserved and exhibit large proportions of tooth marks. Comparisons with a new and existing reference collection combined with an analysis of all recovered faunal remains suggest the accumulation reflects the action of young red foxes. This is supported by the fact that, although these young individuals leave teeth mark, they do not necessarily consume all parts of medium-sized prey species, including the European hamster. Conversely, the remains of smaller rodents, such as microtine, show distinct patterns of digestion and tooth marks. Carnivore bone accumulations from scats are generally poorly preserved; however, our results demonstrate prey size plays a major role, both qualitatively and quantitatively, in skeletal representation, bone preservation, and bone surface modifications. The present paper underlines the need for more diversified taphonomic reference collections based on an integrative approach designed to evaluate multi-taxa accumulations.

Keywords: burrow; fossil record; multi-taxa approach; small carnivore accumulations; small mammals; European hamster; taphonomy

1. Introduction

Small- and medium-sized mammal fossil remains are regularly preserved in both archaeological and palaeontological sites and can result from series of complex processes. Over the last 40 years, numerous taphonomic studies have investigated faunal assemblages and associated bone modifications in order to characterise each possible scenario responsible for these accumulations (e.g., [1–7], including bone collectors [8–10], human predation and consumption [11–19], and attritional or accidental-natural mortality [20]. Among these potential scenarios, while accumulations by mammalian carnivores are regularly cited [3,21–25], there is a paradoxical lack of detailed analyses, and studies are often limited to specific questions. In a recent synthesis, Denys and Cochard [24] listed a number of taphonomic studies based on reference collections; however, the comparative material for each predator is generally small and limited to few taxa. For instance, no reference collections are currently available for the least weasel (*Mustela nivalis*) or stoat (*Mustela erminea*), despite these species being specialist predators who capture small prey in their tunnels, and store food in reserves [26]. Beyond the predators involved, reference collections do not routinely include sufficiently representative numbers of individuals or remains, often less than 10 individuals [27–30]. Furthermore, reference collections that intend to characterise accumulations of small-sized prey equally remain limited. For instance, among the eight listed by Denys and Cochard [24] for red foxes (*Vulpes vulpes*), only one concerns rodents [31], with the rest focusing on larger prey species, such as lagomorpha and/or birds [32–36]. This is important as prey size has a major influence on the traces left on bone remains [3,33,37–39]. Studies focusing on the taphonomy of small mammal remains are also rare [40–45], especially for European species [18,31,38,46], and often mix rodent species of different sizes [44]). Although the remains of small- and medium-sized mammals from archaeological sites do not necessarily result from the same processes, it is critical to develop multi-taxa approaches to decipher taphonomic issues [47].

Recent rescue excavation in Ittenheim (Alsace, France) recovered numerous small mammals from a burrow, including the European hamster, associated with small numbers of leporid and microtine (voles, wood mice) remains. Here, we explore whether this accumulation reflects attritional mortality or the action of carnivores based on data from three new reference collections of small mammal remains, two recovered from fox or badger dens (Bettant I and VIII, Ain, France) and one from a rabbit warren (Les Six Chemins, Charente, France).

2. Material and Methods

2.1. Description of Sites

Ittenheim is located in the Bas-Rhin region (Figure 1) and was the subject of rescue excavations carried out by the French National Institute for Preventive Archaeological Research (INRAP) in 2019 as part of a highway development project [48]. A deep sounding of 7 m was sunk in order to better contextualise fossil faunal remains recovered from the site's Pleistocene levels. Very few remains of large mammals were collected throughout the sequence, whereas no lithic material was identified [48]. Chronology of Ittenheim is still preliminary, mainly based on sedimentary correlations with local sequences. This loess sequence could have recorded discontinuously, due to many erosion processes, a time interval beginning from at least 350 ky (and possibly from 400 ky). The upper part of this loess deposit is attributed to the Weichselian glaciation (MIS 4-2), above a first paleosol correlated to the Eemian in-terglacial (MIS 5) [48]. Yet, no reliable OSL date have been obtained up to now. Approximately 3 m below the surface (172 m a.s.l.), in level 3, a burrow containing a large quantity of small mammal remains was identified in the loess deposits. No radiocarbon date on this bone material has been realized. The infill of the burrow was water sieved using two superimposed meshes (2.0 and 0.5 mm).

Bettant I and VIII are two morden spoil heaps from red fox (*Vulpes vulpes*) and badger (*Meles meles*) burrows, both of which yielded large faunal assemblages (approximately 5000 remains each). The remains of medium-sized mammals and *Talpa* sp. were studied as part of a previous taphonomic analysis [37,49]. For the present study, the remains of additional small mammals (rat, hedgehog, shrew, and vole) were collected following the water sieving of more than 500 litres of sediments from Bettant I and about 600 litres from Bettant VIII sediment using a 1.8 mm mesh, and analysed.

Les Six Chemins (LSC) is a morden rabbit warren in southwest France (Gensac-la-Pallue, France), which was excavated in 2015 using modern archaeological methods in order to provide taphonomic criteria for the identification of attritional death signatures in fossil accumulations [20,50] The remains of several small mammals demonstrate that the Les Six Chemins warren has supported trogloxene species. For this study, we recorded taphonomic data from small mammal bones recovered from all sedimentary units [20], totalling 2346 litres of sediments, wet-sieved with a 1.6 mm mesh. The warren lies approximately 100m from a fox den (*Vulpes vulpes*), where the remains of 15 rabbits were recovered, including seven with carnivore marks (tooth marks), in loose anatomical connection within a 40 m radius of the den. While this association of warrens and fox dens is relatively common [51,52], no visible evidence of the presence of foxes or other predators was noted in the excavated area.

Figure 1. (**A**) Maps showing the locations of Ittenheim, Bettant, and Les Six Chemins; (**B**) excavations at Ittenheim; and (**C**,**D**) the Ittenheim burrow.

2.2. Taxonomic Identification

As recently demonstrated (e.g., [37–39]), prey size does matter, and is a key factor for interpreting the traces left on bones by predators. Accordingly, bone remains were assigned to four size categories: (1) leporids and fox; (2) hamster, rat, muskrat, and hedgehog; (3) larger voles (*Arvicola amphibius* and *A. sapidus*); and (4) smaller voles and mice, including *Microtus* sp., *Clethrionomys glareolus*, and *Apodemus* sp. Species were identified with reference to modern vertebrate collections curated at the Biogéosciences Laboratory of the Université Bourgogne Franche-Comté and at the PACEA Laboratory of the University of Bordeaux, France. The age-at-death of fox remains was estimated using the data previously published by Harris [53], which is based on tooth eruption and the closure of bone epiphysis. For hare, we used the same criteria as in Pelletier et al. [54], which are mainly based on long-bone epiphysis fusion.

2.3. Anatomical Representation, Bone Breakage, and Surface Modifications

Skeletal part representation was calculated primarily from the principal long bones: humerus (HUM), radius (RAD), and ulna (ULNA) for the forelimb; and the femur (FEM) and tibia (TIB) for the hindlimb; as well as the coxal bone (PELVIS), hemi-mandibles (MAND), maxillary (MAX), and vertebrae (VERT). Radial, carpal, metacarpal, tarsal, metatarsal, phalange, and rib fragments were excluded when calculating the number of identified skeletal elements (NISP) to simplify the approach and make the analysis more efficient (see [30] for more details). The relative abundance of skeletal elements [3,55] was calculated using the following formula:

$$Ri = MNEi/(MNI \times E) \times 100 \quad (1)$$

In which Ri is the relative abundance of element i, MNEi is the minimum number of skeletal elements i, MNI is the minimum number of individuals based on the greatest number of any single element in the assemblage, and E is the number of elements i in the prey skeleton.

In addition, the proportions of skeletal elements were evaluated using the two ratios originally proposed by Andrews [3] but slightly modified in order to ensure statistical rigor in calculating the confidence intervals [20,30,56] that take into account discarded bone elements:

$$-\mathbf{AN/PO\%} = (\text{HUM} + \text{ULNA})/(\text{HUM} + \text{ULNA} + \text{FEM}+\text{TIB}) \times 100 \quad (2)$$

where **AN** represents the number of bones from the anterior limbs, and **PO** is the number of bones of the posterior limbs:

$$-\mathbf{CRA/POSTCRA\%} = \text{MAND} \times 2/(\text{MAND} \times 2 + \text{HUM} + \text{FEM}) \times 100 \quad (3)$$

where **CRA** refers to the cranial elements and **POSTCRA** to the postcranial elements:

$$-\mathbf{Z/E\%} = (\text{TIB} + \text{ULNA})/(\text{TIB} + \text{ULNA} + \text{FEM} + \text{HUM}) \times 100 \quad (4)$$

where **Z** represents zygopodia (tibiae and ulnae) and **E** is stylopodia (femora and humeri).

All calculations produced values between 0 and 100. For instance, the absence of cranial or postcranial elements would produce a **CRA/POSTCRA%** value of 0 or 100, respectively, whereas the **CRA/POSTCRA%** would be around 50 when similar numbers of these different elements were recorded. These modifications allow a 95% confidence interval to be calculated from a binomial test.

The degree of bone breakage was calculated to assess overall assemblage fragmentation:

$$(\text{NISPi complete long bones})/(\text{total NISPi long bones}) \times 100 \quad (5)$$

for each element (i).

Evidence for digestion was recorded on the lower and upper incisors; first lower molars of Arvicolinae; distal humerus; proximal femur, radius, ulna, and tibia; and the acetabulum. The degree of digestion evident on teeth were assessed based on the categories described by Andrews [3] and Fernández-Jalvo et al. [57]: null (0), light (1), moderate (2), heavy (3), and extreme (4). Isolated molars and incisors were not analysed separately from those still in situ in the mandible or maxillary, as the proportion of in situ teeth is highly dependent on the methods used to recover remains from archaeological contexts (i.e., water sieving with baths or continuous water flow). For all other skeletal elements, evidence for digestion was recorded as either present or absent.

Four types of tooth marks were recorded: (1) pitting, which corresponds to compressed oval areas without perforation; (2) punctures, which are areas of compression that pierced the cortical surface of the bone; (3) grooves/scoring in the form of channels associated with carnivore tooth marks; and (4) notches, which are semicircular to curved marks on a fracture edge. The location and character (multiple or bilateral) were also recorded for each bone. The proportion of tooth marks (Toothmark%) was calculated as:

$$(\text{NISPi long bones and coxal bone with tooth marks})/(\text{total NISPi long bones and coxal}) \times 100 \quad (6)$$

2.4. Statistical Analyses

All statistical analyses were conducted in R v.3.6.3 [58] and are detailed in the Supplementary Materials. It includes the data of Ittenheim, and the three reference collections (Bettant I, VIII and Les Six Chemins) from this work, as well as many data as possible from previously published reference collections of dens and small carnivores consuming rodents and rabbits. Few data from pellets of diurnal and nocturnal birds of prey have been also included. Confidence intervals were generated using the probability of success from the *binom.test* function in R. Correspondence analyses (CAs) were calculated using 'missMDA' [59] and 'FactomineR' [60] packages. Loss of basipods, scapula, and ribs results from numerous non-predator-related factors, including post-depositional alterations or recovery protocols. As such, these elements were removed from the CA analysis. As the assumptions of multivariate analysis of variance (MANOVA) were not fully met, we performed a non-parametric multivariate analysis of variance (NP-MANOVA) based on Bray–Curtis dissimilarity using the *adonis* function from the R 'vegan' package [61]. This was designed to test the null hypotheses of differences between modern carnivore reference collections according to prey species (rabbit vs. rodent) and origin (i.e., scat, no digested elements, den). A post-hoc multilevel pairwise analysis based on Bray–Curtis distances was also performed using the *pairwise.adonis* function in the pairwiseAdonis package [62]. The R script is available in the Supplementary Materials and at https://github.com/AurelienRoyer/Ittenheim (accessed on 16 July 2021).

3. Results

3.1. Ittenheim

A total of 2686 remains were determined from the Ittenheim burrow, representing at least nine distinct species (Table 1). The several hundred hare remains from Ittenheim were assigned to mountain hare or brown hare based on the size of the incisors and the morphology of the upper second premolars in order to differentiate this species from brown hare [63–67]). As only the remains of mountain hare (*Lepus timidus*) were identified, and taking into consideration the homogeneity of the assemblage, all hare remains were attributed to the mountain hare. Most hare remains were from juvenile individuals (i.e., less than four months; see [54,68]). Two red fox (*Vulpes vulpes*) individuals were identified, an approximately three and a four-month-old fox cub. The presence of a single baculum (length = 25 mm) indicates one of the two cubs was male. Compared to data published by Harris [53], its morphology and length are consistent with a 6-week-old fox cub. Based on the epiphysation of the metapods, the second individual could be less than 15 weeks old. Tooth marks, and to a lesser extent traces of digestion, were recorded on both leporid and canid remains (Table 2). Two elements exhibited evidence of digestion, one belonging to a hare, the other an unidentified medium-sized mammal (fox or hare). Tooth marks were recorded on nine hare remains and single fox mandible. Most tooth marks consisted of small depressions (pittings) on bone surfaces, primarily those of long bones. One element displayed scoring and another a fracture edge with a crenellated morphology.

Table 1. Number of remains (NISP), minimal number of individuals (MNI), and minimal number of elements (MNE) by taxa from Ittenheim.

Taxon	NISP	MNE	MNI
Lepus timidus	108	68	3
Vulpes vulpes	53	45	2
Lagomorpha/Vulpes	10	-	-
Cricetus cricetus	2358	2279	64
Apodemus cf. *sylvaticus*	4	4	2
Apodemus/mus	2	2	1
Muridae	25	25	1
Arvicola cf. *amphibius*	8	8	1
Microtus agrestis	4	4	3
Microtus arvalis/agrestis	5	5	3
Arvicolidae	87	85	-
Talpa europaea	7	7	1
Suinae	1	1	1
Bird	2	2	1
Indéterminés	12		-
Total	2686	2535	83

Table 2. Conservation of skeletal elements from Ittenheim by taxa (leporid and fox) and NISP bones showing digestion and tooth marks by anatomical element. %CB: Complete bones; %RA: Relative abundance; NISP CO: number of identified specimens complete.

	Leporid (MNI = 3)							Fox (MNI = 2)						
Anatomy	**NISP**	**MNE**	**NISP CO**	**% CB**	**% RA**	**NISP Digested**	**NISP Bones with Tooth Marks**	**NISP**	**MNE**	**NISP CO**	**% CB**	**% RA**	**NISP Digested**	**NISP Bones with Tooth Marks**
Mandible	-	-	-	-	-	-	-	3	3	2	66.7	75.0	0	1
Maxillary	2	2	0	0.0	33.3	-	0	0	0	-	-	0.0	-	-
Cranial	5	-	0	0.0	-	-	0	7	1	1	14.3	-	0	0
Incisors	1	1	1	100.0	5.6	-	0	2	2	1	50.0	8.3	0	0
Molars	9	8	4	44.4	12.1	1	0	1	1	0	0.0	5.0	0	0
Other teeth	-	-	-	-	-	-	-	2	2	2	100.0	50.0	0	0
Scapula	-	-	-	-	-	-	-	-	-	-	-	-	-	-
Humerus	1	1	0	0.0	16.7	-	0	1	1	1	100.0	25.0	0	0
Radius	-	-	-	-	-	-	0	-	-	-	-	-	-	-
Ulna	1	1	0	0.0	16.7	-	1	3	2	2	66.7	50.0	0	0
Femur	8	3	2	25.0	50.0	-	4	1	1	0	0.0	25.0	0	0
Tibia	9	3	5	55.6	50.0	-	3	-	-	-	-	-	-	-
Pelvis	1	1	0	0.0	16.7	-	1	-	-	-	-	-	-	-
Calcaneum	5	5	5	100.0	83.3	-	0	-	-	-	-	-	-	-
Talus	3	3	3	100.0	50.0	-	0	1	1	1	100.0	25.0	-	0
Carpals/tarsals	7	7	7	100.0	9.7	-	0	1	1	1	100.0	25.0	0	0
Patella	1	1	1	100.0	16.7	-	-	-	-	-	-	-	-	-
Sesamoïd	-	-	-	-	-	-	0	1	1	1	100.0	-	-	0
Ribs	2	2	2	100.0	2.6	-	0	3	3	3	100.0	5.8	0	0
Vertebra	1	1	1	100.0	0.9	-	0	16	8	0	0.0	9.8	0	0
Sternum	-	-	-	-	-	-	0	1	1	1	100.0	50.0	-	0
Metapodial	34	22	23	67.6	40.7	-	0	7	7	3	42.9	19.4	0	0
Phalanx	21	20	20	95.2	19.6	-	0	10	10	6	60.0	9.6	0	0
Baculum	-	-	-	-	-	-	0	1	1	1	100.0	50.0	0	0
Total	111	81				1	9	61	46				0	1

Nearly 92% of the remains (n = 2019) from Ittenheim can be attributed to hamsters (*Cricetus cricetus*), corresponding to a minimum of 64 individuals. All individuals show teeth wear with young adults, adults, and old ones. While all elements of the skeleton are represented, cranial and dental remains are by far the most abundant (MNI was based on hemi-maxillae, Table 3). Only one complete skull was recorded, with the other 45 being broken at the level of the occipital. Conversely, long bones and pelvis are underrepresented, with an average survival rate of 61%. Despite this, the remains are generally well preserved, with completeness values ranging from 38 to 65%. Only the pelvis bones are more heavily fragmented, which is primarily due to the fragility of the pubic symphysis.

No traces of digestion were observed on hamster incisors, radius, ulna, tibia, or coxal, and only 2.0 and 5.7% of femur and humerus, respectively, bear evidence of digestion (Figures 2 and 3). In contrast, on average, 24% of the hamster remains bear tooth marks (Table 3), primarily the humerus and pelvis (40% of all remains). Among the various carnivore traces observed on long bones, grooves are rare, observed on only two bones (Figure 2C). Notches are evident on more than 50% of bones, while puncture and pitting were observed on 44 and 32%, respectively, of bones. Approximately 65% of bones with carnivore traces exhibit only one form of alteration (Figure 4); bilateral marks are evident in more than 10% of cases, with the majority occurring on the pelvis and humerus (Figures 5 and 6). Trends in the location of the marks are also evident; for example, all pits observed on the ulna are located in the proximal portion of the diaphysis, while punctures are concentrated on the head (Figure 4). In six cases, osteal tissue was removed from the olecranon (Figure 4).

Small mammals from Ittenheim include at least one European mole (*Talpa europaea*) individual and seven voles, the water vole (*Arvicola amphibius*) and field vole (*Microtus agrestis*); and four individuals of muridae, including the wood mouse (*Apodemus sylvaticus*) and possibly *Mus* sp., although the two teeth potentially belonging to this species are heavily altered by digestion, precluding a definitive species attribution. Although small mammal remains are limited, they show distinct patterns, unlike the hamster remains. Skeletal profiles differ as a function of poorer bone preservation and a high degree of fragmentation. On average, more than 65% of small mammal bones and more than 70% of teeth exhibit traces of digestion, some reaching degrees 3 and 4 (Tables 3 and 4, Figures 2G and 3C,D). Again, contrary to the hamster remains, tooth marks are rare and primarily occur on the acetabulum (Figure 5F). However, not all small mammal remains depict the same tendencies; for example, mole remains, which exhibit neither traces of digestion nor tooth marks. Finally, four passerine bird bones were recorded, all bearing digestion damage.

Table 3. Conservation of skeletal elements from Ittenheim by taxa (hamster, *Arvicola* sp., and small rodents) and percentages of digestion and tooth marks by anatomical element. %CB: Complete bones; %RA: Relative abundance; x: observation of digestion recorded but not quantified; *: observations based only on first lower molars.

	Hamster (MNI = 64)					Small Voles and Mice (MNI = 7)					*Talpa europaea* (NMI = 1)				
Anatomy	**MNE**	**% CB**	**% RA**	**% Digestion**	**% Bones with Tooth Marks**	**MNE**	**% CB**	**% RA**	**% Digestion**	**% Bones with Tooth Marks**	**MNE**	**% CB**	**% RA**	**NISP Di-gested**	**NISP Bones with Tooth Marks**
Mandible	119	47	93.0	-	13.1	5	20	35.7	-	0	0	-	-	-	-
Maxillary	125	-	97.7	-	-	2	0	14.3	-	-	0	-	-	-	-
Incisors	219	-	85.5	0	-	26	-	92.9	72.7	-	-	-	-	-	-
Molars	544	-	70.8	-	-	64	-	76.2	85.7 *	-	-	-	-	-	-
Scapula	28	0	21.9	-	-	-	-	-	-	-	1	100	50	0	0
Humerus	74	32	57.8	5.7	40.2	5	0	35.7	60	0	1	100	50	0	0
Radius	72	52	56.3	1.2	10.7	1	0	7.1	100	0	-	-	-	-	-
Ulna	67	40	52.3	0	17.0	5	20	35.7	60	0	1	100	50	0	0
Femur	93	56	72.7	2.9	23.1	7	43	50.0	28.6	14.3	2	100	100	0	0
Tibia	78	53	60.9	0	21.6	2	0	14.3	100	0	1	0	50	0	0
Pelvis	85	8	66.4	0	39.8	7	0	50.0	57.1	57.1	1	100	50	0	0
Calcaneum	62	-	48.4	x	-	-	-	-	-	-	-	-	-	-	-
Talus	56	-	43.8	-	-	-	-	-	-	-	-	-	-	-	-
Carpals/tarsals	-	-	-	-	-	-	-	-	-	-	0	-	-	-	-
Vertebra	397	-	20.7	x	x	22	-	-	x	-	-	-	-	-	-
Metapodial	-	-	-	-	-	-	-	-	-	-	-	-	-	-	-
Phalanx	-	-	-	x	-	-	-	-	-	-	-	-	-	-	-
Total	2019					146					7				

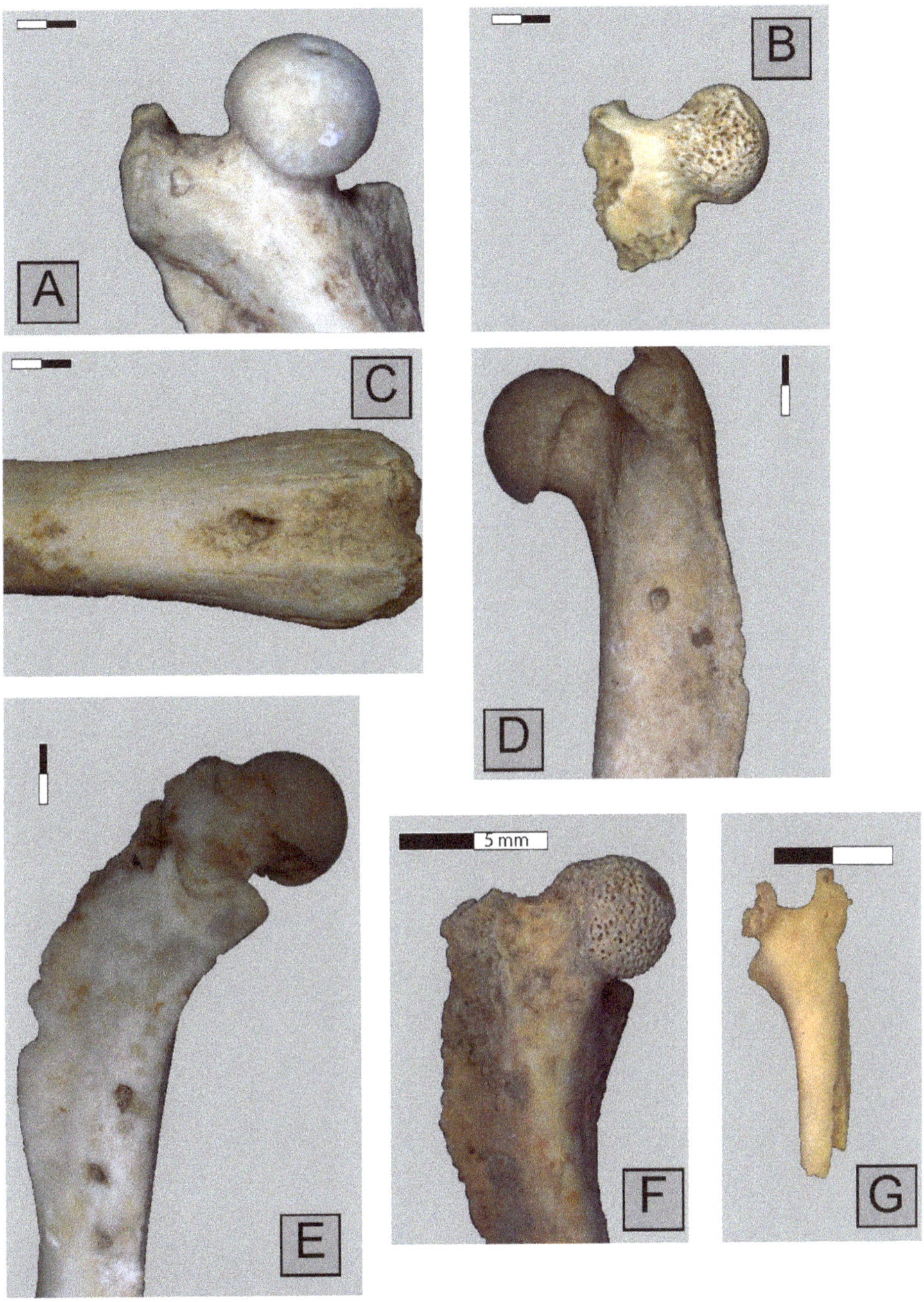

Figure 2. Hamster femur with a single tooth impact (**A**,**C**,**D**), multiple impacts (**E**), and digestion marks (**B**,**F**). Digested Microtus sp. femur (**G**). Scale bar represents 2 mm, unless another value is indicated.

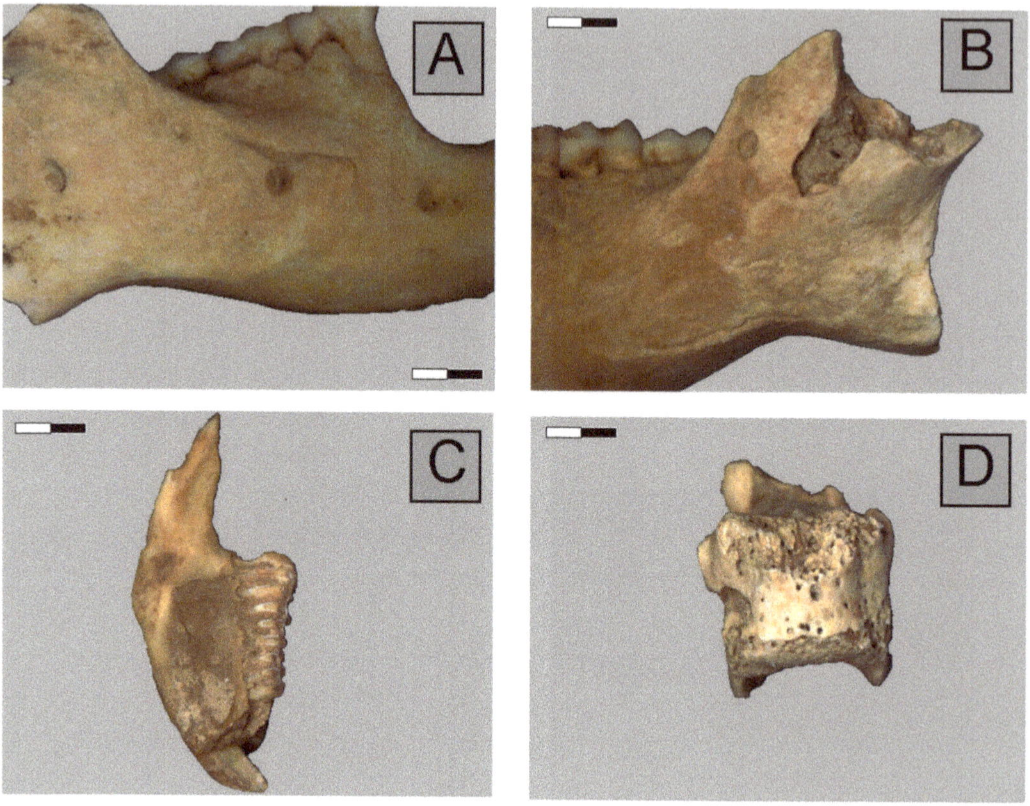

Figure 3. Hamster mandibles from Ittenheim with punctures (**A**,**B**), notch (**B**), and multiple pits (**A**). Digested *Microtus arvalis/agrestis* mandible (**C**) and vertebrae of a small rodent (**D**). Scale bar represents 2 mm.

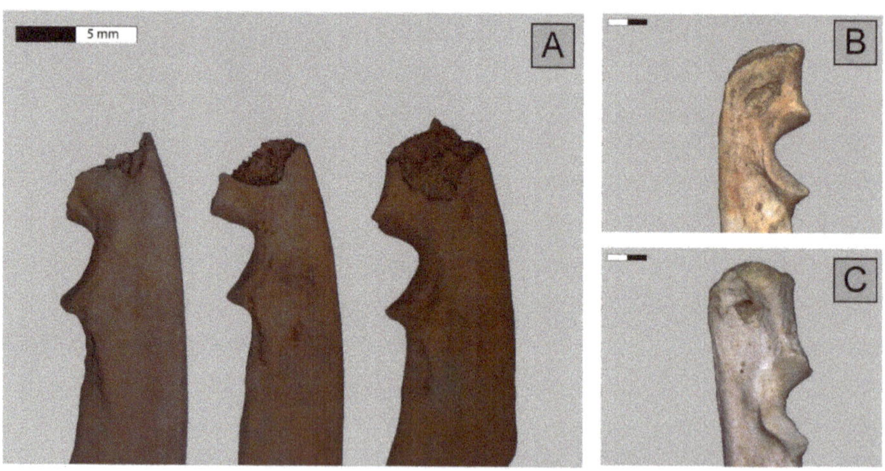

Figure 4. *Cont.*

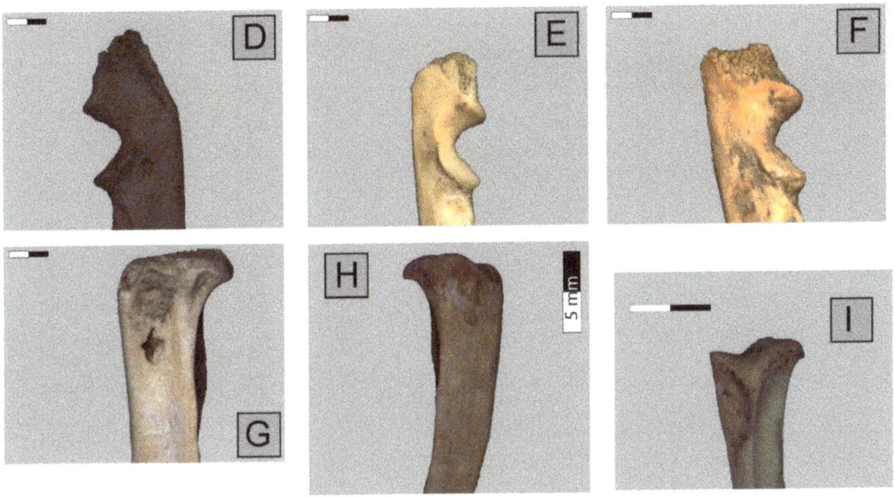

Figure 4. Ittenheim hamster ulna with osteal tissues removed from the olecranon (**A**,**D**,**E**,**F**), pits (**A**), and punctures (**B**,**C**). Ittenheim hamster tibia with punctures (**G**) or pits (**H**). Digested tibia of *Arvicola* sp. (**I**). Scale bar represents 2 mm, unless another value is indicated.

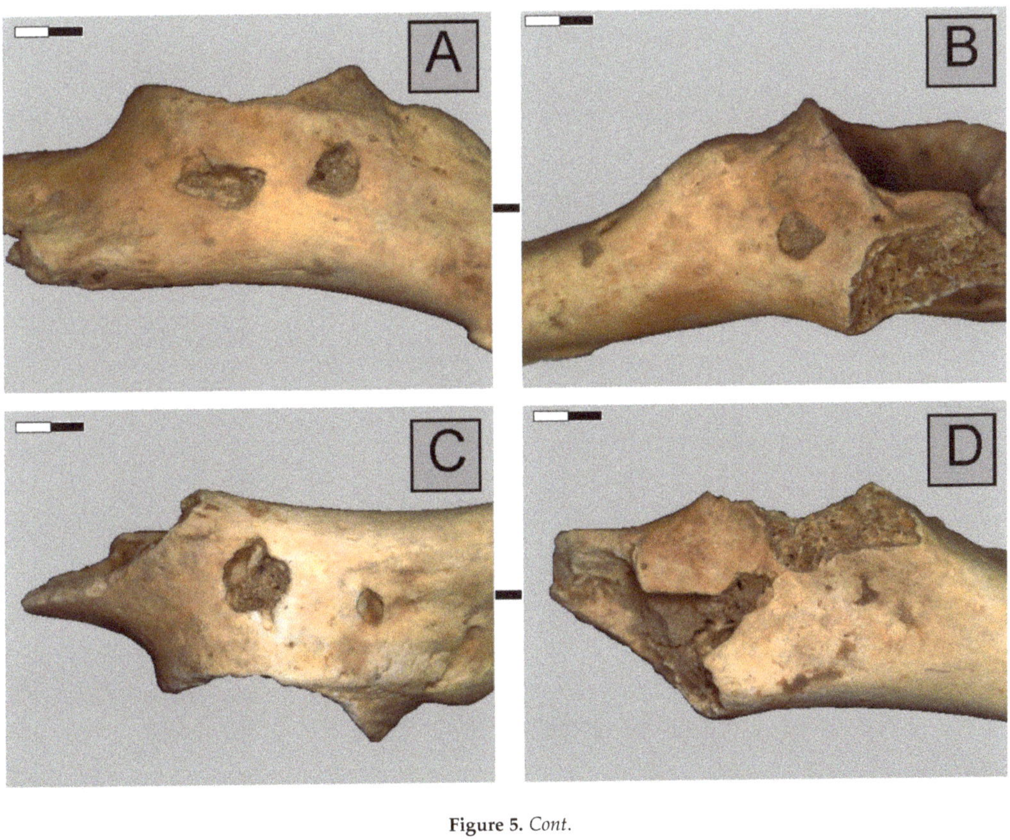

Figure 5. *Cont.*

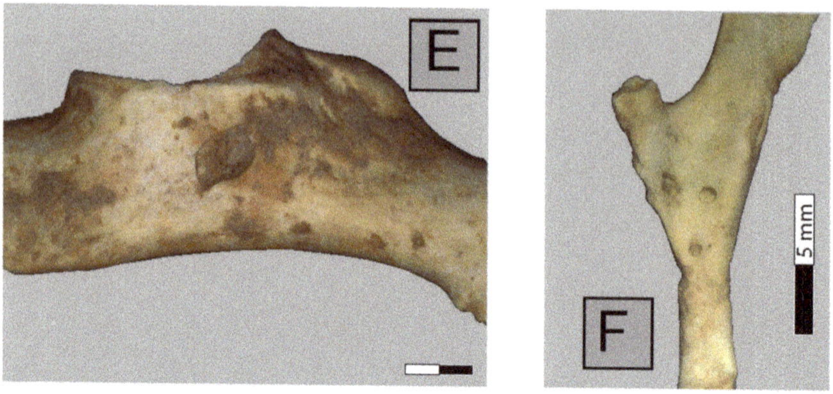

Figure 5. Ittenheim hamster coxal bones exhibiting bilateral and multiple tooth marks (**A–E**). Pelvis bones of *Arvicola* sp. (**F**) from Ittenheim with tooth marks. Scale bar represents 2 mm, unless another value is indicated.

Table 4. Degree of digestion by elements, sites, and prey sizes. Counts are expressed in NISP. I inf: lower incisors, I sup: upper incisors, m1 inf: lower first molars, FEM = femur heads.

Site	Size	Element	Digestion 0	1	2	3	4	Total
Bettant I	*Ondatra, Rattus* and *Erinaceus*	I inf	4	1	0	0	0	5
		I sup	2	2	0	0	0	4
		m1 inf	3	1	1	0	0	5
		FEM	4	-	-	-	-	4
	Arvicola sp.	I inf	4	1	0	0	0	5
		I sup	2	0	0	0	0	2
		m1 inf	4	3	1	0	0	8
		FEM	0	2	0	0	0	2
	Small voles and mice	I inf	58	34	8	7	0	107
		I sup	43	26	6	3	1	79
		m1 inf	64	37	23	5	4	133
		FEM	10	3	0	0	0	13
Bettant VIII	*Rattus* and *Erinaceus*	I inf	3	-	-	-	-	3
		I sup	6	-	-	-	-	6
		m1 inf	1	-	-	-	-	1
		FEM	1	-	-	-	-	1
	Arvicola sp.	I inf	2	0	0	1	0	3
		I sup	3	1	0	0	0	4
		m1 inf	2	2	4	3	1	12
		FEM	1	-	-	-	-	1
	Small voles and mice	I inf	29	16	6	3	3	57
		I sup	20	17	2	3	0	42
		m1 inf	36	20	18	12	3	89
		FEM	3	2	2	1	0	8
Les Six chemins	*Rattus rattus*	I inf	1	-	-	-	-	1
		I sup	0	-	-	-	-	0
		m1 inf	2	-	-	-	-	2
		FEM	2	-	-	-	-	2
	Small voles and mice	I inf	32	3	1	2	1	39
		I sup	28	0	0	1	0	29
		m1 inf	18	6	4	0	0	28
		FEM	6	1	0	0	0	7

I inf: lower incisors, I sup: upper incisors, m1 inf: lower first molars, FEM = femur heads.

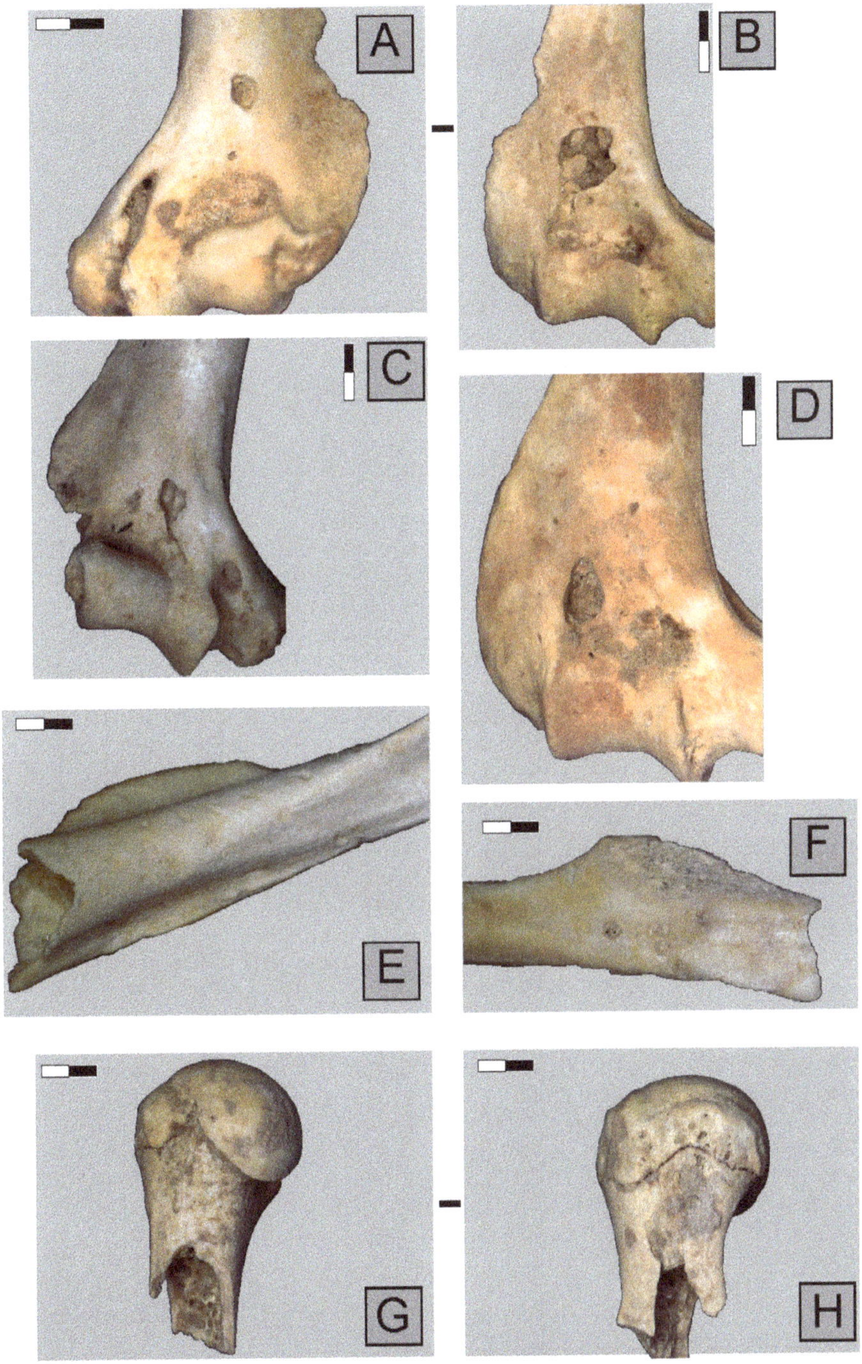

Figure 6. Punctures (**A–D**,**F**), pits (**E**,**F**), and bilateral impact marks (**A**,**B**) with notches (**G**,**H**) on hamster humerus. Scale bar represents 2 mm.

3.2. *Bettant I, a* Meles/Vulpes *den*

Bettant I produced a total of 1091 identifiable bone elements, representing *Rattus rattus, Ondatra zibethicus, Erinaceus europaeus, Arvicola amphibius, Microtus arvalis, Microtus agrestis, Clethrionomys glareolus*, and *Apodemus sylvaticus*. The first three species correspond to larger rodents, which are well preserved, represented by more than 40% complete bones, including a complete rat skull (Table 5). While digestion marks are only present in the form of a slightly altered first lower molar of *Ondatra zibethicus* (Tables 4 and 5), tooth marks are frequent on all long bones, affecting more than 40% of the remains. These marks are primarily pits observable across the bone surface, and are accompanied by two long bones with punctures and one with a single notch.

Of the 113 bone and teeth remains, five individuals of *Arvicola amphibius* were identified at Bettant I. These remains are both poorly preserved and highly fragmentary. Digestion marks are frequent on bones and teeth, and only four bones show tooth marks (pits and punctures).

A total of 895 remains of smaller rodents from Bettant I were identified, representing at least 73 individuals. Less than 10% of bones bear tooth marks, primarily pits followed by punctures and a few notches, although the latter are difficult to reliably determine due to the small size of the remains. More than half of all long bones exhibit traces of digestion. Interestingly, Bettant I also yielded teeth from at least 16 newborn *Microtus arvalis/agrestis* individuals, without tooth wear, but of which half show signs of digestion.

3.3. *Bettant VIII, a* Meles/Vulpes *den*

At least five *Rattus rattus* or *Erinaceus europaeus* were identified from Bettant VII. Slight traces of digestion were observed on the humerus and femur. Multiple pits and punctures are frequent on long bones (Table 6), with the pelvis the most affected by tooth marks, particularly around the acetabulum, sometimes in the form of opposed marks. Half of all identified notches also concern this bone.

Numerous individuals of *Arvicola amphibius* and *Arvicola sapidus* were recognized at Bettant VIII, comprising 145 bones and teeth. Although no complete innominate was recovered, this bone is the most abundant element. All *Arvicola* sp. remains are highly fragmented, showing a low level of completeness, and are poorly preserved, as is the case with the radius, ulna, and mandibles. Less than half of the long bones show no carnivore traces (i.e., digestion, teeth marks). The proportions of digested bones or those with tooth marks are relatively high, varying from 20 to 75% for the most abundant long bones, with several exhibiting both forms of modification. More than 20% of long bones bear notches, and tooth marks are primarily represented by pits, occasionally associated with notches.

A total of 768 remains of Arvicolidae and Muridae were identified from Bettant VIII, including *Apodemus flavicollis, Apodemus sylvaticus, Clethrionomys glareolus, Eliomys quercinus, Microtus agrestis, Microtus arvalis*, and *Microtus (Terricola) subterraneus*. These bones are slightly more complete compared to those from *Arvicola* sp. Like the *Arvicola* sp. remains, these bones bear both evidence of digestion and tooth marks, affecting 34 and 14% of long bones, respectively. Most tooth marks are pits, and punctures are rare (<15% of tooth marks). Clear notches are observed in less than 10% of remains, although several are somewhat uncertain. Numerous teeth show evidence of rounding and corrosion due to digestion, with a non-negligible proportion exhibiting second and third degree traces of digestion (Table 4).

3.4. *Les Six Chemins, a Rabbit Warren*

The small number of *Rattus rattus* (n = 26) remains from Les Six Chemins come from three individuals, including bones of a very young individual with un-erupted molars. Remains are moderately well preserved, fairly complete, and bear no evidence of digestion or tooth marks (Table 7).

Table 5. Conservation of skeletal elements from Bettant I by taxa (*Ondatra*, *Rattus* and *Erinaceus*, *Arvicola* sp., and small rodents) and percentage of digestion and tooth marks by anatomical element. %CB: Complete bones; %RA: Relative abundance; *: observations based only on first lower molars; **: observations based only on first lower molars of *Ondatra zibethicus*.

	Ondatra, Rattus, and *Erinaceus* (MNI = 5)					*Arvicola* sp. (MNI = 5)					Small Voles and Mice (MNI = 73)				
Anatomy	**MNE**	**%CB**	**%RA**	**%Digestion**	**%Bones with Tooth Marks**	**MNE**	**%CB**	**%RA**	**%Digestion**	**%Bones with Tooth Marks**	**MNE**	**%CB**	**%RA**	**%Digestion**	**%Bones with Tooth Marks**
Mandible	5	50.0	50.0	-	16.7	0	-	0.0	-	-	50	21.2	54.3	-	8.0
Maxillary	6	-	60.0	-	-	1	-	6.3	-	-	18	-	19.6	-	-
Incisors	7	-	35.0	33.3	-	7	-	21.9	11.1	-	186	-	101.1	45.7	-
Molars	16	-	26.7	33.3 **	-	29	-	30.2	50.0 *	-	225	-	40.8	52.2*	-
Scapula	3	0.0	30.0	-	x	1	-	6.3	-	-	22	-	23.9	-	-
Humerus	5	100.0	50.0	0.0	40.0	2	0.0	12.5	0.0	50	49	24.3	53.3	65.3	1.4
Radius	2	50.0	20.0	0.0	100.0	2	0.0	12.5	100.0	0.0	14	64.3	15.2	7.1	14.3
Ulna	3	66.7	30.0	-	33.3	0	-	0.0	-	-	42	44.2	45.7	42.9	2.4
Femur	4	100.0	40.0	0.0	25.0	6	25.0	37.5	100.0	12.5	46	23.2	50.0	51.1	8.9
Tibia	5	80.0	50.0	0.0	25.0	6	28.6	37.5	50.0	25.6	37	30.4	40.2	52.9	14.8
Pelvis	5	16.7	50.0	0.0	50.0	3	0.0	18.8	50.0	0.0	35	2.6	38.0	53.8	10.3
Calcaneum	1	-	10.0	-	-	-	-	-	-	-	-	-	-	-	-
Talus	0	-	0.0	-	-	-	-	-	-	-	-	-	-	-	-
Carpals/tarsals	-	-	-	-	-	-	-	-	-	-	-	-	-	-	-
Vertebra	21	-	14.0	-	-	56	-	37.3	-	-	171	-	12.4	-	-
Metapodial	-	-	-	-	-	-	-	-	-	-	-	-	-	-	-
Phalanx	-	-	-	-	-	-	-	-	-	-	-	-	-	-	-
Total	83					113					895				

Table 6. Conservation of skeletal elements from Bettant VIII by taxa (*Rattus* and *Erinaceus*, *Arvicola* sp., and small rodents) and percentage of digestion and tooth marks by anatomical element. %CB: Complete bones %RA: Relative abundance; x: observation of digestion was recorded but not quantified; *: observations based only on first lower molars.

	Rattus and *Erinaceus* (MNI = 5)					*Arvicola* sp. (MNI = 8)					Small Voles and Mice (MNI = 46)				
Anatomy	**MNE**	**%CB**	**%RA**	**%Digestion**	**%Bones with Tooth Marks**	**MNE**	**%CB**	**%RA**	**%Digestion**	**%Bones with Tooth Marks**	**MNE**	**%CB**	**%RA**	**%Digestion**	**%Bones with Tooth Marks**
Mandible	2	50.0	20.0	-	0.0	2	50	12.5	-	0.0	50	14.8	54.3	-	13.0
Maxillary	1	-	10.0	-	-	0	-	0.0	-	-	13	-	14.1	-	-
Incisors	9	-	45.0	0.0	-	7	-	21.9	33.3	-	100	-	54.3	50.5	-
Molars	5	-	8.3	-	-	45	-	46.9	83.3 *	-	170	-	30.8	59.6 *	-
Scapula	3	0.0	30.0	-	-	3	-	18.8	-	-	14	-	15.2	-	-
Humerus	2	0.0	20.0	50.0	0.0	7	9.1	43.8	57.1	45.5	38	19.2	41.3	60.5	5.7
Radius	2	100.0	20.0	0.0	50.0	2	100.0	12.5	0.0	0.0	7	57.1	7.6	28.6	0.0
Ulna	0	-	0.0	-	-	3	33.3	18.8	33.3	0.0	19	21.1	20.7	50.0	10.5
Femur	3	20.0	30.0	33.3	40.0	7	8.3	43.8	42.9	75.0	38	18.2	41.3	59.5	31.8
Tibia	1	0.0	10.0	0.0	100.0	7	8.3	43.8	100.0	41.7	30	13.9	32.6	70.0	22.2

Table 6. *Cont.*

	Rattus and *Erinaceus* (MNI = 5)					*Arvicola* sp. (MNI = 8)					Small Voles and Mice (MNI = 46)				
Anatomy	MNE	%CB	%RA	%Digestion	%Bones with Tooth Marks	MNE	%CB	%RA	%Digestion	%Bones with Tooth Marks	MNE	%CB	%RA	%Digestion	%Bones with Tooth Marks
Pelvis	8	12.5	80.0	0.0	87.5	13	0.0	81.3	28.6	26.7	19	0.0	20.7	43.5	30.4
Calcaneum	0	-	0.0	-	-	-	-	-	-	-	-	-	-	-	-
Talus	0	-	0.0	-	-	-	-	-	-	-	-	-	-	-	-
Carpals/tarsals	-	-	-	-	-	-	-	-	-	-	-	-	-	-	-
Vertebra	7	-	4.7	-	x	33	-	13.8	-	-	236	-	17.1	-	-
Metapodial	-	-	-	-	-	-	-	-	-	-	-	-	-	-	-
Phalanx	-	-	-	-	x	-	-	-	-	-	-	-	-	-	-
Total	43					129					734				

Table 7. Conservation of skeletal elements from Les Six Chemins by taxa (*Rattus rattus* and small rodents) and percentage of digestion and tooth marks by anatomical element. %CB: Complete bones; %RA: Relative abundance; *: observations based only on first lower molars.

	Rattus rattus (MNI = 3)					Small Voles and Mice (MNI = 21)				
Anatomy	MNE	%CB	%RA	NISP Digested	NISP Bones with Tooth Marks	MNE	%CB	%RA	%Digestion	%Bones with Tooth Marks
Mandible	3	33.3	50.0	-	0	29	60.0	69.0	-	0.0
Maxillary	1	-	16.7	-	-	13	-	31.0	-	-
Incisors	1	-	8.3	0	-	69	-	82.1	12.5	-
Molars	7	-	19.4	-	-	65	-	25.8	35.7 *	-
Scapula	0	-	0.0	-	-	2	-	-	-	-
Humerus	1	0.0	16.7	-	0	10	28.6	23.8	16.7	0.0
Radius	2	50.0	33.3	0	0	11	72.7	26.2	0.0	0.0
Ulna	1	100.0	16.7	0	0	3	100.0	7.1	0.0	0.0
Femur	4	50.0	66.7	0	0	14	36.8	33.3	20.0	0.0
Tibia	0	-	0.0	-	-	19	54.2	45.2	0.0	0.0
Pelvis	1	0.0	16.7	0	0	11	9.1	26.2	44.4	18.2
Calcaneum	0	-	0.0	-	-	-	-	-	-	-
Talus	0	-	0.0	-	-	-	-	-	-	-
Carpals/tarsals	-	-	-	-	-	-	-	-	-	-
Vertebra	5	-	5.6	-	-	16	-	-	-	-
Metapodial	-	-	-	-	-	-	-	-	-	-
Phalanx	-	-	-	-	-	-	-	-	-	-
Total	26					275				

Small rodents from the Les Six Chemins warren include at least 15 individuals of trogloxene species, including the common vole (*Microtus arvalis*), field vole (*Microtus agrestis*), bank vole (*Clethrionomys glareolus*), and muridae (*Apodemus* sp. and *Mus* sp.). Unlike the rats from Les Six Chemins, the remains of these small rodents are fragmented or generally lack most skeletal elements, particularly the ulna. Tooth marks are rare and are only evident on the acetabulum. Digestion marks affect up to 45 and 36% of the pelvis and molars, respectively, and to a lesser extent, incisors, femur, and humerus (Table 7), suggesting a proportion were introduced to the warren by carnivores. Finally, two fragmented passeriform vertebrae were identified.

3.5. Comparison with Other Carnivore Reference Collections

We performed correspondence analyses that included as many reference collections as possible for small carnivores that consume rodents and rabbits, as well as several from dens and pellets from rapacious birds (Figure 7A,B, see Supplementary Table S1 for more details). This correspondence analysis is based on anatomical representation. The first two axes explain more than 65% of the total inertia, the third 15.7%. The first axis is characterized by a major contribution of the mandibles (50.9%), tibia (21.7%), and radius (12.4%) and is relatively well correlated with **CRA/POSTCRA%** (Pearson's correlation: $r = -0.36$, $t = -3.16$, $df = 65$, p-value $= 0.002$). The second axis opposes posterior limb bones, the femur (19.3%) and pelvis (19.0%), and anterior limb bones, the scapula (16.5%) and ulna (24.2%), and is highly correlated with **AN/PO%** ($r = 0.77$; $t = 9.91$, $df = 67$, p-value < 0.00001), while the third is highly negatively correlated with **Z/E%** ($r = -0.73$; $t = -8.71$, $df = 67$, p-value < 0.00001). Reference collections are distributed over a large area of the factorial space of the first two axes, reflecting significant variation in skeletal profiles, particularly those of canidae predators. Nonetheless, the ordination space appears strongly influenced by prey categories, with minimal overlap between rabbits and rodents, which are primarily associated with smaller carnivore reference collections. This separation is likely connected to distinct proportions of mandibles and zigopodia between rodents and leporids. Rodent scats and reference collections comprising non-ingested leporids reference collections occupy opposite ends of axis 1. Reference collections of nocturnal rapacious by prey size tend to cluster together and are distinct from leporids and rodents, while diurnal birds of prey are separate. The Ittenheim hamsters fall in the middle of the factorial space defined by first two axes, relatively close to both the Ittenheim small rodents and the Bettant I rats. Bettant VIII small rodents overlap with scats as well as the Les Six Chemins small rodents and rats. Rats and *Arvicola* sp. from Bettant VIII are closer to rabbit assemblages accumulated by *Bubo* sp. and to the cache of Iron Age small mustelids described by Vigne et al. [26] and characterized by high proportions of femur and tibia. These reference collections occupy opposite ends of the second axes of *Talpa europaea* remains.

The **CRA/POSTCRA%, AN/PO%**, and **Z/E%** values (Figure 7C,D) show similar trends to those obtained from CA analyses, including considerable variability amongst scats. The same separation according to prey size (rabbits vs. rodents) and the origin of the material (i.e., scat, den, non-ingested) is also evident; for example, rabbits tend to display **CRA/POSTCRA%** values lower than 50, reflecting an over-representation of postcranial elements, or scats tend to have a **Z/E%** values lower than 50 and an overabundance of stylopodia elements. Hamsters and rodents from Ittenheim fall both within the variability of carnivore scats with rodent prey, as well as small rodents and rats from Bettant I and small rodents from Bettant VIII (Figure 7C,D). Only rats and *Arvicola* sp. from Bettant VIII tend to be associated with rabbit prey reference collections. An NP-MANOVA test returned low not statically significant p-values, except for the interaction (NP-MANOVA interaction term $p \approx 0.01$) between the prey (rodents and leporids) and their origins (den, scat, non-ingested, natural death). While these results depict some general trends, in particular for non-ingested elements according to pairwise comparisons, the high degree of variability prevents any clear separation with the other categories.

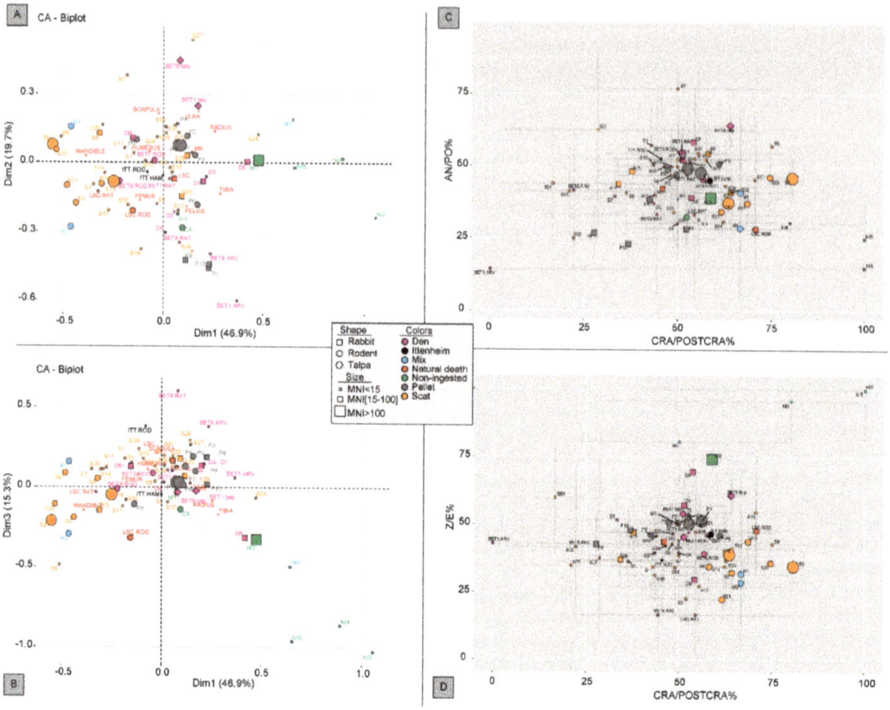

Figure 7. Correspondence analysis biplot of carnivore bone accumulations and reference collections from this study (**A**) dimensions 1 and 2, and (**B**) 1 and 3. In red, coordinates of each bone used in the analysis. Scatter plots of AN/PO% and CRA/POSTCRA% (**C**) and Z/E% and CRA/POSTCRA% (**D**) from the carnivore reference collections. The light grey in scatter plots represents the calculated 95% confidence intervals. For biplots and scatter plots, reference codes are listed in the supplementary data and correspond to: this study = ITT.HAM—Hamster of Ittenheim; ITT—ROD—small rodents; LSC.ROD—small rodents of Les Six Chemins; LSC.RAT—rat, BET1.RAT—rats, muskrats and hedgehogs from Bettant I, BET1.*ARV*—*Arvicola* sp., BET1.ROD—small rodents, BET8.RAT—rats and hedgehogs from Bettant VIII, BET8.ARV—Arvicola sp., BET8.ROD—small rodents; from the literature = BET1.talp—[37], BET8.talp—[37], LSC—rabbit warren [20], P = pellets, P1—[3], P2—[69], P3—[69], P4—[69], P5—[3], P6—[70], P7—[30], S = scats, S1—[3], S2—[3], S3—[3], S4—[3], S5—[3], S6—[3], S7—[3], S8—[44], S9—[46], S10—[40], S11—[40], S12—[40], S13—[43], S14—[71], S15—[41], S16—[42], S17—[72], S18—[38], S19—[73], S20—[73], S21—[74], S22—[17], S23—[75], S24—[45], S25—[34], S26—[34], S27—[35], S28—[35], S29—[76], S30—[73], M = Mix, M1—[38], M2—[38], CA = Cache, CA—[26], MN = Natural death, MN—[77], D = Den, D1—[33], D2—[37], D3—[37], D4—[78], D5—[79], D6—[80], NI = Non—ingested, NI1—[81], NI2—[82], NI3—[82], NI4—[35], M3—[35], NI5—[17], S31—[17], S32—[83], P8—[84], P9—[84], P10—[70]. Shapes represents different prey (squares for rabbits, circles for rodents, and hexagons for talpa), size varies according to the number of prey according to three size groups ([1–15[, [15–100[, ≥100), and colours illustrate the different origins (Den—purple, Ittenheim—black, Mix—blue, Natural Death—red, Non—ingested elements—green, Pellets—grey and Scat—orange).

Figure 8A shows the percentage of digested femur alongside the percentage of digested incisors according to the different predator categories defined by Andrews [3]. Whether for rabbits or rodents, clear differences are evident for carnivore prey recovered from scats, with high proportions of digested and non-ingested remains or those from natural death assemblages. The hamsters from Ittenheim fall squarely in this latter category. The other reference collections fall between these two extremes and correspond to category 2, 3, and 4 predators. Small rodent assemblages from Ittenheim, Bettant I, and Bettant VIII are similar to those produced by category 3 and 4 predators, whereas rats from Bettant I and VIII are closer to predators from categories 1 and 2. Conversely, *Arvicola* sp. from Bettant I stand out from the others, given a high proportion of digested femur and low numbers of digested incisors, which may be partially due to the small number of individuals. Small rodents from Ittenheim are characterized by slightly more digested incisors compared to femurs. When the percentage of digested incisors is compared to the percentages of tooth marks, an inverse correlation can be seen between these two taphonomic signatures (Figure 8B). A first group comprises scat and is characterized by high proportions of digestion teeth and low (or no) numbers of tooth marks. Den, non-ingested reference collection, and Ittenheim are all characterized by higher proportions of tooth marks and lower proposition of bone with signs of digestion. Smaller rodents have intermediate values for both tooth and digestion marks. European hamsters from Ittenheim show a distinct pattern characterized by the absence of signs of digestion on all incisors and relatively high proportions of tooth marks.

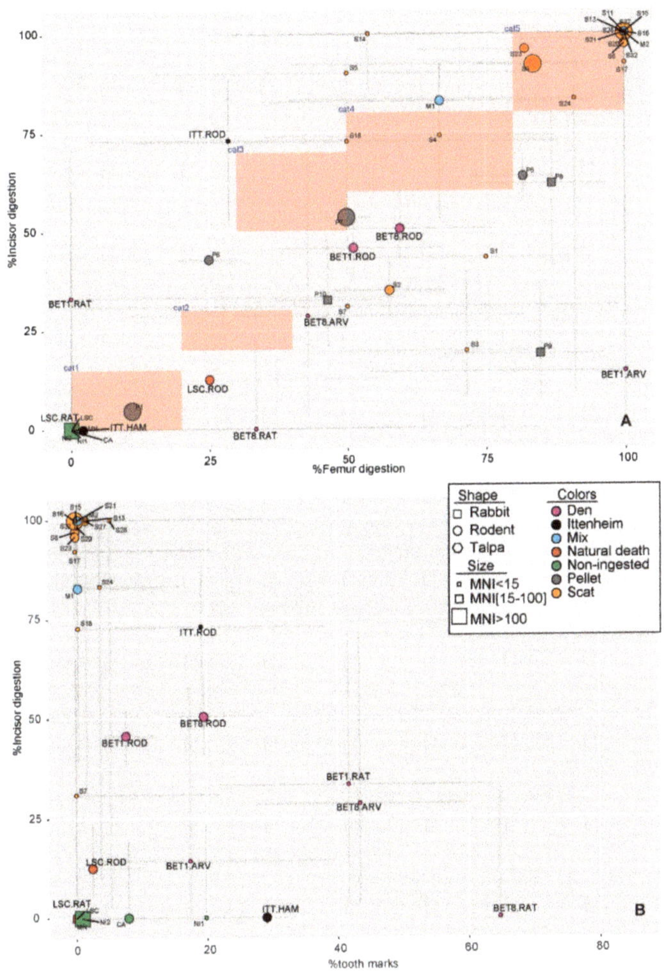

Figure 8. Scatter plot of percentages of digested incisors (**A**) and femur heads, (**B**) and tooth marks, from carnivore reference collections and those built for this study. Corresponding codes are listed in Figure 7 and in the supplementary data. As in Figure 7, shapes represent different prey (squares for rabbits, circles for rodents, and hexagons for Talpa), size varies according to the number of prey according to three size groups ([1–15[, [15–100[, ≥100), and colours illustrate the different origins (Den—purple, Ittenheim—black, Mix—blue, Natural Death—red, Non ingested elements-green, Pellets—grey and Scat—orange). The light grey lines represent the 95% confidence interval.

4. Discussion

4.1. An Integrative Approach for Reconstructing the Taphonomic History of the Ittenheim Accumulation

The small and medium mammal bone accumulations from Ittenheim depict limited species diversity comprising less than 10 species. While no radiocarbon dates are currently available, the presence of interglacial fauna suggests a Holocene age, which would be consistent with the possible presence of *Mus* sp. [85]. However, the identification of *Lepus timidus*, which is no longer present in the region, but whose date of last occurrence is still unknown, also attests to a non-modern accumulation. The Ittenheim accumulation is remarkable due to the high number of hamster (*Cricetus cricetus*) remains, representing at least 64 individuals, associated with the remains of young leporids, fox cubs, and numerous small rodents. Given the context of the site, a faunal assemblage accumulated by rapacious birds can safely be ruled out, leaving only two other possibilities: a catastrophic death related to the collapse of a burrow that was also occupied by trogloxene species or a carnivore accumulation

The hypothesis of catastrophic death related to the collapse of the burrow would be consistent with the fact that hamsters are burrowing mammals. However, this moderately sized rodent (weighing between 150 and 600 g according to the season) is a solitary mammal [86], with each individual occupying a unique burrow, the one exception being nursing females [87,88]. A female can give birth to more than 10 young, which leave the burrow following weaning. Hamsters dig one small burrow during the summer and a deeper, more complex one during the winter, which can reach depths of up to two meters and can include, for example, several food storage and waste disposal areas [86]. The diameter of the burrow's tunnels varies between approximately 5 and 10 cm. Although they have distinct behavioural characteristics compared to rabbits, hamster burrows are likely to display some traits that overlap with rabbit warrens [20]. A similar accumulation for denning hamsters would be characterized by numerous juvenile and immature individuals, an under-representation of postcranial remains, no digestion or tooth marks, and a low proportion of fragmented bones. However, apart from the excellent preservation of the remains, none of these criteria were encountered for the Ittenheim assemblage. The important numbers of tooth marks, reaching almost 30%, amongst the highest observed for the rodent and rabbit carnivore accumulations, undoubtedly attests to a trogloxene carnivore contribution to the accumulation. Furthermore, no young hamster individuals were recorded. As hamsters are solitary mammals, the high number of adult hamster individuals argues against a natural death assemblage, which should theoretically concern only a single individual, as in the case of Combe-Cullier [77]. While the Ittemheim hamsters being accumulated by carnivores seems the most parsimonious hypothesis, the question remains of which of its main predators is responsible: a small predator, such as *Mustela erminea*, *Mustela nivalis*, *Martres foina*, or *Martres martres*; or a larger one, such as *Meles meles* or *Vulpes vulpes*.

One defining feature of the Ittenheim hamster assemblage is its overall good preservation and low proportion of digested elements, indicating a large part of the assemblage did not pass through a predator's digestive tract. Independent of the carnivore, all scats clearly contain a large proportion of digested remains, whereas non-ingested elements generally bear only tooth marks (Figure 8). Although there is an under-representation of limb bones compared to cranial bones, which would be in agreement with a carnivore accumulation, the skeletal profile is not entirely consistent with the non-ingested assemblages, falling more within the large variability of scats and dens.

Small mustelids are known to cache excess prey for later consumption (e.g., [89,90]). Generally, these predators tend to dispatch their prey by targeting the occipital and brain, eating the latter promptly. Consequently, these caches can be characterized by large numbers of prey, including numerous skulls with occipital breaks. In the case of the Ittenheim hamster accumulation, numerous fragmented skulls lack their occipital region. Impacts evident on the bones are also similar to those described by Vigne et al. [26]

for *Mus/Apodemus, Eliomys*, and *Arvicola* remains, characterized by numerous small perforations in, for example, the olecranon area. However, the proportions of remains at Ittenheim with these types of marks are much greater, affecting more than 20% of the long bones. Tooth marks observed on hamster remains are relatively similar to the ones observed on rodent remains from Bettant I and VIII. Their distribution also follows a similar pattern as those observed for rabbits from the fox den of Les Rochers-de-Villeneuve [33], with tooth marks preferentially located on mandibles, girdles, and long bones with partially removed bony eminences, such as the olecranon. Vigne et al. [26] interpret these small perforations to reflect predation by small carnivores, such as weasels or stoats, which would be consistent with the small quantity of associated weasel remains. Such perforations can also be produced by fox or badger cubs, as demonstrated, for example, by Mallye et al. [37] for the bird and mole remains from the Bettant dens. Though similar, tooth marks were nevertheless relatively rare amongst mole remains, being evident on about 5% of remains compared to 20–30% for larger vertebrate remains. Similar patterns were observed for both small and large birds, demonstrating the clear impact of prey size on patterns of bone modification [37]. Campmas et al. [38] showed that larger rodents (*Ctenodactylus gundi* weighing between 200 and 300 g) exhibit similar patterns (digestion and fragmentation) to hyraxes and leporids from scats of small and medium carnivores, leading them to propose a weight limit of prey between small and medium mammal preys of around 300 g. Our analysis of the Bettant remains is in good agreement with these observations as well as a clear relation between proportions of tooth marks and prey size. The larger rodents (rats, muskrats) and hedgehogs are characterised by good bone preservation, as reflected in the presence of a complete rat skull, low proportions of digested bones, and large proportions of bones with tooth marks. On the contrary, smaller rodents are represented by only a few complete bones, small quantities of tooth marks (mainly pits), and numerous traces of digestion on incisors, molars, and bones. Intermediate rodents from *Arvicola* sp. present an intermediate pattern, with both high proportions of digestion and tooth marks. Prey size therefore has a direct impact on the proportions of bone surface modifications that varies according to predator size; for example, young foxes cannot swallow larger prey, such as hamsters. The distinct patterns between these two prey sizes reflect different ways in which predators consume prey, shredding the larger ones and swallowing the smaller.

Finally, the Ittenheim accumulation shows similar patterns to those from Bettant. Both are characterised by a faunal spectrum comprising small and medium mammals, as well as other vertebrates, such as birds. Leporid remains, including young individuals, bear limited traces of carnivore damage, primarily traces of gnawing. Larger rodents and hedgehogs exhibit high proportions of tooth marks and low proportions of digestion damage, whereas smaller ones show an opposite pattern. This being the case, the accumulation of small- and medium-sized mammals at Ittenheim could be interpreted as reflecting a red fox (*Vulpes vulpes*) den, an interpretation consistent with the presence of a small quantity of fox cub remains, as recovered from the Bettant den. The only exception is *Talpa europaea*, which, given the absence of carnivore modification, could represent a natural death. In summary, the Ittenheim bone accumulation could be the result of foxes occupying a burrow. Ethological data show that female foxes occupy the den during denning and the weaning period of juveniles, from March to the end of summer. After the lactation period, foxes bring whole prey to the cubs in the den before they are strong enough to follow their mother during hunting episodes [91,92]. The accumulation of hamsters, other rodents, and hares could reflects these hunting episodes. Furthermore, given the age at death of the fox cubs, Ittenheim could have been occupied during the late spring and early summer. During this period, rodent populations experience a jump in birth rate and a significant increase in population size. Considering that the density of some hamster populations during the summer vary between 300 and 500 individuals per hectare [86], the Ittenheim accumulation resulting from a single occupation of the burrow by a fox family cannot be ruled out.

4.2. Implications for New Reference Collections and Fossil Material

Small carnivores constitute a potential accumulation agent of small and medium mammal bones [3,24,25,31,32,35,37,38,73,75,76,78,81,93–95]. Identifying the role played by small carnivores in fossil-bearing contexts remains challenging. While numerous taphonomic studies have focused on establishing reference collections for each of these mammal accumulators in order to identify taphonomic criteria susceptible to differentiating them, several studies have highlighted difficulties in identifying predators to species, suggesting identifications should remain at the genus level [96,97]. While statistically, the Ittenheim accumulation does not overlap with any particular reference collection, the results of our integrated systemic approach are most consistent with a fox den. The results of our study suggest two major avenues for future research: (1) the importance of contextual data and the integration of all types of remains, and (2) the need for more diversified reference collections that reflect the complexity of fossil accumulations.

We emphasise the importance of the discovery context and the integration of all recovered remains when analysing fossil materials. Our interpretation of the Ittenheim hamster remains was only possible by the identification of the burrow during excavations and the presence of leporid and fox remains, and to a lesser extent those of passeriform birds. The presence of young fox remains, which are regularly found in carnivore dens [37,49,78] supports the hypothesis of the Ittenheim burrow being a fox den. However, predator remains in fossil accumulation do not necessarily indicate the predator responsible for its formation [3], as they also could be the prey or just visiting roommates. Moreover, other trogloxene vertebrates take advantage of pre-existing burrows, such as the wood mouse [78] or muskrat, leading to difficulties in interpreting their presence in faunal assemblages. For instance, the partly digested juvenile remains of *Microtus arvalis/agrestis* from Bettant I pose questions as to their origin. In the case of the Les Six Chemins warren, the presence of neonate rats clearly indicates this rodent used the burrow alongside an additional small predator that introduced the remains of other vertebrate species, as demonstrated by the presence of bird bones and digested small rodent remains. This once again underlines the importance of including multiple multi-spectrum reference collections when reconstructing the formation history of bone accumulations [80,98].

Mammalian carnivores are the most taphonomically destructive predators, compared to diurnal raptors and owls [3,96,97]. Digestion marks and high proportions of fragmented bones are frequently considered as one of the main elements identifying carnivore bone accumulations [3], especially as tooth marks were mainly recorded on bones from canid scats. However, this paradigm has primarily been derived from currently available reference collections. Most taphonomic reference collections concern carnivore scats and clearly demonstrate how bones are modified as they pass through the digestive tracts of carnivores, sometimes becoming heavily altered and fragmented [3]. Non-ingested elements exhibit higher proportions of tooth marks, including those consumed by felids [33,35,45,81], which could be slightly more important during the denning season due to the presence of infants [79]. As mentioned above, beyond issues related to the reliance on modern references to evaluate taphocoenosis [99], without taking into the account the complexity of fossilization processes and attendant post-depositional changes, carnivore bone accumulations could be more complex than just remains coming from a defecation area, particularly in burrows. Burrows are complex structures that can fulfil multiple functions: temporary refuge, long term shelters, or food storage areas [100]. Remains recovered in burrows can therefore reflect natural deaths, stored or discarded food storage, or a mix of all these different origins. The Ittenheim and Bettant carnivore accumulations show clear taphonomical patterns distinct from those of scats. Furthermore, given the complexity of living, burrows can be used differently by multiple animals, as was the case for the Les Six Chemins warrens. Relying only on data from scat to understand fossil accumulation may run the risk of producing erroneous interpretations, especially when variability in carnivore bone accumulations is not incorporated. For example, most neotaphonomic experiments and reference collections did not take into account the variability [29,100–107] that can be observed between individuals [3,27,99,108] or reflect population-level variability [3,75,105,109]). Moreover, several experimental studies have involved captive animals [17,35,76,79,81,93,110], leading Gidna et al. [111] and Comay and Dayan [97] to caution against the inappropriate use of actualistic data derived from inadequate contexts, which can lead to erroneous analogies and, subsequently, unreliable interpretations of prehistoric faunal assemblages. More diverse statistically representative reference collections in terms of the number of individuals or types of remains are necessary to reliably incorporate both the natural diversity and taphonomic implications of distinct contexts in analyses [26,37,78,112,113]. Furthermore, these new reference collections should be based on an integrative approach designed to evaluate multi-taxa accumulations [30,47,70,98].

5. Conclusions

Small carnivores constitute potential accumulation agents of both small- and medium-sized mammal bones. Our analysis of taphonomic modifications on bones from three modern and one fossil accumulation demonstrates three important aspects that need to be taken into account when evaluating fossil accumulations: (1) prey size, as tooth marks are more frequent on larger prey while traces of digestion occur in larger proportions on smaller ones; (2) the discovery context and the need to include all elements of the faunal spectrum in order to reconstruct the full history of bone accumulations; and (3) the fact that den and burrow reference collections clearly differ from those built from carnivore scat. Our data illustrate the need for more taphonomic reference collections and not only ones based on scats or from feeding carnivores in zoos. Finally, further efforts are needed to multiply modern reference collections of carnivore scats to include larger samples reflecting prey sizes and origins in order to better characterise the variability of these different types of accumulations.

Supplementary Materials: The following are available online at https://www.mdpi.com/2571-550X/4/4/41/s1 and at https://github.com/AurelienRoyer/Ittenheim, Table S1: supplementary data. R script S1: R script.

Author Contributions: Conceptualization: A.R., J.-B.M., M.P. and S.G. Formal analysis: A.R., J.-B.M., M.P. and S.G. Writing: A.R., J.-B.M., M.P. and S.G. All authors have read and agreed to the published version of the manuscript.

Funding: This study was partially supported by the project HARCGLOB (AAP 2020 Région Bourgogne Franche–Comté).

Institutional Review Board Statement: Not applicable.

Informed Consent Statement: Not applicable.

Data Availability Statement: All references including the data used for this study are available at https://github.com/AurelienRoyer/Ittenheim, accessed on 16 July 2021.

Acknowledgments: Authors are grateful to the two anonymous reviewers and to the editors of this special issue (J. Rofes, J. Ochoa and E. Stoetzel) for their comments and corrections that greatly improved our manuscript.

Conflicts of Interest: The authors declare no conflict of interest.

References

1. Mellet, J.S. Scatological origin of microvertebrate fossil accumulation. *Science* **1974**, *185*, 349–350. [CrossRef] [PubMed]
2. Korth, W.W. Taphonomy of Microvertebrate fossil assemblages. *Ann. Carnegie Mus.* **1979**, *48*, 235–285.
3. Andrews, P. *Owls, Caves and Fossils*; Natural History Museum Publication: London, UK, 1990.
4. Denys, C. Introduction générale. Taphonomie des microvertébrés: Méthodologie et état de l'art: Micromammifères, oiseaux et problèmes taphonomiques. *Artefacts* **1994**, *9*, 195–206.
5. Lyman, R.L. *Vertebrate Taphonomy*; Cambridge University Press: Cambridge, MA, USA, 1994.
6. Stahl, P.W. The recovery and interpretation of microvertebrate bone assemblages from archaeological contexts. *J. Archaeol. Method Theory* **1996**, *3*, 31–75. [CrossRef]

7. Lyman, R.L. Actualistic neotaphonomic research on bone modifying animal species: An analysis of the literature. *Palaios* **2018**, *33*, 542–554. [CrossRef]
8. Shipman, P.; Walker, A. Bone-collecting by harvesting ants. *Paleobiology* **1980**, *6*, 496–502. [CrossRef]
9. Brain, C. Porcupines as bone collectors in African caves. In *The Hunters or the Hunted*; University of Chicago Press: Chicago, IL, USA, 1981; pp. 109–117.
10. Leoni, F.; Boudadi-Maligne, M.; Mallye, J.B.; Beauval, C.; Faivre, J.P. Renards chapardeurs et amas de boucherie en pleine nature. Quantification des modifications et incidence sur les analyses archéozoologiques. In *Relations Hommes—Canidés de la Préhistoire aux Périodes Modernes*; Boudadi-Maligne, M., Mallye, J.-B., Eds.; Ausonius éditions; Collection DAN@: Pessac, France, 2020; pp. 69–94.
11. Stahl, P.W. On small mammal remains in archaeological context. *Am. Antiq.* **1982**, *47*, 822–829. [CrossRef]
12. Fiedler, L.A. Rodents as a food source. In Proceedings of the Fourteenth Vertebrate Pest Conference, Sacramento, CA, USA, 6–8 March 1990; p. 30.
13. Simonetti, J.A.; Cornejo, L.E. Archaeological evidence of rodent consumption in central Chile. *Lat. Am. Antiq.* **1991**, *2*, 92–96. [CrossRef]
14. Henshilwood, C.S. Identifying the collector: Evidence for human processing of the Cape dune mole-rat, Bathyergus suillus, from Blombos Cave, southern Cape, South Africa. *J. Archaeol. Sci.* **1997**, *24*, 659–662. [CrossRef]
15. Stiner, M.C.; Munro, N.D.; Surovell, T. The Tortoise and the Hare. Small-Game Use, the Broad Spectrum Revolution, and Paleolithic Demography. *Curr. Anthropol.* **2000**, *41*, 39–79. [CrossRef]
16. Cochard, D.; Brugal, J.-P. Importance des fonctions de sites dans les accumulations paléolithiques de léporidés. In *Petits Animaux et Sociétés Humaines. Du Complément Alimentaire aux Ressources Utilitaires, Actes Des XXIVème Rencontres Internationales D'archéologie et D'histoire d'Antibes*; Brugal, J.-P., Desse, J., Eds.; APDCA: Antibes, France, 2004; pp. 217–230.
17. Armstrong, A. Small mammal utilization by Middle Stone Age humans at Die Kelders Cave 1 and Pinnacle Point Site 5-6, Western Cape Province, South Africa. *J. Hum. Evol.* **2016**, *101*, 17–44. [CrossRef]
18. Mallye, J.-B.; Laroulandie., V.; Avec la Collaboration de Bonnet-Jacquement, P.; Langlais, M.; Man-Estier, E.; Vercoutère, C.; Paillet, P. L'utilisation des petits gibiers au Tardiglaciaire: Le cas de Rochereil (collection Jude, Dordogne). In Proceedings of the Animal Symbolisé—Animal Exploité, Du Paléolithique à la Protohistoire, Actes du 141eme Congrès du CTHS, Rouen, Paris, France, 11–15 April 2016; Costamagno, S.C., Dupont, O., Dutour, L., Gourichon, D.V., Eds.; CTHS: Cherrybrook, Australia, 2018.
19. Pelletier, M.; Desclaux, E.; Brugal, J.P.; Texier, P.J. The exploitation of rabbits for food and pelts by last interglacial Neandertals. *Quat. Sci. Rev.* **2019**, *224*, 105972. [CrossRef]
20. Pelletier, M.; Brugal, J.P.; Cochard, D.; Lenoble, A.; Mallye, J.B.; Royer, A. Identifying fossil rabbit warrens: Insights from a taphonomical analysis of a modern warren. *J. Archaeol. Sci. Rep.* **2016**, *10*, 331–344. [CrossRef]
21. Mondini, M. Taphonomic action of Foxes in Puna Rockshelters: A case study in Antofagasta de la Sierra (Province of Catamarca, Argentina). In *Ethnoarchaeology of Andean South America*; International Monographs in Prehistory. Ethnoarchaeology Series; Kuznar, L.A., Ed.; Berghahn Books: New York, NY, USA, 2001; Volume 4, pp. 266–295.
22. Mondini, M. Accumulation of small and large vertebrates by carnivores in Andean South America. In *Petits Animaux et Sociétés Humaines. Du Complément Alimentaire aux Ressources Utilitaires. XXIVè Rencontres Internationales D'archéologie et D'histoire d'Antibes*; Brugal, J.-P., Desse, J., Eds.; APDCA: Antibes, France, 2004; pp. 483–487.
23. Mondini, M. Use of Rockshelters by carnivores in the Puna. Implications for hunter-gatherer archaeology. *Before Farming* **2005**, *2*, 1–25. [CrossRef]
24. Denys, C.; Cochard, D. Taphonomie des restes ingérés par les petits carnivores. In *TaphonomieS*; Brugal, J.P., Ed.; EAC, Editions des Archives Contemporaines: Paris, France, 2017; pp. 369–388.
25. Lloveras, L. Taphonomy of carnivores: Understanding archaeological small prey accumulations. In *Themes in Old World Zooarchaeology, From the Mediterranean to the Atlantic*; Albraella, U., Detry, C., Gabriel, S., Ginja, C., Pires, A.E., Tereso, J.P., Eds.; Oxbow Books: Oxford, UK, 2021; pp. 13–27.
26. Vigne, J.D.; Méniel, P.; Lambot, B.; Bonani, G. Un assemblage d'ossements de micromammifères dans une cache de petit mustélidé de l'âge du fer (Acy-Romance, Ardennes, France): Micromammifères, oiseaux et problèmes taphonomiques. *Artefacts* **1994**, *9*, 207–215.
27. Williams, J.P. Small Mammal Deposits in Archaeology: A Taphonomic Investigation of *Tyto alba* (Barn Owl) Nesting and Roosting Sites. Ph.D. Thesis, University of Sheffield, Sheffield, UK, 2001.
28. Denys, C. Taphonomy and experimentation. *Archaeometry* **2002**, *44*, 469–484. [CrossRef]
29. Denys, C. Des référentiels en taphonomie des petits vertébrés: Bilan et perspectives. In *Taphonomie des Petits Vertébrés: Référentiels et Transferts aux Fossiles*; Laroulandie, V., Mallye, J.-B., Denys, C., Eds.; BAR Publishing: Oxford, UK, 2011; Volume 2269, pp. 20–21.
30. Royer, A.; Montuire, S.; Gilg, O.; Laroulandie, V. A taphonomic investigation of small vertebrate accumulations produced by the snowy owl (*Bubo scandiacus*) and its implications for fossil studies. *Palaeogeogr. Palaeoclimatol. Palaeoecol.* **2019**, *514*, 189–205. [CrossRef]
31. Andrews, P.; Nesbit Evans, E.M. Small mammal bone accumulations produced by mammalian carnivores. *Paleobiology* **1983**, *9*, 289–307. [CrossRef]
32. Sanchís, A.S. Los restos de *Oryctolagus cuniculus* en las tafocenosis de *Bubo bubo* y *Vulpes vulpes* y su aplicación a la caracterización del registro faunístico arqueológico. *Saguntum* **2000**, *32*, 31–49.

33. Cochard, D. Etude taphonomique des léporidés d'une tanière de renard actuelle: Apport d'un référentiel à la reconnaissance des accumulations anthropiques. *Rev. Paléobiol.* **2004**, *23*, 659–673.
34. Sanchís, A.S.; Pascual, J.B. Análisis de las acumulaciones óseas de una guarida de pequeños mamíferos carnívoros (Sitjar Baix, Onda, Castellón): Implicaciones arqueológicas. *Archaeofauna* **2011**, *20*, 47–71.
35. Lloveras, L.; Moreno-García, M.; Nadal, J. Feeding the foxes: An experimental study to assess their taphonomic signature on leporid remains. *Int. J. Osteoarchaeol.* **2012**, *22*, 577–590. [CrossRef]
36. Cohen, B.F.; Kibii, J.M. Taphonomy of a novel small carnivore: Experimental analysis of honey badger (*Mellivora capensis*) modifications on leporid prey. *Archaeol. Anthropol. Sci.* **2019**, *11*, 2309–2327. [CrossRef]
37. Mallye, J.-B.; Cochard, D.; Laroulandie, V. Accumulations osseuses en périphérie de terriers de petits carnivores: Les stigmates de prédation et de fréquentation. *Ann. Paléontol.* **2008**, *94*, 187–208. [CrossRef]
38. Campmas, E.; Stoetzel, E.; Denys, C. African carnivores as taphonomic agents: Contribution of modern coprogenic sample analysis to their identification. *Int. J. Osteoarchaeol.* **2018**, *28*, 237–263. [CrossRef]
39. Rufà, A.; Laroulandie, V. Prey size as a critical factor for bird bone taphonomy in Eagle Owl (*Bubo bubo*) pellets. *Sci. Rep.* **2019**, *9*, 19200. [CrossRef]
40. Matthews, T. Taphonomic characteristics of micromammals predated by small mammalian carnivores in South Africa: Application to fossil accumulations. *J. Taphon.* **2006**, *4*, 143–161.
41. Montalvo, C.I.; Pessino, M.E.M.; Bagatto, F.C. Taphonomy of the bones of rodents consumed by Andean hog-nosed skunks (*Conepatus chinga*, Carnivora, Mephitidae) in central Argentina. *J. Archaeol. Sci.* **2008**, *35*, 1481–1488. [CrossRef]
42. Montalvo, C.I.; Bisceglia, S.; Kin, M.S.; Sosa, R.A. Taphonomic analysis of rodent bone accumulations produced by Geoffroy's cat (*Leopardus geoffroyi*, Carnivora, Felidae) in Central Argentina. *J. Archaeol. Sci.* **2012**, *39*, 1933–1941. [CrossRef]
43. Montalvo, C.I.; Vezzosi, R.I.; Kin, M.S. Taphonomic analysis of rodent bones from *Lontra longicaudis* (Mustelidae, Carnivora) scats in fluvial environments. *Mastozool. Neotrop.* **2015**, *22*, 319–333.
44. López, J.M.; Rosi, M.I.; Tabeni, S.; Bender, B.; Chiavazza, H. Taphonomic analysis of small mammal bone remains preyed upon by wildcats (Carnivora: Felidae) from the central Monte Desert (Mendoza, Argentina). *Boreas* **2017**, *46*, 282–293. [CrossRef]
45. Scheifler, N.A.; Álvarez, M.C.; Rafuse, D.J.; Kaufmann, C.A.; Massigoge, A.; González, M.E.; Gutiérrez, M.A. Taphonomic signature of Geoffroy's cat (*Leopardus geoffroyi*) on small sized preys: A comparative study of ingested and non-ingested leporid bones. *J. Archaeol. Sci. Rep.* **2020**, *31*, 102340. [CrossRef]
46. Marin-Monfort, M.D.; García-Morato, S.; Olucha, R.; Yravedra, J.; Piñeiro, A.; Barja, I.; Andrews, P.; Fernández-Jalvo, Y. Wildcat scats: Taphonomy of the predator and its micromamal prey. *Quat. Sci. Rev.* **2019**, *225*, 106024. [CrossRef]
47. Denys, C.; Stoetzel, E.; Andrews, P.; Bailon, S.; Rihane, A.; Huchet, J.B.; Fernández-Jalvo, Y.; Laroulandie, V. Taphonomy of Small Predators multi-taxa accumulations: Palaeoecological implications. *Hist. Biol.* **2018**, *30*, 868–881. [CrossRef]
48. Lefranc, P.; Griselin, S.; Schneider, N.; Arbogast, R.-M.; Chenal, F.; Gebhardt, A.; Jodry, F.; Mallye, J.-B.; Moine, O.; Royer, A.; et al. Ittenheim Bas-Rhin. A355-Contournement Ouest de Strasbourg—Tronçon 3—Site 3.4. In *der Flies et Musau*; La Formation d'un Versant Puis son Occupation en Rive Droite de la Vallée du Musaubach, une Histoire Dynamique du Paléolithique, au Néolithique et Jusqu'à L'époque Moderne; Rapport de Fouille INR AP; SRA Grand Est: Vandœuvre-lès-Nancy, France, 2021.
49. Mallye, J.-B. Les Restes de Blaireau en Contexte Archéologique: Taphonomie, Archéozoologie et Eléments de Discussion des Séquences Préhistoriques. Ph.D. Thesis, The University of Bordeaux 1, Bordeaux, France, 2007.
50. Pelletier, M.; Royer, A.; Holliday, T.W.; Discamps, E.; Madelaine, S.; Maureille, B. Rabbits in the grave! Consequences of bioturbation on the Neandertal "burial" at Regourdou (Montignac-sur-Vézère, Dordogne). *J. Hum. Evol.* **2017**, *110*, 1–17. [CrossRef]
51. Biadi, F.; Le Gall, A. *Le Lapin de Garenne. Vie Gestion et Chasse d'un Gibier Authentique*; Office National de la Chasse Hatier: Paris, France, 1993.
52. Mulder, J.L.; Wallage-Drees, J.M. Red fox predation on young rabbits in breeding burrows. *Neth. J. Zool.* **1979**, *29*, 144–149. [CrossRef]
53. Harris, S. Age determination in the Red fox (*Vulpes vulpes*)—An evaluation of technique efficiency as applied to a sample of suburban foxes. *J. Zool.* **1978**, *184*, 91–117. [CrossRef]
54. Pelletier, M.; Desclaux, E.; Mallye, J.B.; Crégut-Bonnoure, E. Identifying the accidental-natural mortality of leporids in the archaeological record: Insights from a taphonomical analysis of a pitfall without evidence of human presence. *J. Quat. Sci.* **2020**, *35*, 677–694. [CrossRef]
55. Dodson, P.; Wexlar, D. Taphonomic investigations of owl pellets. *Paleobiology* **1979**, *5*, 275–284. [CrossRef]
56. Royer, A.; Lécuyer, C.; Montuire, S.; Escarguel, G.; Fourel, F.; Mann, A.; Maureille, B. Late Pleistocene (MIS 3–4) climate inferred from micromammal communities and $\delta^{18}O$ of rodents from Les Pradelles, France. *Quat. Res.* **2013**, *80*, 113–124. [CrossRef]
57. Fernández-Jalvo, Y.; Andrews, P.; Denys, C.; Sesé, C.; Stoetzel, E.; Marin-Monfort, D.; Pesquero, D. Taphonomy for taxonomists: Implications of predation in small mammal studies. *Quat. Sci. Rev.* **2016**, *139*, 138–157. [CrossRef]
58. R Core Team. *R: A Language and Environment for Statistical Computing. v3.6.3.*; R Core Team: Vienna, Austria, 2020.
59. Josse, J.; Husson, F. missMDA: A package for handling missing values in multivariate data analysis. *J. Stat. Softw.* **2016**, *70*, 1–31. [CrossRef]
60. Husson, F.; Josse, J.; Le, S.; Mazet, J.; Husson, M.F. Package 'FactoMineR'. *R Package* **2016**, *96*, 698.

61. Oksanen, J.; Blanchet, F.; Friendly, M.; Kindt, R.; Legendre, P.; Minchin, P.R.; O'hara, R.B.; Simpson, G.L.; Solymos, P.; Stevens, M.H.H.; et al. Vegan: Community Ecology Package, R package, v. 2.4-6. 2016.
62. Martinez Arbizu, P. pairwiseAdonis: Pairwise multilevel comparison using adonis. R package version 0.0.1. 2017.
63. Donard, E. *Recherche sur le Léporinés Quaternaires (Pléistocène Moyen et Supérieur, Holocène*; Université de Bordeaux: Bordeaux, France, 1982; Volume 1, 161p.
64. Koby, F.-E. Contribution au diagnostic ostéologique différentiel de *Lepus timidus* Linné et *L. europaeus* Pallas. *Verh. Nat. Ges.* **1959**, *70*, 19–44.
65. Koby, F.-E. Contribution à la connaissance des lièvres fossiles, principalement de ceux de la dernière glaciation. *Verh. Nat. Ges.* **1960**, *71*, 149–173.
66. Morel, P.; Müller, W. *Hauterive-Champréveyres: Un Campement Magdalénien au Bord du lac de Neuchâtel. Étude Archéozoologique*; Archéologie Neuchâteloise: Neuchâtel, France, 1997; 149p.
67. Müller, W. *Le Site Magdalénien de Monruz, 3 Acquisition, Traitement et Consommation des Ressources Animales*; Archéologie Neuchâteloise: Neuchâtel, France, 2013.
68. Flux, J.E.C. Life history of the Mountain hare (*Lepus timidis scoticus*) in north-east Scotland. *J. Zool.* **1970**, *161*, 75–123. [CrossRef]
69. Matthews, T. South African micromammals and predators: Some comparative results. *Archaeometry* **2002**, *44*, 363–370. [CrossRef]
70. Lebreton, L.; Bailon, S.; Guillaud, E.; Testu, A.; Perrenoud, C. Multi-taxa referential of a modern Eurasian Eagle-Owl (*Bubo bubo*) aerie. *J. Archaeol. Sci. Rep.* **2020**, *32*, 102417. [CrossRef]
71. Gómez, G.N.; Kaufmann, C.A. Taphonomic analysis of *Pseudalopex griseus* (Gray, 1837) scat assemblages and their archaeological implications. *J. Taphon.* **2007**, *5*, 59–70.
72. Denys, C.; Kowalski, K.; Dauphin, Y. Mechanical and chemical alterations of skeletal tissues in a recent Saharian accumulation of faeces from *Vulpes rueppelli* (Carnivora, Mammalia). *Acta Zool. Cracovensia* **1992**, *35*, 265–283.
73. Schmitt, D.N.; Juell, K.E. Toward the identification of coyote scatological faunal accumulations in archaeological contexts. *J. Archaeol. Sci.* **1994**, *21*, 249–262. [CrossRef]
74. Montalvo, C.I.; Pessino, M.E.; González, V.H. Taphonomic analysis of remains of mammals eaten by pumas (*Puma concolor* Carnivora, Felidae) in central Argentina. *J. Archaeol. Sci.* **2007**, *34*, 2151–2160. [CrossRef]
75. Coll, D.G.; Montalvo, C.I.; Fernández, F.J.; Pia, M.V.; Mondini, M. Actualistic taphonomic study of the rodents digested by the Achala culpeo fox (*Lycalopex culpaeus smithersi*) in the highlands of central Argentina. *Boreas* **2021**, *50*, 1146–1160. [CrossRef]
76. Lloveras, L.; Moreno-García, M.; Nadal, J. Taphonomic analysis of leporid remains obtained from modern Iberian lynx (*Lynx pardinus*) scats. *J. Archaeol. Sci.* **2008**, *35*, 1–13. [CrossRef]
77. Royer, A.; Sécher, A.; Langlais, M. A Brief Note on the Presence of the Common Hamster during the Late Glacial Period in Southwestern France. *Quaternary* **2018**, *1*, 8. [CrossRef]
78. Arilla, M.; Rufà, A.; Rosell, J.; Blasco, R. Small carnivores' cave-dwelling: Neo-taphonomic study of a badger (*Meles meles*) sett and its archaeological implications. *Hist. Biol.* **2020**, *32*, 951–965. [CrossRef]
79. Rodríguez-Hidalgo, A.; Saladié, P.; Marín, J.; Canals, A. Expansion of the referential framework for the rabbit fossil accumulations generated by Iberian lynx. *Palaeogeogr. Palaeoclimatol. Palaeoecol.* **2015**, *418*, 1–11. [CrossRef]
80. Rodríguez-Hidalgo, A.; Sanz, M.; Daura, J.; Sánchez-Marco, A. Taphonomic criteria for identifying Iberian lynx dens in quaternary deposits. *Sci. Rep.* **2020**, *10*, 7225. [CrossRef] [PubMed]
81. Álvarez, M.C.; Kaufmann, C.A.; Massigoge, A.; Gutiérrez, M.A.; Rafuse, D.J.; Scheifler, N.A.; González, M.E. Bone modification and destruction patterns of leporid carcasses by Geoffroy's cat (*Leopardus geoffroyi*): An experimental study. *Quat. Int.* **2012**, *278*, 71–80. [CrossRef]
82. Rodríguez-Hidalgo, A.; Lloveras, L.; Moreno-García, M.; Saladié, P.; Canals, A.; Nadal, J. Feeding behaviour and taphonomic characterization of non-ingested rabbit remains produced by the Iberian lynx (*Lynx pardinus*). *J. Archaeol. Sci.* **2013**, *40*, 3031–3045. [CrossRef]
83. Lloveras, L.; Nadal, J.; Fullola, J.M. Distinguishing the taphonomic signature of wolves from humans and other predators on small prey assemblages. *Sci. Rep.* **2020**, *10*, 8030. [CrossRef]
84. Lloveras, L.; Moreno-García, M.; Nadal, J. The eagle owl (*Bubo bubo*) as a leporid remains accumulator: Taphonomic analysis of modern rabbit remains recovered from nests of this predator. *Int. J. Osteoarchaeol.* **2009**, *19*, 573–592. [CrossRef]
85. Cucchi, T.; Vigne, J.D.; Auffray, J.C. First occurrence of the house mouse (*Mus musculus domesticus* Schwarz & Schwarz, 1943 in the Western Mediterranean: A zooarchaeological revision of subfossil occurrences. *Biol. J. Linn. Soc.* **2005**, *84*, 429–445.
86. Quéré, J.P.; Le Louarn, H. Les rongeurs de France: Faunistique et biologie. In *3e Edition Revue et Augmentée*; Editions Quae: Versailles, France, 2011.
87. Górecki, A. Energy flow through the common hamster population. *Acta Theriol.* **1977**, *22*, 25–66. [CrossRef]
88. Nechay, G. *Status of Hamsters Cricetus Cricetus, Cricetus Migratorius, Mesocricetus Newtoni, and Other Hamster Species in Europe Nature and Environment*; Council of Europe: Strasburg, France, 2000; Volume 106.
89. Delattre, P. *La Belette (Mustela Nivalis, Linnaeus, 1766) et L'hermine (Mustela Erminea, Linnaeus, 1758)*; Société Française pour l'étude et la Protection des Mammifères: Paris, France, 1987.
90. Labrid, M. Labrid, M. La Martre (*Martes Martes*, Linnaeus, 1758. In *Encyclopédie des Carnivores de France*; Société française pour l'Étude et la Protection des Mammifères: Bourges, France, 1986; pp. 1–22.

91. Artois, M. *Le Renard Roux: Vulpes Vulpes Linnaeus, 1758*; Société Française Pour l'Étude et la Protection des Mammifères: Bourges, France, 1989; 90p.
92. Meia, J.-S. *Le Renard: Description, Comportement, vie Sociale, Mythologie, Observation*; Delachaux et Niestlé: Paris, France, 2011; 183p.
93. Lloveras, L.; Thomas, R.; Cosso, A.; Pinyol, C.; Nadal, J. When wildcats feed on rabbits: An experimental study to understand the taphonomic signature of European wildcats (*Felis silvestris silvestris*). *Archaeol. Anthropol. Sci.* **2018**, *10*, 449–464. [CrossRef]
94. Mondini, M. Carnivore taphonomy in South America: A review of actualistic studies and their implications in the southern Neotropics. *Hist. Biol.* **2018**, *30*, 774–785. [CrossRef]
95. Montalvo, C.I.; Fernández, F.J. Review of the actualistic taphonomy of small mammals ingested by South American predators. Its importance in the interpretation of the fossil record. *Publicación Electrónica De La Asoc. Paleontológica Argent.* **2019**, *19*, 18–46. [CrossRef]
96. Terry, R.C. Inferring predator identity from skeletal damage of small-mammal prey remains. *Evol. Ecol. Res.* **2007**, *9*, 199–219.
97. Comay, O.; Dayan, T. Taphonomic signatures of owls: New insights into micromammal assemblages. *Palaeogeogr. Palaeoclimatol. Palaeoecol.* **2018**, *492*, 81–91. [CrossRef]
98. Laroulandie, V.; Royer, A.; Mallye, J.B.; Bailon, S.; Danger, M. Les petits vertébrés de Peyrazet et les chasseurs-collecteurs du Magdalénien supérieur. In *La Grotte-Abri de Peyrazet (Creysse, Lot, France) au Magdalénien*; Originalité Fonctionnelle d'un Habitat des Derniers Chasseurs de Rennes du Quercy, CNRS, Supplément à Gallia Préhistoire 43; Langlais, M., Laroulandie, V., Eds.; CNRS: Paris, France, 2021; pp. 141–152. Available online: https://hal.archives-ouvertes.fr/hal-03202072 (accessed on 1 November 2021).
99. Kowalski, K. Some problems of the taphonomy of small mammals. In *International Symposium Evolution, Phylogeny and Biostratigraphy of Arvicolids Rohanov, May 1987*; Fejfar, O., Heinrich, W.-D., Eds.; Czech Geological Survey: Prague, Czech Republic, 1990; pp. 285–296.
100. Butler, D.R. *Zoogeomorphology: Animals as Geomorphic Agents*; Cambridge University Press: Cambridge, MA, USA, 1995.
101. Denys, C. Nouveaux critères de reconnaissance des concentrations de microvertébrés d'après l'étude des pelotes de chouettes du Botswana (*Afrique australe*). Bulletin du Muséum national d'histoire naturelle. *Sect. A Zool. Biol. Écologie Anim.* **1985**, *7*, 879–933.
102. Hoffman, R. The contribution of raptorial birds to patterning in small mammal assemblages. *Paleobiology* **1988**, *14*, 81–90. [CrossRef]
103. Denys, C.; Dauphin, Y.; Rzebik-Kowalska, B.; Kowalski, K. Taphonomic study of Algerian owl pellet assemblages and differential preservation of some rodents: Palaeontological implications. *Acta Zool. Crac.* **1996**, *39*, 103–116.
104. Denys, C.; Dauphin, Y.; Fernández-Jalvo, Y. Apports biostratigraphiques et paléoécologiques de l'étude taphonomique des assemblages de micromammifères. Bilan et perspectives. *Geobios* **1997**, *30*, 197–206. [CrossRef]
105. Saavedra, B.; Simonetti, J.A. Small mammal taphonomy: Intraspecific bone assemblage comparison between South and North American barn owl, *Tyto alba*, populations. *J. Archaeol. Sci.* **1998**, *25*, 165–170. [CrossRef]
106. Cochard, D. Discussion sur la variabilité intraréférentiel d'accumulations osseuses de petits prédateurs. *Ann. Paléontologie* **2008**, *94*, 89–101. [CrossRef]
107. Rudzik, S.M.; Fernández, F.J.; Carrera, J.D. Taphonomic analysis of micromammal remains from striped owl (*Pseudoscops clamator*) pellets in northeastern Buenos Aires province, Argentina: Implications for archaeological sites formation. *Int. J. Osteoarchaeol.* **2015**, *25*, 550–563. [CrossRef]
108. Raczyński, J.; Ruprecht, A.L. The effect of digestion on the osteological composition of owl pellets. *Acta Ornithol.* **1974**, *14*, 25–36.
109. Andrews, P.; Fernández-Jalvo, Y. Seasonal variation in prey composition and digestion in small mammal predator assemblages. *Int. J. Osteoarchaeol.* **2018**, *28*, 318–331. [CrossRef]
110. Cohen, B.F.; Kibii, J.M. Experimental study of bone modification by captive caracal (Caracal caracal); a model for fossil assemblage analysis. *J. Archaeol. Sci.* **2015**, *62*, 92–104. [CrossRef]
111. Gidna, A.; Yravedra, J.; Domínguez-Rodrigo, M. A cautionary note on the use of captive carnivores to model wild predator behavior: A comparison of bone modification patterns on long bones by captive and wild lions. *J. Archaeol. Sci.* **2013**, *40*, 1903–1910. [CrossRef]
112. Rafuse, D.J.; Kaufmann, C.A.; Gutiérrez, M.A.; González, M.E.; Scheifler, N.A.; Álvarez, M.C.; Massigoge, A. Taphonomy of modern communal burrow systems of the Plains vizcacha (*Lagostomus maximus*, Chinchillidae) in the Pampas region of Argentina: Implications for the fossil record. *Hist. Biol.* **2019**, *31*, 517–534. [CrossRef]
113. Scheifler, N.A.; Massigoge, A.; Kaufmann, C.A.; Rafuse, D.J.; González, M.E.; Gutiérrez, M.A. Modern Bone Distribution in the Pampas of Argentina: Taphonomic Implications for the Regional Archaeological Record. In *Actualistic Taphonomy in South America. Topics in Geobiology*; Martínez, S., Rojas, A., Cabrera, F., Eds.; Springer: Cham, Switzerland, 2020; Volume 48, pp. 193–220.

quaternary

MDPI

Article

The Microvertebrates of Shanidar Cave: Preliminary Taphonomic Findings

Emily Tilby *, Preston Miracle and Graeme Barker

Department of Archaeology, University of Cambridge, Cambridge CB2 3DZ, UK; ptm21@cam.ac.uk (P.M.); gb314@cam.ac.uk (G.B.)
* Correspondence: emct3@cam.ac.uk

Abstract: Shanidar Cave, Iraqi Kurdistan, is one of the most important Palaeolithic sites in Southwest Asia. This is due to the long sequence of hominin occupation of the cave and the discovery of multiple Neanderthal individuals from the original Solecki excavations (1951–1960) and recent excavations (2014 to present). Preliminary taphonomic analyses of the microvertebrate assemblage were undertaken to understand the factors affecting assemblage formation and accumulation, and this paper presents the first results of these analyses. All contexts display a high proportion of fragmentation, with a slight decrease in breakage towards the base of the sequence. Black staining and root etching were observed in a similar pattern, present in most contexts but with an increase in the lower levels. A significant proportion of the microvertebrate remains examined displayed light traces of digestion, indicating some contribution to the assemblage by predators. The results are consistent with wider palaeoecological records that indicate relatively warm, wet conditions at the base of the sequence and cooler, drier conditions at the top.

Keywords: microvertebrates; taphonomy; palaeoecology; Shanidar Cave

1. Introduction

Climate change and resultant environmental shifts have long been associated with major population and cultural turnover in human history. Understanding the nature of these shifts, the exact duration, rapidity, and amplitude, can provide a comprehensive understanding of the situations faced by early humans. Local environmental records are particularly valuable when investigating human responses, as they reflect the specific conditions and resulting challenges faced by humans. Micromammals provide particularly good local, high-resolution environmental records due to their restricted habitat range, short generational turnover time, and ubiquity in the zooarchaeological record [1].

Taphonomy is "the study of the transition, in all details, of organics from the biosphere into the … geological record" [2]. Taphonomic processes may distort and alter the archaeological and palaeontological record, potentially producing spurious interpretations and conclusions if not accounted for. Understanding these taphonomic processes is particularly important in analyses that are based on tracking change through a sequence (as is the case in palaeoenvironmental studies), as changes due to shifts in the taphonomic process may mimic or mask environmental changes. In the case of microvertebrates the effect of avian predators is particularly important, as the hunting range and prey preferences may produce biases in the eventual microfaunal community recorded and resultant palaeoenvironmental reconstruction. The aim of this preliminary study was to carry out an initial taphonomic analysis of the microvertebrate assemblage of Shanidar Cave. The results of this analysis will be the baseline for future taxonomic, taphonomic, and palaeoenvironmental studies of the site.

1.1. The Site

Shanidar Cave (36°50′ N, 44°13′ E) is a karstic cave located in the Baradost Mountains in the Zagros foothills of northwest Iraqi Kurdistan, approximately 2.5 km from the Greater Zab River, a tributary of the Tigris (Figure 1). The cave is at an elevation of 745 m, with a large, south-facing mouth measuring about 25 m in width and 8 m in height. The cave is slightly wider than it is long. It extends back approximately 40 m and is 53 m wide at its maximum point, resulting in a cave floor of about 1200 m^2 [3,4]. The Zagros Mountains extend from southeastern Turkey down towards the Persian Gulf, the main massif forming the modern boundary between Iraq and Iran. They are a series of karstic formations, and as such multiple caves are present in the region. Many of these caves have provided both archaeological and paleoclimatic information on the region; along with Shanidar Cave other key archaeological sites in the region include Yafteh Cave, Bisitun Cave, Kaldar Cave, and Kobeh Cave [5].

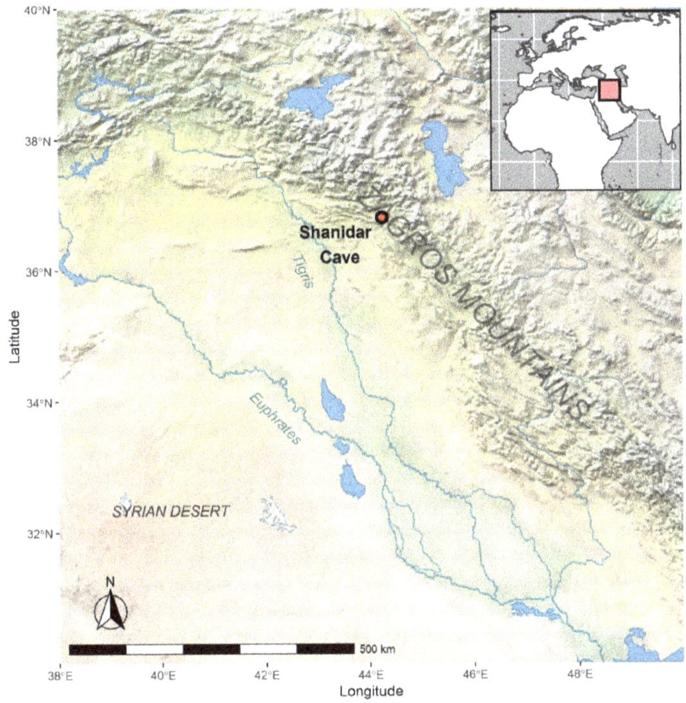

Figure 1. Map showing the location of Shanidar Cave. Map created in R using Natural Earth raster.

1.1.1. Excavations 1951–1960

The cave was initially excavated by Ralph Solecki in a series of four field seasons through the 1950s and has yielded one of the key Palaeolithic sequences for the region [4,6–12]. The original excavations resulted in a 14 m-deep trench roughly in the centre of the cave floor. During the original excavations the remains of nine Neanderthal individuals were recovered from the Mousterian Layer D, with remains from a 10th, immature individual recovered during reanalysis of the faunal remains in 2007 [4,13,14]. These remains attracted significant interest due both to their biology and to the inferences drawn from them about Neanderthal behaviour, particularly Shanidar I and Shanidar IV. Shanidar I shows signs of severe trauma and may have been partially incapacitated during his life, suggesting a degree of care for the wounded or disabled among Neanderthals [15–17]. Shanidar IV is an individual who appears to have been deliberately buried with flowers, which has significant implications for our understanding of Neanderthal mortuary practices [16,18,19]. This interpretation has been challenged by various authors, including Sommer [20], who suggested that the clusters of flowers may be the result of jirds burrowing in the sediments around the burial.

Although the focus of the original excavations was the recovery of a deep cultural sequence, with the Neanderthal remains an unexpected bonus, palaeoenvironmental studies of the flora (pollen) and fauna were also carried out. The former was analysed by Arlette Leroi-Gourhan [12,18] and the later primarily by Dexter Perkins [21], though additional work was carried out by Mary Evins [22,23] on the Mousterian faunal assemblage that was excluded from the original analyses. The original faunal analyses mostly focused on larger mammals, though short reports on the molluscs and microvertebrates were included as appendices to Evins' thesis. Microvertebrate taxa reported here included *Hemiechinus auratus*, *Talpa* sp., *Miniopterus schreibersi*, *Taphozous* sp., *Ellobius* sp., *Meriones* sp. and *Spalax* sp.

1.1.2. Recent Excavations

All material studied in this research derives from the recent excavations at Shanidar Cave led by Graeme Barker. The project began with preliminary work in 2014, and excavations commenced in 2015. The recent excavations have two main aims. The first is to locate and date the cultural layers and Neanderthal burial locations described by Solecki using modern methods. The new dates will produce a high-resolution sequence of occupation of the cave, securely establishing the presence of Neanderthals in the landscape at set points in time. The second aim is to collect a range of palaeoecological data to provide a local record of climate and environment. Although, as discussed above, some palaeoenvironmental analyses were carried out in the original excavations, certain groups such as the microvertebrates were largely excluded due to the difficulty in finding appropriate reference collections. The recent excavations offer an opportunity to redress this absence. Although the recent excavations are processing a far smaller amount of material than the original excavations, it is hoped that the application of modern techniques and methods will provide a rich, well-dated palaeoenvironmental sequence for the cave [24]. The re-opened excavations are ongoing with faunal, shell, and botanical samples taken for use as climate proxies and isotope analyses of shell and faunal material planned. To date, work at the cave has uncovered further remains attributed to Shanidar V from the area defined by Solecki as Layer D, as well as an incised shell object from the base of the Baradostian Layer C [25,26]. The most notable discovery has been the torso and skull of an additional Neanderthal individual in the immediate vicinity of the original "Flower Burial" [27]. The discovery of the new hominin remains has expanded the aims and methodology of the overall project. The remains offer an opportunity to excavate and analyse the Neanderthal remains and their burial contexts in much higher detail than was possible during the original excavations, potentially providing robust insights into Neanderthal mortuary behaviour [28].

1.1.3. Stratigraphy and Dating of the Cave

In the original excavations, Solecki recognised four major layers or cultural phases in the 14 m sequence of the cave: Layer A (modern to Neolithic), Layer B1 (Proto-Neolithic), Layer B2 (Mesolithic), Layer C (Upper Palaeolithic Baradostian), and Layer D (Middle Palaeolithic Mousterian) [4]. Multiple rockfall events occurred throughout the sequence in addition to mudflow events, and as a result the stratigraphy is highly complex. His radiocarbon dates indicated that Layer A dates from approximately 7000 BP to the present, Layer B from approximately 12,000 BP to 10,600 BP, Layer C from 35,540 BP to 28,700 BP, and that Layer D ended around 46,900 BP. The start date of the sequence is uncertain, though at the time Solecki estimated that it could be as old as 100,000 BP on the basis of sedimentation rates in the cave [4]. It should be noted that a significant hiatus in the occupation of the cave was recorded between Layers B2 and C, and between C and D. In addition to the radiocarbon dates, obsidian samples from Layers B and C were taken for obsidian hydration dating [29], though Solecki did not feel that the dates obtained for Layer C using this method were reliable [4].

In 2017 Becerra-Valdivia et al. published a paper reanalysing the dates obtained for the original excavations alongside new and original radiocarbon dates for other Palaeolithic sites in the region. Bayesian modelling of these dates was used to suggest a transition to the Upper Palaeolithic between 45,000 to 40,250 cal BP in the region, and a transition between 43,200 to 39,600 cal BP for Shanidar Cave specifically (i.e., from Solecki's Layer D to Layer C) [5]. Dating of material from the recent excavations is ongoing, with charcoal AMS dates and preliminary OSL dates broadly agreeing with the original Solecki dates but without indicating any substantial hiatus between Layers C and D. Altogether the current set of available dates suggests that the full cave sequence extends back at least 100,000 years, which corresponds to marine isotope stage 5.

1.2. Taphonomy and Microvertebrate Studies

Microvertebrates are a valuable component of the zooarchaeological assemblage, as they offer unique perspectives on human–environment interactions that cannot always be gleaned from other sources [30]. They offer particular advantages in the study of human migration and sedentism through the study of commensal microvertebrates [31,32]. The other major use of microvertebrate assemblages is in palaeoenvironmental reconstruction (e.g., [33–37]). The abundance, small range, and specific ecological niches of microvertebrates mean that they are able to offer a much more localised, high-resolution palaeoclimatic record than other proxies.

The selection of prey by predators may create biases in the observed micromammal community composition. Information on which predators are responsible for depositing the micromammals in an archaeological assemblage, and whether there is change in the predator, may have large implications for any resulting palaeoecological reconstruction. Taphonomic studies aim to provide insights into any possible predatory action in the accumulation of the assemblage, as well as provide information on other taphonomic processes that may be related to environmental conditions within the cave. For example, black discolouration of the microvertebrate remains may be indicative of burning or manganese staining. The former may be caused by human activity in the cave and the latter may be indicative of increased humidity in the cave or the presence of standing water [38]. Root etching of the bones is indicative of relatively high levels of vegetation and plant growth in an area, which may in turn suggest warmer or more humid environmental conditions that are more conducive to plant growth. Abrasion of the microvertebrate remains due to transportation and mixing of sediments also provides important information on the assemblage formation. Abrasion and digestion can produce similar effects in microvertebrate remains, though they can be distinguished from each other on the basis of the location and smoothness of any rounding [39].

The most common way in which microfaunal remains are accumulated is through predation [40,41]. As different predators have different dietary preferences and prey selectivity, the resulting species presence and proportional abundance may be more of a reflection of predator activity than the actual microfaunal community. The predator responsible for assemblage accumulation therefore acts as a taphonomic filter. A substantial body of work dedicated to the effects of predation on microfaunal remains exists, with key work carried out by Mayhew [42] and Andrews [40] focusing on the taphonomic signals of different predators. The different feeding habits and digestive environments of different predators result in different degrees of breakage and digestion of the microfaunal skeletons. Digestion has been particularly thoroughly studied using teeth, and as a result the intensity of digestion of the enamel and dentine of teeth can be linked to different predator types [39,41,43–45]. Once the digestion category is known it may be possible to deduce the predator responsible using evidence from the larger mammal and avian faunal remains. Given different predators' preferences, this can then be used to determine the strength of any taphonomic or ecological signal in an assemblage composition [40]. If a predator preferentially feeds on a subset of the microfaunal community, this produces stronger taphonomic than ecological patterning in the community composition, and any environmental reconstruction based on this will be less reliable. If a predator is more generalist in its feeding habits, the community composition in the predator diet and resulting archaeological assemblage are much more likely to be representative of the actual community. In this case, the ecological signal is stronger than the taphonomic signal and we can be more confident in any resulting palaeoecological reconstruction [44].

2. Materials and Methods

2.1. Sampling and Processing of Material

All the material used in this study is from the recent excavations at Shanidar Cave, with all microfaunal samples prior to those of the 2019 excavation season examined, including seasons 2015, 2015A, 2016, 2016A, 2017, and 2018. The sections and trench walls from the Solecki excavations were exposed and cleaned and then a series of sample columns were cut through the length of the sequence. The location of these sample columns is indicated in Figure 2. Where possible, sample columns were cut with a dimension of 30 by 30 cm, though in many cases smaller diameters were cut due to the large amount of rockfall in the sequence. In addition to the sample columns, in the 2018 excavation season material was collected from two open-plan excavation areas, also shown in Figure 2. For the purposes of this analysis the upper open-plan excavation area is referred to as SH18GS and the lower SH19GS. Overall, on the basis of the dates from the recent excavations, the Solecki excavations, and the Becerra-Valdivia reanalysis, it is thought that the material examined here dates from around 30,000 years ago to around 85,000 years ago. This corresponds to marine isotope stages 5a to 3.

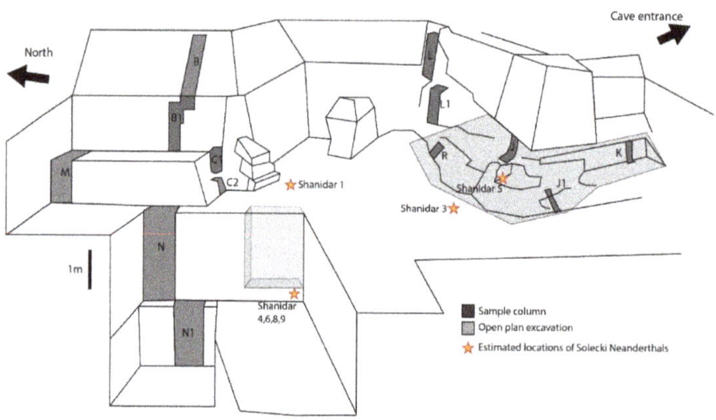

Figure 2. Schematic of recent excavations showing the locations of the sample columns and open-plan excavation areas. The northern sequence combines samples from Sample Columns B, B1, M, N, and N1 and the lower open-plan excavation area. SH19GS samples all come from the lower open-plan excavation area, shown in the figure as the lower light-grey rectangle. The southern sequence combines samples from Sample Columns L, L1, J, J1, and R and the upper open-plan excavation area. SH18GS samples all come from the upper open-plan excavation area, shown in the figure as the upper light-grey area. Diagram created by E. Hill.

In the field, the volume of excavated bulk material was recorded and then material was wet sieved through a 1 mm mesh. Once dry, materials were separated by dry sieving into three fractions of approximately >10 mm, >3 mm, and >0.3 mm. Following this, the shells, lithics, and bones (all mostly fragments) were hand sorted from each fraction into material type. All faunal material from the excavations is held on loan from the Kurdistan Directorate of Antiquities at the Department of Archaeology in Cambridge. On the return to the UK, the faunal sample bags were sorted again in the Grahame Clark

Zooarchaeology Laboratory with the aid of a Leica MZ 75 light microscope. Micro- and macrofauna were analysed separately and so were initially sorted apart from one another. Within the microfaunal fraction, indeterminate bones were separated from the rest of the assemblage and the remainder was sorted into the major taxonomic groups (e.g., birds, reptiles, fish, mammals). The analysis focused on the micromammals, and within this group bones were sorted according to element. Overall, 2592 microfaunal remains collected from the recent excavation were examined. The preliminary micromammal faunal list is given in Table 1.

Table 1. Preliminary microvertebrate faunal list from Shanidar Cave.

Order	Family	Subfamily	Species
Rodentia			
	Cricetidae		
		Arvicolinae	
			Microtus socialis s.l.
			Ellobius sp.
			Arvicola cf. *amphibius*
		Cricetinae	
			Cricetulus sp.
			Mesocricetus sp.
	Muridae		
		Murinae	
			Apodemus (*Sylvaemus*) sp.
			Mus cf. *musculus*
			Micromys sp.
		Gerbillinae	*Meriones* cf. *persicus*
	Calomyscidae		
			Calomyscus sp.
Eulipotyphla			
	Soricidae		
		Crocidurinae	
			Crocidura cf. *suaveolens*

2.2. Taphonomic Methods

Breakage patterns and the degree of digestive etching on microfaunal remains can be used as an indicator of the predator responsible for assemblage accumulation. Surface digestion of the specimens, recorded following the categories set out by Andrews [40] and expanded in subsequent work by Fernandez-Jalvo et al. [44], was recorded for both postcranial (humeri and femora) and dental material. Diagrams and descriptions of humeri and femora digestion from Jenkins aided categorisation of this material [46]. The categories were as follows: None, Light, Moderate, Heavy, and Extreme. Corrosion of bones and teeth can be produced by other processes, such as chemical weathering, but in the case of the Shanidar Cave assemblage the larger fauna and the mollusc assemblages showed no preliminary evidence of corrosion. In addition to this, the corrosion of the molars originated at the salient angles of the tooth, rather than at random spots. Chemical corrosion leaves randomly dispersed points of corrosion on the tooth surface [44]. This suggests that any corrosion observed on the microfauna is due to digestion rather than other processes. Data on breakage were also collected for these two groups, and, where relevant, additional information on other taphonomic marks such as burning and manganese staining was also collected. Studies of anatomical representation can provide useful additional information on the predator or post-depositional processes [40]. However, fully sorted postcranial data are not available for all contexts at this stage. These data, and their analyses, will be included in future work.

The data are presented as an overall assemblage dataset, and then disaggregated into separate stratigraphic units in order to investigate whether taphonomic processes changed through the sequence. As there is a degree of overlap between the sample columns and open-plan excavation areas, a new stratigraphic framework of a northern and southern sequence was developed to integrate these areas. Many of the individual contexts (i.e., within-layer sediment units identified in the field) were very small, and so, where possible, data from individual contexts have been combined into single sequence units in the northern and southern sequences. The northern sequence brings together samples from the back of the cave, encompassing contexts from Sample Columns B, B1, M, N, and N1 and the SH19GS open-plan excavation area, shown on the left side of Figure 2. The remaining areas, Sample Columns J, J1, L, and L1 and the SH18GS open-plan excavation area, are located towards the entrance of the cave and were grouped together into the southern sequence, shown in the right side of Figure 2. Contexts were grouped together into new units for the northern and southern sequences and assigned sequence numbers to indicate the relative position of each unit, with units numbered from top to bottom. Smaller sequence numbers, e.g., N01 and S03, represent samples from the upper stratigraphic levels that correspond to later dates, and larger sequence numbers, e.g., N66 and S43, represent lower stratigraphic levels that correspond to earlier dates. A small number of contexts that were very isolated were not linked to the main northern and southern sequences.

In order to maximise the amount of usable data in the timeframe available, the taphonomic analysis focused on two element groups, the dentition and proximal long bones (humerus and femur). Breakage and other taphonomic marks were recorded for all taxa, but the dental taphonomic analyses used only the arvicoline molars, as this group was present in nearly all contexts and in large numbers. A small subsample of murid data was collected (see Supplementary Table S1), but given the very small and uneven sample sizes is not included in the digestion data presented and discussed. Breakage data considered all soricid teeth and all rodent molars, whereas the digestion data only considered arvicoline molars. All femora were considered for digestion. As unfused femoral heads could be mistaken for digestive marks, particular care was taken in distinguishing between these two categories. Digestion marks were also visible on the greater and lesser trochanters and particular attention was likewise paid to these in cases where the femoral head was unfused. Digestion and breakage were recorded for all available humeri and femora, which are generally among the most commonly preserved postcranial elements in the assemblage and for which there are available comparative digestion data. With the exception of the talpid humerus, there are few diagnostic differences between the different micromammal postcrania, and so these results were not disaggregated by taxa. In the case of dental breakage, a tooth was recorded as incomplete if part of the occlusal surface or a cusp was absent. Absolute frequencies for all data are provided in the Supplementary Materials.

3. Results

3.1. Breakage

3.1.1. Humeri and Femora Results

Following the guidelines in Andrews [40], breakage of the humeri and femora was recorded as being present or absent, and if breakage had occurred, whether the proximal, shaft, distal, or a combination of these sections was present. The vast majority of humeri and femora surveyed were broken: 93% of the total surveyed. The data are shown broken down by element and portion present in Table 2. Approximately equal numbers of humeri and femora were surveyed, 336 to 343, respectively, and both showed similar levels of breakage, with 92.3% of humeri broken compared to 91.5% of femora. In terms of the portion present, most of the humeri were either a distal or a distal and shaft fragment, and the opposite was true of the femora, with most consisting of a proximal or proximal and shaft fragment.

As the humeri and femora sample size was smaller than the dentition sample, the data were very coarse when broken down by approximate area of the assemblage (sample column or cluster), particularly in the smaller upper sample columns. The sample size per area is shown in Table 3, demonstrating that certain sample columns had very small humeri and femora samples, in particular Sample Column J, from which no humeri or femora were recovered. As would be expected from the overall data, breakage in all areas was very high, ranging from 85.3 to 100% of all specimens showing signs of breakage. It is notable that some of the lowest proportions of breakage were observed in the lower sample columns and clusters: Sample Column N1, the lowest area of the excavations, had the highest number of complete specimens. Sample Columns B, B1, J, J1, L, and L1 are all located in the upper levels and in section appeared to have higher levels of rockfall, characteristics that may explain the highly fragmented nature of the assemblage.

Table 2. Overall humeri and femora fragmentation data: The data are disaggregated by element and portion present.

Element	Portion Present	Number of Specimens Examined	Proportion (%) [1]
Humerus	Proximal epiphysis	20	6.0
	Proximal epiphysis and shaft	10	3.0
	Shaft	39	11.6
	Distal epiphysis and shaft	67	20.0
	Distal epiphysis	174	51.8
	Complete	26	7.7
Femur	Proximal epiphysis	176	51.3
	Proximal epiphysis and shaft	95	27.7
	Shaft	24	7.0
	Distal epiphysis and shaft	7	2.0
	Distal epiphysis	12	3.5
	Complete	29	8.5
Unidentified	Shaft	3	100

[1] "Proportion": the proportion of each fragment type relative to the overall number of fragments for each element type.

Table 3. Postcranial fragmentation data, disaggregated by sample column or area.

Sample Column or Area	Femora		Humeri	
	NISP	% Broken [1]	NISP	% Broken [1]
Area SH18GS	165	90.9	163	94.5
Area SH19GS	69	92.8	70	88.6
Sample Column J	4	100.0	0	-
Sample Column J1	2	100.0	4	100.0
Sample Column L	1	100.0	2	100.0
Sample Column L1	3	100.0	3	100.0
Sample Column B	1	100.0	0	-
Sample Column B1	42	97.6	6	100.0
Sample Column M	42	92.9	38	94.7
Sample Column N	14	85.7	30	93.3
Sample Column N1	113	91.2	20	85.0
Total	317	92.4	336	92.9

[1] "% Broken" refers to the proportion of humeri and femora specimens examined that were broken.

When broken down by context, 74% of contexts had no complete specimens, though it must be noted that this may be partially due to the small sample size of contexts. Contexts with no complete specimens had an average sample size of 3.34 in comparison to 5.9 in contexts with at least one complete specimen, with an average sample size of 5.9. Contexts with greater sample sizes than average had a mean breakage of 92.3% in comparison to 94.6% in those with a below-average sample size. Whilst there was a large number of smaller contexts with no complete specimens, there did not seem to be a significant increase in the number of complete specimens as sample size increased.

In order to investigate whether there were any changes in the taphonomic processes throughout, the sequence data were grouped into northern and southern sequences. Overall, the proportion of broken humeri and femora remained high throughout both sequences (see Supplementary Figures S1–S4). A possible reduction in breakage was visible between sequence units N34 to N38 and S36 onwards.

3.1.2. Teeth Results

Given the contrasting morphology of different teeth within and between taxa, tooth breakage was recorded as simply present or absent. Whether the teeth were loose or present in a tooth row was also recorded. Very few teeth were found in a tooth row: Only 175 or 9.5% of the teeth recorded were present in the mandible or maxilla. In terms of fragmentation of the teeth themselves, 12% of in situ teeth and 34.8% of isolated teeth were broken. It should be noted that given the way in which teeth and bones were picked from the sieved material and then sorted in the lab, fragmented teeth, particularly single cusps of the microtine molars, were perhaps more likely to be missed and not kept due to their resemblance to the bone "dust" that was ubiquitous in the assemblage, so it might be expected that as sampling effort (and sample size) increases, more of the tiny individual teeth and toothrow fragments would be recovered. In this case, the same team was responsible for the initial picking of all teeth, and one individual was responsible for all lab sorting of the microfaunal remains. By keeping these conditions constant throughout the samples, it is hoped that sorting effort was maintained at a constant level throughout the assemblage. Examination of the data suggests that it is not a lack of sampling effort that led to the lack of teeth found in tooth rows: Even the largest sample area, SH18GS, had only 7.3% teeth in situ (see Table 4). The area with the highest proportion of teeth still present in a tooth row was SH19GS, with still only 17.3% of specimens containing toothrows.

Table 4. Number of teeth examined in each sample column or area.

Sample Column or Area	In Situ		Isolated Teeth	
	NISP	% Broken	NISP	% Broken
Sample Column B	5	20.0	30	53.3
Sample Column B1	1	0.0	20	35
Sample Column J	4	0.0	31	51.6
Sample Column J1	0	-	8	50
Sample Column L	0	-	8	50
Sample Column L1	2	0.0	52	55.8
Sample Column M	15	0.0	283	44.2
Sample Column N	7	14.3	127	31.5
Sample Column N1	17	5.9	122	19.7
Area SH18GS	58	14.7	728	36.0
Area SH19GS	56	14.3	267	21.0
Total	175	12.0	1676	34.8

In addition to data on whether teeth were present in the jaw, data were also collected on the breakage of the teeth. Given the different types of teeth surveyed, the data were not broken down by portion of the tooth present, but instead whether the tooth in question was complete. Overall, 32% of teeth were incomplete across the whole assemblage, with these data shown broken down by excavation area in Table 4.

The proportion of incomplete isolated teeth ranged from 19.7% in Sample Column N1 to 55.8% in Sample Column L1, a fairly large range. In the toothrow dataset, breakage was generally lower, ranging from 0 in several sample columns to 20% in Sample Column B. As discussed above, it is possible that increased sampling effort may be a factor leading to greater recovery of fragmentary teeth. However, this does not seem to be the case, as there did not appear to be a strong correlation between sample size and the proportion of specimens examined that were incomplete. This indicates that the higher proportion of incomplete specimens in certain areas is likely to be a genuine indication of higher disturbance to the sediments or other taphonomic processes, rather than greater sampling effort resulting in increased numbers of incomplete specimens being recovered.

As with the humeri and femora, the dental breakage patterns were also examined in sequence. The initial data were complicated by the large number of sequence units with 100% or 0% breakage. This was due to their relatively small sample sizes, with only one or two specimens examined. The data were filtered to remove extremely small samples ($n < 5$), producing a clearer picture of the variation through the sequence. Bar charts displaying the proportion of broken isolated teeth in each sequence unit are presented in the Supplementary Materials (Figures S5 and S6), but overall there appeared to be a slight increase in the proportion of teeth that were broken over time in both sequences. This was particularly evident at the base of the northern sequence from unit N64 to N36, where all units except N51 had breakage levels of less than 37.5%.

3.1.3. Comparison between the Postcranial and the Dental Breakage Patterns

Postcranial and tooth breakage patterns were compared to investigate whether any parts of the sequence showed similar patterns of breakage in both elements. There did not appear to be any particularly strong correlation between the two datasets, though the absence of postcranial data for certain parts of the sequence, where no humeri or femora were recovered, makes comparisons throughout the whole sequence hard. The postcrania were generally much more fragmented than the teeth, which is to be expected given the more delicate nature of these elements. Something that did seem to be evident from the fragmentation and breakage data is that there was a slight decline in the proportion of broken teeth, loose teeth, and broken postcrania in the lower sample columns (Sample Column N1 and Area SH19GS) and lower sections of the northern sequence.

3.2. *Digestion*

3.2.1. Humeri and Femora Results

Digestion on the long bones is generally most evident at the epiphyses. A subset of the overall postcranial dataset was taken, with only the distal epiphysis of the humerus and the proximal epiphysis of the femur examined for signs of digestion. Overall, 67.7% of the humeri and femora surveyed showed no signs of digestion and a further 26.5% only light. Although some specimens with higher digestion intensities were recorded, they accounted for only 5.8% of the whole humeri and femora assemblage.

These data can be broken down further by element, as shown in Table 5. The data are presented as independent proportions of the overall humeri and femora assemblage and as a cumulative proportion. Similar numbers of each element were examined: 343 femora and 336 humeri. The data here show that generally lower degrees of digestion were observed on humeri than on femora, with 82.1% of the humeri showing no signs of digestion in comparison to 53.4% of the femora. It should also be noted that no specimens displaying heavy or extreme digestion were observed in the humeri. Similar numbers of each element were examined—343 femora and 336 humeri—so we can be confident that these patterns are not an artefact of relatively small sample size in one of the elements. It is possible that the distal epiphysis of the humerus is generally denser, and more robust to digestion than the proximal epiphysis of the femur. An additional explanation is that the proximal epiphysis of the femur fuses later than the distal humerus, and so unfused or fusing epiphyses were mistaken for or exacerbated by digestion.

Table 5. Observed humeri and femora digestion across the assemblage. Data presented disaggregated by element.

Element	Observed Digestion	NISP	Proportion (%)
Femur	None	183	53.4
	Light	125	36.4
	Moderate	26	7.6
	Heavy	7	2
	Extreme	2	0.6
Humerus	None	276	82.1
	Light	56	16.7
	Moderate	4	1.2

The results of the digestion intensity analysis for the humeri and femora assemblage are presented in Figures 3 and 4, respectively. Overall, the majority of samples in each sequence displayed relatively low digestion intensities, with the majority of each sequence unit showing either no or light digestion. In the southern sequence, no humeri or femora samples displaying extreme digestion were observed, and only one heavily digested specimen. It was difficult to discern any sequence units with particularly distinct digestion patterns, although some areas of both sequences seemed to have relatively higher proportions of more severe digestion. These areas also had relatively small sample sizes.

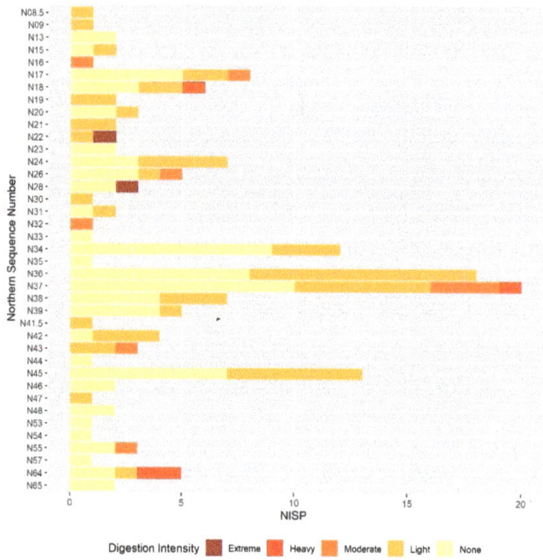

(**A**)

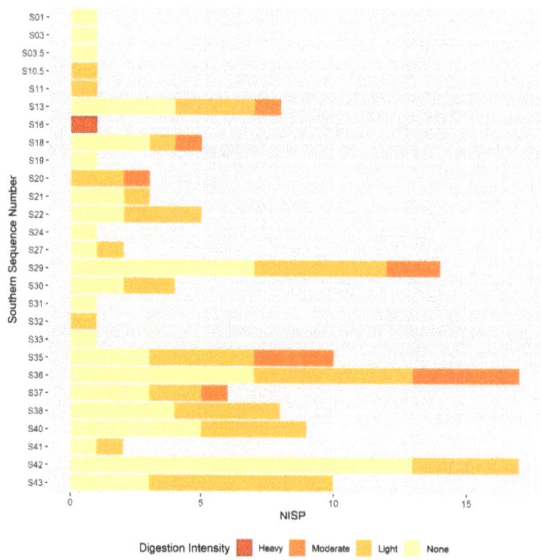

(**B**)

Figure 3. Observed femora digestion intensity in the northern (**A**) and southern (**B**) sequence. NISP: number of identifiable specimens.

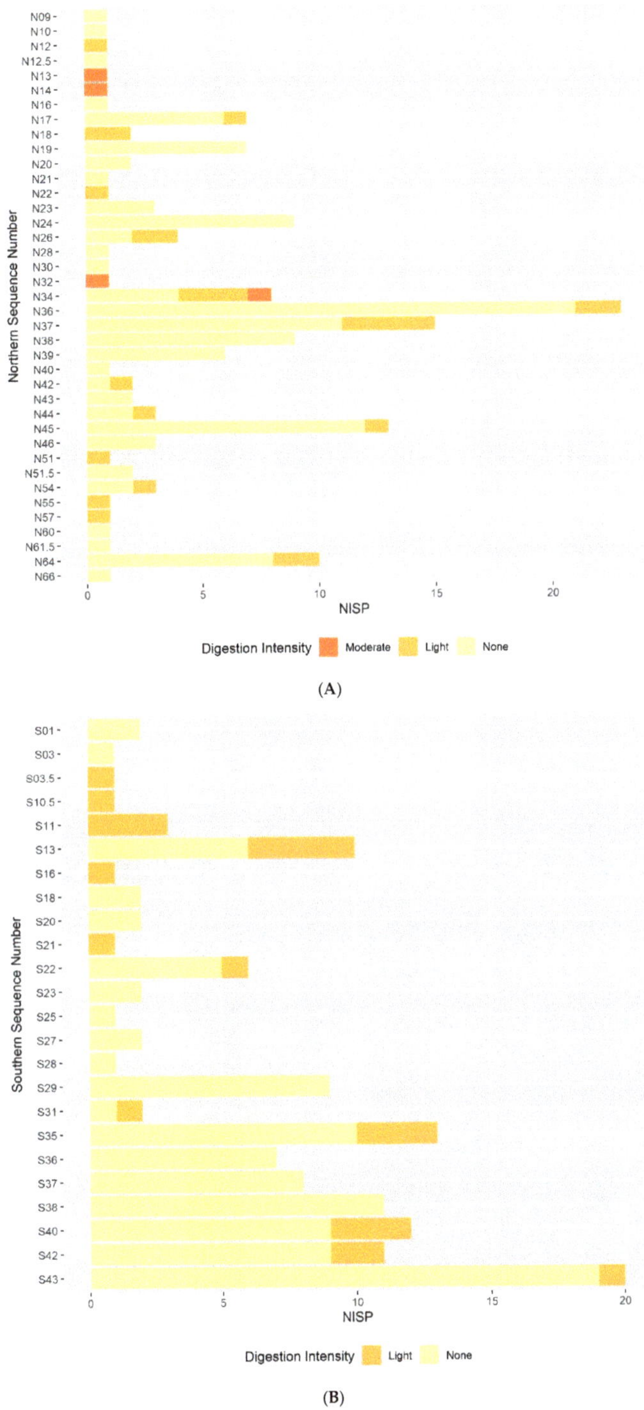

Figure 4. Observed humeri digestion intensity in the northern (**A**) and southern (**B**) sequence. NISP: number of identifiable specimens.

3.2.2. Molar Results

A large number of molars were too incomplete to discern whether any digestion marks were present, or the molar was obscured by large amounts of sediment. In these cases, digestion was recorded as N/A (not applicable) and not included in these results. Examples of molars displaying the different digestion intensities are shown in Figure 5.

(a)

(b)

(c)

(d)

(e)

Figure 5. Arvicoline molars from the Shanidar Cave assemblage displaying the different categories of digestion intensity: (**a**) Molar displaying no digestion marks. (**b**) Molar displaying light digestion marks. (**c**) Molar displaying moderate digestion marks. (**d**) Molar displaying heavy digestion marks. (**e**) Molar displaying extreme digestion marks. Note the manganese staining in Figure 5d,e.

A breakdown of the overall assemblage molar digestion data is shown in Table 6. Overall, more severe levels of digestion intensity were very low: 80% of the isolated molars surveyed had no signs of digestion and 13.2% light evidence, with less than 8% of molars displaying signs of more intense digestion than this. The in situ molars also displayed even lower levels of digestion intensity than the isolated molars, which is to be expected given that isolated molars are generally more exposed and digested. Given the very small sample size of in situ molars, they are excluded from the datasets disaggregated by sample column, unit, or area.

Table 6. Overall assemblage observed digestion of molars.

Observed Digestion	Isolated Molars		In Situ Molars	
	NISP	Proportion (%)	NISP	Proportion (%)
None	968	80.0	25	89.3
Light	162	13.2	2	7.1
Moderate	62	5.1	1	3.6
Heavy	24	2.0	0	0
Extreme	10	0.8	0	0
Total	1226	-	28	-

When these data are broken down further into broad sample columns or area groups, a little more variation in the observed digestion intensities can be seen (Table 7). In all areas except in Sample Column L, over 70% of molars showed no signs of digestion, and in all areas except in Sample Column L and Sample Column B1, over 90% of molars showed light or no signs of digestion. This variation in digestion intensity may be partially explained by the extreme variations in sample size; in Sample Column L, only four molars were examined for digestion and in Sample Column B1, 14.

Table 7. Overall isolated arvicoline molar digestion intensities observed in each area or sample column. The observed digestion is presented as a proportion of each sample column or area.

Sample Column	Observed Digestion (%)					Total NISP
	None	Light	Moderate	Heavy	Extreme	
Sample Column B	70.0	20.0	5.0	5.0	0.0	20
Sample Column B1	85.7	0.0	14.3	0.0	0.0	14
Sample Column J	90.5	4.8	0.0	0.0	4.8	21
Sample Column J1	83.3	16.7	0.0	0.0	0.0	6
Sample Column L	25.0	50.0	0.0	25.0	0.0	4
Sample Column L1	80.0	20.0	0.0	0.0	0.0	20
Sample Column M	75.3	14.4	7.2	2.1	1.0	194
Sample Column N	81.7	8.5	1.4	4.2	4.2	71
Sample Column N1	86.1	11.1	2.8	0.0	0.0	72
Area SH18GS	76.6	14.4	5.9	2.5	0.7	611
Area SH19GS	86.5	10.4	3.1	0.0	0.0	193
Total	79.0	13.2	5.1	2.0	0.8	1226

The effect of sample size is evident when the assemblage is broken down further into the northern and southern sequence units, presented in Figures 6 and 7. Broadly, most sequence units showed mostly no, or light, digestion, though there are a few that appeared to show higher proportions of more severe digestion. For example, in N34, S02, and S30, at least 50% of all specimens showed signs of digestion, including moderate or more severe digestion, but all three of these sequence units had very small sample sizes, of fewer than five specimens. There was perhaps a decrease in digestion intensity towards the base of the northern sequence and the top of the southern sequence, with a greater proportion of molars showing no digestion. However, these two areas also coincided with relatively small sample sizes, which may be a significant biasing factor.

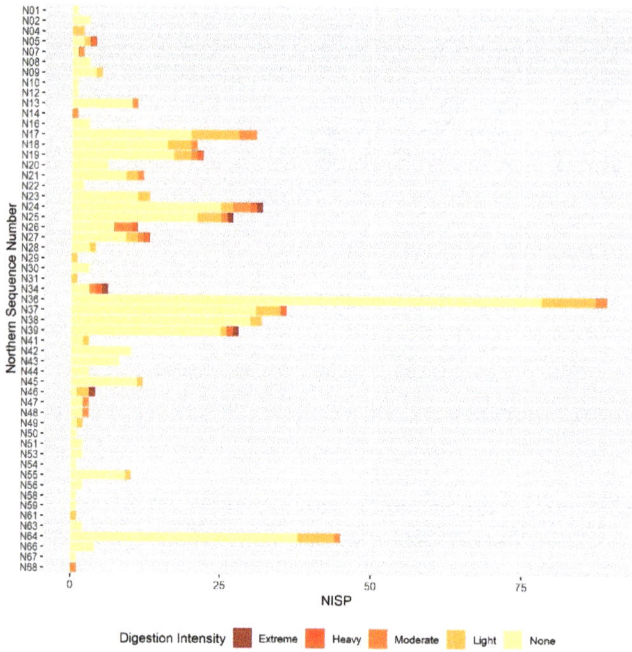

Figure 6. Observed isolated molar digestion intensities in the northern sequence. NISP: number of identifiable specimens.

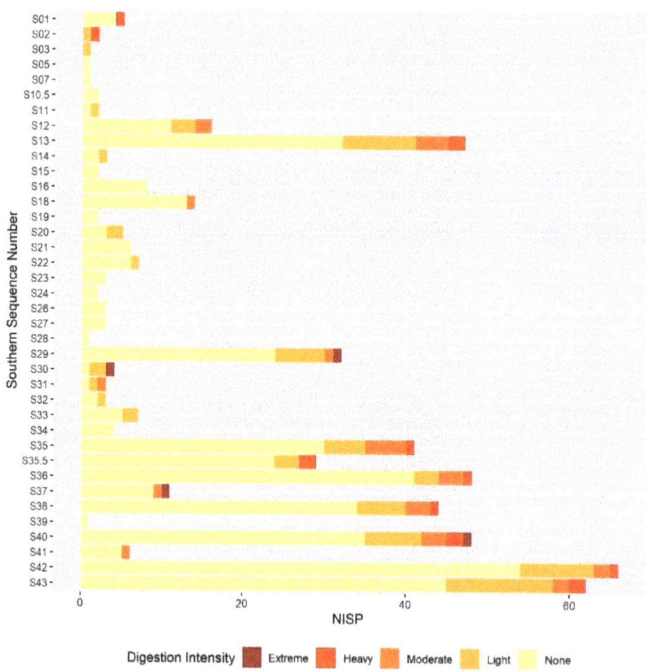

Figure 7. Observed isolated molar digestion intensities in the southern sequence. NISP: number of identifiable specimens.

3.2.3. Comparison between the Postcranial and the Molar Digestion Patterns

Overall, the digestion data from both molars and postcrania displayed similar patterns of no or relatively low digestion levels. In both cases, certain sample columns or areas and certain sequence units did appear to have relatively high proportions of more severe digestion, but it is difficult to determine whether this is a genuine signal, or simply the effect of small sample size. Humeri and femora remains did display slightly lower proportions of elements with no observed digestion compared to the molars, though similar levels of elements showed light or no digestion. It is unlikely that this disparity indicates anything unusual about the taphonomic processes affecting the humeri and femora material versus the molars in the Shanidar Cave assemblage, as the former has widely been observed to display digestion more readily than molars in other assemblages.

A possible confounding factor in the recorded molars and humeri and femora digestion intensities is the increased fragility of more heavily digested molars. The erosion of the stronger enamel layers leaves the molars more susceptible to fragmentation, which in turn makes the element less likely to survive to be recovered during the excavation and sorting process. As such, it is possible that certain samples displayed artificially low levels of digestion due to the difference in the recoverability of severely digested elements.

3.3. *Other Taphonomic Marks*

Although the more systematic data collection was aimed at investigating breakage patterns and digestion, data on other taphonomic marks were also collected. Burning was recorded on 58 teeth specimens, five humeri, and four femora specimens. On occasion, burnt teeth, humeri, or femora appeared to be the result of being in close proximity to hearths, with all specimens showing signs of burning, but more commonly burnt specimens appeared sporadically across the sequence in context with unburnt material, likely as a result of fluvial or aeolian processes leading to the dispersal of burnt remains. A commonly observed taphonomic mark was manganese oxide staining or spotting of teeth. This was observed on 89 tooth specimens across all levels of the sequence, though there did appear to be a higher proportion of teeth showing this marking in the lower part of Sample Column N and Sample Column N1. Manganese spotting has been attributed to a number of sedimentary processes, including the deposition of organic matter and relatively damp conditions [38,45,47]. Root etching on larger faunal remains and larger microvertebrate postcranial fragments was also observed, and this again was more common in the lowest layers of the sequence.

Although the analysis of larger fauna was outside the scope of the project, the highly fragmentary nature of the zooarchaeological assemblage meant that micro and macro remains were generally bagged together, giving the opportunity to briefly examine these larger fragments when sorting. As a result, although the data collection of large mammal bones was not systematic, it was occasionally possible to record the presence of rodent gnawing on fragments of bone, as seen in Figure 8. Although data collection here was not systematically completed, this finding is significant, as it indicates that rodents inhabited the cave at certain times in the past. Here we can see how hominins may have disrupted the ecology of the cave by their presence. Faunal data from the original excavations indicates that large numbers of ungulates were being consumed in the cave: Over 90% of the faunal assemblage studied by Perkins was *Ovis orientalis*, *Capra hircus aegagrus*, and *Cervus elaphus* [21]. It is possible that the presence of discarded food remains in the cave may have encouraged the presence of rodents and other scavenging species. This also has implications for the interpretation of other taphonomic data from the cave. Within Andrews' framework, for example, the absence of digestion marks could be taken as evidence of certain species of avian predator: *Tyto alba*, for example, leaves no or only light digestion marks. The presence of gnawed bones may indicate that these undigested specimens are from rodent specimens that inhabited the cave and died a more "natural" death, though it is equally possible that these rodents were temporary visitors to the site and died elsewhere.

Figure 8. Detail of rodent gnawing on ungulate long bone fragment.

4. Discussion

The results of this preliminary analysis suggest that in most respects the taphonomic processes affecting the microvertebrate assemblage remain constant throughout the sequence. Certain individual sequence units do appear to have particularly distinct digestion or breakage patterns, but these are generally extremely small units. If these extremely small sequence units are excluded, the overall picture of the assemblage is one of high fragmentation and low digestion intensity throughout the sequence. The only exception to this continuity is the slight decline in fragmentation in the lower contexts that are dated to the earlier phases of the sequence. The majority of humeri and femora are still broken, and most teeth are still loose from the jaw but there is a small reduction in the overall proportion in each case. This is best illustrated in the teeth breakage data, with the proportion of teeth broken in the lowest sample column and area half that of the proportion in the highest sample columns. There are multiple possible causes of the high fragmentation rates, including rockfall, trampling, predation, and sediment movement. Although some predators, such as diurnal raptors and mammalian carnivores, can produce very high levels of fragmentation [45], they are unlikely to be the cause in this case. Predators that produce high levels of fragmentation generally leave more intensive digestive marks on the teeth, and this does not seem to be the case in the Shanidar Cave assemblage. Indeed, many of the incomplete specimens examined display no traces of digestion.

Extensive rockfall is visible throughout most of the assemblage, particularly in the upper layers, and these high energy rockfall events are likely to be responsible for most of the observed fragmentation. Geomorphological analyses to determine the exact nature and causes of these repeated rockfall events are ongoing. One current explanation is that they are associated with colder, arid periods, as they are generally absent from the lowest layers of the sequence [27]. Relatively warmer, humid conditions are also indicated by the presence of increased manganese dioxide staining in these layers. These lower layers are present at the bottom of the northern sequence only, as the southern sequence does not extend as deep. This is significant because the northern sequence is located towards the back of the cave, further away from the entrance than the southern sequence. The presence of manganese staining and root etching indicates that the warm and humid conditions extended quite far into the cave, enough to support some plant growth. Although dating for the lowest layers is still ongoing, preliminary dates suggest that these sediments below the lower Neanderthal cluster are older than 70,000 years BP. This would place these sediments in the range of Marine Isotope Stage 5a, a period of relatively warmer conditions. Further palaeoecological analyses of the microvertebrate assemblage and other environmental samples are needed to confirm these warmer conditions, but these results present an intriguing starting point.

Results from the digestion analyses also show no large-scale changes in digestion intensity, or at least none that can be distinguished from others due to the small sample size. In the molars and, to a slightly lesser degree, the humeri and femora, the majority of specimens show no signs of digestion, with the next most commonly occurring digestion category of light. The presence of occasional more heavily digested specimens indicates that at least one predator is responsible for at least part of the assemblage accumulation, and that this predator is likely to be a category 1 or 2 predator following Andrews' classification of different predators' modification of microvertebrate teeth, humeri and femora [40], though further analyses of the incisors and other microvertebrate taxa will be needed to confirm this. The proportion of digested molars does not closely match the proportions seen in modern predators, except for the Eurasian eagle owl (*Bubo bubo*), the tawny owl (*Strix aluco*), and possibly the barn owl (*Tyto alba*), long-eared owl (*Asio otus*), and short-eared owl (*Asio flammeus*) [44]. These species can be found in present-day Iraq; *B. Bubo* and *T. alba* are resident year-round, whereas *S. alucois*, *A. otus*, and *A. flammeus* are much rarer and present only in the winter [48]. None of these species are reported in Hesse's preliminary list of microvertebrates reported in Appendix B of Evins' 1981 report [22]. It should be noted that absence of physical evidence of the predator responsible is not necessarily evidence of absence, and that we would not generally expect the remains of the predator responsible to be found among the prey fossils. The only bird of prey found was a kestrel, *Falco* cf. *tinnunculus* [22]. If *Falco* was responsible for assemblage accumulation, a much higher proportion of digested molars would be expected (~53%), and those that were digested would be categorized as heavily digested. *S. aluco* also produces teeth that would be categorized as heavily digested, and so is unlikely to be the primary contributor to the assemblage. *B. bubo* produces a lower proportion of teeth displaying any digestion, though those teeth still tend to be more heavily digested than those examined in the Shanidar Cave assemblage [44]. *B. bubo* nests in rocky habitats, cliffs, and small caves, and hunts over open habitats, very similar to the karstic landscape around Shanidar Cave. Attempts were made during the excavation seasons to collect owl pellets in order to understand the local avian predator community and diet. Unfortunately, no pellets have been recovered to date, though it is possible that some may be recovered in future excavation seasons. A motion-activated trail camera was placed in the cave during the 2018 excavation season. It recorded images of a larger owl species visiting the cave, but the image quality was too low to allow specific identification. Preliminary taphonomic analysis of the small mammal assemblage of Kaldar Cave (approximately 500 km northwest of Shanidar Cave) found a high proportion of digested elements in the assemblage, and also considered *S. aluco* and *B. bubo* to be possible contributors to the assemblage [49,50]. The high proportion of arvicoline teeth in the assemblage may be indicative of certain predators with a preference for this taxon. Several avian predators have been reported as preferring arvicolids, in particular *Asio otus*, *A. flammeus*, and *Strix nebulosa* [40]. The latter is not currently found in the region, but both *A. otus* and *A. flammeus* are known to visit the region in winter [48].

Although no owl pellets were collected from the local area of Shanidar Cave, a recent taphonomic study of *T. alba* was carried out in Birecik, Turkey, approximately 550 km west of Shanidar Cave [51]. The study found that the *T. alba* prey assemblage was dominated by *Meriones tristrami*, followed by *Mus musculus*. Species from both genera were also recovered from the Shanidar Cave assemblage (see Table 1), and a similarly low level of digestion intensity was observed in both assemblages. *Microtus* dominates the Shanidar Cave assemblage, and was not recovered from the Birecik *T. alba* prey assemblage. *Microtus* is known to be taken by *T. alba* in the wider region; a 2006 study of *T. alba* pellets in the Beit She'an Valley (approximately 890 km southwest of Shanidar Cave) found that *Microtus* accounted for 19.9% of the total prey MNI, with *Meriones* accounting for 32.3% and *Mus*

37.3% [52]. The difference between the two studies is unlikely to be due to season of capture, as the Birecik pellets were collected in June and the Beit She'an in July. The Beit She'an study noted a large difference in recovery of *Microtus* between different study sites that was not observed in *Mus* or *Meriones*, and so it is possible that the lack of *Microtus* in the Birecik study is due to a lack of vole populations in the study area. *T. alba* is a category 1 predator that produces absent or minimal digestion of the microfaunal remains. This broadly matches the pattern seen in the Shanidar Cave assemblage. Occasional specimens with more intense digestion have been recorded in the Shanidar Cave assemblage, though as demonstrated by Williams [53], there is variation in digestion intensity between different *T. alba* individuals depending on age and other factors.

The results of the taphonomic analysis with respect to predation are not conclusive. The digestion patterns, landscape, and predators present in the region today suggest that the eagle owl is the most likely predator responsible for assemblage accumulation out of the avian species listed above, though the digestion intensities reported in the literature are slightly higher than those observed here. It should be recognised that the assemblage is being analysed over very large temporal ranges, and it is possible that other predators occasionally contributed to the assemblage at different points, which may explain the occasional teeth and bones displaying more intense digestion. As the agent of accumulation is uncertain, it is difficult to be confident that the community composition observed in any palaeoecological analyses is a genuine ecological rather than taphonomic signal. The taxonomic identification of microvertebrate specimens is outside the scope of this paper, but generally speaking, arvicolines dominate the assemblage. Arvicolines are known to dominate the diet of several avian predators [44], but can also reach very high population densities in the landscape, and so it is difficult to know whether their presence is due to predator bias. Further studies of predator distribution, hunting behaviour, and preferences in the region are needed to resolve this problem.

Occasional animal burrows have been observed during excavation. Some of these are likely to have occurred during and following the 1950s excavations, when the trench walls were more accessible. Burrowing taxa are present in the preliminary faunal list, for example, *Microtus*, *Meriones*, and *Ellobius*, and may be responsible for some of these burrows, though many of these taxa more commonly construct burrows in open areas rather than within caves. Repeated rockfall throughout the sequence has resulted in very rocky sediments that may be unsuitable for some burrowing rodent taxa. Further data on the shape, depth and, where possible, age, of these burrows, and the palynological data from their sediments, will help to determine which are modern and which taxa may be responsible.

5. Conclusions

The preliminary taphonomic results presented here are indicative of relatively constant taphonomic processes throughout the length of the sequence at Shanidar Cave exposed in the new excavations, spanning from MIS 5a to MIS 3. There are no extreme changes in taphonomy, though there does appear to be a gradual increase in rockfall damage towards the upper layers alongside a decrease in manganese staining and root etching. This slight shift is suggestive of relatively wet, warm conditions in the lower levels. Early faunal data appear to support this, with increased proportions of taxa found in wetland and more vegetated habitats found in these lower levels. Further palaeoecological analyses will need to be conducted to confirm this. In terms of the agent of accumulation, digestion of the teeth is rare and generally of a low intensity throughout the sequence. Several avian predator species are recorded as being present in the region, but current records of the digestion intensity left on microvertebrate teeth and bones by these species does not closely match those observed in the Shanidar Cave assemblage. Further surveys of the hunting behaviour of avian predators of the region are needed to identify the agent of accumulation of microvertebrates in this assemblage. Further taxonomic and taphonomic work on the Shanidar Cave microvertebrate assemblage, focusing in particular on the non-arvicolines and postcranial elements, is required to produce a more complete account of the taphonomic history of the cave.

Supplementary Materials: The following are available online at https://www.mdpi.com/article/10.3390/quat5010004/s1, Figure S1. The proportion of femora that were broken in the northern sequence, Figure S2. The proportion of humeri that were broken in the northern sequence, Figure S3. The proportion of femora that were broken in the southern sequence, Figure S4. The proportion of humeri that were broken in the southern sequence, Figure S5. Proportion of isolated teeth broken by sequence unit in the northern sequence, Figure S6. Proportion of isolated teeth broken by sequence unit in the southern sequence, Table S1: Murid digestion intensity, Table S2: Sample column or area humerus data, Table S3: Sample column or area femur data, Table S4: Northern sequence femur data, Table S5: Northern sequence humerus data, Table S6: Southern sequence femur data, Table S7: Southern sequence humerus data, Table S8: Northern sequence arvicoline isolated molar digestion data, Table S9: Southern sequence arvicoline isolated molar digestion data, Table S10: Northern sequence arvicoline in situ molar digestion data, Table S11: Southern sequence arvicoline in situ molar digestion data, Table S12: Sample column or area dentition breakage data, Table S13: Northern sequence dentition breakage data, Table S14: Southern Sequence dentition breakage data, Table S15: Sample column or area arvicoline isolated molar digestion data, Table S16: Sample column or area arvicoline in situ molar digestion data.

Author Contributions: Conceptualization, E.T.; methodology, E.T.; software, E.T.; formal analysis, E.T.; investigation, E.T.; resources, E.T., P.M. and G.B.; writing—original draft preparation, E.T.; writing—review and editing, E.T., P.M. and G.B.; visualization, E.T.; supervision, P.M. and G.B.; project administration, E.T. and G.B.; funding acquisition, E.T., P.M. and G.B. All authors have read and agreed to the published version of the manuscript.

Funding: Doctoral research was funded by the Natural Environment Research Council, grant number NE/L002507/1. The overall Shanidar Cave Project has received funding from the Leverhulme Trust (Research Grant RPG-2013-105), the Rust Family Foundation, the British Academy, the Wenner-Gren Foundation, the Society of Antiquaries, the McDonald Institute of Archaeological Research at the University of Cambridge, and the Natural Environment Research Council's Oxford Radiocarbon Dating Facility (grant NF/2016/2/14). The ongoing dating programme has also been supported by the European Research Council under the European Union's Seventh Framework Programme (FP7/2007–2013)/ERC grant agreement number 324139 "PalaeoChron" awarded to Tom Higham, University of Oxford.

Data Availability Statement: Data are available within this article, in the supplementary materials and on request from the corresponding author.

Acknowledgments: We wish to thank the Kurdistan Regional Government for the original invitation to G.B. to plan new excavations at Shanidar Cave, and the Kurdistan General Directorate of Antiquities for permission to conduct these excavations and to study the finds, as well as for invaluable logistical support. In particular, we thank the General Director Mala Awat and his successor Kaifi Mustafa Ali, the Director for the Soran District Abdulwahab Suleiman, and the on-site inspectors from the Soran District Directorate of Antiquities, Dlshad Abdulmutalb and Jeghir Khalil. E.T. wishes to thank Jessica Rippengal and Catherine Kneale for technical and laboratory support, Ruairidh Macleod for reviewing drafts and coding support, Evan Hill for providing Figure 2 and useful discussion, and past and present members of the Grahame Clark Zooarchaeology Laboratory. We thank our colleagues of the Shanidar Cave Project field team for help and support in the field and post-excavation, particularly Chris Hunt, Tim Reynolds, Paul Bennett, Lucy Farr, Emma Pomeroy, Evan Hill, James Holman, Ross Lane, Jessica Twyman, and Andreas Nymark. We also wish to thank the general editors, in particular E. Stoetzel, and three anonymous reviewers for their support, useful comments, and suggestions.

Conflicts of Interest: The authors declare no conflict of interest.

References

1. Stahl, P.W. The recovery and interpretation of microvertebrate bone assemblages from archaeological contexts. *J. Archaeol. Method Theory* **1996**, *3*, 31–75. [CrossRef]
2. Lyman, R.L. *Vertebrate Taphonomy*; Cambridge Manuals in Archaeology; Cambridge University Press: Cambridge, UK, 1994. ISBN 978-0-521-45840-5.
3. Fiacconi, M.; Hunt, C.O. Pollen taphonomy at Shanidar Cave (Kurdish Iraq): An initial evaluation. *Rev. Palaeobot. Palynol.* **2015**, *223*, 87–93. [CrossRef]
4. Solecki, R.S. Prehistory in Shanidar Valley, Northern Iraq. *Science* **1963**, *139*, 179–193. [PubMed]
5. Becerra-Valdivia, L.; Douka, K.; Comeskey, D.; Bazgir, B.; Conard, N.J.; Marean, C.W.; Ollé, A.; Otte, M.; Tumung, L.; Zeidi, M.; et al. Chronometric investigations of the Middle to Upper Paleolithic transition in the Zagros Mountains using AMS radiocarbon dating and Bayesian age modelling. *J. Hum. Evol.* **2017**, *109*, 57–69. [CrossRef]
6. Solecki, R.S. Notes on a brief archaeological reconnaissance of cave sites in the Rowanduz district of Iraq. *Sumer* **1952**, *8*, 37–48.
7. Solecki, R.S. A palaeolithic site in the Zagros Mountains of Northern Iraq, report on a sounding at Shanidar Cave: Part I. *Sumer* **1952**, *8*, 127–192.
8. Solecki, R.S. A palaeolithic site in the Zagros Mountains of Northern Iraq, report on a sounding at Shanidar Cave: Part II. *Sumer* **1953**, *9*, 60–93.
9. Solecki, R.S. Shanidar Cave, a Paleolithic site in Northern Iraq. In *Annual Report of the Smithsonian Institution, 1954*; Annual Report of the Smithsonian Institution; Smithsonian Institution: Washington, DC, USA, 1955; pp. 389–425.
10. Solecki, R.S. Three adult Neanderthal skeletons from Shanidar Cave, Northern Iraq. In *Annual Report of the Smithsonian Institution, 1959*; Annual Report of the Smithsonian Institution; Smithsonian Institution: Washington, DC, USA, 1960; pp. 603–635.
11. Solecki, R.S. New anthropological discoveries at Shanidar, Northern Iraq. *Trans. N. Y. Acad. Sci.* **1961**, *23*, 690–699. [CrossRef]
12. Solecki, R.S.; Leroi-Gourhan, A. Palaeoclimatology and archaeology in the Near East. *Ann. N. Y. Acad. Sci.* **1961**, *95*, 729–739. [CrossRef]
13. Cowgill, L.W.; Trinkaus, E.; Zeder, M.A. Shanidar 10: A Middle Paleolithic immature distal lower limb from Shanidar Cave, Iraqi Kurdistan. *J. Hum. Evol.* **2007**, *53*, 213–223. [CrossRef] [PubMed]
14. Trinkaus, E. *The Shanidar Neandertals*; Academic Press: New York, NY, USA, 1983.
15. Dettwyler, K.A. Can paleopathology provide evidence for "compassion"? *Am. J. Phys. Anthropol.* **1991**, *84*, 375–384. [CrossRef] [PubMed]
16. Solecki, R.S. *Shanidar: The Humanity of Neanderthal Man*; Allen Lane The Penguin Press: London, UK, 1971; ISBN 978-0-7139-0306-5.
17. Trinkaus, E.; Zimmerman, M.R. Trauma among the Shanidar Neandertals. *Am. J. Phys. Anthropol.* **1982**, *57*, 61–76. [CrossRef] [PubMed]
18. Leroi-Gourhan, A. The flowers found with Shanidar IV, a Neanderthal burial in Iraq. *Science* **1975**, *190*, 562–564. [CrossRef]
19. Solecki, R.S. The implications of the Shanidar cave Neanderthal flower burial. *Ann. N. Y. Acad. Sci.* **1977**, *293*, 114–124. [CrossRef]
20. Sommer, J.D. The Shanidar IV 'flower burial': A re-evaluation of Neanderthal burial ritual. *Camb. Archaeol. J.* **1999**, *9*, 127–129. [CrossRef]
21. Perkins, D. Prehistoric fauna from Shanidar, Iraq. *Science* **1964**, *144*, 1565–1566. [CrossRef]
22. Evins, M. A Study of the Fauna from the Mousterian Deposits at Shanidar Cave, Northeastern Iraq. Master's Thesis, University of Chicago, Chicago, IL, USA, 1981.
23. Evins, M. The fauna from Shanidar Cave: Mousterian wild goat exploitation in Northeastern Iraq. *Paléorient* **1982**, *8*, 37–58. [CrossRef]
24. Reynolds, T.; Boismier, W.; Farr, L.; Abdulmutalb, D.; Barker, G. New investigations at Shanidar Cave, Iraqi Kurdistan. In *The Archaeology of the Kurdistan Region of Iraq and Adjacent Regions*; Kopanias, K., MacGinnis, J., Eds.; Archaeopress: Oxford, UK, 2016; pp. 369–372. ISBN 978-1-78491-394-6.
25. Hunt, C.O.; Hill, E.A.; Reynolds, T.; Abdulmutalb, D.; Farr, L.; Lane, R.; Szabó, K.; Barker, G. An incised shell object from Baradostian (Early Upper Palaeolithic) layers in Shanidar Cave, Iraqi Kurdistan. *J. Archaeol. Sci. Rep.* **2017**, *14*, 318–322. [CrossRef]
26. Pomeroy, E.; Mirazón Lahr, M.; Crivellaro, F.; Farr, L.; Reynolds, T.; Hunt, C.O.; Barker, G. Newly discovered Neanderthal remains from Shanidar Cave, Iraqi Kurdistan, and their attribution to Shanidar 5. *J. Hum. Evol.* **2017**, *111*, 102–118. [CrossRef]
27. Pomeroy, E.; Bennett, P.; Hunt, C.O.; Reynolds, T.; Farr, L.; Frouin, M.; Holman, J.; Lane, R.; French, C.; Barker, G. New Neanderthal remains associated with the 'flower burial' at Shanidar Cave. *Antiquity* **2020**, *94*, 11–26. [CrossRef]
28. Pomeroy, E.; Hunt, C.O.; Reynolds, T.; Abdulmutalb, D.; Asouti, E.; Bennett, P.; Bosch, M.; Burke, A.; Farr, L.; Foley, R.; et al. Issues of theory and method in the analysis of Paleolithic mortuary behavior: A view from Shanidar Cave. *Evol. Anthropol.* **2020**, *29*, 263–279. [CrossRef]
29. Friedman, I.; Smith, R.L. Part I, the development of the method. *Am. Antiq.* **1960**, *25*, 476–493. [CrossRef]
30. Brothwell, D.R.; Jones, R. The relevance of small mammal studies to archaeology. In *Research Problems in Zooarchaeology*; Routledge: London, UK; Institute of Archaeology: London, UK, 1978; pp. 47–57.
31. Cucchi, T.; Papayianni, K.; Cersoy, S.; Aznar-Cormano, L.; Zazzo, A.; Debruyne, R.; Berthon, R.; Bălășescu, A.; Simmons, A.; Valla, F.; et al. Tracking the Near Eastern origins and European dispersal of the Western house mouse. *Sci. Rep.* **2020**, *10*, 8276. [CrossRef] [PubMed]

32. O'Connor, T. *Animals as Neighbors: The Past and Present of Commensal Species*; Michigan State University Press: East Lansing, MI, USA, 2013.
33. Avery, D.M. Micromammals and paleoenvironmental interpretation in Southern Africa. *Geoarchaeology* **1988**, *3*, 41–52. [CrossRef]
34. Belmaker, M.; Bar-Yosef, O.; Belfer-Cohen, A.; Meshveliani, T.; Jakeli, N. The environment in the Caucasus in the Upper Paleolithic (Late Pleistocene): Evidence from the small mammals from Dzudzuana cave, Georgia. *Quat. Int.* **2016**, *425*, 4–15. [CrossRef]
35. Belmaker, M.; Hovers, E. Ecological change and the extinction of the Levantine Neanderthals: Implications from a diachronic study of micromammals from Amud Cave, Israel. *Quat. Sci. Rev.* **2011**, *30*, 3196–3209. [CrossRef]
36. Maul, L.C.; Smith, K.T.; Barkai, R.; Barash, A.; Karkanas, P.; Shahack-Gross, R.; Gopher, A. Microfaunal remains at Middle Pleistocene Qesem Cave, Israel: Preliminary results on small vertebrates, environment and biostratigraphy. *J. Hum. Evol.* **2011**, *60*, 464–480. [CrossRef] [PubMed]
37. Smith, K.T.; Maul, L.C.; Flemming, F.; Barkai, R.; Gopher, A. The microvertebrates of Qesem Cave: A comparison of the two concentrations. *Quat. Int.* **2016**, *398*, 233–245. [CrossRef]
38. López-González, F.; Grandal-d'Anglade, A.; Vidal-Romaní, J.R. Deciphering bone depositional sequences in caves through the study of manganese coatings. *J. Archaeol. Sci.* **2006**, *33*, 707–717. [CrossRef]
39. Fernández-Jalvo, Y.; Andrews, P.; Sevilla, P.; Requejo, V. Digestion versus abrasion features in rodent bones. *Lethaia* **2014**, *47*, 323–336. [CrossRef]
40. Andrews, P. *Owls, Caves and Fossils*; The University of Chicago Press: Chicago, IL, USA, 1990; ISBN 978-0-565-01118-5.
41. Fernández-Jalvo, Y.; Scott, L.; Andrews, P. Taphonomy in palaeoecological interpretations. *Quat. Sci. Rev.* **2011**, *30*, 1296–1302. [CrossRef]
42. Mayhew, D.F. Avian predators as accumulators of fossil mammal material. *Boreas* **1977**, *6*, 25–31. [CrossRef]
43. Denys, C.; Stoetzel, E.; Andrews, P.; Bailon, S.; Rihane, A.; Huchet, J.B.; Fernández-Jalvo, Y.; Laroulandie, V. Taphonomy of small predators multi-taxa accumulations: Palaeoecological implications. *Hist. Biol.* **2018**, *30*, 868–881. [CrossRef]
44. Fernández-Jalvo, Y.; Andrews, P.; Denys, C.; Sesé, C.; Stoetzel, E.; Marin-Monfort, D.; Pesquero, D. Taphonomy for taxonomists: Implications of predation in small mammal studies. *Quat. Sci. Rev.* **2016**, *139*, 138–157. [CrossRef]
45. Fernández-Jalvo, Y.; Andrews, P. *Atlas of Taphonomic Identifications: 1001+ Images of Fossil and Recent Mammal Bone Modification*; Springer: New York, NY, USA, 2016.
46. Jenkins, E.L. *Unwanted Inhabitants? The Microfauna from Çatalhöyük and Pınarbası*; VDM: Saarbrücken, Germany, 2009. ISBN 978-3-639-11213-9.
47. Marín Arroyo, A.B.; Landete Ruiz, M.D.; Vidal Bernabeu, G.; Seva Román, R.; González Morales, M.R.; Straus, L.G. Archaeological implications of human-derived manganese coatings: A study of blackened bones in El Mirón Cave, Cantabrian Spain. *J. Archaeol. Sci.* **2008**, *35*, 801–813. [CrossRef]
48. Porter, R.; Salim, M.; Ararat, K. A provisional checklist of the birds of Iraq. *Marsh Bull.* **2010**, *5*, 56–95.
49. Bazgir, B.; Ollé, A.; Tumung, L.; Becerra-Valdivia, L.; Douka, K.; Higham, T.; Van Der Made, J.; Picin, A.; Saladié, P.; López-Garciá, J.M.; et al. Understanding the emergence of modern humans and the disappearance of Neanderthals: Insights from Kaldar Cave (Khorramabad Valley, Western Iran). *Sci. Rep.* **2017**, *7*, 43460. [CrossRef]
50. Rey-Rodríguez, I.; López-García, J.-M.; Blain, H.-A.; Stoetzel, E.; Denys, C.; Fernández-García, M.; Tumung, L.; Ollé, A.; Bazgir, B. Exploring the landscape and climatic conditions of Neanderthals and Anatomically Modern Humans in the Middle East: The rodent assemblage from the Late Pleistocene of Kaldar Cave (Khorramabad Valley, Iran). *Quat. Sci. Rev.* **2020**, *236*, 106278. [CrossRef]
51. Rey-Rodríguez, I.; Stoetzel, E.; López-García, J.M.; Denys, C. Implications of modern barn owl pellets analysis for archaeological studies in the Middle East. *J. Archaeol. Sci.* **2019**, *111*, 105029. [CrossRef]
52. Charter, M.; Izhaki, I.; Meyrom, K.; Motro, Y.; Leshem, Y. Diets of barn owls differ in the same agricultural region. *Wilson J. Ornithol.* **2009**, *121*, 378–383. [CrossRef]
53. Williams, J.P. Small Mammal Deposits in Archaeology: A Taphonomic Investigation of Tyto Alba (Barn Owl) Nesting and Roosting Sites. Ph.D. Thesis, University of Sheffield, Sheffield, UK, 2001.

quaternary

MDPI

Article

Unravelling the Taphonomic Stories of Bird Bones from the Middle Pleistocene Layer VIII of Grotte Vaufrey, France

Anna Rufà [1,2,*] and Véronique Laroulandie [2]

1 ICArEHB—Interdsciplinay Center for Archaeology and the Evolution of Human Behaviour, Campus de Gambelas, Universidade do Algarve, 8005-139 Faro, Portugal

2 CNRS, UMR 5199—PACEA, Université de Bordeaux, Bâtiment B8, Allée Geoffroy Saint Hilaire, 33615 Pessac, France; veronique.laroulandie@u-bordeaux.fr

* Correspondence: arufabonache@gmail.com

Abstract: In recent years, several studies have significantly changed our knowledge concerning the use of birds by Neanderthals. However, what remains to be clarified is the geographical and chronological variability of this human behaviour. The present case study provides new information on this topic/debate. The Grotte Vaufrey was discovered during the 1930s and was excavated during different periods. Work carried out by J.-P. Rigaud during the 1980s motivated many multidisciplinary studies in the cave, but accurate studies were not focused on avian remains. In this work, we provide new data on the bird remains from layer VIII (MIS 7), which is the richest among all the sequences and which has an important Mousterian component. Corvids are predominant in the assemblage and are associated with medium-sized birds and small Passeriformes, among others. Most of the remains present modern fractures, which hinder taphonomic interpretation. However, some alterations associated with raptor or mammalian carnivore activities, together with the anatomical representation and age profile, suggest a non-human accumulation of the majority of the bird remains, especially in the case of corvids that naturally died in the cave. However, at least some bones show evidence of anthropic activity, suggesting the occasional use of large- and medium-sized birds by human populations.

Keywords: birds; zooarchaeology; taphonomy; natural origin; carnivores; cut marks; Neanderthal; Middle Paleolithic

1. Introduction

The use of birds by past human communities has been a widely discussed subject during the last two decades, which has contributed to our knowledge of the evolution of past human behaviour. The discoveries published about birds processed by humans in the pre-Upper Palaeolithic period have prompted the development of zooarchaeological research on avian remains. Until the 1990s, faunal analyses mainly focused on large herbivores, which are considered more nutritionally profitable than small prey [1]. Thus, birds, as other small animal remains, had commonly passed unnoticed from accurate taphonomic studies under the general assumption that they were introduced to the archaeological record by non-anthropogenic inputs. Nevertheless, the use of fast-running (e.g., rabbits and hares) or quick-flying prey (birds) by human groups could not only be linked to dietary purposes [2] but also to the non-edible products they can offer—including feathers, furs, tendons, or bones [3–12]. This fact makes birds attractive to humans as they can provide elements to protect, raw materials to produce tools, or other aesthetic or symbolic components related to human culture. Thus, there is increasing awareness of the necessity of identifying accumulator agents of small prey remains in archaeological sites.

Determining the origin of an accumulation is challenging, as different actors could have occupied archaeological sites, mixing the waste of their activities with those performed by humans. Such palimpsests are difficult to tackle as it is not always possible to differentiate events during the excavation process. Many neo-taphonomic studies have been conducted to create a referential corpus that could help distinguish small prey accumulators in archaeological sites, such as those produced by mammalian carnivores or diurnal and nocturnal raptors. In the case of small fast-running prey, the referential framework has been widely developed in southwestern Europe, where these animals are abundant [13–19]. Studies on quick-flying animals have mostly focused on predators such as diurnal and nocturnal birds of prey [20–43]. The modifications that mammalian carnivores can produce on bird accumulations are still barely explored [31,34,38,42,44].

Additionally, it must be taken into account that some accumulations could be partially or totally the consequence of natural accumulations where any predator—neither human nor non-human—intervened. Caves and shelters are places of nesting and roosting for many bird species that could die inside. In a hypothetical scenario in which natural death takes place, complete skeletons are expected to be found in articulation, as they are not modified by predators. Nevertheless, this situation is ideal, as only carcasses from individuals that are quickly covered by sediment can maintain their original position after death [45]. Prior to and during burial, carcasses are exposed to predators who can scavenge on them [46–48], as well as to other natural processes that could alter bone surfaces [24,49–52]. Even after being buried, many postdepositional alterations could modify the original position of the remains, hindering possible interpretations when recovering and studying them [53,54].

To distinguish these natural accumulations from those of predatory origin, different studies were conducted. In 1987, Ericson proposed that the proportion of wing and leg elements could be a guide for identifying the possible anthropogenic origin of an assemblage. He argues that anthropogenic bird accumulations tend to have wing elements underrepresented, while in natural accumulations, the proportion is similar or wings tend to be overrepresented [55]. In line with this argument, some work has already been published suggesting the anthropogenic origin of avian accumulations in the base of anatomical representation [56,57]. However, Ericson's model was criticized and some authors suggested that factors like functional anatomy or the mode of locomotion should be taken into account when assessing an assemblage, as they could influence on bone density and, as a consequence, lead to differential preservation of bones [58–60]. The bone density hypothesis is nowadays only partially accepted; it has been proven that anthropogenic accumulations can present an equal or higher proportion of wing elements. Besides, other cultural factors could influence skeletal part representation [61–63]. Therefore, anatomical representation is no longer used as a unique criterion to determine bone accumulations' origin, and data should be contrasted with other evidence. The presence of taphonomic traits related to predatory activities is currently the most effective way to diagnose a predator's intervention on avian remains. The presence of chewing and gnawing marks, digestive damage, cut marks, or burning are indicators of these actions [64,65]. However, human activities involving birds are not always easy to interpret, as these animals can also be processed without using tools. In order to detect possible anthropogenic activities, some work has been performed to distinguish other distinctive traits that could be attributed to humans, such as the presence of human teeth marks [66], peeling resulting from bone bending [42,67,68] or wrenches and holes related to the disarticulation of the elbow joint [69].

The capacity of prehistoric humans to capture and use bird products is conceived of as an important acquired ability because it proves that they are versatile enough to adapt to different conditions and maximize the resources present in their environment for different purposes [3,70–73]. This is part of what is called "complex behaviour." Thus, there is growing interest in exploring when and why this complex behaviour appeared. Currently, we know that the evidence of bird use by pre-modern human populations has its highest expression during the Late Middle Palaeolithic (MIS 5–3) [3–6,8–10,74–76]. However, the first current data existing on anthropic bird processing goes back to the

Early Pleistocene. The most ancient evidence was documented at the 1.2 Ma level TE9a of the Sima del Elefante site (Spain), where a large bird radius bearing cut marks was identified [77]. At the Dursunlu site (Turkey), a tarsometatarsus of a large-sized bird also had cut marks in the context of 0.9 Ma [78]. Other rare evidence was documented at the Achelo-Yabroudian Lower Palaeolithic of Quesem Cave (Israel), where three bones presented cut marks on their surfaces and a humerus of starling (*Sturnus* sp.) had possible human modifications related to chewing activities. One of the cut-marked bones was a swan (*Cygnus* sp.) carpometacarpus, which also bore peeling on its proximal area [12]. Five bones were also documented at the Middle Pleistocene of the Gran Dolina site (four at unit TD10-1 and one at TD10-2) [79,80]. At the site of Payre (MIS 8–5), bird use was proposed because of the presence of some feather barbules on a stone tool [81], even though the taphonomic study of bird remains does not confirm this hypothesis [82]. In more recent chronologies, anthropogenic activity on Alpine chough (*Pyrrhocorax graculus)* bones was suggested in the "Acheulean cabin" of the Lazaret in France [56], even if no direct evidence on bones was confirmed. At the same site, a rock dove (*Columba livia)* cut-marked humerus was found at UA24 [83].

Apart from this evidence, a remarkable case was documented at Bolomor Cave (Spain), where bird use seemed to occur systematically during the Middle Pleistocene (MIS 9–5e). The processing of bird body parts in this site was confirmed along the sequence (levels VI, XI, XII and XVII), affecting a variety of bird bones from both aquatic and terrestrial species [79,84,85]. During these ancient periods, the procurement of birds seemed to be mostly related to their use as food. Nevertheless, the handling of inedible products possibly involved in symbolically mediated behaviour and aesthetical expressions is documented in some cases [12], in particular at the end and after the MIS 5 [3–10,74–76].

In this line, we would like to explore the bird bone assemblage at Grotte Vaufrey (Dordogne, France), analysing part of the remains recovered from J.-P. Rigaud's excavations.

La Grotte Vaufrey

The Grotte Vaufrey, or Grotte XV, is one of the 22 caves conforming to the karstic system at Falaise du Conte (Cénac-et-Saint-Julien, Dorgodgne, France), a cliff overlooking the Céou River, at its confluence with the Dordogne river (Figure 1). The cave has a large entrance and is composed of a 22-metre-long principal room, NE oriented, which joins with a little chamber of 4 × 3 metres. Raymond Vaufrey discovered the cave in 1930 and conducted the first archaeological excavations in the site. Two test pits were performed during this period, confirming the archaeological potential of the site. However, the works were dropped out. The Direction des Antiquités Préhistoriques closed the entrance of the cave to protect it from destruction and plundering suffered since the end of Vaufrey's excavation. In 1969, J.-P. Rigaud restarted the works in the cave, which lasted until 1982 and uncovered fifteen archaeo-paleontological layers (I–XV). After this work, Texier refined the stratigraphy by identifying five lithostratigraphic units that included the previously stated layers. From them, twelve archaeological layers remained (I–XII), while the others only had a paleontological component [86,87].

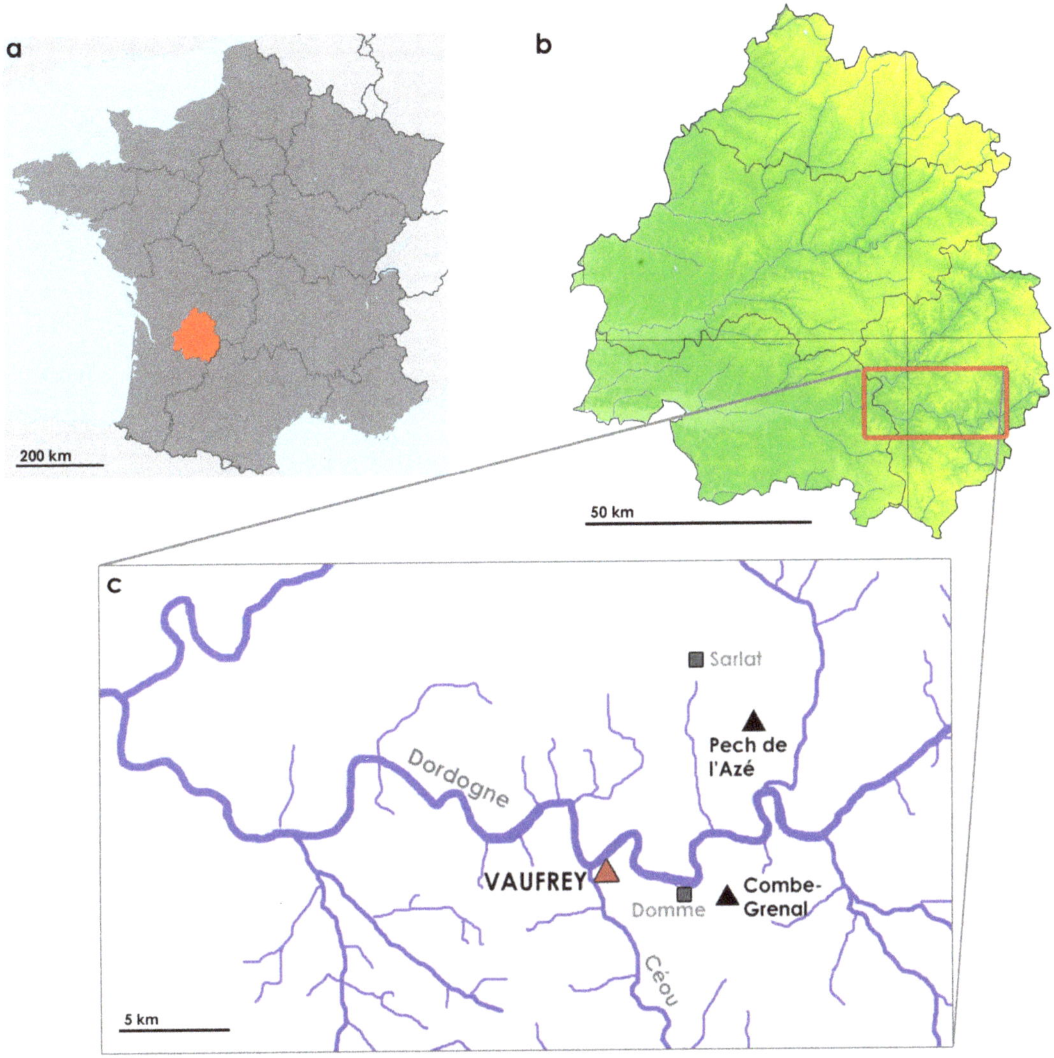

Figure 1. Location of Grotte Vaufrey (or Grotte XV) within Dordogne area. (**a**) Map of France with the location of the Dordogne region marked in red; (**b**) Dordogne region with the specific area of interest along the Dordogne river; (**c**) location of Grotte Vaufrey, at Céou and Dordogne confluence.

Considering chronology, Vaufrey's sequence has commonly provided inconsistent results [86–91]. The first dating was performed using the U-Th technique on speleothems, providing a chronological range from 246 ± 76 ka BP (layer X) to 74 ± 18 ka BP (layer I). However, the extremely low content of uranium and the high presence of detrital impurity only delivered some estimations [86,91]. Delpech, using biostratigraphy, and Texier, based on the presence of cryosols and radiometric data, tried to estimate the chronology of different strata, situating the formation of unit I during MIS 4. The chronology for the lowermost layers was more inconsistent, proposing variability in dating between MIS 7 and MIS 11 for layers IX to XI [86,88,91]. New TT-OSL dating performed in the cave by Hernandez and colleagues suggests that the sequence spreads from MIS 14 (545 ± 47 ka BP) at layer XIII to a minimum age of 109 ± 7 ka BP (layer II) [89]. No new absolute dates were made for layer I, as it was attributed to MIS 4, based on paleontological data, which is consistent with the environmental data available [89,91]. Considering cultural assignations, layer I contains the Quina Mousterian industry. Layers from II–VIII are assigned to the Typical Mousterian, with a significant Levallois component. In layers IX to XII, Levallois products are limited or absent, and an increasing number of tools are typically assigned to the Acheulean [86]. Faunal remains show a wide diversity of species all along the sequence, where ungulates stand out: red deer (*Cervus elaphus*), roe deer (*Capreolus capreolus*), reindeer (*Rangifer tarandus*), horse (*Equus ferus*), aurochs (*Bos primigenius*), steppe bison (*Bison priscus*), Bonal tar (*Hemitragus* cf. *bonali*), chamois (*Rupicapra rupicapra*), and wild boar (*Sus scrofa*), among others. Leporid remains are abundant and comprise an important component of the faunal assemblage [92]. Carnivores are also present: dhole (*Cuon* sp.), fox (*Vulpes/Alopex*), wolf (*Canis lupus*), cave lynx (*Lynx spelaea*), bear (*Ursus* sp.), and the European badger (*Meles meles*).

During Rigaud's excavations, many multidisciplinary studies were performed in the cave. As a result, a monograph dedicated to the cave was published in 1988 [86]. According to it, layer VIII is the richest among all the archaeo-stratigraphic sequence. Its formation took place during a temperate and humid period, probably from MIS 7 [89]. This layer preserves abundant Mousterian lithic industry from different raw materials that were collected in a radius of 50 km. The faunal assemblage is mainly composed of ungulates, where red deer stands out. Tahr, horse, chamois, and roe deer are present to a lesser extent. Rabbits (*Oryctolagus cuniculus*) are numerous and are probably accumulated by the action of the dhole [92]. Several carnivore remains were also discovered belonging to dhole, fox, wolf, lynx, bear, and badger [86,93]. First analyses carried out by Binford on the ungulate remains from the cave proposed that human activity in layer VIII was mainly related to scavenging activities [86]. This hypothesis was later criticized and rejected by Grayson and Delpech, who proposed that these animals were hunted and that the subsistence strategies observed from faunal remains pointed towards hominid–carnivore alternation, where humans performed short but repeated occupations [93]. In addition to mammal remains, fish and bird remains were recovered during the excavation [42,94]. In the case of the birds, their study was first assigned to P. Vilette for a paleontological determination. However, this work was not finished. After him, one of us (VL) reviewed the corvid remains from unit VIII, taking them as an example of the natural contribution of birds in archaeological assemblages [42,95]. No further studies have been conducted on the bird assemblage until now. Thus, the purpose of this study was to analyse bird bones from layer VIII (those previously studied by VL and those from other bird species) from a taphonomic point of view. Different objectives were pursued: (a) to find out other possible accumulations of birds in the site to see the different taphonomic stories that could have happened in the cave and (b) to search for traces of human activities so that they could help to unveil the role of humans in this assemblage. These data will provide information on the occupation dynamics of layer VIII and will help to understand the role of small animals, such as birds, in archaeological accumulations.

2. Materials and Methods

Bird remains from Grotte Vaufrey were reviewed in December 2019 during a stay at Musée National de Préhistoire (MNP) at Les Eyzies-de-Tayac, where the material is currently housed. The whole bird bone assemblage (from layers I to XI) was analysed and identified anatomically and taxonomically at the family, genus, or species level by using both osteological atlases [96–99] and the osteological bird collection from the PACEA (De la Préhistoire à l'Actuel: Culture, Environnement et Anthropologie) laboratory, Université de Bordeaux. No bird remains underneath layer XI were documented. When detailed determination was not possible, the bones were classified by size and followed the same criteria as mentioned below. The degree of ossification of cortical tissue was used to determine immature individuals [54]. The sex determination was occasionally possible by the presence of medullary bone in some long bones. The medullary bone is a granular dense tissue of calcium formed within the bones during the laying period [54,100,101]. The remaining bones were quantified according to the number of identified specimens (NISP), the minimum number of elements (MNE), and the minimum number of individuals (MNI).

At this point, further analyses were performed only for the layer VIII collection, as it was the richest in the number of remains to obtain detailed results. The percentage of relative abundance (%RA) was computed to assess possible biases in the assemblage [102]. This formula was applied by dividing the MNE found for each bone category by the total number of elements expected to be found, based on the MNI represented, and is expressed as a percentage. After the initial identification, the remains were classified, taking into account six larger categories according to the bird's weight [40]: size 1 (<50 g); size 2 (51–150 g); size 3 (151–250 g); size 4 (251–500 g); size 5 (501–1500 g); and size 6 (>1501 g). This allowed for the calculation of different indexes with more precision. The wing-to-leg ratio and the proximal-to-distal element ratio were calculated for the principal size categories and for the medium-sized (msz) Corvidae group to assess possible differences in anatomical representation. Medium-sized corvids were the only taxa considered in addition with its size category, as other taxa were not representative enough to establish reliable statistics. The wing-to-leg ratio was calculated by dividing wing remains (humeri, ulnae, and carpometacarpi) by the sum of the wing and leg specimens (femora, tibiotarsi, and tarsometatarsi) [55]. The proximal-to-distal element ratio was calculated by dividing the proximal bones (scapulae, coracoides, humeri, femora, and tibiotarsi) by the sum of the proximal and distal bones (ulnae, radii, carpometacarpi, and tarsometatarsi). Additionally, the proximal and distal fragment ratio was estimated for long bones from size 4, as the number of remains was large enough to compute it. This ratio is obtained by dividing the total proximal parts (whole bones and proximal ends) by the sum of proximal and distal parts (whole bones and distal ends) [30]. These ratios are expressed as a percentage and allowed us to assess possible biases in anatomical representation. Bone fractures were classified by green, dry, and modern fractures. Green fractures are usually distinguished by the presence of oblique angles and smooth edges, while dry and modern fractures usually have rough and irregular edges. In addition, modern fractures can be identified because the edges do not present other post-depositional modifications at the fracture points, and the fractures usually preserve the internal colouration of the bone [42,103].

All layer-VIII bird bone surfaces were analysed using an Euromex stereomicroscope (Nexius Zoom NZ 1902-P) with magnification up to 45×. Bones were observed to search for any kind of modification related to the different taphonomic agents involved in their accumulation. Mechanical modifications produced by mammalian carnivores or raptors (tooth marks/beak and claw marks) were observed in the form of pits and punctures [64,104], and their location (proximal/distal end, shaft) and distribution (isolated, concentrated, unilateral, bilateral) were considered. Chemical alterations produced by the ingestion of the bones were documented, distinguishing among the five degrees of corrosion stated by Andrews [105]: 0—no corrosion; 1—light; 2—moderate; 3—strong; and 4—extreme. The distribution of damage along the bone was also considered (proximal/distal end or fracture edge). On the other hand, cut marks have been documented as anthropic modifications on bones [106,107]. The distribution (isolated, concentrated), orientation (oblique, longitudinal, transversal), morphology (straight, curved), and dimensions were registered. In addition, the degrees of colouration were documented for burned bones from 0 (no burned, no specific colouration) to 5 (calcined, white) [65]. A small selection of remains was borrowed from the MNP to perform further analyses using an environmental scanning electron microscope (ESEM) Zeiss EVO operated in a low vacuum and with magnification up to 100×.

3. Results

3.1. Spectrum and Population Structure

More than two thousand specimens, organised among eleven different layers, compose the bird assemblage of Grotte Vaufrey (Table A1). Layer VIII is the most representative of the number of remains, with 1181 bones (55.5% of the total assemblage). The assemblage is composed of at least fifteen bird species, where corvids are the most abundant (Table 1). The Corvidae family is represented by at least two different species (*Corvus corone/frugilegus* and *Pyrrhocorax graculus*), which together represent 59.4% of the total MNI of the assemblage (MNI = 60). Most of the corvid remains were attributed to Alpine chough (*P. graculus*). Other elements could not be assigned to a specific corvid species because they belong to young individuals or because distinctive characters are absent on the bones. Thus, firstly, the birds were classified separately as "medium-sized Corvidae." However, some of these remains complement the absences perceived for the Alpine chough group and have a similar size. Therefore, they could belong to the same group. For this reason, we assumed that part of the non-assigned corvid remains were choughs, and they were finally included within the same category. To avoid further confusion in this field, all corvid remains that have the same size and morphology as the Alpine chough (this species included) were classified as "medium-sized Corvidae."

The medium-sized Corvidae group comprised 57.4% of the total MNI (n = 58), followed by other medium and small Passeriformes (8.9% and 6.9% of the total individuals, respectively). Seven individuals are partridges (*Perdix* sp.) that could be assigned to the species *P. perdix* or *P. palaeoperdix*. Six more individuals were classified as pigeons of the species *Columba livia/oenas* (6% of the MNI). The remaining species are only represented by one or two MNI. One medium-sized Charadriiformes was included in the individuals counted, as it could not be included in any other species (it is larger than the snipe—*Gallinago* sp.—identified in the same layer). The grouse family was documented by three identified elements. One of them corresponded to the rock ptarmigan (*Lagopus muta*), while the others were included in the *Lagopus* sp. group because non-distinctive criteria were detected. Nevertheless, *Lagopus* sp. remains were counted as another different individual, as their size does not fit with the *L. muta*, and they would belong to another *Lagopus* individual.

If only the size of the specimens is considered, the size 4 category is the most important in MNI, representing nearly three-quarters of the total assemblage (Table 2). Within this category, 17.8% are immature individuals, but nearly all belong to medium-sized corvids (n = 15; 28.8% of medium-sized corvid remains). Five long bones from the medium-sized Corvidae group also contained medullary bone, indicating the presence of females who died during the laying period.

Table 1. NISP and MNE present at layer VIII of Grotte Vaufrey. MNE indicated inside the brackets. MNI indicated in the last column, specifying the number of immature (im) individuals. Bek: beak; man: mandible; ste: sternum; fur: furcula; cor: coracoid; ver: vertebra; sca: scapula; syn: synsacrum; hum: humerus; rad: radius; uln: ulna; cmc: carpometacarpus; whp: wing phalanx; fem: femur; tib: tibiotarsus; tmt: tarsometatarsus; pph: posterior phalanx (talon excluded); tal: talon. Lsz: large size; msz: medium size; ssz: small size.

NISP (MNE)	bek	man	ste	fur	cor	ver	sca	syn	hum	rad	uln	cmc	wph	fem	tib	tmt	pph	tal	Total	MNI
Anas crecca																1 (1)			1 (1)	1
Tadorna tadorna							1 (1)					1 (1)			2 (2)				4 (4)	2
Anatinae																2 (2)			2 (2)	-
Perdix perdix/palaeoperdix					10 (10)		3 (3)		3 (3)	1 (1)	2 (2)	3 (3)			1 (1)	4 (4)			27 (27)	6 + 1 im = 7
Lagopus muta																1 (1)			1 (1)	1
Lagopus sp.					1 (1)							1 (1)							2 (2)	1
Phasianidae					2 (2)						1 (1)	1 (1)			3 (3)				7 (7)	-
Columba livia/oenas			1 (1)		3 (3)		1 (1)		5 (5)	3 (3)	2 (2)	5 (5)	2 (2)		1 (1)	1 (1)			24 (24)	4 + 2 im = 6
Gallinago sp.					1 (1)														1 (1)	1
Charadriiformes msz					1 (1)				2 (2)										3 (3)	1
Charadriiformes ssz									1 (1)			1 (1)							2 (2)	-
Aquila chrysaetos																	2 (1)		1 (1)	1
Buteo sp.																	1 (1)		1 (1)	1
Accipitridae														1 (1)			1 (1)	2 (2)	4 (4)	-
Accipitridae/Strigidae								1 (1)									1 (1)		2 (2)	-
Falco tinnunculus												1 (1)				2 (2)			3 (3)	2
Bubo bubo																	2 (2)	1 (1)	3 (3)	1
Strigidae												1 (1)							1 (1)	-
Corvus corone/frugilegus											3 (3)	2 (2)		1 (1)					6 (6)	2
Corvidae msz					74 (65)		45 (42)		106 (51)	18 (12)	132 (70)	135 (96)	37 (37)	52 (34)	95 (63)	71 (43)			765 (513)	43 + 15 im = 58
Passeriformes msz					2 (2)				21 (12)	1 (1)	8 (8)	2 (2)		1 (1)	1 (1)	3 (3)			39 (30)	9
Passeriformes ssz					4 (4)				6 (5)		14 (10)	12 (12)		3 (2)	4 (3)	10 (9)			53 (45)	7
Aves lsz													1 (1)			3 (3)			4 (4)	-
Aves msz	2 (2)	2 (2)	3 (3)	4 (3)	4 (3)	9 (9)		6 (6)	5 (2)	1 (1)	4 (4)	1 (1)	2 (2)	1 (1)	13 (8)	6 (3)	134 (132)	17 (17)	214 (199)	-
Total	2 (2)	2 (2)	4 (4)	4 (3)	107 (93)	9 (9)	50 (47)	7 (7)	151 (81)	24 (18)	167 (100)	167 (127)	42 (42)	59 (40)	120 (82)	105 (72)	141 (138)	20 (20)	1181 (886)	101

3.2. Anatomical Representation and Completeness of the Assemblage

In general, long bones are the most rendered. Ulnae and carpometacarpi are the best represented in categories from size 1 to size 5, presenting remarkable values in size 1 (%RA 71.4 and 85.7, respectively) and size 4 (58.1% and 66.9%, respectively). Tarsometatarsi are well documented in size 1 (66.2%), while in size 4, ulnae and carpometacarpi are followed by humeri (49.3%), tibiotarsi (48.6%), and coracoides (48%). Tarsometatarsi are also well represented in this latter category, but with values under 38% (Figure 2). In other categories (sizes 2, 3, 5, and 6), the values of relative abundance are low. However, it should be taken into account that these groups have few specimens, which makes the percentages difficult to interpret. Considering the wing-to-leg ratio, wing elements dominate, and the values calculated for the main categories (size 1: 65.3%; size 2: 86.8%; size 4: 62%; Corvidae msz: 61.1%) differed significantly from the expected value, based on the MNI, except for size 1 (Table 3). As for the proximal to distal elements ratio, the calculated values (size 1: 32.1%; size 2: 64.3%; size 4: 52.4%; Corvidae msz: 51.1%) are not different from the expected ratio except for size 1, where distal remains are significantly more numerous than proximal ones.

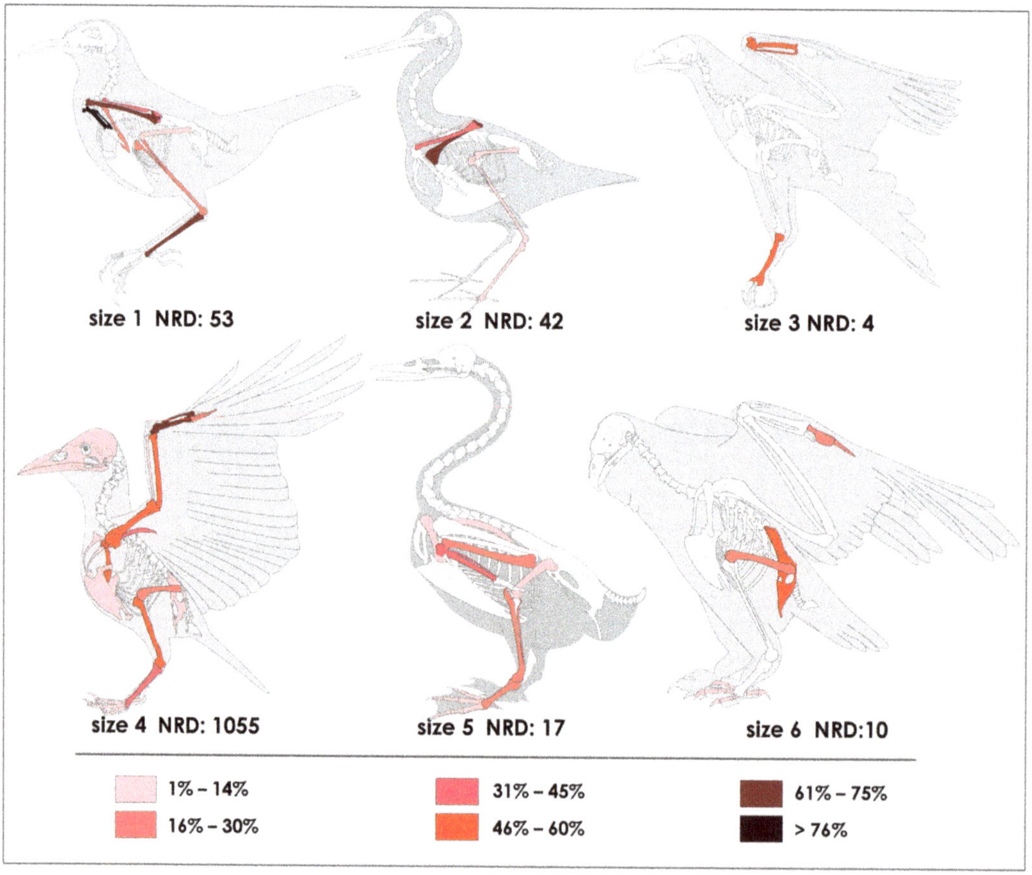

Figure 2. Percentage of relative abundance of bird remains at Grotte Vaufrey layer VIII classified by size. NRD: number of remains determined. See Materials and Methods for "size" definition.

The degree of fragmentation in layer VIII is high (79.9% of the remains), which means that only 237 of the total 1181 remains were complete. This trend is true for all size categories present in the assemblage. Sizes 3 and 6 present higher completeness, but the low number of remains (four and ten, respectively) led us to consider this result with caution. The most complete elements are phalanges—including wing phalanges, posterior phalanges, and talons. In size 4, which is the most representative in terms of the number of remains, wing phalanges raised 63.4% of complete elements, while pedal phalanges and talons had values of 83.6% and 84.2% of completeness, respectively. The small dimensions, morphology, and robustness of these bones probably facilitate its preservation.

Nevertheless, the values of fragmentation obtained at Grotte Vaufrey layer VIII were difficult to assess, as most of the fractures observed on bones are modern fractures, caused during the process of excavation or later. This was particularly evident when observing fragmentation on long bones; 81.5% of the total fracture edges revised present modern fractures, and they could not be analysed from a zooarchaeological point of view. Only 13.8% of the bone fractures were produced when the bone was still fresh. The low proportion of ancient fractures hinders the calculation of significant fracture indexes.

The calculation of proximal ends with respect to the distal ends of long bones show that, in general terms, the proximal and distal ends are both well preserved. Only in the case of tibiotarsi, there is a clear predominance of distal ends (27.6%). In some cases, there is a slight preponderance of proximal (e.g., omal coracoides, with values of 61.9%) or distal ends (e.g., radii—40%, and tarsometatarsi—41.6%).

3.3. *Bone Surface Modifications*

3.3.1. Traces of Non-Human Activity

Only 4.4% of the bones present evidence of alteration (n = 52) resulting from non-human predators. Most of the damage is from digestion, and only five elements present mechanical modifications. The latter occur on a humerus of a medium Passeriformes, a carpometacarpus of a common kestrel (*Falco tinnunculus*), and three elements of medium-sized corvids (Table 2). Teeth/beak and claw mechanical damage (pits, punctures, or beak/claw marks) are isolated and unilateral (Figure 3). The only exception is a corvid tibiotarsus, which has five pits concentrated on its proximal end at the anterior face.

Table 2. Predatory-induced damage on bones at layer VIII of Grotte Vaufrey and its final percentage of representation.

		Pits/Punctures	Digestion	Cut Marks	Burning
Size 1	Passeriformes ssz		1		
Size 2	Passeriformes msz	1	2		
Size 3	*Falco tinnunculus*	1			
Size 4	*Perdix perdix/palaeoperdix*		5		
	Phasianidae		1		
	Accipitridae		1		
	Corvidae msz	3	21		3
	Aves msz		10		
Size 5	*Tadorna tadorna*		1		
	Corvus corone/frugilegus		1		
	Aves lsz		1		
Size 6	*Aquila chrysaetos*		1		
	Accipitridae		1	1	
	Bubo bubo		1		
Total		5 (0.4%)	47 (4%)	1 (0.1%)	3 (0.3%)

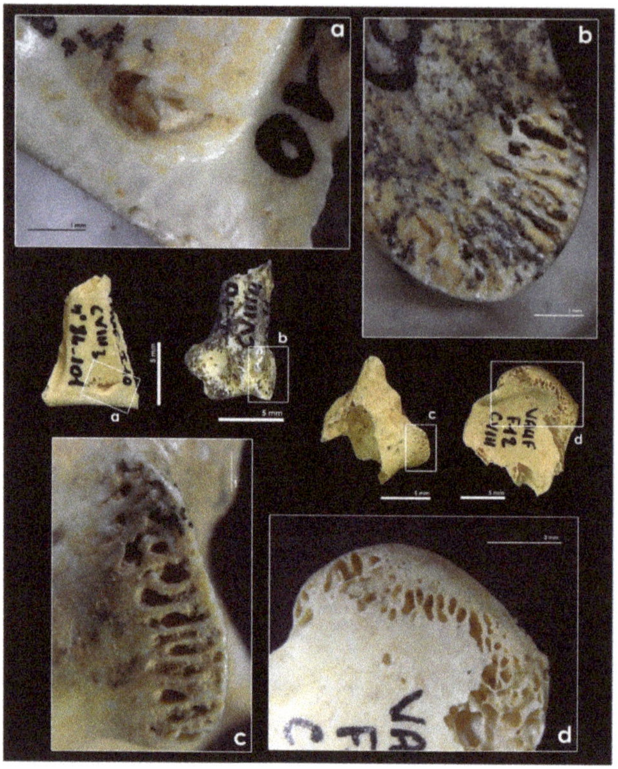

Figure 3. Non-human alterations observed at Grotte Vaufrey layer VIII: pit (**a**) and digestive damage (**b–d**). (**a**) Left sternal coracoid of *P. graculus*. (**b**) Left distal medium-sized *Corvidae ulna*. (**c**) Right proximal ulna of *C. corone/frugilegus*. (**d**) Left humerus of *P. graculus*.

Digestive damage is attested in all size categories, except for size 3, and the number of affected remains is low (4% of altered remains). Size 4 is the most representative, affecting different skeletal elements from different species: medium-sized corvids, phasianids, and an Accipitridae. Alterations were slight (grade 1) on most of the digested remains (85.1%). Only seven remains (14.9%) have moderate damage (grade 2), and no strong or extreme degrees were documented. Digestion mainly affects the proximal and distal ends of long bones (Table 3), while it is less extended at fracture edges—only detected on two coracoides and a radius.

It is important to remark that, although there are few digested remains, if we look at the species affected, the remains of corvids—which are nesting species in caves and shelters—have a low percentage of bones digested, representing 2.7% of the total remains of medium-sized corvids. On the other hand, if we focus on the non-habitual inhabitants of these areas (including Phasianidae, Anatidae, and Charadriiformes), we see that the percentage of digested elements is much higher, reaching values of 14%. A statistical Z-test (Z-value = 4.233) was applied between these two groups, providing a p-value < 0.001 and confirming that the difference is significant between them.

Table 3. Digestion grades observed at layer VIII of Grotte Vaufrey by skeletal elements. cor: coracoid; hum: humerus; rad: radius; uln: ulna; cmc: carpometacarpus; fem: femur; tib: tibiotarsus; tmt: tarsometatarsus; pph: posterior phalanx (talon excluded); tal: talon.

	Element	Grade 1	Grade 2	Total Digested
Size 1	cmc	1		1
Size 2	hum	2		2
Size 4	cor	3		3
	hum	2		2
	uln	5	2	7
	rad	2		2
	fem	4	2	6
	tib	5		5
	tmt	4		4
	pph	7	1	8
	tal	1		1
Size 5	tib	1		1
	tmt	1		1
	uln	1		1
Size 6	pph	1	2	3
Total		40	7	47

3.3.2. Traces of Human Activity

Anthropogenic activity was documented on four avian remains (0.3% of the assemblage). Three posterior elements of medium-sized corvids present burning damage. They correspond to a femur and two tibiotarsi, which have homogeneous brown and black colourations over their entire surface, a consequence of fire exposure.

Apart from these, a group of cut marks was identified on a shaft fragment of the femur of a large diurnal raptor (Figures 4 and 5). As it is a small fragment, it was not possible to make a taxonomical determination. However, the curvature of the bone and the positioning of the anterior intermuscular line—straight with respect to the longitudinal axis of the bone—indicate that it is a taxon belonging to the Accipitridae family. No other bird of similar size has these characteristics. This fragment, broken into two parts, probably during the excavation process, is 3.3 cm long and 1.6 cm wide. The fragment could not be joined together because of recent damage.

Two groups of three and five incisions can be observed on the largest portion of the fragment. The first group is composed of deep parallel striae and seems to be the result of the same action. Next to them, another group of five incisions is present. The deepest incision in this group slightly overlapped with one of the previous incisions described and could be interpreted as an extension of the previous mark. However, this group of striations was more oblique than the first group and possibly indicates another gesture.

The second and smallest portion of the fragment also has two short and parallel cut marks on its surface. They are deep and oblique. This notwithstanding, the fragment was marked after the excavation with its reference number. Even if the reference is removed, the marks are more difficult to observe.

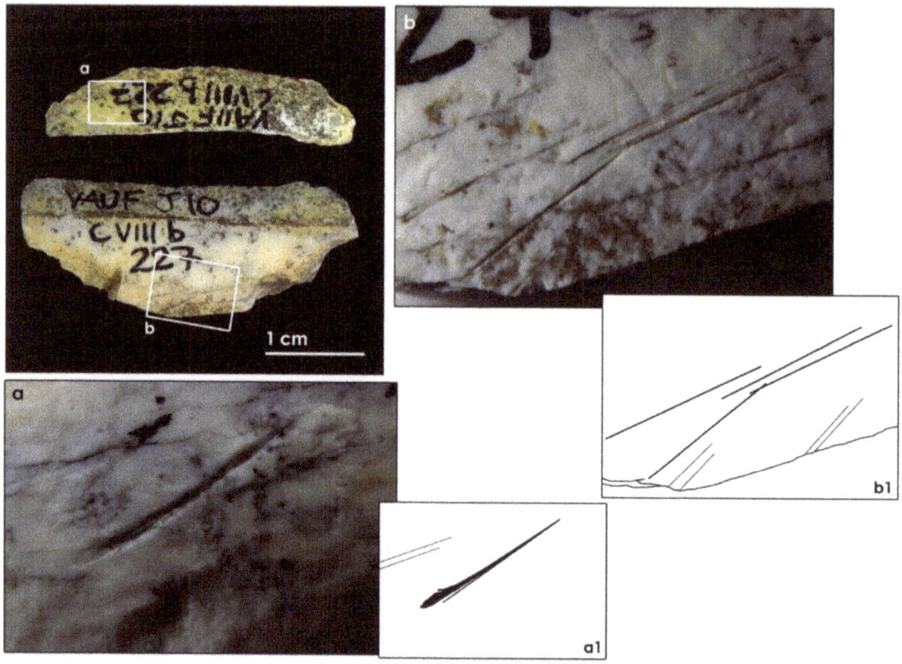

Figure 4. Photo of the large raptor femur shaft with the location (**a**,**b**) and the scheme (**a1**,**b1**) of the cut marks.

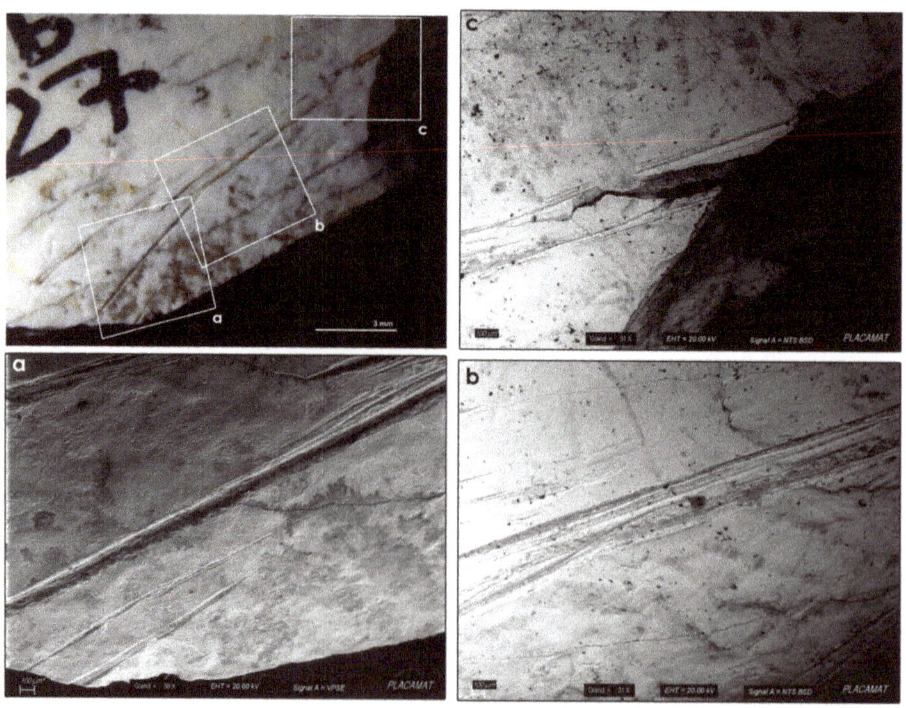

Figure 5. Detail of the cut marks observed on the large raptor femur (**a**–**c**).

4. Discussion

In archaeological sites, many inputs can be involved in the accumulation of bone remains. Determining their origin is sometimes an arduous task, but it allows us to understand the dynamics of occupation, as well as the functionality of these places [104, 108]. Similarly, the presence of some agents at a site does not necessarily imply that they were the accumulators of the assemblage as a whole, and many factors must be taken into account when assessing the possible causes of accumulation. The bird bone assemblage recovered from Grotte Vaufrey layer VIII provides new data on a complex taphonomic story at the site. This accumulation has a mixed origin in which the natural dynamics of accumulation themselves play a prominent role.

First, the results of the present study show that medium-sized corvids are the predominant category in Vaufrey's assemblage, where the Alpine chough is the only identified species. In the preliminary study carried out on Alpine chough remains at the site [42,95], it was determined that the origin of their accumulation was due to the natural death and deposition of these birds on the site's soil. These inferences were established by the significant presence of immature individuals, the anatomical representation of the assemblage, and the absence of predator marks on the remains. Our results corroborate previously established hypotheses for corvids while introducing some nuances.

According to several studies, the mortality rate of chicks during the breeding season is high among Alpine choughs [109,110]. Thus, as corvids are frequent inhabitants of caves and shelters, it is not surprising that some of them die in these spaces, including adult individuals but also a high percentage of juveniles. The existence of some corvid remains with medullary bone indicates that some females died at the site during the laying period [54,100], which supports the hypothesis that the cavity was used as a nesting site for this taxa.

It is theoretically expected that remains generated by natural deaths should be found in anatomical connection or associated together. At Vaufrey, there are no data from Rigaud's excavation about the spatial distribution of the bird remains, probably because most of the elements were recovered during the sieving process [86]. Thus, if there were any anatomical associations, they have not been documented. Nevertheless, no important biases in the anatomical representation of the assemblage, except those that can be explained by small size (phalanx, carpals) and weakness (sternum, skull, pelvis), suggest other possible agents intervening in the assemblage. The wing-to-leg ratio indicates a predominance of wing elements among Corvidae. This value support Ericson's hypothesis [55], arguing that an equal amount or predominance of wing bones is often associated with natural accumulations. Moreover, the proximal-to-distal ratio is close to the expected value [30].

Although there is a percentage of corvid bones with possible tooth/beak and claw marks (0.4%) and digestive alterations (2.7%), it is marginal. If the whole accumulation had been generated by a mammalian carnivore or a raptor, we would expect a higher percentage of remains with mechanical modifications (in the case of non-ingested remains) or a higher proportion of digestions (for the ingested ones) [20–44]. In general terms, it is stated that nocturnal birds of prey tend to ingest their prey. As a consequence of this process, an important number of remains present digestive corrosion in light or moderate degrees. They could also have mechanical modifications in percentages from 0.1% to 7.7%, depending on the prey [20–22,39–42]. In the case of diurnal birds of prey, if the prey is ingested, digestive damage is intense or extreme to most of the bones. Mechanical modifications are not frequent and when present they are under 1.2% [22,25,35,36,42]. On the contrary, on non-ingested diurnal raptor waste, digestive damage is absent, while mechanical modifications are frequent, in percentages that can reach up to 35.1%, depending on the species [23,25,26,28,33,35,36,42,43]. Concerning mammalian carnivores, studies have mainly focused on non-ingested remains, where mechanical modifications usually raise values of 48%. The mastication process is intense on long bones, with an abundance of multiple and bilateral pits and punctures. Crenulated edges are common, mainly affecting the bone extremities and causing the loss of part of the bone in these areas. Ingested remains produced by mammalian carnivores are characterized by intense bone corrosion [31,34,38,42,44,47].

We also observed that corvid remains are less affected by non-human predator marks than other species that do not inhabit the karst (including related size class species such as the partridges). This result supports the hypothesis that both groups have different taphonomic stories.

Based on the above, all the arguments mainly point to an accumulation of natural origin, to which an unknown predator might have contributed either by bringing a corvid carcass or by scavenging some of the remains already deposited in the cavity [45,46]. The same explanation could be applied to other possible cave-dwelling bird species present in the assemblage, such as small Passeriformes or pigeons [111,112]. Nevertheless, none of these remains show evidence of predation, which supports this argument.

Second, apart from corvids, other taxa not inhabiting caves and shelters also compose the avifaunal assemblage at Vaufrey layer VIII. These species, such as phasianids, anatids, or Charadriiformes, usually occupy open spaces in grassy and scrubby areas (phasianids) or spaces close to aquatic resources (anatids and Charadriiformes) [113,114]. Therefore, the fact that these taxa are found at Vaufrey indicates that they were introduced into the cavity by an external agent. In this group of prey, mechanical modifications to the remains were not observed. However, the proportion of digested bones is higher than that documented for corvids, and all the corrosions are slight. The small number of damaged remains prevents us from identifying a potential accumulator, but the absence of anthropogenic marks on bones and the digestive damage observed point to a raptor or a small mammalian carnivore.

During the Upper Palaeolithic and even the Late Middle Palaeolithic period, certain medium-sized taxa, such as phasianids and anatids, were recurrent human prey [75,115–121]. However, Vaufrey layer VIII is a clear example that this is not always the case. Our results differ from those obtained at other sites of similar chronologies and thus confirm that the use of birds in ancient chronologies is not systematised. Many factors can influence their exploitation. In fact, the recurrent use of medium-sized birds for this period has only been confirmed in the case of Bolomor Cave [84].

Actually, other birds that are not typically consumed by humans have direct or indirect anthropogenic modifications at Vaufrey layer VIII. A few corvid bones present colourations related to bone exposure to fire. Despite being the most abundant taxon at the layer, there are no other finds with evidence of anthropic activity among them. This suggests that the presence of these burned bones could be accidental, which means it is not linked to direct human activity on the bones themselves. Although no combustion structures have been recognised in layer VIII, Binford pointed in his zooarchaeological analysis to the presence of some burned macrofauna bones as an indicator of the use of fire in the cavity [86]. Therefore, the occasional existence of hearths in the cave cannot be discarded. It would be possible that some bird bones already deposited on the soil of the site had contact with the hearths, thus producing thermoalterations on some remains [65,122,123].

The raptor cut-marked bone recovered at Vaufrey is the only bone with direct evidence of human activity. The fact that this is the bone of a large bird of prey is of remarkable importance, as in Eurasia, there are a few birds of prey found in the Middle Palaeolithic with clear evidence of human processing. The peculiarity of this finding has two noteworthy aspects. The first is the association of the remains with butchery activities linked to meat removal. The second is related to the age of the find itself. Although the use of diurnal raptors during the Middle Palaeolithic period is a known phenomenon, much of the existing evidence is limited to the presence of striae on wing bones or phalanges/talons [3–7], which have little or no meat content. This has been interpreted as the use of large raptors to obtain non-edible products, such as feathers or claws. However, the cut marks observed at Vaufrey are on a femur, a bone from the fleshy area of the leg.

Furthermore, the location of the cut marks, on the shaft, suggests activities related to meat procurement.

Defleshing activities potentially linked to obtaining food on raptor bones are rare and restricted to MIS 3 [5,74,76,124,125], with the exception of Cova Negra (Table 4). In this site, a scapula of lesser kestrel (*Falco naumanni*) has striae on its shaft [76]. At Axlor, a golden eagle (*Aquila chrysaetos*) femur and tibiotarsus have striae in the proximal and distal areas, respectively, which have been attributed to the procurement of meat [74]. At Les Fieux, a golden eagle femur is cut-marked in the proximal and articular areas of the bone [124,125]. Apart from these cases, a kestrel distal humerus also presents longitudinal striae in Le Noisetier [5], but it is difficult to interpret whether they are associated with meat or feather procurement [126]. At unit IV of Gorham's Cave (Gibraltar), a femur of *Gyps* with marks on its proximal end and a tibiotarsus of a black kite (*Milvus migrans*) with cut marks on the shaft were probably used for food. However, some other wing bones of golden eagle and red kite (*Milvus milvus*) in the site bearing cut marks have been interpreted as the possible procurement of feathers [4]. Thus, the Vaufrey femur could be one of the oldest raptor remains associated with the procurement of edible products to date.

Table 4. Sites with evidence of possible raptor use for food. mid (middle), prox (proximal), dist (distal), ant (anterior), post (posterior), lat (lateral), art (articular). hum: humerus; sca: scapula; fem: femur; tib: tibiotarsus. FR: France; ES: Spain; GIB: Gibraltar (United Kingdom).

Site	Layer/Unit	Chronology	Taxon	Element	Striae Location	Reference
Vaufrey, FR	VIII	MIS 7	Accipitridae	fem	mid shaft	Present study
Cova Negra, ES	IIIb	MIS 5-3	*Falco naumanni*	sca	mid shaft	[76]
Axlor, ES	IV V	MIS 3 MIS 3	*Aquila chrysaetos* *Aquila chrysaetos*	fem tib	prox end (post and art) dist end (ant)	[74]
Les Fieux, FR	K	MIS 3	*Aquila chrysaetos*	fem	prox end (post, lat and art)	[125]
Le Noisetier, FR	-	MIS 3	*Falco* sp.	hum	dist end (ant)	[5]
Gorham's, GIB	IV	MIS 3	*Gyps melitensis/fulvus* *Milvus migrans*	fem tib	mid shaft (caudal) prox shaft (lat)	[4]

In this sense, we can ask what could have driven humans to exploit this type of prey. The find did not seem to have occurred during a period of food shortage, as there are other much larger prey at the site—mainly ungulates—with evidence of anthropogenic activity [93]. Likewise, as attested by the numerous remains of corvids and other birds found at the site, other avian resources were available and of potential interest. Taking into account that animals such as corvids and pigeons nest in cave environments, they would have been easily captured, especially during egg-laying and the breeding of chicks, when adults stay in the nest [127,128]. However, none of these taxa appeared to interest humans in Vaufrey.

Large raptors are attractive animals for humans, impressive, and often associated with symbolism and power [129], which could lead to their capture. Regular or occasional scavenging birds—including large Accipitriformes—when feeding, gorge themselves to the point that they are unable to fly after the meal. Humans could have taken advantage of this behavior to capture these birds of prey without too much difficulty [130]. Although this scenario could have occurred at Vaufrey, we have only one undetermined Accipitridae element bearing cut marks, which limits the interpretation scope.

Besides the femur fragment identified, several posterior phalanges of Accipitriformes were identified at the site, some of which show light or moderate digestion. No particular human interest in these anatomical elements is known at Vaufrey, unlike those documented in several more recent Middle Palaeolithic assemblages [5–9,124]. The data obtained demonstrate the necessity of accurate taphonomic analysis and contribute to a better knowledge of past human relationships with birds.

5. Conclusions

The bird assemblage from layer VIII at Vaufrey is a clear example of complex taphonomic story in which there are natural contributions mixed with those of other agents. Corvids are the most represented taxa in the assemblage, and they seem to be mainly deposited in the site naturally, possibly with other species used to inhabit karstic or rocky environments. Predator intervention on these remains is almost non-existent or could be the result of scavenging.

Contrary to corvids, other bird species that are not cave-dwelling inhabitants (e.g., partridges) seem to have been brought to the site by external agents, such as small mammalian carnivores or birds of prey. However, the small number of remains recovered does not allow accurate identification of the predator.

It is noteworthy that despite the scarce presence of anthropic activity on the remains, the raptor bone with cut marks recovered in the cavity is one of the oldest recorded in Europe so far and is not directly associated with possible aesthetic or symbolic purposes. This case in itself demonstrates that the populations inhabiting Europe during MIS 7 were capable of exploiting a wide variety of animal resources. Nevertheless, it also indicates that the use of birds in ancient choronologies is not systematic, and other factors beyond resource availability need to be considered.

Author Contributions: Conceptualization, A.R. and V.L.; methodology, A.R.; software, A.R.; validation, A.R. and V.L.; formal analysis, A.R.; investigation, A.R. and V.L.; resources, A.R.; data curation, A.R. and V.L.; writing—original draft preparation, A.R. and V.L.; writing—review and editing, A.R. and V.L.; visualization, A.R. and V.L.; supervision, A.R. and V.L; project administration, A.R.; funding acquisition, A.R. All authors have read and agreed to the published version of the manuscript.

Funding: This research was funded by by the IdEx University of Bordeaux Investments for the Future Program (ANR No.–10–IDEX-03-02). A. Rufà is a beneficiary of the Individual Call to Scientific Employment Stimulus—3rd Edition promoted by the Portuguese FCT (reference: 2020.00877.CEECIND). She also participates in the Spanish MINECO/FEDER project PID2019-103987GB-C31; the Generalitat de Catalunya projects CLT009/18/00053, CLT009/18/00054 and CLT009/18/00055; and the AGAUR project 2017 SGR 836.

Institutional Review Board Statement: Not applicable.

Informed Consent Statement: Not applicable.

Data Availability Statement: The data presented in this study are fully available in the article and Appendix A.

Acknowledgments: We would like to acknowledge the scientific staff of the MNP—Les Eyzies-de-Tayac for their kind support during the revision of the materials. We also would like to thank Jean-Baptiste Mallye for his comments and suggestions on the Vaufrey fossil collection. We also acknowledge the PLACAMAT laboratory for allowing us to perform the ESEM analyses using their microscope. Finally, we would like to thank the anonymous reviewers for their helpful comments on a previous draft of the manuscript and to the editors for giving us the opportunity to participate in this special volume.

Conflicts of Interest: The authors declare no conflict of interest.

Appendix A

Table A1. Synthetic table with the NISP, MNE, and MNI for each archaeological layer of Grotte Vaufrey. NISP (MNE); * = presence of immatures; # = presence of females (medullary bone). Lsz: large size; msz: medium size; ssz: small size. NRD: number of remains determined. The classification by size has been included considering six categories according to bird's weight (see Section 2). Note that the MNI is very low in all the layers, with the exception of layer VIII.

Layers	Size Category	I	II	III	IV	V	VI	VII	VIII	IX	X	XI	Total
MNI		15	21	11	38	5	13	16	101	10	8	6	159
Anas crecca	4	-	-	-	-	-	-	-	1 (1)	-	-	-	1 (1)
Tadorna tadorna	5	-	-	-	-	-	-	-	4 (4)	-	-	-	4 (4)
Anatinae	4, 5	1 (1) #	-	-	-	-	1 (1)	1 (1) *	2 (2) *	1 (1)	1 (1)	-	7 (7)
Anseriformes	5	-	-	-	-	-	-	-	-	-	1 (1)	-	1 (1)
Coturnix coturnix	2	-	1 (1)	-	-	-	-	-	-	-	-	-	1 (1)
Perdix perdix/palaeoperdix	4	1 (1)	2 (2)	-	6 (6)	1 (1)	1 (1)	-	32 (27) *	1 (1)	2 (2)	-	46 (41)
Lagopus muta	4	-	-	-	-	-	-	-	1 (1)	-	-	-	1 (1)
Lagopus sp.	5	-	-	-	-	-	-	-	2 (2)	-	-	-	2 (2)
Phasianidae	4	2 (2) *	2 (2)	1 (1)	5 (5)	-	3 (3)	2 (2)	8 (7) *	-	-	2 (2)	25 (24)
Columba livia/oenas	4	-	4 (4)	-	34 (29) *	-	3 (3)	5 (5)	28 (24) *	6 (6)	6 (6)	2 (2)	88 (79)
Gallinago sp.	2	-	-	-	-	-	-	-	1 (1)	-	-	-	1 (1)
Charadriiformes msz	4	-	-	-	3 (3)	-	-	-	3 (3)	1 (1)	-	-	7 (7)
Charadriiformes ssz	2	-	-	-	-	-	1 (1)	-	2 (2)	-	-	-	3 (3)
Aquila chrysaetos	6	-	-	-	-	-	-	-	2 (1)	-	-	-	2 (1)
Buteo sp.	5	-	-	-	-	-	-	-	1 (1)	-	-	-	1 (1)
Accipitridae	4, 6	-	6 (6)	-	4 (4)	-	-	4 (4)	4 (4)	1 (1)	-	1 (1)	20 (20)
Accipitridae/Strigidae	6	-	-	-	-	-	-	-	2 (2)	-	-	-	2 (2)
Falco tinnunculus	3	-	-	-	-	-	-	-	3 (3)	-	-	-	3 (3)
Bubo bubo	6	-	-	-	-	-	-	-	3 (3)	-	-	-	3 (3)
Strigidae	3	-	-	-	-	-	-	-	1 (1)	-	-	-	1 (1)
Corvus corone/frugilegus	5	1 (1)	-	-	-	-	-	-	6 (6)	-	-	-	7 (7)
Corvus monedula	4	-	1 (1)	-	1 (1)	-	-	-	-	-	-	-	2 (2)
Pyrrhocorax pyrrhocorax	4	-	-	-	1 (1)	-	-	-	-	1 (1)	-	-	2 (2)
Pyrrhocorax graculus	4	13 (9) *	97 (69) *	33 (29) *	218 (167) *#	20 (17)	78 (58) *	99 (70) *#	-	25 (20) *#	10 (9)	-	946 (738)
Corvidae msz	4	-	3 (3) *	1 (1)	5 (5) *	1 (1)	-	-	765 (513) *#	3 (3) *	-	1 (1)	426 (273)
Passeriformes msz	2	4 (4)	10 (10)	1 (1)	15 (15) *	1 (1)	4 (4)	5 (5)	39 (30)	3 (3)	1 (1)	1 (1)	84 (75)
Passeriformes ssz	1	34 (33)	7 (7)	1 (1)	22 (21)	3 (3)	3 (3)	11 (11)	53 (45)	-	1 (1)	4 (4)	139 (129)
Aves lsz	5, 6	-	1 (1)	-	1 (1)	-	-	-	4 (4)	-	2 (2)	-	8 (8)
Aves msz	4	1 (1)	10 (10) *	-	28 (28)	5 (5)	16 (16)	13 (13)	214 (199)	3 (3)	3 (3)	-	293 (178)
Aves ssz	1	-	-	-	1 (1)	-	-	-	-	-	-	-	1 (1)
NISP (NRD) Total		57 (52)	144 (116)	37 (33)	344 (287)	31 (28)	110 (90)	140 (111)	1181 (886)	45 (40)	27 (26)	11 (11)	2127 (1647)

Table 2. Representation of Grotte Vaufrey layer VIII remains by size. The classification of remains by size has been evaluated considering the bird's weight (see Section 2). NRD: number of remains determined. Bek: beak; man: mandible; ste: sternum; fur: furcula; cor: coracoid; ver: vertebra; sca: scapula; syn: synsacrum; hum: humerus; rad: radius; uln: ulna; cmc: carpometacarpus; whp: wing phalanx; fem: femur; tib: tibiotarsus; tmt: tarsometatarsus; pph: posterior phalanx (talon excluded); tal: talon.

NRD	Size 1 (MNI = 7)	Size 2 (MNI = 10)	Size 3 (MNI = 2)	Size 4 (MNI = 74)	Size 5 (MNI = 6)	Size 6 (MNI = 2)	Total (MNI = 101)	% Size 1	% Size 2	% Size 3	% Size 4	% Size 5	% Size 6
bek				2			2	0.0	0.0	0.0	0.2	0.0	0.0
man				2			2	0.0	0.0	0.0	0.2	0.0	0.0
ste				4			4	0.0	0.0	0.0	0.4	0.0	0.0
fur				4			4	0.0	0.0	0.0	0.4	0.0	0.0
cor	4	3		99	1		107	7.5	7.1	0.0	9.4	5.9	0.0
ver				9			9	0.0	0.0	0.0	0.9	0.0	0.0
sca				49	1		50	0.0	0.0	0.0	4.6	5.9	0.0
syn				6		1	7	0.0	0.0	0.0	0.6	0.0	10.0
hum	6	22		123			151	11.3	52.4	0.0	11.7	0.0	0.0
rad		1		23			24	0.0	2.4	0.0	2.2	0.0	0.0
uln	14	8		142	3		167	26.4	19.0	0.0	13.5	17.6	0.0
cmc	12	3	2	146	4		167	22.6	7.1	50.0	13.8	23.5	0.0
wph				41		1	42	0.0	0.0	0.0	3.9	0.0	10.0
fem	3	1		53	1	1	59	5.7	2.4	0.0	5.0	5.9	10.0
tib	4	1		113	2		120	7.5	2.4	0.0	10.7	11.8	0.0
tmt	10	3	2	86	4		105	18.9	7.1	50.0	8.2	23.5	0.0
pph				134	1	6	141	0.0	0.0	0.0	12.7	5.9	60.0
tal				19		1	20	0.0	0.0	0.0	1.8	0.0	10.0
Total	53	42	4	1055	17	10	1181	4.5	3.6	0.3	89.3	1.4	0.8

Table 3. Wing-to-leg (WL) and proximal-to-distal (PD) bone ratios, considering the total observed and expected remains. The classification by size was evaluated considering the bird's weight (see Section 2). The ratios were only calculated for those categories for which the number of remains was large enough to obtain reliable results. A Z-test was calculated to see if the differences between the observed and expected elements are significant. * Significant. Msz: medium size.

	Observed			Expected (Based on the MNI)				
	Wing	Leg	Ratio WL	Wing	Leg	Ratio WL	Z-Test	*p*-Value
Size 1	32	17	65.30	42	42	50	1.713905298	$0.08 < p < 0.09$
Size 2	33	5	86.8	60	60	50	4.022140631	$p < 0.01$ *
Size 4	411	252	61.9	444	444	50	4.69715692	$p < 0.01$ *
Corvidae msz	373	218	63.1	348	348	50	4.723087674	$p < 0.01$ *
	Proximal	Distal	Ratio PD	Prox	Distal	Ratio PD	Z-Test	*p*-Value
Size 1	17	36	32.1	70	56	55.55	−2.869431268	$p < 0.01$ *
Size 2	27	15	64.3	100	80	55.55	1.02966479	$0.30 < p < 0.31$
Size 4	437	397	52.4	740	592	55.55	−1.435548179	$0.15 < p < 0.16$
Corvidae msz	372	356	51.1	580	464	55.55	−1.851106305	$0.06 < p < 0.07$

References

1. Stiner, M.C.; Munro, N.D.; Surovell, T.A. The tortoise and the hare. Small game use, the broad-spectrum revolution, and Paleolithic demography. *Curr. Anthropol.* **2000**, *41*, 39–73. [CrossRef]
2. Hockett, B.; Haws, J.A. Nutritional ecology and the human demography of Neandertal extinction. *Quat. Int.* **2005**, *137*, 21–34. [CrossRef]
3. Peresani, M.; Fiore, I.; Gala, M.; Romandini, M.; Tagliacozzo, A. Late Neandertals and the intentional removal of feathers as evidenced from bird bone taphonomy at Fumane Cave 44 ky B.P., Italy. *Proc. Natl. Acad. Sci. USA* **2011**, *108*, 3888–3893. [CrossRef]
4. Finlayson, C.; Brown, K.; Blasco, R.; Rosell, J.; Negro, J.J.; Bortolotti, G.R.; Finlayson, G.; Sánchez Marco, A.; Pacheco, F.G.; Rodríguez-Vidal, J.; et al. Birds of a feather: Neanderthal exploitation of raptors and corvids. *PLoS ONE* **2012**, *7*, e45927.

5. Morin, E.; Laroulandie, V. Presumed symbolic use of diurnal raptors by Neanderthals. *PLoS ONE* **2012**, *7*, e32856. [CrossRef] [PubMed]
6. Romandini, M.; Peresani, M.; Laroulandie, V.; Metz, L.; Pastoors, A.; Vaquero, M.; Slimak, L. Convergent evidence of eagle talons used by late Neanderthals in Europe: A further assessment on symbolism. *PLoS ONE* **2014**, *9*, e101278.
7. Radovčić, D.; Sršen, A.O.; Radovčić, J.; Frayer, D.W. Evidence for Neandertal jewelry: Modified white-tailed eagle claws at krapina. *PLoS ONE* **2015**, *10*, e0119802. [CrossRef] [PubMed]
8. Fiore, I.; Gala, M.; Romandini, M.; Cocca, E.; Tagliacozzo, A.; Peresani, M. From feathers to food: Reconstructing the complete exploitation of avifaunal resources by Neanderthals at Fumane cave, unit A9. *Quat. Int.* **2016**, *421*, 134–153. [CrossRef]
9. Majkić, A.; Evans, S.; Stepanchuk, V.; Tsvelykh, A.; d'Errico, F. A decorated raven bone from the Zaskalnaya VI (Kolosovskaya) Neanderthal site, Crimea. *PLoS ONE* **2017**, *12*, e0173435. [CrossRef] [PubMed]
10. Rodríguez-Hidalgo, A.; Morales, J.I.; Cebrià, A.; Courtenay, L.A.; Fernández-Merchena, J.L.; García-Argudo, G.; Marín, J.; Saladié, P.; Soto, M.; Tejero, J.-M.; et al. The châtelperronian Neandertals of Cova Foradada (Calafell, Spain) used Iberian imperial eagle phalanges for symbolic purposes. *Sci. Adv.* **2019**, *5*, eaax1984. [CrossRef] [PubMed]
11. Pelletier, M.; Desclaux, E.; Brugal, J.-P.; Texier, P.-J. The exploitation of rabbits for food and pelts by Last Interglacial Neandertals. *Quat. Sci. Rev.* **2019**, *224*, 105972. [CrossRef]
12. Blasco, R.; Rosell, J.; Sánchez-Marco, A.; Gopher, A.; Barkai, R. Feathers and food: Human-bird interactions at Middle Pleistocene Qesem Cave, Israel. *J. Hum. Evol.* **2019**, *136*, 102653. [CrossRef]
13. Cochard, D. Les Leporidés Dans la Subsistence Paléolithique du Sud de la France. Ph.D. Dissertation, Université de Bordeaux, Bordeaux, France, 2004.
14. Yravedra, J. Implicaciones tafonómicas del consumo del lagomorfos por búho real (*Bubo bubo*) en la interpretación de los yacimientos arqueológicos. *An. Murcia* **2006**, *22*, 33–47.
15. Lloveras, L.; Moreno-García, M.; Nadal, J. Feeding the foxes: An experimental study to assess their taphonomic signature on leporid remains. *Int. J. Osteoarchaeol.* **2012**, *22*, 577–590. [CrossRef]
16. Lloveras, L.; Moreno-García, M.; Nadal, J. Assessing the variability in taphonomic studies of modern leporid remains from Eagle Owl (*Bubo bubo*) nest assemblages: The importance of age of prey. *J. Archaeol. Sci.* **2012**, *39*, 3754–3764. [CrossRef]
17. Sanchis, A. *Los Lagomorfos del Paleolítico en la Vertiente Mediterránea Ibérica. Humanos y Otros Predadores como Aentes de Aporte y Alteración de los Restos Óseos en Yacimientos Arqueológicos*; Serie de Trabajos Varios 115; Museo de Prehistoria de Valencia, Diputación de Valencia: Valencia, Spain, 2012; p. 271.
18. Rodríguez-Hidalgo, A.; Lloveras, L.; Moreno-García, M.; Saladié, P.; Canals, A.; Nadal, J. Feeding behaviour and taphonomic characterization of non-ingested rabbit remains produced by the iberian lynx (*Lynx pardinus*). *J. Archaeol. Sci.* **2013**, *40*, 3031–3045. [CrossRef]
19. Rodríguez-Hidalgo, A.; Saladié, P.; Marín, J.; Canals, A. Expansion of the referential framework for the rabbit fossil accumulations generated by Iberian lynx. *Palaeogeogr. Palaeoclimatol. Palaeoecol.* **2015**, *418*, 1–11. [CrossRef]
20. Bochenski, Z.M.; Tomek, T.; Boev, Z.; Mitev, I. Patterns of bird bone fragmentation in pellets of the Tawny Owl (*Strix aluco*) and the Eagle Owl (*Bubo bubo*) and their taphonomic implications. *Acta Zool. Crac.* **1993**, *36*, 313–328.
21. Bochenski, Z.M.; Tomek, T. Patterns of bird bone fragmentation in pellets of the Long-eared Owl *Asio otus* and its taphonomic implications. *Acta Zool. Crac.* **1994**, *37*, 177–190.
22. Bochenski, Z.M. Preliminary taphonomic studies on damage to bird bones by Snowy Owls *Nyctea scandiaca*, with comments on the survival of bones in palaeontological sites. *Acta Zool. Crac.* **1997**, *40*, 279–292.
23. Bochenski, Z.M.; Korovin, V.A.; Nekrasov, A.E.; Tomek, T. Fragmentation of bird bones in food remains of imperial eagles (*Aquila heliaca*). *Int. J. Osteoarchaeol.* **1997**, *7*, 165–171. [CrossRef]
24. Bochenski, Z.M.; Tomek, T. Preservation of bird bones: Erosion versus digestion by owls. *Int. J. Osteoarchaeol.* **1997**, *7*, 372–387. [CrossRef]
25. Bochenski, Z.M.; Huhtala, K.; Jussila, P.; Pulliainen, E.; Tornberg, R.; Tunkkari, P.S. Damage to bird bones in pellets of Gyrfalcon *Falco rusticolus*. *J. Archaeol. Sci.* **1998**, *25*, 425–433. [CrossRef]
26. Bochenski, Z.M.; Huhtala, K.; Sulkava, S.; Tornberg, R. Fragmentation and preservation of bird bones in food remains of the Golden Eagle Aquila chrysaetos. *Archaeofauna* **1999**, *8*, 31–39.
27. Bochenski, Z.M.; Nekrasov, A.E. The taphonomy of sub-Atlantic bird remains from Bazhukovo III, Ural Mountains, Russia. *Acta Zool. Crac.* **2001**, *44*, 279–292.
28. Laroulandie, V. Damage to Pigeon long bones in pellets of the Eagle Owl *Bubo bubo* and food remains of Peregrine Falcon, *Falco peregrinus*: Zooarchaeological implications. *Acta Zool. Crac.* **2002**, *45*, 331–339.
29. Bochenski, Z.M.; Tornberg, R. Fragmentation and preservation of bird bones in uneaten food remains of the Gyrfalcon *Falco rusticolus*. *J. Archaeol. Sci.* **2003**, *30*, 1665–1671. [CrossRef]
30. Bochenski, Z.M. Owls, diurnal raptors and humans: Signatures on avian bones. In *Biosphere to Lithosphere: New Studies in Vertebrate Taphonomy, Proceedings of the 9th Conference of the International Council of Archaeozoology, Durham, UK, 23–28 August 2002*; O'Connor, T., Ed.; Oxbow Books: Oxford, UK, 2005; pp. 31–45.
31. Mallye, J.B.; Cochard, D.; Laroulandie, V. Accumulations osseuses en périphérie de terriers de petits carnivores: Les stigmates de prédation et de fréquentation. *Ann. Paleontol.* **2008**, *94*, 187–208. [CrossRef]

32. De Cupere, B.; Thys, S.; Van Neer, W.; Ervynck, A.; Corremans, M.; Waelkens, M. Eagle Owl (*Bubo bubo*) pellets from Roman Sagalassos (SW Turkey): Distinguishing the prey remains from nest and roost sites. *Int. J. Osteoarchaeol.* **2009**, *19*, 1–22. [CrossRef]
33. Bochenski, Z.M.; Tomek, T.; Tornberg, R.; Wertz, K. Distinguishing non human predation on birds: Pattern of damage done by the white-tailed eagle *Haliaetus albicilla*, with comments on the punctures made by the golden eagle *Aquila chrysaetos*. *J. Archaeol. Sci.* **2009**, *36*, 122–129. [CrossRef]
34. Monchot, H.; Gendron, D. Les restes d'une tanière de renard en milieu périglaciare (Île Digges, rive sud du détroit d'Hudson, Canada). In *Taphonomie des Petits Vertébrés: Référentiels et Transferts aux Fossiles. Actes de la Table Ronde du RTP Taphonomie, Talence 20–21 Octobre 2009*; BAR International Series 2269; Laroulandie, V., Mallye, J.-B., Denys, C., Eds.; Archaeopress: Oxford, UK, 2011; pp. 65–76.
35. Lloveras, L.; Nadal, J.; Moreno-García, M.; Thomas, R.; Anglada, J.; Baucells, J.; Martorell, C.; Vilasís, D. The role of the Egyptian Vulture (*Neophron percnopterus*) as a bone accumulator in cliff rock shelters: An analysis of modern bone nest assemblages from North-eastern Iberia. *J. Archaeol. Sci.* **2014**, *44*, 76–90. [CrossRef]
36. Lloveras, L.; Thomas, R.; Lourenço, R.; Caro, J.; Dias, A. Understanding the taphonomic signature of Bonelli's Eagle (*Aquila fasciata*). *J. Archaeol. Sci.* **2014**, *49*, 455–471. [CrossRef]
37. Royer, A.; Montuire, S.; Gilg, O.; Laroulandie, V. A taphonomic investigation of small vertebrate accumulations produced by the snowy owl (*Bubo scandiacus*) and its implications for fossil studies. *Palaeogeogr. Palaeoclimatol. Palaeoecol.* **2019**, *514*, 189–205. [CrossRef]
38. Rodríguez-Hidalgo, A.; Saladié, P.; Marín, J.; Canals, A. Bird-bone modifications by Iberian lynx: A taphonomic analysis of non-ingested red-legged partridge remains. *Quat. Int.* **2016**, *421*, 228–238. [CrossRef]
39. Alonso, G.; Rufà, A.; Arilla, M.; Blasco, R. Taphonomic signature of the Eurasian eagle-owl (*Bubo bubo*) on the avian accumulation of Cau del Duc (Lleida, Spain). *Hist. Biol.* **2020**, *32*, 1320–1333. [CrossRef]
40. Rufà, A.; Laroulandie, V. Prey size as a critical factor for bird bone taphonomy in Eagle Owl (*Bubo bubo*) pellets. *Sci. Rep.* **2019**, *9*, 19200. [CrossRef]
41. Lebreton, L.; Bailon, S.; Guillaud, E.; Testu, A.; Perrenoud, C. Multi-taxa referential of a modern Eurasian Eagle-Owl (*Bubo bubo*) aerie. *J. Archaeol. Sci. Rep.* **2020**, *32*, 102417.
42. Laroulandie, V. Taphonomie et archéozoologie des oiseaux en grotte: Applications aux Sites Paléolithiques du Bois-Ragot (Vienne), de Combe Saunière (Dordogne) et de la Vache (Ariège). Ph.D. Dissertation, Université de Bordeaux I, Bordeaux, France, 2000.
43. Wertz, K.; Tornberg, R.; Bochenski, Z.M. The taphonomy of medium-sized grouse in food remains of the northern goshawk *Accipiter gentilis*, compared with damage done by man and other predators. *Int. J. Osteoarchaeol.* **2021**, *31*, 188–195. [CrossRef]
44. Rufà, A.; Laroulandie, V. Carnívoros y carroñeros. Una nueva visión sobre la depredación de los lobos sobre las aves carroñeras y su implicación en el registro arqueológico. *ArkeoGazte* **2020**, *10*, 127–140.
45. Nicholson, R.A. Bone degradation, burial medium and species representation: Debunking the myths, an experiment-based approach. *J. Archaeol. Sci.* **1996**, *23*, 513–533. [CrossRef]
46. Oliver, J.S.; Graham, R.W. A catastrophic kill of ice-trapped coots; time-averaged versus scavenger-specific disarticulation patterns. *Paleobiology* **1994**, *20*, 229–244. [CrossRef]
47. Leoni, F.; Boudadi-Maligne, M.; Mallye, J.-B.; Beauval, C.; Faivre, J.-P. Renards chapardeurs et amas de boucherie en pleine nature. Quantification des modifications et incidence sur les analyses archéozoologiques. In *Relations Hommes—Canidés de La Préhistoire Aux Périodes Modernes*; Mallye, J.-B., Boudadi-Maligne, M., Eds.; Ausonius Editions: Bordeaux, France, 2020; Collection DAN@3; pp. 69–93.
48. Beasley, J.C.; Olson, Z.H.; Devault, T.L. Ecological Role of Vertebrate Scavengers. In *Carrion Ecology, Evolution, and Their Applications*; Benhow, M.E., Tomberlin, J.K., Tarone, A.M., Eds.; USDA National Wildlife Research Center—Staff Publications: Lincoln, NE, USA, 2015; Volume 1745, pp. 107–127.
49. Trapani, J. Hydrodynamic sorting of avian skeletal remains. *J. Archaeol. Sci.* **1998**, *25*, 477–487. [CrossRef]
50. Schiffer, M. *Formation Processes of the Archaeological Record*; University of New Mexico Press: Albuquerque, NM, USA, 1987.
51. Behrensmeyer, A.K.; Stayton, C.T.; Chapman, R.E. Taphonomy and ecology of modern avifaunal remains from Amboseli Park, Kenya. *Paleobiology* **2003**, *29*, 52–70. [CrossRef]
52. Cruz, I. Avian and mammalian bone taphonomy in southern continental Patagonia: A comparative approach. *Quat. Int.* **2008**, *180*, 30–37. [CrossRef]
53. Bochenski, Z.M.; Wertz, K.; Tomek, T. What happens to dead birds left in a cave and its vicinity for a quarter of a century? A taphonomic experiment. *Hist. Biol.* **2018**, *30*, 855–867. [CrossRef]
54. Serjeantson, D. *Birds*; Cambridge University Press: New York, NY, USA, 2009.
55. Ericson, G.P. Interpretations of Archaeological Bird Remains: A Taphponomic Approach. *J. Archaeol. Sci.* **1987**, *14*, 65–75. [CrossRef]
56. Lumley, H. *Une Cabane Acheuléenne Dans la Grotte du Lazaret (Nice)*; Mémoires de la Société Préhistorique Française—CNRS: Paris, France, 1969.
57. Mourer-Chauviré, C. Las aves del sitio de ocupación achelense de Aridos-1 (Arganda, Madrid). In *Ocupaciones Achelenses en el Valle del Jarama (Arganda-Madrid)*; López, N., Pérez, A., Santonja, M., Eds.; Diputación Provincial, Servicios de Extensión Cultural y Divulgación: Madrid, Spain, 1980; pp. 145–160.
58. Livingston, S.D. The taphonomic interpretation of avian skeletal part frequencies. *J. Archaeol. Sci.* **1989**, *16*, 537–547. [CrossRef]

59. Cruz, I. La representación de partes esqueléticas de aves. Patrones naturales e interpretación arqueológica. *Archaeofauna* **2005**, *14*, 69–81.
60. Broughton, J.M.; Mullins, D.; Ekker, T. Avian resource depression or intertaxonomic variation in bone density? A test with San Francisco Bay avifaunas. *J. Archaeol. Sci.* **2007**, *34*, 374–391. [CrossRef]
61. Bovy, K.M. Differential avian skeletal part distribution: Explaining the abundance of wings. *J. Archaeol. Sci.* **2002**, *29*, 965–978. [CrossRef]
62. Bovy, K.M. Why so many wings? A re-examination of avian skeletal part representation in the south-central Northwest Coast, USA. *J. Archaeol. Sci.* **2012**, *39*, 2049–2059. [CrossRef]
63. Lefèvre, C.; Laroulandie, V. Avian skeletal part representation: A case study from offing 2, A Hunter-Gatherer-Fisher site in the Strait of Magellan (Chile). *Int. J. Osteoarchaeol.* **2014**, *24*, 256–264. [CrossRef]
64. Lyman, R.L. *Vertebrate Taphonomy*; Cambridge University Press: New York, NY, USA, 1994.
65. Stiner, M.; Kuhn, S.; Weiner, S.; Bar-Yosef, O. Differential burning, recrystallization, and fragmentation of archaeological bone. *J. Archaeol. Sci.* **1995**, *22*, 223–237. [CrossRef]
66. Landt, M.J. Tooth marks and human consumption: Ethnoarchaeological mastication research among foragers of the Central African Republic. *J. Archaeol. Sci.* **2007**, *34*, 1629–1640. [CrossRef]
67. White, T.D. *Prehistoric Cannibalism at Mancos 5MTUMR-2346*; Princenton Univeristy Press: Princenton, NJ, USA, 1992.
68. Laroulandie, V. Anthropogenic versus non-anthropogenic bird bone assemblages: New criteria for their distinction. In *Biosphere to Lithosphere: New Studies in Vertebrate Taphonomy, Proceedings of the 9th Conference of the International Council of Archaeozoology, Durham, UK, 23–28 August 2002*; O'Connor, T., Ed.; Oxbow Books: Oxford, UK, 2005; pp. 25–30.
69. Laroulandie, V.; Costamagno, S.; Cochard, D.; Mallye, J.-B.; Beauval, C.; Castel, J.C.; Ferrié, J.G.; Gourichon, L.; Rendu, W. Quand désarticuler laisse des traces: Le cas de l'hyperextension du coude. *Ann. Paleontol.* **2008**, *94*, 287–302. [CrossRef]
70. McBrearty, S.; Brooks, A.S. The revolution that wasn't: A new interpretation of the origin of modern human behavior. *J. Hum. Evol.* **2000**, *39*, 453–563. [CrossRef]
71. D'Errico, F. The invisible frontier. A multiple species model for the origin of behavioral modernity. *Evol. Anthropol.* **2003**, *12*, 188–202. [CrossRef]
72. Villa, P.; Roebroeks, W. Neandertal Demise: An Archaeological Analysis of the Modern Human Superiority Complex. *PLoS ONE* **2014**, *9*, e96424. [CrossRef]
73. Fiorenza, L.; Benazzi, S.; Henry, A.G.; Salazar-García, D.C.; Blasco, R.; Picin, A.; Wroe, S.; Kullmer, O. To meat or not to meat? New perspectives on neanderthal ecology. *Am. J. Phys. Anthropol.* **2015**, *156*, 43–71. [CrossRef]
74. Gómez-Olivencia, A.; Sala, N.; Núñez-Lahuerta, C.; Sanchis, A.; Arlegi, M.; Rios-Garaizar, J. First data of Neandertal bird and carnivore exploitation in the Cantabrian Region (Axlor; Barandiaran excavations; Dima, Biscay, Northern Iberian Peninsula). *Sci. Rep.* **2018**, *8*, 10551. [CrossRef]
75. Romero, A.J.; Díez, J.C.; Brugal, J.P. Aves de caza. Estudio tafonómico y zooarqueológico de los restos avianos de los niveles Musterienses de Pié Lombard (Alpes-Maritimes, Francia). *Munibe Antropol.* **2017**, *68*, 73–84. [CrossRef]
76. Martínez Valle, R.; Guillem Calatayud, P.M.; Villaverde Bonilla, V. Bird consumption in the final stage of Cova Negra (Xátiva, Valencia). *Quat. Int.* **2016**, *421*, 85–102. [CrossRef]
77. Huguet, R.; Saladié, P.; Cáceres, I.; Díez, C.; Rosell, J.; Bennàsar, M.; Blasco, R.; Esteban-Nadal, M.; Gabucio, M.J.; Rodríguez-Hidalgo, A.; et al. Successful subsistence strategies of the first humans in south-western Europe. *Quat. Int.* **2013**, *295*, 168–182. [CrossRef]
78. Güleç, E.; White, T.; Kuhn, S.; Özer, I.; Sağr, M.; Yilmaz, H.; Howell, F.C. The lower pleistocene lithic assemblage from Dursunlu (Konya), Central Anatolia, Turkey. *Antiquity* **2009**, *83*, 11–22. [CrossRef]
79. Blasco, R.; Rosell, J.; Fernández Peris, J.; Arsuaga, J.L.; Bermúdez de Castro, J.M.; Carbonell, E. Environmental availability, behavioural diversity and diet: A zooarchaeological approach from the TD10-1 sublevel of Gran Dolina (Sierra de Atapuerca, Burgos, Spain) and Bolomor Cave (Valencia, Spain). *Quat. Sci. Rev.* **2013**, *70*, 124–144. [CrossRef]
80. Rodríguez-Hidalgo, A.; Saladié, P.; Ollé, A.; Arsuaga, J.L.; Bermúdez de Castro, J.M.; Carbonell, E. Human predatory behavior and the social implications of communal hunting based on evidence from the TD10.2 bison bone bed at Gran Dolina (Atapuerca, Spain). *J. Hum. Evol.* **2017**, *105*, 89–122. [CrossRef] [PubMed]
81. Hardy, K.; Moncel, M.H. Neanderthal use of fish, mammals, birds, starchy plants and wood 125–250,000 years ago. *PLoS ONE* **2011**, *6*, e23768. [CrossRef]
82. Rufà, A.; Blasco, R.; Roger, T.; Moncel, M.-H. What is the taphonomic agent responsible for the avian accumulation? An approach from the Middle and early Late Pleistocene assemblages from Payre and Abri des Pêcheurs (Ardèche, France). *Quat. Int.* **2016**, *421*, 46–61. [CrossRef]
83. Roger, T. L'avifaune du Pléistocène Moyen et Supérieur du Bord de la Méditerranée Européenne: Orgnac 3, Lazaret (France), Caverna delle Fate, Arma delle Manie (Italie), Kalamakia (Grèce), Karain E (Turquie). Paléontologie, Taphonomie et Paléoécologie. Ph.D. Dissertation, Muséum National d'Histoire Naturelle, Paris, France, 2004.
84. Blasco, R.; Fernández Peris, J. Middle Pleistocene bird consumption at level XI of Bolomor Cave (Valencia, Spain). *J. Archaeol. Sci.* **2009**, *36*, 2213–2223. [CrossRef]
85. Blasco, R.; Fernández Peris, J. Small and large game: Human use of diverse faunal resources at level IV of Bolomor Cave (Valencia, Spain). *C.R. Palevol* **2012**, *11*, 265–282. [CrossRef]

86. Rigaud, J.-P. *La Grotte Vaufrey à Cénac et Saint-Julien (Dordogne): Paléoenvironnements, Chronologie et Activités Humaines*; Mémoires de la Société Préhistorique Française XIX; Société Préhistorique Française: Paris, France, 1988.
87. Guadelli, J.-L.; Rigaud, J.-P. Les sites du Massif du Conte (Dordogne). In *Les Cahiers de la Vallée de la Couze*; CIRC: Montferrand-du-Périgord, France, 1995; pp. 1–17.
88. Texier, J.-P. *Historie Géologique de Sites Préhistoriques Classiques du Périgord: Une Vision Actualisée. La Micoque, la Grotte Vaufrey, le Pech de lAzé I et II, la Ferrassie, l'Abri Castanet, le Flageolet, Laugerie Haute*; Éditions du Comité des Travaux Historiques et Scientifiques: Paris, France, 2009.
89. Hernandez, M.; Mercier, N.; Rigaud, J.P.; Texier, J.P.; Delpech, F. A revised chronology for the Grotte Vaufrey (Dordogne, France) based on TT-OSL dating of sedimentary quartz. *J. Hum. Evol.* **2014**, *75*, 53–63. [CrossRef] [PubMed]
90. Garralda, M.D.; Maureille, B.; Rigaud, J.-P.; Vandermeersch, B. La molaire néandertalienne de la grotte Vaufrey (Dordogne, France). *Bull. Mem. Soc. Anthropol. Paris* **2004**, *16*, 189–199. [CrossRef]
91. Delpech, F. Rapport sur les travaux de biostratigraphie effectuées en 1999. In *Litho- et Biostratigraphie de Quelques Sites Paléolithique de Référence du Périgord*; Texier, J.-P., Delpech, F., Rigaud, J.-P., Eds.; Programme Collectif de Recherche, Service Régional de l'Archéologie d'Aquitaine: Paris, France, 1999.
92. Cochard, D. Caractérisation des apports de Léporidés dans les sites paléolithiques et application méthodologique à la couche VIII de la grotte Vaufrey. In *Un Siècle de Construction du Discours Scientifique en Préhistoire, des Idées d'hier, Proceedings of the XXVIe Congrès Préhistorique de France—Avignon, 21–25 Septembre 2004*; Société Préhistorique Française: Paris, France, 2007; Volume II, pp. 467–480.
93. Grayson, D.K.; Delpech, F. The evidence for the Middle Palaeolithic Scavenging from Couche VIII, Grotte Vaufrey (Dordogne, France). *J. Archaeol. Sci.* **1994**, *21*, 359–375. [CrossRef]
94. Le Gall, O. Ichtyophagie et Pêche Préhistoriques. Quelques Données de l'Europe Occidentale, Thèse de Doctorat D'état. Ph.D. Dissertation, Université de Bordeaux I, Bordeaux, France, 1999.
95. Laroulandie, V. Alpine chough Pyrrhocorax graculus from Pleistocene sites between Pyrenees and Alps: Natural versus cultural assemblages. In *Birds in Archaeology, Proceedings of the 6th Meeting of the ICAZ Bird Working Group in Groningen, Groningen, The Netherlands, 23–27 August 2008*; Prummel, W., Zeiler, J.T., Brinkhuizen, D.C., Eds.; Barkhuis: Groningen, The Netherlands, 2010; pp. 219–232.
96. Tomek, T.; Bochenski, Z.M. *The Comparative Osteology of European Corvids (Aves: Corvidae), with a Key to the Identification of Their Skeletal Elements*; Institute of Systematics and Evolution of Animals, Polish Academy of Sciences: Krákow, Poland, 2000.
97. Bochenski, Z.M.; Tomek, T. *A Key for the Identification of Domestic Bird Bones in Europe: Preliminary Determination*; Institute of Systematics and Evolution of Animals, Polish Academy of Sciences: Krákow, Poland, 2009.
98. Cohen, A.; Serjeantson, D. *A Manual for the Identification of Bird Bones from Archaeological Sites*; Archetype Publications Ltd.: London, UK, 1996.
99. Kessler, J.E. Evolution and skeletal characteristics of European owls. *Ornis Hung.* **2017**, *25*, 65–103. [CrossRef]
100. Whitehead, C.C. Overview of bone biology in the egg-laying hen. *Poult. Sci.* **2004**, *78*, 1033–1041. [CrossRef]
101. Canoville, A.; Schweitzer, M.H.; Zanno, L.E. Systemic distribution of medullary bone in the avian skeleton: Ground truthing criteria for the identification of reproductive tissues in extinct Ave metatarsalia. *BMC Evol. Biol.* **2019**, *19*, 71. [CrossRef]
102. Dodson, P.; Wexlar, D. Taphonomic investigations of owl pellets. *Paleobiology* **1979**, *5*, 275–284. [CrossRef]
103. Steadman, D.W.; Plourde, A.; Burley, D.V. Prehistoric Butchery and Consumption of Birds in the Kingdom of Tonga, South Pacific. *J. Archaeol. Sci.* **2002**, *29*, 571–584. [CrossRef]
104. Binford, L.R. *Bones. Ancient Men and Modern Myths*; Academic Press Inc.: New York, NY, USA, 1981.
105. Andrews, P. *Owls, Caves and Fossils*; University of Chicago Press: London, UK, 1990.
106. Shipman, P. Applications of scanning electron microscopy to taphonomic problems? In *The Research Potential of Anthropological Museum Collections*; Cantwell, A.M.E., Griffin, J.R., Rothschild, N.R., Eds.; Annals of the New York Academy of Sciences: New York, NY, USA, 1981; pp. 357–386.
107. Shipman, P.; Rose, J. Early hominid hunting, butchering and carcass-processing behaviors: Approches to the fossil record. *J. Anthropol. Archaeol.* **1983**, *2*, 57–98. [CrossRef]
108. Brugal, J.-P.; Díez-Lomana, C.; Huguet, R.; Michel, P.; Rosell, J. Karstic cavities, natural bone accumulations and discrete human activities in the European Palaeolithic: Some case studies. In *Paleolithic Zooarchaeology in Practice*; BAR International Series 1564; Haws, J.A., Hockett, B.S., Brugal, J.-P., Eds.; Archaeopress: Oxford, UK, 2006; pp. 1–12.
109. Bonham, P.F. Studies of less familiar birds. 157. Chough and Alpine Chough. *Brit. Birds* **1970**, *63*, 28–32.
110. Delestrade, A.; Stoyanov, G. Breeding biology and survival of the Alpine Chough *Pyrrhocorax graculus*. *Bird Study* **1995**, *42*, 222–231. [CrossRef]
111. Murton, R.K. Natural selection and the breeding seasons of the Stock Dove and Wood Pigeon. *Bird Study* **1966**, *13*, 311–327. [CrossRef]
112. Hetmański, T. The timing of fledging and annual post-fledging survival of juvenile Feral Pigeons, *Columba livia*, in a city area (Pomerania, NW Poland). *Pol. J. Ecol.* **2007**, *55*, 367–375.
113. Gill, F.; Donsker, D. *IOC World Bird List (v11.1)*; IOC: Baton Rouge, LA, USA, 2021. [CrossRef]
114. Del Hoyo, J. *All the Birds of the World*; Lynx Edicions: Barcelona, Spain, 2020.
115. Diez, C.; Sánchez, A.; Moreno, V. Grupos avicaptores del Tardiglaciar: Las aves de Berroberría. *Munibe Antropol.* **1995**, *47*, 3–22.

116. Louchart, A.; Soave, R. Changement d'ampleur de l'exploitation des oiseaux entre le Magdalénien et l'Azilien: L'exemple du Taï 2 (Brome). *Quaternaire* **2002**, *13*, 297–312. [CrossRef]
117. Laroulandie, V. Bird exploitation pattern: The case of Ptarmigan Lagopus sp. in the Upper Magdalenian site of La Vache (Arihge, France). In *Feathers, Grit and Symbolism: Birds and Humans in the Ancient Old and New Worlds*; Grupe, G., Peters, J., Eds.; Verlag Marie Leidorf: Rahden, Germany, 2005; pp. 165–178.
118. Marco, A.S.; Quesada, C.C. Avian wings as ornaments in the Magdalenian? *Archaeofauna* **2010**, *19*, 133–139.
119. Lloveras, L.; Salazar, R.; García-Argüelles, P.; Nadal, J. Birds and Epipalaeolithic hunter-gatherers in northeast Iberia. The case of the Balma del Gai site. *Quat. Int.* **2020**, *543*, 25–33. [CrossRef]
120. Lloveras, L.; Garcia, L.; Marqueta, M.; Maroto, J.; Soler, J.; Soler, N. The role of birds in Upper Palaeolithic sites: Zooarchaeological and taphonomic analysis of the avian remains from Arbreda Cave (Serinyà, northeast Iberia). *Quat. Int.* **2020**, in press. [CrossRef]
121. Goffette, Q.; Germonpré, M.; Lefèvre, C.; Brecko, J.; Goemaere, E.; Rots, V. Bird bones from Trou de Chaleux and the human exploitation of birds during the late Magdalenian in Belgium. *J. Archaeol. Sci. Rep.* **2020**, *29*, 102096. [CrossRef]
122. Spenneman, D.H.R.; Colley, S.M. Fire in a pit: The effects of burning of faunal remains. *Archaeozoologia* **1989**, *3*, 51–64.
123. Bennett, J.L. Thermal alteration of buried bone. *J. Archaeol. Sci.* **1999**, *26*, 1–8. [CrossRef]
124. Laroulandie, V.; Faivre, J.-P.; Gerbe, M.; Mourre, V. Who brought the bird remains to the Middle Palaeolithic site of Les Fieux (Southwestern, France)? Direct evidence of a complex taphonomic story. *Quat. Int.* **2016**, *421*, 116–133. [CrossRef]
125. Gerbe, M.; Thiebaut, C.; Mourre, V.; Bruxelles, L.; Coudenneau, A.; Jeannet, M.; Laroulandie, V. Influence des facteurs environnementaux, économiques et culturels sur les modalités d'exploitation des ressources organiques et minérales par les Néandertaliens des Fieux (Miers, Lot). In *Transitions, Ruptures et Continuités en Préhistoire, Actes du XXVIIème Congrès Préhistorique de France, Bordeaux-les Eyzies 31 Mai–5 Juin 2010*; Jaubert, J., Fourment, N., Depaepe, P., Eds.; Société Préhistorique Française: Paris, France, 2014; Volume 2, pp. 257–279.
126. Blasco, R.; Cochard, D.; Colonese, A.C.; Laroulandie, V.; Meier, J.; Rufà, A.; Tassoni, L.; Thompson, J.C. Small animal use by Neanderthals. In *Updating Neanderthals: Understanding Behavioural Complexity in the Late Middle Palaeolithic*; Benazzi, S., Rivals, F., Romagnoli, F., Eds.; Elsevier Monographs: Amsterdam, The Netherlands. (In press)
127. Juste, J.; Fa, J.E.; Perez del Val, J.; Castroviejo, J. Market dynamics of bushmeat species in Equatorial Guinea. *J. Appl. Ecol.* **1995**, *32*, 454–467. [CrossRef]
128. Blasco, R.; Finlayson, C.; Rosell, J.; Marco, A.S.; Finlayson, S.; Finlayson, G.; Negro, J.J.; Pacheco, F.G.; Vidal, J.R. The earliest pigeon fanciers. *Sci. Rep.* **2014**, *4*, 5971. [CrossRef] [PubMed]
129. Finlayson, S.; Finlayson, G.; Guzman, F.G.; Finlayson, C. Neanderthals and the cult of the Sun Bird. *Quat. Sci. Rev.* **2019**, *217*, 217–224. [CrossRef]
130. Finlayson, S.; Finlayson, C. The birdmen of the Pleistocene: On the relationship between Neanderthals and scavenging birds. *Quat. Int.* **2016**, *421*, 78–84. [CrossRef]

quaternary

MDPI

Article

Human Activities, Biostratigraphy and Past Environment Revealed by Small-Mammal Associations at the Chalcolithic Levels of El Portalón de Cueva Mayor (Atapuerca, Spain)

Juan Rofes [1,2,3,*], Amaia Ordiales [4], Eneko Iriarte [5], Gloria Cuenca-Bescós [6], María Ángeles Galindo-Pellicena [7,8,9], Amalia Pérez-Romero [5], José Miguel Carretero [5] and Juan Luis Arsuaga [9]

1 Archaeological Studies Program, University of the Philippines-Diliman, Albert Hall, Lakandula St., Quezon City 1101, Philippines
2 Archéozoologie, Archéobotanique: Sociétés, Pratiques et Environnements (AASPE, UMR 7209), Sorbonne Universités, Muséum National d'Histoire Naturelle, CNRS, CP56, 55 rue Buffon, F-75005 Paris, France
3 National Museum of the Philippines, P. Burgos Drive, City of Manila 1000, Philippines
4 Independent Researcher, Barrio Ergoien 60C, 48113 Gamiz-Fika, Spain; amaiaocb@gmail.com
5 Laboratorio de Evolución Humana, Departamento de Historia, Geografía y Comunicación, Universidad de Burgos, Edificio I+D+i, Plaza de Misael Bañuelos s/n, 09001 Burgos, Spain; eiriarte@ubu.es (E.I.); apromero14@gmail.com (A.P.-R.); jmcarre@ubu.es (J.M.C.)
6 Aragosaurus-IUCA-Earth Sciences Department, University of Zaragoza, C/Pedro Cerbuna 12, 50009 Zaragoza, Spain; cuencag@unizar.es
7 Museo Arqueológico Regional, Plaza de Las Bernardas s/n, Alcalá de Henares, 28801 Madrid, Spain; mangeles.galindo@fgua.es
8 Fundación General de la Universidad de Alcalá de Henares, C/Imagen 3, Alcalá de Henares, 28801 Madrid, Spain
9 Centro Mixto UCM-ISCIII de Evolución y Comportamiento Humanos, C/Monforte de Lemos 5, Pabellón 14, 28029 Madrid, Spain; jlarsuaga@isciii.es
* Correspondence: jcrofes@up.edu.ph

Abstract: The Chalcolithic levels of El Portalón de Cueva Mayor (Atapuerca, Burgos, Spain) offer a good opportunity to test whether the small-mammal contents of different archaeo-stratigraphical units may be useful to characterize them as independent entities. With that purpose, we studied representative samples of small-mammal remains from the two main contexts identified: the Early Chalcolithic (EC) funerary context and the Late Chalcolithic (LC) habitat/stabling context, with the latter comprising three different archaeological units according to their origin, namely prepared floors, activity floors and stabling surfaces or fumiers. Following the distribution of taxa in their respective contexts, we performed several statistical tests to check for significant discrepancies between archaeological units. The exclusive presence of certain taxa, together with the statistical difference in relative taxonomic ratios, points to the integrity and unpolluted condition of the EC context. The interspersed arrangement of the different LC context's units made them prone to inter-pollution as they are not statistically different. The unexpected presence of *Pliomys lenki* and *Chionomys nivalis* in the prepared floors evidences their Upper Pleistocene allochthonous origin. The EC levels of El Portalón contribute the first Holocene records of nine taxa in the Sierra de Atapuerca. An environment dominated by woodland, shrubland and wet meadows, with moderate presence of grassland, inland wetlands and rocky areas, is inferred from the small-mammal association of the EC levels.

Keywords: small mammals; archaeological contexts; Chalcolithic; biostratigraphy; past environment; Holocene; Sierra de Atapuerca; Iberian Peninsula

1. Introduction

The fossil remains of small vertebrates are frequently preserved in archaeological and paleontological contexts such as cave and rock shelter deposits, sometimes in association with human remains and artefacts. Contrary to their larger counterparts, which are usually a product of human selection, the small-vertebrate accumulations mainly result from digestion or storing by their predators (i.e., birds of prey and small carnivores) or natural death. Despite unavoidable filters due to specific predators [1–3], they reflect local biocenoses reasonably well. The presence or absence and relative abundance of small-mammal species can thus be used as proxies for biochronology (e.g., References [4, 5]) and for the reconstruction of both past environments (e.g., References [6–10]) and biogeographical histories (e.g., References [11,12]).

Such interpretations are based on the working hypothesis of a strict contemporary association between small vertebrates and archaeological remains, implying good archaeological coherence of the analysed assemblages. There are exceptions to this hypothesis, i.e., some post-depositional processes can affect the stratigraphy of a given site by altering the integrity of the deposits [13,14] and, therefore, the subsequent scientific interpretations. For instance, caves and rock shelters are propitious places for small- and medium-sized animals to nest or build burrows, which may significantly modify and pollute faunal assemblages [15–17]. The tiny size of small-vertebrate remains favors percolations into underlying layers along stratigraphic sequences, and this is a known phenomenon in given archaeological deposits [18].

The Chalcolithic levels of El Portalón de Cueva Mayor (Atapuerca, Burgos, Spain) offer a unique opportunity to test whether the small-mammal contents of different archaeo-stratigraphical units may be useful to characterize them as independent entities (for biostratigraphical, biogeographical and paleo-environmental interpretations), and, if so, to what extent. For this purpose, we depart from a working hypothesis opposite to the one previously mentioned: there was (some degree of) pollution among units. Is this true? Moreover, even if so, is it still possible to distinguish and characterize archaeo-stratigraphical units/contexts according to their respective small-mammal contents?

To answer these questions, we have studied representative samples of small-mammal remains from the two main archaeo-stratigraphical contexts identified in the Chalcolithic levels of El Portalón: the Early Chalcolithic (EC) funerary context and the Late Chalcolithic (LC) habitat/stabling context, with the latter comprising three different archaeological units according to their origin, namely prepared floors, activity floors and stabling surfaces or fumiers. Following the taxonomical identification, we performed several statistical tests to check for significant discrepancies between contexts and archaeological units. Relevant biostratigraphic and environmental insights are also given.

Preliminary results of this work were presented elsewhere [19], but they were substantially revised and improved here.

2. Materials and Methods

2.1. The Site

El Portalón is located at the current entrance to the Cueva Mayor-Cueva del Silo karst system of the Sierra de Atapuerca (Burgos, Spain) (Figure 1). The latter is internationally well-known for its Pleistocene hominin-bearing localities [20–23], but also provides a rich stratigraphic record from the Holocene [24–31]. The relevance of El Portalón relies on its long human occupation, documented from the Paleolithic to historic times (from ca. 35,205 yr cal BP to 796 yr cal BP) [24,26–29]. El Portalón site alone currently surpasses 200 m^2. However, this space is part of a much larger endokarstic archaeological area that comprises the room known as Salón del Coro or Galería Principal (which is part of El Portalón site itself) and the Galería del Sílex (Figure 1b,c).

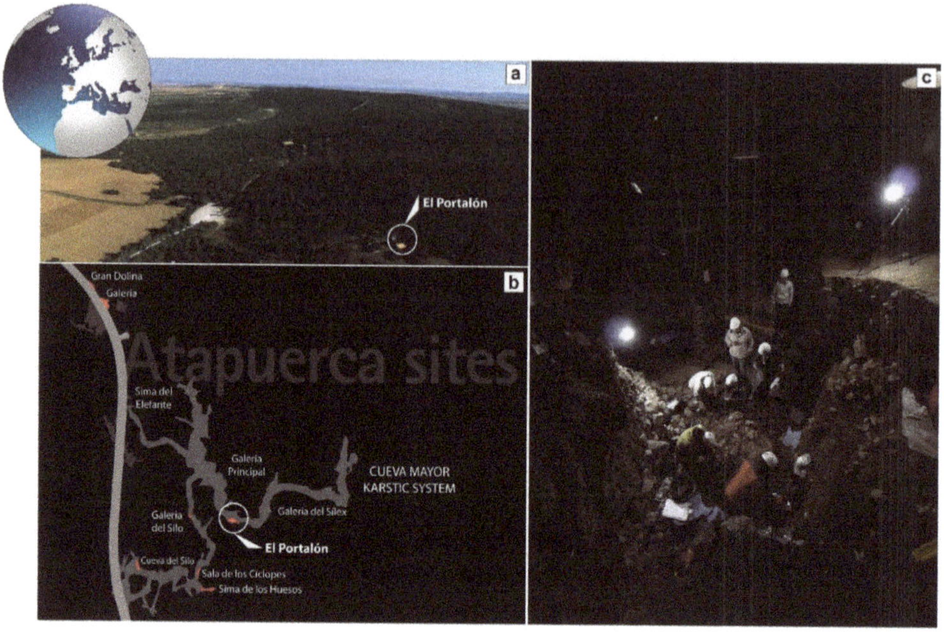

Figure 1. Location of El Portalón de Cueva Mayor archaeological site. (**a**) Location of the site in the Sierra de Atapuerca; (**b**) location of the site at the entrance of the Cueva Mayor-Cueva del Silo karstic system (plan view), with the latter containing numerous Lower Pleistocene to Holocene archaeo-paleontological sites (in red); (**c**) general view of the works during the excavation of the early Chalcolithic funerary tumulus (ca. 5000 yr cal BP).

The currently known stratigraphic sequence of El Portalón exceeds 10 m and is divided into 11 stratigraphic units grouped into two sedimentary units (Figure 2). There is a basal sedimentary unit from the Upper Pleistocene with abundant microfauna [32]; however, evidence of macrofauna and human presence is scarce. The second sedimentary unit comprises the Holocene and is divided into ten stratigraphic units [26]. The cultural affiliation of the Holocene units based on the material records and supported by a large radio-chronological set provide evidence for occupations in the Middle Ages (9th–12th Centuries CE), the Roman Age (2nd Century BCE to 2nd Century CE), the Iron Age I (2360–2740 yr cal BP), Bronze Age (2876–4250 yr cal BP), Chalcolithic (4100–5572 yr cal BP) and Neolithic (5745–7270 yr cal BP) [26].

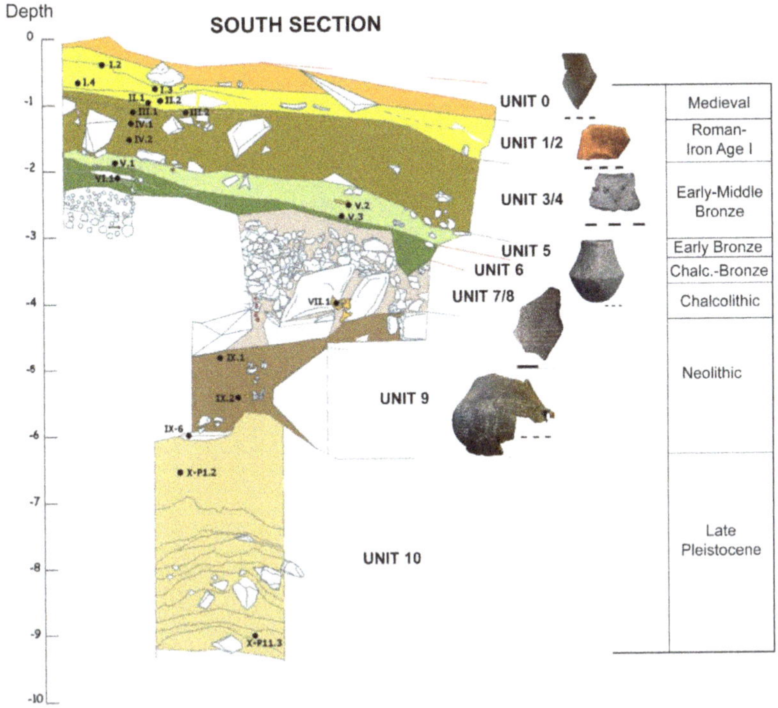

Figure 2. Stratigraphy of El Portalón de Cueva Mayor archaeological site (south profile). Chalcolithic units 7 and 8 represent a Pre-Bell-Beaker funerary context (stone tumulus) (unit 8) and a late Pre-Bell-Beaker and Bell-Beaker stabling/habitational context (unit 7).

This study focuses on small-mammal samples from the Chalcolithic archaeo-stratigraphic levels (Figure 2).

2.2. *Archaeo-Stratigraphic Units and Chronology*

The Chalcolithic stratigraphic units are divided into two phases. The older corresponds to an Early Chalcolithic or Pre-Bell-Beaker funerary context, whereas the Late Pre-Bell-Beaker and Bell-Beaker phase is characterized by a herding and habitat context. The data in this section mainly follow Pérez-Romero et al. [29].

2.2.1. Early Chalcolithic (Pre-Bell-Beaker) Funerary Context (UE79)

The burial phase comprises a tumular stacking of decimetric limestone clasts of approximately 8 m in diameter and 2 m high in the center. It seems to have an oval shape and is made up of limestone blocks with sizes varying from 40 to 10 cm, and locally they appear to form upward-fining sequences.

The tumular structure was built by a progressive accumulation of limestone clasts in an aggrading (vertical) and prograding manner over a basal surface defined by a floor covered with pottery "pavements", on which there are numerous circular small fire pits filled with partially combusted charcoal fragments (UE79) (Figure 3d). Associated with the fire pits, nearly complete small ceramic bowls are found, along with remains of immature individuals of domestic fauna, mainly lambs, in anatomical connection. Among the limestone clasts and sometimes lying on the floor, different human remains are found, defining funerary contexts with some archaeological elements typical of grave goods (Figure 3e).

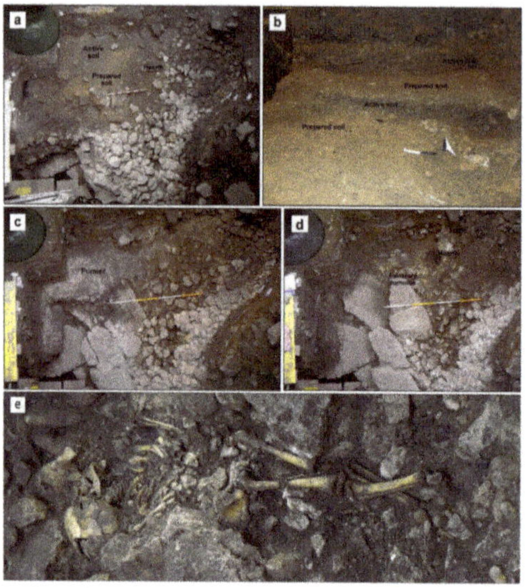

Figure 3. Different archaeological contexts of the Chalcolithic levels of El Portalón de Cueva Mayor site: (**a**) General view during the excavation of a prepared floor and active soils from the stabling phase. Note the overlapping disposition of these units in relation to the limestone blocks of the funerary tumulus from the previous phase; (**b**) detailed view of the alternation of prepared clay floors (orange) and active soils (dark) from the habitat/stabling context; (**c**) partially excavated white ash layer (fumier) covering the funerary tumulus; (**d**) general view of the funerary tumulus; note the basal large limestone blocks which originated from a collapse of the cave roof; (**e**) an example of the human remains contained in the funerary tumulus corresponding to an intact burial of a 6-year-old child.

The characteristics of the funerary context suggest this to be the result of repetitive burial activity following a similar funerary pattern over time. Each event appears to have partially disturbed a previous funerary context, with possible collapse of a large part of the roof of the cavity during this period. There is a minimum of eight buried individuals. Furthermore, the later habitation and stable use of the cave seems to have contributed to the disturbance of some of the funerary structures. Under this scenario, human bone remains are a common occurrence, and grave-goods elements often appear scattered among the limestone blocks of the tumular structure [29,33].

2.2.2. Late Chalcolithic (Late Pre-Bell-Beaker and Bell-Beaker) Habitat and Stabling Context

Prepared floors (UE23A): They are composed of orange-brown silty/clayey levels of less than 10 cm thick incorporating small limestone fragments. They include fragmentary archaeological remains (e.g., charcoal, bone, pottery, etc.), mainly at the top. These floors are interpreted as built anthropic floors made of endokarstic silts and clays extracted from the internal galleries of the cave where the pits are still visible. They were likely intended to adequate moist areas for various human activities (living, cooking, etc.). They usually contain anthropic structures, such as hearths and postholes, and appear interspersed with gray sediments belonging to "activity floors" (Figure 3a,b).

Activity floors (UE23): Grayish clayey units formed by the remains of human activities that took place on them (ashes, charcoal, bone fragments, pottery, vegetal remains, etc.) mixed with fine sediments, trampled and normally deposited above the prepared floors or materials originated from other uses, as stabling (fumiers). They have a variable thickness, from few centimetres to decimetres (Figure 3a,b).

Stabling surfaces or fumiers (Layers 214B, 214A, 214, 212, 220, 216): They appear as fine alternating horizontal (millimetric to centimetric) layers with abundant carbonaceous remains (blackish) and ashes (whitish). They represent burned remains of vegetal layers (herbaceous and branches) and manure from livestock stabling areas. These units appear either attached, laterally equivalent or interspersed with the prepared floors and the activity floors (Figure 3c).

Thus, we observe a compartmentalization and an alternation of the uses of the cave space.

2.2.3. Chronology

Nine different cereal seed and bone remains from Chalcolithic archaeo-stratigraphic units from the funerary and stabling contexts studied in this work have been radiocarbon dated [26,29,34] (Table 1). More in detail, four human samples and two faunal remains from the funerary context and two cereal seeds (*Triticum* sp.) and one bone sample from the stabling context were radiocarbon dated using accelerator mass spectrometry (AMS) at Beta Analytic Inc. (Miami, Florida) [34]. Calibration to years cal BP was made by using Oxcal v4.4 software based on the IntCal20 radiocarbon age calibration curve [35].

Table 1. Radiocarbon dates of the Chalcolithic levels of El Portalón de Cueva Mayor sites. Con., conventional; cal, calibrated.

Sample	Chalcolithic Cultural Context	Laboratory Code	Con. Age (yr BP)	Age (yr cal BP)	Reference
Human tibia	Bell-Beaker (UE4)	Beta-269494	3900 ± 40	4423–4158	[26]
Triticum sp. seed	Pre-Bell Beaker/Stabling (UE209)	Beta-347579	4230 ± 30	4859–4646	[26]
Triticum sp. seed	Pre-Bell Beaker/Stabling (UE212a)	Beta-347580	4280 ± 30	4957–4732	[29]
Human adult proximal foot phalange	Pre-Bell Beaker/Funerary (UE79)	Beta-368290	4280 ± 30	4957–4732	[34]
Human tooth	Pre-Bell Beaker/Funerary (UE80)	Beta-337300	4300 ± 30	4960–4830	[34]
Bos sp.	Pre-Bell Beaker/Funerary (UE80)	Beta-337299	4380 ± 30	5042–4860	[29]
Human adult thoracic vertebra	Pre-Bell Beaker/Funerary (UE79)	Beta-368289	4400 ± 30	5214–4862	[34]
Ovicaprine tooth	Pre-Bell Beaker/Funerary (UE79)	Beta-197389	4440 ± 50	5285–4874	[29]
Human fragment of a right tibial shaft	Pre-Bell Beaker/Funerary (UE79)	Beta-368285	4460 ± 40	5294–4886	[34]

For a comprehensive chronology/biochronology of the different sites comprising the Atapuerca archaeo-paleontological complex, we refer the reader elsewhere [4,5]. Further recent references may enrich this general frame [36–39].

2.3. *Samples*

The studied assemblage includes 12,867 elements, of which 2325 were identified either to the family, genus and/or species level. To obtain the samples, the sediment from the different archaeo-stratigraphic units of the Chalcolithic levels of the site was water-screened, using a stack of sieves of decreasing mesh size (5, 2 and 0.5 mm). The small mammals were collected from residue coarser than 0.5 mm in the following proportions: 112 liters from the funerary context, 84 liters from the prepared floors, 90 liters from the activity floors and 91,75 liters from the fumiers.

Fossils were sorted, classified, counted and studied with the aid of a binocular microscope (Nikon SMZ-U; 7x, 20x, and 40x magnifications). Most of the elements are teeth, isolated mandibles, skull fragments and postcranial bones.

The light-to-moderate gastric digestion and scant breakage observed in the small-vertebrate remains indicate that the bones were likely accumulated by an avian predator of category 1, such as a barn owl (*Tyto alba*), which is an opportunistic rather than a selective hunter [2,3]. However, there are certain instances of heavy-to-extreme modification, which means categories 4 to 5 on the Fernández-Jalvo et al. [3] scale. In these cases, the agents of deposition were most probably small-mammalian carnivores and/or diurnal birds of prey.

2.4. Taxonomic Identification and Quantification

The species/taxon of each small-mammal remain was identified based on cranial and post-cranial diagnostic elements: isolated teeth for the Murinae, Gliridae, Sciuridae and Arvicolinae; mandibles, maxillae and isolated teeth for the Soricidae and Chiroptera; and post-cranial skeleton for the Talpidae. The taxonomic classification follows well-known References [40,41].

The absolute number of elements per taxon was expressed as the number of identified specimens (NISP). The relative ratios of fossil species were established with the minimum number of individuals (MNI), also used as a quantitative measure to evaluate differences among archaeo-stratigraphic units and for paleo-environmental estimations. To determine the MNI, we consider a diagnostic anatomical element, taking into account laterality, e.g., first lower molar in arvicolines and humerus in Talpidae.

2.5. Statistical Analysis

Differences in small-mammal relative taxonomic ratios among samples from different archaeo-stratigraphical units/contexts were assessed by an analysis of the variance (ANOVA) on the MNI, with subsequent pairwise t-tests between samples, namely Tukey's (Copenhaver-Holland 1988), Mann–Whitney and Dunn's post hoc, with the latter two with Bonferroni corrections. Statistical analyses were performed by using PAST software v. 3.21 [42].

2.6. Paleo-Environmental Estimations

The information on the habitat affinities of the small-mammal taxa identified at the EC levels of El Portalón was taken from the IUCN Red List of Threatened Species [41] and the bibliography therein; it is summarized in Table 2. The habitats of the wild large mammals mentioned in the Discussion come from the same source. We apply an actualist approach; that is, being a site with a relatively recent chronology, all the fauna are extant, which means that there is little doubt in the species–habitat correlations.

Table 2. Habitat affinities (in green) of small-mammal taxa identified at the early Chalcolithic levels of El Portalón de Cueva Mayor. Fo, forest; Sh, shrubland; Gr, grassland; WI, wetland (inland); RA, rocky areas including caves.

Taxa	Fo	Sh	Gr	WI	RA
Crocidura suaveolens		X	X		X
Sorex gr. *araneus-coronatus*	X	X		X	
Sorex minutus	X	X	X	X	
Neomys fodiens	X		X	X	
Talpa sp.	X	X	X		
Miniopterus schreibersii	X	X	X		X
Myotis gr. *myotis-blythii*	X	X	X		X
Arvicola sapidus				X	
Microtus (*T.*) *duodecimcostatus*		X			
Microtus (*T.*) *lusitanicus*	X				
Microtus agrestis	X	X	X	X	
Microtus arvalis	X	X	X		
Apodemus gr. *sylvaticus-flavicollis*	X	X	X		
Eliomys quercinus	X				X
Sciurus vulgaris	X				
Oryctolagus cuniculus		X	X		

3. Results

The complete association of small mammals from the Chalcolithic levels of El Portalón includes five eulipotyphlans (*Crocidura suaveolens*, *Sorex* gr. *araneus-coronatus*, *S. minutus*, *Neomys fodiens* and *Talpa* sp.), two chiropterans (*Miniopterus shreibersii* and *Myotis* gr. *myotis-blythii*), nine rodents (*Arvicola sapidus*, *Pliomys lenki*, *Chionomys nivalis*, *Micro-*

tus (Terricola) duodecimcostatus, M. (T.) lusitanicus, M. agrestis, M. arvalis, Apodemus gr. *sylvaticus-flavicollis, Eliomys quercinus*, and *Sciurus vulgaris*) and one lagomorph (*Oryctolagus cuniculus*). The detailed list of species, together with the NISP and NMI values per archaeo-stratigraphic unit and context, is shown in Table 3.

Table 3. Number of identified specimens (NISP) and minimal number of individuals (MNI) for the Chalcolithic levels of El Portalón de Cueva Mayor.

Chrono-Cultural Period	**Early Chalcolithic**		**Late Chalcolithic**					
Context	**Funerary**		**Habitat/Stabling**					
Floor			**Prepared**		**Activity**		**Fumier**	
Taxa	**NISP**	**MNI**	**NISP**	**MNI**	**NISP**	**MNI**	**NISP**	**MNI**
Crocidura suaveolens	281	83	31	6	46	11	67	25
Sorex gr. *araneus-coronatus*	326	79	19	3	67	16	104	38
Sorex minutus	4	3	0	0	0	0	2	2
Neomys fodiens	3	1	0	0	0	0	0	0
Talpa sp.	10	4	0	0	0	0	7	3
Miniopterus schreibersii	9	5	1	1	4	3	0	0
Myotis gr. *myotis-blythii*	27	6	2	1	13	5	5	2
Arvicola sapidus	7	4	1	1	0	0	0	1
Pliomys lenki	0	0	2	1	0	0	0	0
Chionomys nivalis	0	0	2	1	0	0	0	0
Microtus (T.) duodecimcostatus	15	10	2	2	2	2	7	6
Microtus (T.) lusitanicus	49	33	2	1	7	5	14	9
Microtus agrestis	58	36	9	6	16	11	40	25
Microtus arvalis	95	54	12	10	14	9	37	24
Apodemus gr. *sylvaticus-flavicollis*	555	121	37	9	94	17	143	33
Eliomys quercinus	35	10	3	1	7	3	12	7
Sciurus vulgaris	2	2	1	1	1	1	0	0
Oryctolagus cuniculus	8	3	6	2	2	2	0	0
Totals	1484	454	130	46	273	85	438	175

In taxonomical terms, the EC funerary context differs from the LC habitat/stabling context in the exclusive presence of *N. fodiens* in the former. The prepared floors differ from any other archaeo-stratigraphic unit in the presence of *P. lenki* and *C. nivalis*. Within the habitat/stabling context, the activity floors differ from the other two units in the absence of *A. sapidus*; the fumiers, in turn, differ from the other two in the presence of *S. minutus* and *Talpa* sp. and the absence of *O. cuniculus*, *Sc. vulgaris* and *M. shreibersii*. Figure 4 (EC) and Figure 5 (LC) display diagnostic elements from the different archaeo-stratigraphic units under study, highlighting those taxa which only appear in particular units.

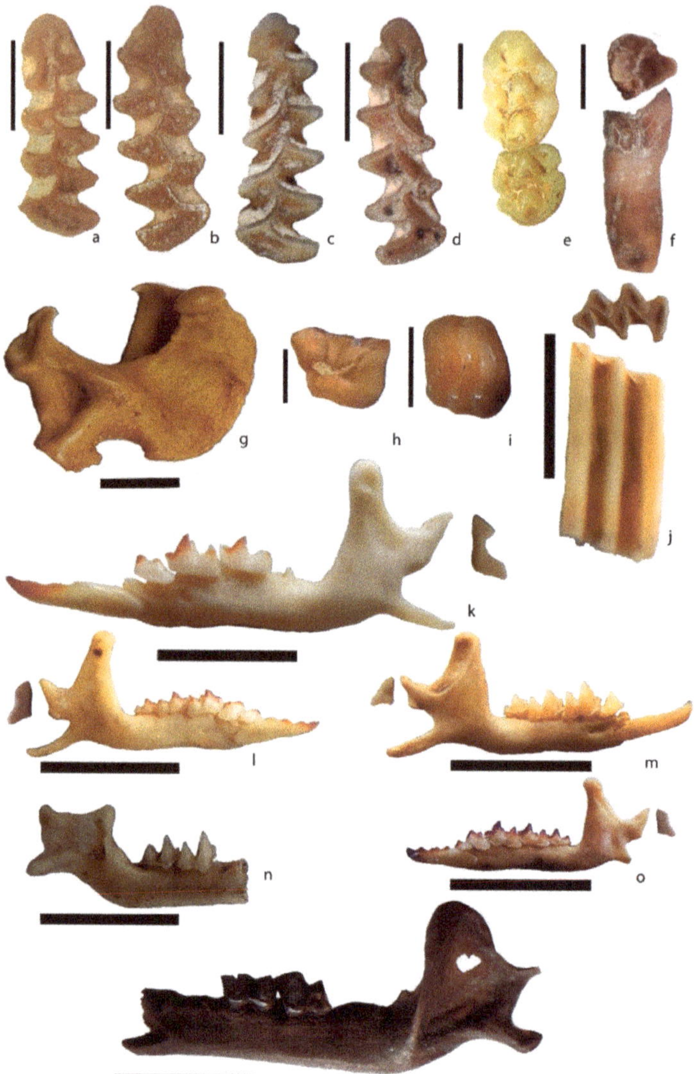

Figure 4. Diagnostic elements of the different taxa present at El Portalón de Cueva Mayor Early Chalcolithic funerary context: (**a**) *Microtus arvalis*, left m1 (occlusal view); (**b**) *M. agrestis*, right m1 (occlusal view); (**c**) *M. (Terricola) duodecimcostatus*, left m1 (occlusal view); (**d**) *M. (T.) lusitanicus*, right m1 (occlusal view); (**e**) *Apodemus* gr. *sylvaticus-flavicollis*, right M1-M2 (occlusal view); (**f**) *Sciurus vulgaris*, left p4 (occlusal and lingual views); (**g**) *Talpa* sp., left humerus (posterior view); (**h**) *Oryctolagus cuniculus*, right I1 (occlusal view); (**i**) *Eliomys quercinus*, left m1 or m2 (occlusal view); (**j**) *Arvicola sapidus*, left M1 (occlusal and buccal views); (**k**) *Neomys fodiens*, left mandible with i1-p4-m1-m2 (lateral view) and condyle detail (posteroventral view); (**l**) *Sorex* gr. *araneus-coronatus* right mandible with i1-a1-p4-m1-m2-m3 (lateral view) and condyle detail (posteroventral view); (**m**) *Crocidura suaveolens*, right mandible with i1-p4-m1-m2-m3 (lateral view) and condyle detail (posteroventral view); (**n**) *Miniopterus schreibersii*, right mandible with m2-m3 (lateral view); (**o**) *Sorex minutus*, left mandible with i1-a1-p4-m1-m2-m3 (lateral view) and condyle detail (posteroventral view); (**p**) *Myotis* gr. *myotis-blythii* left mandible with m1-m2 (lateral view). **Scale bars** of a–f, h, i = 1 mm; g, j–p = 5 mm.

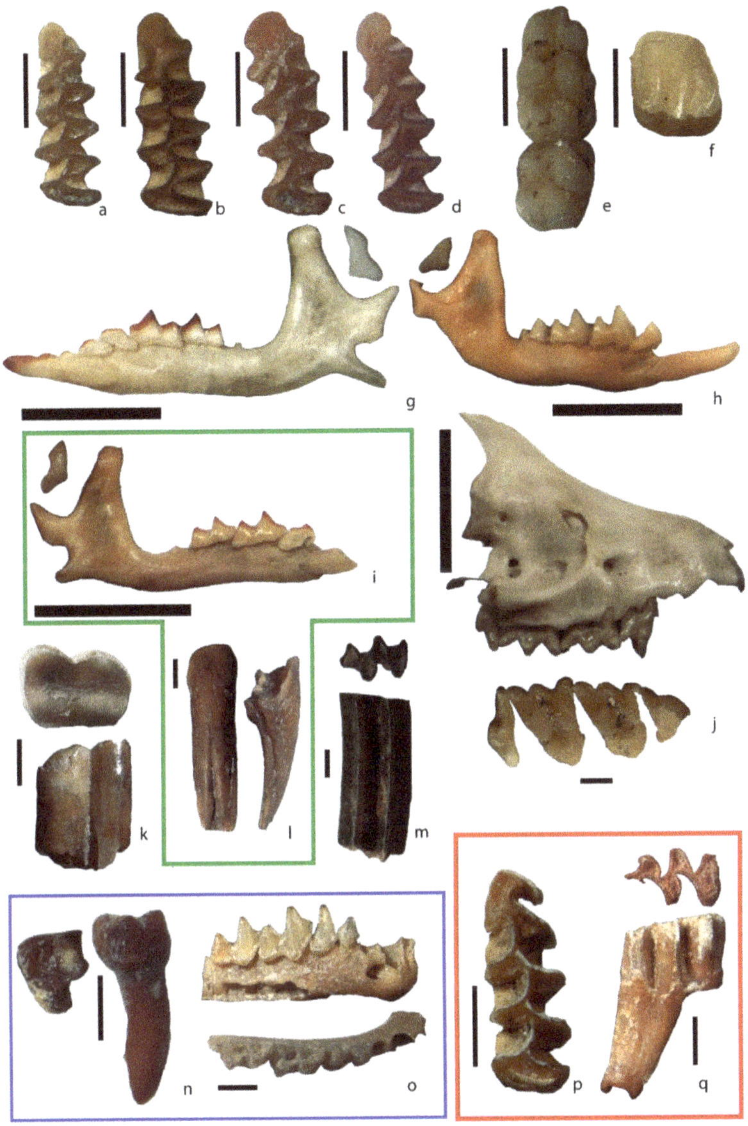

Figure 5. Diagnostic elements of the different taxa present at El Portalón de Cueva Mayor Late Chalcolithic habitational/stabling context, highlighting the taxa which are particular to specific floors: (**a**) *Microtus arvalis*, left m1 (occlusal view); (**b**) *M. agrestis*, left m1 (occlusal view); (**c**) *M. (Terricola) duodecimcostatus*, left m1 (occlusal view); (**d**) *M. (T.) lusitanicus*, left m1 (occlusal view); (**e**) *Apodemus* gr. *sylvaticus-flavicollis*, right m1-m2 (occlusal view); (**f**) *Eliomys quercinus*, left m1 or m2 (occlusal view); (**g**) *Sorex* gr. *araneus-coronatus*, right mandible with i1-a1-p4-m1-m2 (lateral view) and condyle detail (posteroventral view); (**h**) *Crocidura suaveolens*, left mandible with i1-p4-m1-m2-m3 (lateral view) and condyle detail (posteroventral view); (**i**) *Sorex minutus*, right mandible with p4-m1-m2 (lateral view) and condyle detail (posteroventral view); (**j**) *Myotis* gr. *myotis-blythii* right maxilla with P4-M1-M2-M3 (lateral and occlusal views); (**k**) *Oryctolagus cuniculus*, left I1 (occlusal and buccal views); (**l**) Talpa sp., 3rd phalanx (dorsal and lateral views); (**m**) *Arvicola sapidus*, right M2 (occlusal and lingual views); (**n**) *Sciurus vulgaris*, right p4 (occlusal and posterior views); (**o**) *Miniopterus schreibersii*, right mandible with p3-p4-m1-m2 (lateral and occlusal views); (**p**) *Chionomys nivalis*, left m1 (occlusal view); (**q**) *Pliomys lenki*, left m2 (occlusal and buccal views). Red, only in the prepared floors; green, only in the fumiers; blue, only in the prepared and activity floors. Scale bars of a–f, j (lower), k–q = 1 mm; g–j (upper) = 5 mm.

Although the ANOVA test for equal means showed a statistically significant difference among units ($p < 0.05$), the subsequent pairwise t-tests (Tukey's, Mann–Whitney and Dunn's post hoc) revealed that the three units from the LC were not statistically different in their relative taxonomic ratios. That is not the case for the EC Funerary context, which is significantly different from the prepared floors according to all the pairwise t-tests performed ($p = 0.004$; $p = 0.03$; $p = 0.04$, respectively) and significantly different from the activity floors according to Tukey's ($p = 0.01$).

Crossing data from Table 2 (habitat) and Table 3 (MNI), we can state that, at the EC layers, twelve species represent the forest habitat; eleven, the shrubland; ten, the grassland; five, the inland wetland; and four, the rocky areas, including caves.

4. Discussion

Particularly striking is the presence of *Pliomys lenki* in the Chalcolithic layers of El Portalón. The last records of this species in Europe date to the very end of the Younger Dryas, and they come from Southwestern France and the Cantabrian region of the Iberian Peninsula [5,43]. However, several diagnostic traits (i.e., presence of roots, absence of cement in the re-entrant angles of the crown and large size in relation to the other contemporaneous rooted arvicoline, i.e., *Myodes glareolus*) indisputably assign the lower molar found at the prepared floors to *P. lenki* (Figure 5q). The red pigment of the teeth and the general shape and size of a soricine mandible found at the funerary context (Figure 4k) identify it as belonging to *Neomys fodiens* and not to its smaller relative *N. anomalus*.

The presence of *P. lenki* and *Chionomys nivalis* confirm the allochthonous endokarstic nature of the clay sediments used in the prepared floors, biostratigraphically attributable to the Upper Pleistocene [4,5,43]. *Chionomys nivalis* has been previously recorded for the latest Pleistocene levels of El Portalón [32], but the last evidences of *P. lenki* in Atapuerca are from the Middle Pleistocene (i.e., Galería-Cueva de los Zarpazos [4,5]). Nevertheless, as stated before, the latest records of the species in Spain come from the Cantabrian region, dating back prior to the Pleistocene/Holocene transition (e.g., Laminak and El Mirón [43]).

Two bone remains of wild large mammals were found within the prepared floors: an atlas of *Bos primigenius* (ATP11.UE23a.862) and a distal part of a humerus of *Sus scrofa* (ATP11.UE23a.861) [44–46]. They might be allochthonous as well, considering that the wild large mammalian fauna is very scarce at the LC levels of El Portalón (accounting for only 2% of the identified taxa, otherwise largely dominated by domestic species), and that these two species are not present in the underlying EC levels [44]. We cannot be certain on their attribution to the Upper Pleistocene though, a status which absolute dates may eventually assess.

The prepared floors are statistically different from the EC funerary context in terms of their small-mammal relative taxonomic ratios, and arguably different from the activity floors as well. This, together with the presence of *N. fodiens* and the absence of *P. lenki* and *C. nivalis* in the funerary context, argues in favor of the integrity of this archaeo-stratigraphic unit, which likely remained unpolluted from overlying LC small-mammal remains percolations.

In spite of the Pleistocene components of the prepared floors, this unit is not statistically different from the other two within the compartmentalized LC habitat/stabling context, and inter-pollution among units cannot be ruled out. This is not surprising for layers that appear interspersed one another.

Comparing the EC and LC contexts as whole individual archaeo-stratigraphic entities, we notice that, even if the amount of residue tested for the LC (266.75 liters) is higher than for the EC (112 liters), the NISP and MNI are clearly higher in the latter, while the number of taxa is slightly lower. Presence of Pleistocene allochthonous components in the LC may explain the latter phenomenon but the former may correspond to a more complex scenario: prepared/activity floors and fumiers are all products of human endeavors; consequently, avian predators and small-mammalian carnivores (main agents responsible for the accumulations of small-mammal remains) were not able to enter the cave with the frequency they did when it was used mainly for burying the dead. NISP and MNI at the funerary context are higher than at the habitat/stabling context, and this is likely because humans and stabled animals scared away the predators. When the cave was a sacred space, during the EC, with minimal presence of people, predators were freer to enter. This possibility is supported by the low synanthropic index (35.58%) obtained elsewhere for these levels [47]. Human-induced disturbances on small-mammal palaeocommunities (i.e., bats) are already known in other cave sites of the Atapuerca complex [10].

Cuenca-Bescós et al. [4,48] defined seven biostratigraphic units for the Sierra de Atapuerca sequence (i.e., Faunal Units (FU) 1–7), the uppermost of them, FU 7, corresponding to the latest Pleistocene small-vertebrate association of El Portalón [32]. An updated version of this biostratigraphic series [5] adds two more faunal units, being FU 8 characterized mainly by the first appearance of the species *M. (T.) lusitanicus* at the Chalcolithic levels of El Portalón [19]. Subsequently, Bañuls-Cardona et al. [37] contributed a detailed study of the small-mammal assemblages from the contiguous cave of El Mirador, with a chrono-stratigraphy going from the latest Pleistocene up to the late Holocene. In her Ph.D. thesis, Bañuls-Cardona [47] presented a study of the EC (UE79) and LC (UE85) small-mammal associations of El Portalón, mainly intended for the reconstruction of past climate and environments; this study was based on different samples than our own study.

Considering the small-mammal associations and chronologies presented in these previous works and here, we can state that the early Chalcolithic levels of El Portalón contribute the first Holocene records of *Crocidura suaveolens, Sorex minutus, Neomys fodiens, N. anomalus, Microtus (Terricola) lusitanicus, M. pyrenaicus, Sciurus vulgaris, Rhinolophus ferrumequinum* and *Oryctolagus cuniculus* in the Sierra de Atapuerca, which are the highest occurrences of these taxa in the sequence as well, together with *Miniopterus shreibersii* and *Myotis* gr. *myotis-blythii*. The latter two appear at El Mirador earlier in the Holocene [37]. Faunal Unit 8 of Atapuerca may be redefined accordingly.

Extending these biostratigraphic remarks to the LC levels could be misleading as their archaeo-stratigraphic units present an evident Pleistocene pollution from the prepared floors; same is applicable for an estimation of the past environment based on these materials. However, we can say that the Pleistocene genesis of the prepared floors is supported by the presence of *Chionomys nivalis*, a good marker of cold conditions, and the limited number of *Apodemus sylvaticus* and *Eliomys quercinus* compared to other units, both these taxa representatives of warm and forested environments [6–8,32,49].

If the EC funerary context remained relatively unbiased until the moment of its excavation, as our results appear to show, then we can give some insights on the environment in the vicinity of the cave, while also keeping in mind that most of the small-mammal samples were accumulated by an opportunistic avian predator randomly sampling the surroundings (see Materials and Methods).

Following Table 2, the general balance obtained from this small-mammal association is that of a patchy landscape dominated by forest, shrubland and grassland, but with a moderate representation of inland wetlands and rocky areas. The estimation is not in contradiction with that obtained by Bañuls-Cardona [47] for the same levels; therefore, they should be considered complementary.

The large wild mammals and other vertebrates identified up to now at the EC funerary context constitute around 2% of the total (the latter again dominated by domestic taxa by far), including *Cervus elaphus* (0.99%), *Capreolus capreolus* (0.18%), *Vulpes* (0.09%), *Mustela* sp. (0.18%) and *Chelonia* indet.: (0.45%) [44]. *Cervus elaphus, C. capreolus* and *V. vulpes* currently inhabit woodland, shrubland and grassland zones. *Cervus elaphus* may be also found in rocky areas, and *C. capreolus* and *V. vulpes* are found in inland wetlands.

Altogether, this ecological distribution corresponds to generally warm–temperate conditions, similar to those recorded for equivalent ages at the Cantabrian range sites of El Mirón (upper part [6]) and the level III of Peña Larga [49]. Said level III of Peña Larga is especially relevant here, as it represents a collective burial episode during which the cave was minimally or not used as a household at all. As in El Portalón, the rest of the sequence of Peña Larga (i.e., early Neolithic to middle Bronze Age) shows evidence of domestic use. Moreover, the level II of Peña Larga, attributed to the late Chalcolithic, exhibits several large alternating lenses, some reddish and some whitish, which probably correspond to the occasional burning of barns or areas occupied by animals [49]. This is again very similar to what we observe at the LC habitat/stabling levels of El Portalón.

El Portalón and Peña Larga similarities are a good example of regional environmental conditions framing similar cultural dynamics (e.g., funerary, habitat and stabling) during the Chalcolithic in north Iberia; this deserves further exploration in other long Holocene archaeo-stratigraphic sequences.

The environmental insights contributed by Bañuls-Cardona [47] and here for El Portalón EC levels (ca. 5294–4732 cal yr BP) fill the chronological gap between the middle (7060–5350 cal yr BP) and Late (4330–3070 cal yr BP) Holocene paleoclimatic and paleo-environmental reconstructions carried out for the neighbor sequence of El Mirador [37] at the Sierra de Atapuerca.

Bañuls-Cardona [47] also obtained an environmental estimation for the LC levels of El Portalón, but we recommend caution with these results, as the sampled material comes from a unit (UE85) mistakenly regarded as a prepared floor. The stratigraphic Unit 85 actually corresponds to natural sediments deposited in the site during a short human occupation hiatus, archaeologically sterile, just prior to its funerary use, detected in the stratigraphic unit 79. However, as interpreted by Bañuls-Cardona [47], the small-mammal association and paleo-environmental inferences are very similar to those observed during the funerary context of the EC (UE79).

5. Conclusions

The associations of small mammals of different archaeo-stratigraphical contexts from the Chalcolithic levels of El Portalón de Cueva Mayor site were successfully used here to characterize them as independent entities. The exclusive presence of certain taxa together with the statistical difference in relative taxonomic ratios point to the integrity and unpolluted condition of the EC Pre-Bell-Beaker Funerary context, which likely remained free of percolations from the overlying LC late Pre-Bell-Beaker and Bell-Beaker habitat and stabling context. The compartmentalization and interspersed arrangement of the different units composing the LC context (prepared floors, activity floors and fumiers) made them prone to inter-pollution, as they do not appear to be statistically different in terms of relative taxonomic ratios.

The presence of *Pliomys lenki* and *Chionomys nivalis* in the prepared floors and their sedimentological composition reveal its allochthonous endokarstic nature, being both rodent species that are Upper Pleistocene in age.

The EC levels of El Portalón contribute the first Holocene records of *Crocidura suaveolens, Sorex minutus, Neomys fodiens, N. anomalus, Microtus (Terricola) lusitanicus, M. pyrenaicus, Sciurus vulgaris, Rhinolophus ferrumequinum* and *Oryctolagus cuniculus* in the Sierra de Atapuerca archaeo-paleontological complex, which will lead to the redefinition of the Faunal Unit 8 of Atapuerca.

A patchy environment dominated by woodland, shrubland and grassland, with moderate presence of inland wetlands and rocky areas, is inferred from the small-mammal association of the EC levels of El Portalón. This ecological distribution matches well with warm–temperate conditions recorded in other sites of similar age and cultural affiliation from the neighboring Cantabrian region.

The Upper Pleistocene origin of the prepared floors included in the LC habitat and stabling context argues against the use of its small-mammal association either for biostratigraphy or estimation of past environment.

Author Contributions: Conceptualization, J.R., A.O. and E.I.; methodology, J.R. and A.O.; validation, J.R., A.O., E.I., G.C.-B., M.Á.G.-P., A.P.-R., J.M.C. and J.L.A.; formal analysis and investigation, J.R. and A.O.; resources, E.I., J.M.C. and J.L.A.; data curation, J.R. and A.O.; writing—original draft preparation, J.R., A.O., E.I., G.C.-B. and M.Á.G.-P.; writing—review and editing, all; visualization, J.R., A.O. and E.I.; supervision, J.R.; funding acquisition, all. All authors have read and agreed to the published version of the manuscript.

Funding: This research was supported by the Spanish Ministerio de Ciencia, Innovación y Universidades Project "Geología, Geocronología y Paleobiolología de los yacimientos de la Sierra de Atapuerca VII" (PGC2018-093925-B-C33).

Data Availability Statement: Data are contained within the article.

Acknowledgments: The authors would like to express their gratitude to the El Portalón de Cueva Mayor archaeological site excavation team. M.A.G.-P. has a technical support staff contract (PTA2018-015145-I) at the Fundación General de la Universidad de Alcala de Henares-Museo Arqueológico Regional (MAR) financed by the Spanish Ministerio de Economía, Industria y Competitividad and the MAR. We also would like to thank Quaternary administrative editors for their permanent assistance and the comments of two anonymous reviewers to improve the paper.

Conflicts of Interest: The authors declare no conflict of interest.

References

1. Mellet, J.S. Scatological origin of microvertebrate fossil accumulations. *Science* **1974**, *185*, 349–350. [CrossRef] [PubMed]
2. Andrews, P.O. *Caves and Fossils: Predation, Preservation and Accumulation of Small Mammal Bones in Caves, with an Analysis of the Pleistocene Cave Faunas from Westbury-Sub-Mendip, Somerset, UK*; University of Chicago Press: Chicago, IL, USA, 1990.
3. Fernández-Jalvo, Y.; Andrews, P.; Denys, C.; Sesé, C.; Stoetzel, E.; Marin-Monfort, D.; Pesquero, D. Taphonomy for taxonomists: Implications of predation in small mammal studies. *Quat. Sci. Rev.* **2016**, *139*, 138–157. [CrossRef]
4. Cuenca-Bescós, G.; Rofes, J.; López-García, J.M.; Blain, H.-A.; De Marfá, R.J.; Galindo-Pellicena, M.A.; Bennásar-Serra, M.L.; Melero-Rubio, M.; Arsuaga, J.L.; Bermúdez de Castro, J.M.; et al. Biochronology of Spanish Quaternary small vertebrate faunas. *Quat. Int.* **2010**, *212*, 109–119. [CrossRef]
5. Cuenca-Bescós, G.; Blain, H.-A.; Rofes, J.; López-García, J.M.; Lozano-Fernández, I.; Galán, J.; Núñez-Lahuerta, C. Updated Atapuerca biostratigraphy: Small-mammal distribution and its implications for the biochronology of the Quaternary in Spain. *C. R. Palevol* **2016**, *15*, 621–634. [CrossRef]
6. Cuenca-Bescós, G.; Straus, L.G.; González-Morales, M.R.; García-Pimienta, J.C. The reconstruction of past environments through small mammals: From the Mousterian to the Bronze Age in El Mirón Cave (Cantabria, Spain). *JAS* **2009**, *36*, 947–955. [CrossRef]
7. López-García, J.M.; Berto, C.; Colamussi, V.; Dalla Valle, C.; Lo Vetro, D.; Luzi, E.; Malavasi, G.; Martini, F.; Sala, B. Palaeoenvironmental and palaeoclimatic reconstruction of the Latest Pleistocene-Holocene sequence from Grotta del Romito (Calabria, southern Italy) using the small-mammal assemblages. *Palaeogeogr. Palaeoclimatol. Palaeoecol.* **2014**, *409*, 169–179. [CrossRef]
8. Rofes, J.; Murelaga, X.; Martínez-García, B.; Bailon, S.; López-Quintana, J.C.; Guenaga-Lizasuc, A.; Ortega, L.A.; Zuluaga, M.C.; Alonso-Olazabal, A.; Castaños, J.; et al. The long paleoenvironmental sequence of Santimamiñe (Bizkaia, Spain): 20,000 years of small mammal record from the latest Late Pleistocene to the middle Holocene. *Quat. Int.* **2014**, *339–340*, 62–75. [CrossRef]
9. Rofes, J.; Garcia-Ibaibarriaga, N.; Aguirre, M.; Martínez-García, B.; Ortega, L.; Zuluaga, M.C.; Bailon, S.; Alonso-Olazabal, A.; Jone Castaños, J.; Murelaga, X. Combining Small-Vertebrate, Marine and Stable-Isotope Data to Reconstruct Past Environments. *Sci. Rep.* **2015**, *5*, 14219. [CrossRef]
10. Galán, J.; Núñez-Lahuerta, C.; López-García, J.; Cuenca-Bescós, G. Did humans disturb bats? Exploring the hominin-chiropter interactions in the Sierra de Atapuerca sites (early to Middle Pleistocene, Spain). *Quat. Sci. Rev.* **2019**, *226*, 106018. [CrossRef]
11. Cucchi, T.; Barnett, R.; Martínková, N.; Renaud, S.; Renvoisé, E.; Evin, A.; Sheridan, A.; Mainland, I.; Wickham-Jones, C.; Tougard, C.; et al. The changing pace of insular life: 5000 Years of microevolution in the Orkney vole (*Microtus Arvalis Orcadensis*). *Evolution* **2014**, *68*, 2804–2820. [CrossRef]
12. Rofes, J.; Cucchi, T.; Hanot, P.; Herman, J.; Stephan, P.; Cersoy, S.; Ivan Horácek, I.; Kerr, E.; Allberry, K.; Valenzuela, S.; et al. Postglacial recolonization and Holocene diversification of *Crocidura suaveolens* (Mammalia, Soricidae) on the north-western fringe of the European continent. *Quat. Sci. Rev.* **2018**, *190*, 1–10. [CrossRef]
13. Wood, W.R.; Lee Johnson, D. 9-A Survey of Disturbance Processes in Archaeological Site Formation. In *Advances in Archaeological Method and Theory*; Schiffer, M.B., Ed.; Academic Press: San Diego, CA, USA, 1978; pp. 315–381.
14. Texier, J. A propos des processus de formation des sites préhistoriques/About prehistoric site formation processes. *Paléo, Revue d'Archéologie Préhistorique* **2000**, *12*, 379–386. [CrossRef]
15. Dalland, M.; Carter, S. The evaluation of a prehistoric mound damaged by rabbit burrowing at Maryton Law, Angus. *Tayside Fife Archaeol. J.* **1998**, *4*, 20–30.
16. Mallye, J.-B. Badger (*Meles meles*) remains within caves as an analytical tool to test the integrity of stratified sites: The contribution of Unikoté Cave (Pyrénées-Atlantiques, France). *J. Taphon.* **2011**, *9*, 15–36.

17. Pelletier, M.; Brugal, J.-P.; Cochard, D.; Lenoble, A.; Mallye, J.-B.; Royer, A. Identifying fossil rabbit warrens: Insights from a taphonomical analysis of a modern warren. *JAS Rep.* **2016**, *10*, 331–344. [CrossRef]
18. Rofes, J.; Cersoy, S.; Zazzo, A.; Royer, A.; Nicod, P.-Y.; Laroulandie, V.; Langlais, M.; Pailler, Y.; Leandri, C.; Leandri, F.; et al. Detecting stratigraphical issues using direct radiocarbon dating from small-mammal remains. *J. Quat. Sci.* **2020**, *35*, 505–513. [CrossRef]
19. Ordiales, A.; Rofes, J.; Iriarte, E.; Mardones, V.; Cuenca-Bescós, G.; Carretero, J.M.; Arsuaga, J.L.; García-Ibaibarriaga, N. Small mammals as tools to characterize archaeological contexts in the Chalcolithic of Portalón de Cueva Mayor (Atapuerca, Burgos, Spain). In Proceedings of the 1st Young Natural History Scientists' Meeting, Paris, France, 12–14 February 2014; p. 12.
20. Carbonell, E.; Bermúdez de Castro, J.M.; Arsuaga, J.L.; Díez, J.C.; Rosas, A.; Cuenca-Bescós, G.; Sala, R.; Mosquera, M.; Rodríguez, X.P. Lower Pleistocene hominids and artifacts from Atapuerca TD6 (Spain). *Science* **1995**, *269*, 826–830. [CrossRef]
21. Carbonell, E.; Bermúdez de Castro, J.M.; Parés, J.M.; Pérez-González, A.; Cuenca-Bescós, G.; Olle, A.; Mosquera, M.; Huguet, R.; van der Made, J.; Rosas, A.; et al. The first hominin of Europe. *Nature* **2008**, *452*, 465–470. [CrossRef]
22. Arsuaga, J.L.; Bermúdez de Castro, J.M.; Carbonell, E. (Eds.) The Sima de los Huesos hominid Site. *J. Hum. Evol.* **1997**, *33*, 105–421. [CrossRef]
23. Bermúdez de Castro, J.M.; Carbonell, E.; Arsuaga, J.L. (Eds.) Gran Dolina Site: TD6 Aurora Stratum (Burgos, Spain). *J. Hum. Evol.* **1999**, *37*, 309–700.
24. Clark, G.; Straus, L.G.; Burton, S.S.; Jackson-Clark, V. The North Burgos Archaelogical Survey: An inventory of cultural remains. In *The North Burgos Archaelogical Survey. Bronze and Iron Age archaeology on the Meseta del Norte (Province of Burgos, North-Central Spain) 19, 18 156*; Clark, G., Ed.; Arizona State University, Department of Antropology: Tempe, AZ, USA, 1979.
25. Apellaniz, J.M.; Domingo, D. Estudios sobre Atapuerca (Burgos): II. Los materiales de superficie del Santuario de la Galería del Sílex. *Cuad. Arqueol. Univ. Deusto* **1987**, *10*, 1–342.
26. Carretero, J.M.; Ortega, A.I.; Juez, L.; Pérez-González, A.; Arsuaga, J.L.; Pérez, R.; Ortega, M.C. A Late Pleistocene-Early Holocene archaeological sequence of Portalón de Cueva Mayor (Sierra de Atapuerca, Burgos, Spain). *Munibe* **2008**, *59*, 67–80.
27. Pérez-Romero, A.; Carretero, J.M.; Juez, L.; Ortega, A.I.; Ortega, M.C.; Arsuaga, J.L. Una dobla almohade del Siglo XIII en el yacimiento de El portalón de Cueva Mayor, Sierra de Atapuerca (Burgos). *Nvmisma* **2010**, *254*, 85–106.
28. Pérez-Romero, A.; Carretero, J.M.; Alday, A.; Arsuaga, J.L. La Cerámica Protohistórica e Histórica en el yacimiento de El Portalón de Cueva Mayor, Sierra de Atapuerca, Burgos. *Bol. Soc. Esp. Ceram. Vidr.* **2013**, *52*, 183–193. [CrossRef]
29. Pérez-Romero, A.; Iriarte, E.; Galindo-Pellicena, M.A.; García-González, R.; Rodríguez, L.; Castilla, M.; Francés-Negro, M.; Santos, E.; Valdiosera, C.; Arsuaga, J.L.; et al. An unusual Pre-bell beaker copper age cave burial context from El Portalón de Cueva Mayor site (Sierra de Atapuerca, Burgos). *Quat. Int.* **2017**, *433*, 142–155. [CrossRef]
30. Vergés, J.M.; Allué, E.; Angeluccí, D.E.; Cebriá, A.; Fontanals, M.D.; Mányanos, A.; Montero, S.; Moral, S.; Vaquero, M.; Zaragoza, J. La sierra de Atapuerca durante el Holoceno: Datos preliminares sobre las ocupaciones de la Edad del Bronce en la cueva de Mirador (Ibeas de Juarros, Burgos). *Trab. Prehist.* **2002**, *59*, 107–126. [CrossRef]
31. Vergés, J.M.; Allué, E.; Angeluccí, D.E.; Burjachs, F.; Carrancho, A.; Cebriá, A.; Expósito, I.; Fontanals, M.; Moral, S.; Rodríguez, A.; et al. Los niveles neolíticos de la cueva del Mirador (Sierra de Atapuerca Burgos): Nuevos datos sobre la implantación y el desarrollo de la economía agropecuaria en la submeseta norte. In *Proceedings of the IV Peninsular Neolithic Congres, MARQ, Alicante, Spain*; Hernández, M.S., Soler, J.A., López, J.A., Eds.; Museo Arqueológico de Alicante, Diputación de Alicante: Alicante, Spain, 2008; pp. 418–427.
32. López-García, J.M.; Blain, H.-A.; Cuenca-Bescós, G.; Ruiz-Zapata, M.B.; Dorado-Valiño, M.; Gil-García, M.J.; Valdeolmillos, A.; Ortega, A.I.; Carretero, J.M.; Arsuaga, J.L.; et al. Palaeoenvironmental and palaeoclimatic reconstruction of the Latest Pleistocene of El Portalón Site, Sierra de Atapuerca, northwestern Spain. *Palaeogeogr. Palaeoclimatol. Palaeoecol.* **2010**, *292*, 453–464. [CrossRef]
33. Castilla, M.; Carretero, J.M.; Gracia, A.; Arsuaga, J.L. Evidence of Rickets and/or Scurry in a complete Chalcolithic Child Skeleton from the El Portalón Site (Sierra de Atapuerca, Spain). *JASs Rep.* **2014**, *92*, 1–16.
34. Günther, T.; Valdiosera, C.; Malmström, H.; Ureña, I.; Rodriguez-Varela, R.; Osk Sverrisdóttir, O.; Daskalaki, E.A.; Skoglund, P.; Naidoo, T.; Svensson, E.M.; et al. Ancient genomes link early farmers from Atapuerca in Spain to modern-day Basques. *Proc. Natl. Acad. Sci. USA* **2015**. [CrossRef]
35. Reimer, P.J.; Austin, W.E.N.; Bard, E.; Bayliss, A.; Blackwell, P.G.; Ramsey, C.B.; Butzin, M.; Cheng, H.; Edwards, R.L.; Friedrich, M.; et al. The IntCal20 Northern Hemisphere Radiocarbon Age Calibration Curve (0–55 cal kBP). *Radiocarbon* **2020**, *62*, 725–757. [CrossRef]
36. Arsuaga, J.L.; Gómez-Olivencia, A.; Sala, N.; Martínez-Pillado, V.; Pablos, A.; Bonmatí, A.; Pantoja, A.; Lira, J.; Alcázar de Velasco, A.; Ortega, A.I.; et al. Evidence of paleoecological changes and Mousterian occupations at the Galería de las Estatuas site, Sierra de Atapuerca, northern Iberian plateau, Spain. *Quat. Res.* **2017**, *88*, 345–367. [CrossRef]
37. Bañuls-Cardona, S.; López-García, J.M.; Morales-Hidalgo, J.I.; Cuenca-Bescós, G.; Vergés, J.M. Late Glacial to Late Holocene palaeoclimatic and palaeoenvironmental reconstruction of El Mirador cave (Sierra de Atapuerca, Burgos, Spain) using the small-mammal assemblages. *Palaeogeogr. Palaeoclimatol. Palaeoecol.* **2017**, *471*, 71–81. [CrossRef]
38. Cuenca-Bescós, G.; Rossell, J.; Morcillo-Amo, A.; Galindo-Pellicena, M.A.; Santos, E.; Moya-Costa, R. Beavers (Castoridae, Rodentia, Mammalia) from the Quaternary sites of the Sierra de Atapuerca, in Burgos, Spain. *Quat. Int.* **2017**, *433*, 263–277. [CrossRef]

39. Benito-Calvo, A.; Ortega, A.I.; Pérez-González, A.; Campaña, I.; Bermúdez de Castro, J.M.; Carbonell, E. Palaeogeographical reconstruction of the Sierra de Atapuerca Pleistocene sites (Burgos, Spain). *Quat. Int.* **2017**, *433*, 379–392. [CrossRef]
40. Wilson, D.E.; Reeder, D.-A.M. Mammal Species of the World. In *A Taxonomic and Geographic Reference*; The Johns Hopkins University Press: Baltimore, MD, USA, 2005.
41. IUCN Red List of Threatened Species. Available online: https://www.iucnredlist.org/ (accessed on 23 February 2021).
42. Hammer, Ø.; Harper, D.A.T.; Ryan, P.D. PAST: Paleontological Statistics Software Package for education and data analysis. *Palaeontol. Electron.* **2001**, *4*, 9. Available online: http://folk.uio.no/ohammer/past/ (accessed on 1 October 2018).
43. Cuenca-Bescós, G.; Straus, L.G.; García-Pimienta, J.C.; González Morales, M.R.; López-García, J.M. Late Quaternary small mammal turnover in the Cantabrian Region: The extinction of *Pliomys lenki* (Rodentia, Mammalia). *Quat. Int.* **2010**, *212*, 129–136. [CrossRef]
44. Galindo-Pellicena, M.A. Estudio de la macrofauna de los niveles holocenos del yacimiento de El Portalón (sierra de Atapuerca, Burgos). Ph.D. Thesis, Department of Paleontology, Complutense University of Madrid, Madrid, Spain, 2014.
45. Galindo-Pellicena, M.A.; Arsuaga, J.L.; Carretero, J.M. Pig management in the Chalcolithic and Bronze Age periods at the El Portalón site: Integration in the Iberian context. *Quat. Int.* **2019**, *515*, 138–149. [CrossRef]
46. Galindo-Pellicena, M.A.; Arsuaga, J.L.; Pérez-Romero, A.; Iriarte, E.; DeGaspar, I.; Carretero, J.M. Metrical analysis of bovine bone remains from the Neolithic to the Bronze Age at the El Portalón site (Atapuerca, Burgos) in the Iberian context. *Quat. Int.* **2020**, *566–567*, 211–213. [CrossRef]
47. Bañuls-Cardona, S. Human Impact on Small-Mammals from Late Glacial to Late Holocene of Western Mediterranean Region: New Environmental and Climate Approach. Ph.D. Thesis, Università degli Studi di Ferrara, Ferrara, Italy, 2017.
48. Cuenca-Bescós, G.; García, N. Biostratigraphic succession of the Early and Middle Pleistocene mammal faunas of the Atapuerca cave sites (Burgos, Spain). *Cour. Forschunginst. Senckenberg* **2007**, *259*, 99–110.
49. Rofes, J.; Zuluaga, M.C.; Murelaga, X.; Fernández-Eraso, J.; Bailon, S.; Iriarte, M.J.; Ortega, L.A.; Alonso-Olazabal, A. Palaeoenvironmental reconstruction of the early Neolithic to middle Bronze Age Peña Larga rock shelter (Álava, Spain) from the small mammal record. *Quat. Res.* **2013**, *79*, 158–167. [CrossRef]

quaternary

MDPI

Article

San Josecito Cave and Its Paleoecological Contributions for Quaternary Studies in Mexico

Joaquin Arroyo-Cabrales [1,2,*], Eileen Johnson [2] and J. Alberto Cruz [1]

1 Laboratorio de Arqueozoología, Subdirección de Laboratorios y Apoyo Académico, I.N.A.H., Moneda 16, Col. Centro, Ciudad de México 06060, Mexico; cruzsilvajac@yahoo.com.mx
2 Museum of Texas Tech University, Lubbock, TX 79409-3191, USA; eileen.johnson@ttu.edu
* Correspondence: arromatu5@yahoo.com.mx

Abstract: San Josecito Cave (2250 m elevation) is located nearby Aramberri, Nuevo León, northeastern Mexico, with excavations occurring in 1935–1941 and 1990. It is a paleontological cave and the significance of its faunal data rests in the understanding of the Quaternary ecosystems of the Mexican Plateau and the Southern Plains. This significance is underpinned by a consideration of associated stratigraphic and geochronological data. The fauna is composed of mollusks, amphibians, reptiles, birds, and mammals. More than 30 extinct vertebrate species have been identified, constituting one of the most important Quaternary localities in the Americas. Radiocarbon dates and faunal correlations indicate the excavated deposits represent an interval of time between 45,000 and 11,000 ^{14}C years BP. The current synthesis demonstrates that the previous view of the assemblage as a single local fauna is erroneous and that, instead, several successive local faunas are present within a stratigraphic framework. This finding underscores the need for detailed studies of single localities in building paleoenvironmental models. As a corollary, results point to the necessity of including all vertebrate classes represented from a locality in building those models. In addition, the field and analytical methodologies demonstrate the importance of very detailed paleontological excavations, with precise spatial and temporal controls, to assess the taphonomic history of a locality, construct a stratigraphic and geochronological framework, and infer the paleoecological conditions during the time span considered based on the number of local faunas represented. The recognition of San Josecito Cave as an important Late Pleistocene vertebrate paleontological locality is enhanced with the consideration of its faunal data for paleoenvironment reconstruction and possible contribution to Quaternary paleoclimatic modeling.

Keywords: late Pleistocene; cave stratigraphy and taphonomy; Sierra Madre Oriental; non-analog fauna; paleoenvironmental reconstruction

1. Introduction

Paleoenvironmental research data are a resource that should be managed in a way that provides the entire global-change research community with the tools to find, access, and manipulate quickly the data needed for a particular investigation [1]. The Past Global Changes Project [2] has identified several types of data that need to be given high priority for global change studies. The first priority is the original "raw" or primary paleoclimate, palaeoecological, and palaeoceanographic data, including associated chronological information. These raw data are the foundation for paleoenvironmental research, and the ultimate source of paleoenvironmental information.

Documentation of Quaternary environmental change generally follows two complimentary paths [3]. The first examines proxy records that define the state of climatic and biological systems at various times in the past and provides empirical data necessary for testing and refinement of paleoclimatic models. The second focuses on reconstructing the response of surface systems to climatic and environmental change. This synthesis is concerned primarily with the first pathway.

A fundamental goal of Quaternary vertebrate paleontology is to understand the interrelationships between environmental and faunal changes. Such relationships, implied by the cyclic pattern of change in the Quaternary, are complex and incompletely understood [4]. Since integration of data from individual localities is required, it is critical to assure the best understanding for each of those localities.

Sedimentological analysis of paleoenvironments and the taphonomic study of organic remains preserved in these environments are essential components of paleoecology [5]. The disciplines of sedimentology and taphonomy are necessary for the reconstruction of the original ecological relationships among fossil organisms. Sedimentology provides information on the physical settings where organisms lived and were buried and on interactions between biotic and abiotic components of ancient ecosystems [5].

Taphonomy plays diverse roles in paleobiology including ecological, biogeographical, and evolutionary questions, as well as diagnosing the roles of various taphonomic agents, processes, and circumstances in generating fossil records [6]. Taphonomy provides the methodology to delimit and categorize the conditions and processes leading to the preservation of fossil organisms [7]. This approach illuminates biases that affect paleoecological reconstructions, provides additional information of biological processes represented in a fossil assemblage, and allows the identification of similar (isotaphonomic) assemblages among which ecological comparisons can be made.

During the Late Pleistocene, the North American Southern Plains grasslands appears to have extended from Oklahoma and the Southern High Plains to the Texas Gulf Coastal Plains and into México through the Mexican Plateau. This grassland ecosystem appears to have been a parkland setting with trees primarily restricted to the waterways [8–10]. A detailed biotic record in the northern part of the Southern Plains grasslands for the last 40,000 years is available [8,11–14]. In order to gain a regional perspective, a comparable time-transgressive database for the southernmost biotic section is needed. San Josecito Cave (western flank of the Sierra Madre Oriental; southern Nuevo León, Mexico; Figure 1) is one of two known localities with a great potential for a detailed Late Pleistocene record in this southernmost section [15]. Rancho La Amapola, Cedral, San Luis Potosí, is the other locality (an open-air locality rather than a cave). It has a rich grassland mammal fauna consisting of 7 orders, 18 families, and 33 species [16].

Based on numerous previous taxonomic studies of the cave materials, the San Josecito Cave fauna was considered an example of a disharmonious or non-analog fauna [17,18]. The fauna was composed of elements characterizing different climatic regimes and presumably ecological incompatibles, coexisting nowhere in the world at present [18,19]. The fauna had six new fossil species, comprised of the horned lizard (*Phrynosoma josecitensis*), big-footed turkey (*Meleagis crassipes*), extinct thicker-billed parrot (*Rhynchopsitta phillipsi*), extinct vampire bat (*Desmodus stocki*), dwarf rabbit (*Sylvilagus leonensis*), and spectacled bear (*Tremarctos floridanus*). The fauna also included eight extinct birds and 19 extinct mammals [20–22]. This disharmonious biota could have been maintained by equable climates with reduced seasonal extremes that have no modern analogs [23]. No specific study addressed taphonomic or stratigraphic analyses of the San Josecito Cave materials, and only short notes were made on the possible causes of bone accumulation [24–26].

This synthesis is the first of its kind to bring together the San Josecito Cave data by focusing on three aspects: (1) a consideration of the overall cave fauna that resulted from Stock's [26] work; (2) results of the 1990 renewed excavation and associated analyses; and (3) the significance that the composite stratum 720 (i.e., combined stratum 720 and Stock's 57–58 ft level taxa) faunal data has for contributing to the understanding of the Quaternary ecosystems of the Mexican Plateau, Sierra Madre Oriental, and Southern Plains in general. The 1990 excavation and subsequent analyses undertaken were aimed at: (1) establishing a stratigraphic framework and chronology; (2) determining the taphonomic history of the cave fauna; (3) examining the collection and documentation now housed at the Natural History Museum of Los Angeles County (LACM) from Stock's [26] excavations; and (4) within the established stratigraphic framework, inferring possible paleoenvironmental scenarios in which the cave faunas could have existed. The 1990 stratigraphically-controlled excavation was conducted by the first two authors [27–31]. A working hypothesis was that no drastic climatic change occurred during the deposition of the studied strata in the mid-Wisconsin period (40,000 to 25,000 ^{14}C years BP).

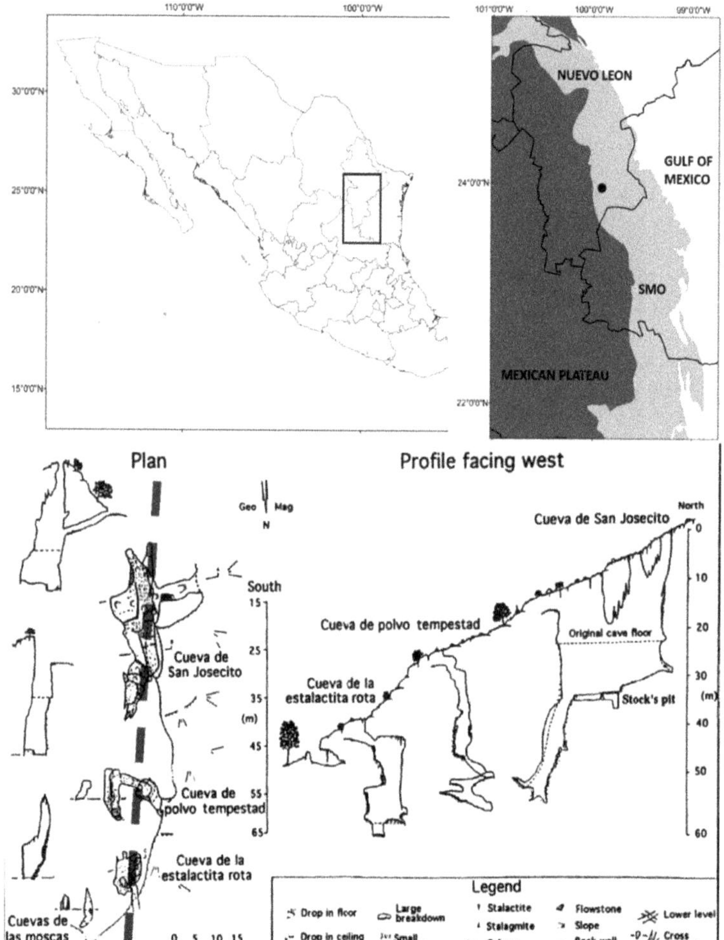

Figure 1. Location of San Josecito Cave (black point) in the Sierra Madre Oriental (SMO), Mexico, with plan and profile (dashed lines) facing west of San Josecito Cave. Modified from [28] and adapted with permission from George Veni, National Speleological Society.

2. Locality Setting

The cave is a single drop, multi-entrance fissure that occurs in folded Late Jurassic or Early Cretaceous limestone [26]. All three now-open natural entrances descend vertically 12 to 30 m to the main room or to a sloping tunnel into the main room (Figure 1). None of the entrances provides easy access today and probably did not in the past; a horizontal or walking entrance probably never existed. This situation implies that the cave acted as a natural pit trap for the larger animals. Smaller animals and birds would have been brought in by raptors or from roosting in the cave. Its use as a den (as proposed by [15,25]) would not have been feasible given the lack of easy access or a walking entrance.

San Josecito Cave consists of only one vertical room tapering to the north, 34 m long and 25 m wide, currently with three vertical shafts. Two of the shafts are high (30 m) above the present cave floor and the third consists of a large upper opening (20 m high) that probably served as the primary trap for the fauna (Figure 1). A separate tunnel descends 41° from the east to a 7 m drop to the floor of the main room. This tunnel entrance is an artificial one opened in the late 1930s to facilitate paleontological excavation. A fourth natural entrance, closed in the late 1930s by local inhabitants, drops about 5 m into the tunnel via a narrow shaft.

A vertical shaft was excavated about 30 m down the south headwall. This shaft was excavated by local people looking for treasure prior to paleontological excavations (Stock's correspondence, 1935, on file at the LACM). Field observations (George Veni, personal communication, 1989) of the shaft indicated the presence of at least 30 m of fossiliferous sediments below the current excavation levels.

On the northern and western walls, thin outgrowths of a calcareous tufa extend the depth of the cave, probably from the original cave floor. The western portion is formed by a boulder-size outgrowth. Enormous amounts of bones are embedded in a hardened matrix on the tufa.

3. Material and Methods

3.1. Stock's Excavations

Under the direction of Stock [26], field crews representing the California Institute of Technology quarried a section of San Josecito Cave between 1935 and 1941. The cave floor was divided into five north-to-south blocks (Figure 2). Workers shovel-excavated in 5 ft levels in each block, removing all sediments (primarily dumped outside the cave entrance) and bone within the cave from the upper 12 m. This work created a new cave floor (the present one) that corresponds with Stock's 55 ft level. At that level, the field crew excavated a test pit in block 2 (southern part of the cave) where they had found the highest concentration of bone. The test pit, now known as Stock's pit, was a 2 × 2 m unit excavated 4.5 m deep. A smaller unit adjacent to (on the east) and contiguous with the larger unit was 1.8 × 1.3 m and excavated 1.8 m deep [28]. The extent of dry screening or screen mesh size was unknown. No wet screening (i.e., sediments water-washed through one or more nested screen or burlap bags) would have been carried out as that technique to recover microvertebrates primarily was a post-WWII development [32].

3.2. The 1990 Excavation and Associated Research

Five steps were undertaken to obtain the basic data for the 1990 work and subsequent study. The first step was the location and reconnaissance of the cave and the establishment of horizontal and vertical controls [30]. Bringing control into the cave proved difficult due to the great distance through rough terrain between the known reference point (INEGI #3146) and the cave entrance. Nevertheless, the datum established at the cave was reliable as it was obtained through two different methods, field survey instrumentation [33] and GPS calculation. Inside the cave, marks on the walls (left by Stock's [26] crew) were found that helped tie the elevation data to the provenience data salvaged from Stock's materials and documentation. The north and west walls of Stock's pit were cleaned and photographed. The deposits showed a great potential for more than 4.5 m of fossiliferous sediments below the current cave floor [28]. Remnants of the upper strata also were located in various parts of the cave.

The second step was the review of most of the materials and available documentation regarding the cave excavations by Cal Tech personnel in 1935–1941 [29,30]. This review proved helpful in reconstructing the history and methodology of those excavations, including the deficiencies with the excavations. A literature search demonstrated the taxonomic importance of the cave fauna [34–36]. Several taphonomic features were found on the bones from the collection that formed the initial analysis categories for the remains recovered in the 1990 excavation.

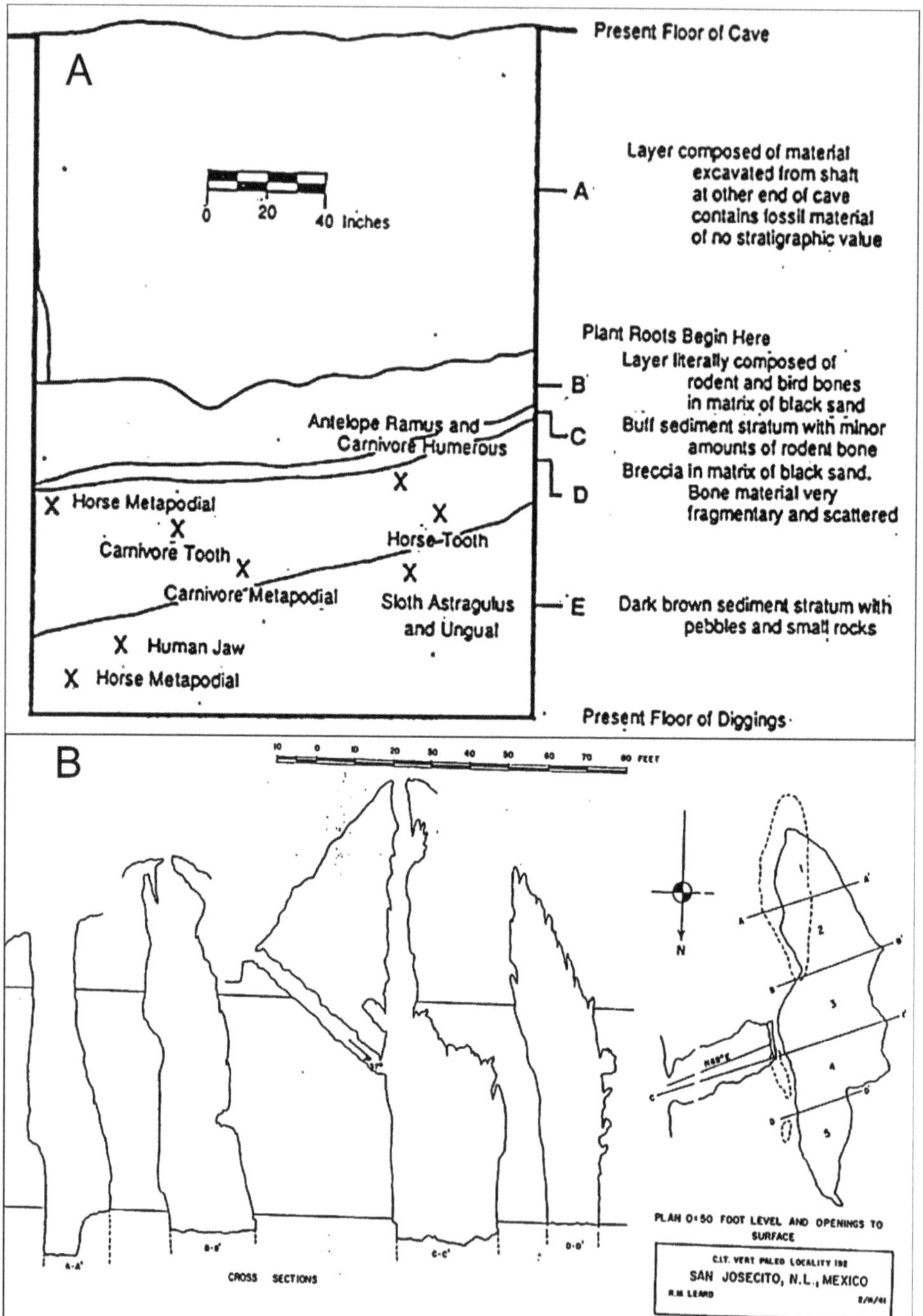

Figure 2. (**A**) Original cave profile indicating the spatial position of bones considered noteworthy (letter of Bode to Stock-Stock, personal correspondence, 19 January 1936). (**B**) Stock´s cave plan and profile showing excavations blocks.

The third step was the 1990 renewed, stratigraphically-controlled excavation to recover high-resolution data that would permit the paleoenvironmental reconstruction of the ecosystems represented by the cave fauna [31]. Four test units were placed. The two main units were adjacent to Stock's pit along the north face. A 1 × 0.3 m unit was excavated stratigraphically to stratum 570 at a depth of 1.26 m. A 50 × 50 cm unit was excavated to stratum 680 at a depth of approximately 50 cm. The third unit (30 × 30 cm; excavated 23 cm deep) was placed in an alcove (known as the Bone Alcove; in Stock's block 5) within upper deposits that corresponded to Stock's 35 ft level. The last unit (20 × 20 cm; excavated 3 cm deep) was placed in another alcove (known as the Bat Alcove; also in Stock's block 5) within upper deposits ~0.5 m above the present floor. Every step was recorded on field forms and photodocumented. All sediments were collected and several samples were sent to different specialists for study.

Several cave sediment samples were taken for insects, macrobotanical and pollen remains, and mollusks. Samples for insects, macrobotanical remains, and mollusks were water-processed through nested fine-mesh screens. Pollen samples were processed using appropriate, standard lab techniques. The sparse results were not informative [28].

In order to obtain high-quality data for the taphonomic and paleoecologic analyses, a new approach was used in the excavations. A minimal taphonomic unit (taphonomic box) was defined as a 10 × 10 cm square with thickness dependent on stratum thickness. This approach was based on the knowledge that the stratigraphic units to be encountered were no more than 10 cm thick (generally much less). Most previous paleontological studies lacked the very detailed data necessary to analyze the small vertebrates within a short span of time and space in accordance with their biology. The methods for statistical manipulation were: number of individual specimens (NISP); minimum number of elements (MNE); minimum number of individuals (MNI); and MNI by taphonomic box (BOX).

Orientation data were recorded for 88 bones found in-situ. As all but one were small bones, only direction was noted; dip and strike were obtained only for the one large bone. These data did not prove useful and have not been analyzed further.

Sediments were washed through nested screens (1.41, 0.96, and 0.56 mm) and remaining concentrates sorted for biological materials. These materials were separated into the different vertebrate classes (Amphibia, Reptilia, Aves, and Mammalia) and for mammals, different mammal sizes (small, medium, and large). The vertebrate remains were identified preliminarily.

The fourth step was the detailed analysis of the bones from stratum 720, including the paleoecology of the identified taxa [28]. Due to the enormous amount of microvertebrate material recovered from the concentrates, a single stratum was examined in detail. Stratum 720 was chosen because it had good radiometric control and the largest amount of bone material within the excavated strata based on sorted concentrate samples. Species identifications were pursued only for the cranial materials. Postcranial materials were identified as general groups and data used for taphonomic analyses and comparisons.

The fifth step was the paleoenvironmental reconstruction of composite stratum 720. The MNI and NISP values [37,38], habitat weighting analysis [39], and correspondence canonical analysis (CCA) in different taxonomic groups (small mammals, birds, and reptiles) and taxonomic groups were used.

Habitat weighting has been calculated considering the MNI and NISP since becauseboth measurements provide paleoenvironmental insights of finer resolution than taxonomic presence alone [38]. MNI is at best an ordinal-scale measurement unit. NISP is likely to provide an ordinal scale measure of taxonomic abundance at best. MNI underestimates the true abundance of some taxa, overestimates others, and accurately estimates the abundance of still others. NISP is likely to do the same, although the taxa the abundances of which are overestimated may not be the same as those that are overestimated by MNI [37].

The habitat weighting method consisted of establishing the present-day habitat distribution of each taxon. Each species received a score for each habitat type proportional to habitat preference; the sum of scores for all habitat types for each species is 1. The calculated values of habitat weighting (with presence-absence data) were multiplied by the fossil abundance (NISP and MNI) to add abundance into the analysis [39]. The vegetation types were coniferous forest, oak forest, xeric scrub, grassland, and tropical forest. The analysis was conducted in R using CCA [40] and vegan packages [41].

4. Results

4.1. Sediments

Stock's [26] (p. 3) observations on the cave sediments were limited. He noted: "The sediments consist for the most part of unconsolidated, grayish cave dirt, generally of a fine consistency. These deposits came from outside the cavern and were either washed in or accumulated through aeolian action. Often thickly interspersed in these beds are fragments of limestone, varying greatly in size, that have fallen from the walls and ceiling."

Cave sediments from different strata defined during the 1990 work were analyzed. Results [42] pointed to a minimal contribution to the cave sediments from external sources (mammal and bird bones) contrary to Stock's [26] notation. A great diversity of constituents from inside, including abiotic and biotic sources, were found in the sediments. This diversity included: (1) variable-sized fragments of primary limestone; (2) residue from dissolution of the limestone; (3) authigenically formed calcite-cemented tufa-like materials with varying degrees of porosity; and (4) apatite-rich material likely formed in guano.

4.2. Stratigraphy and Geochronology

While excavated materials from Stock's [26] work were sent back to California with excavation data, those data are lost at some point in the collection's history. Stock's collection now is only taxonomically important, as other studies are hampered by the lack of a complete stratigraphic record. The few available excavation data are meaningless regarding paleoenvironmental reconstruction. This unfortunate occurrence is quite contrary to Stock's [43,44] concern regarding the importance of recording any available stratigraphic data during paleontological excavations.

Despite that concern, Stock's descriptions of the cave stratigraphy were limited. He recognized four major units (A, B–C, D, E), each many feet thick (Figure 2). Most bone probably came from his lower three units: B-C ("5–6 feet of fine-grained brown sandstone, definitely stratified"); D ("10–15 feet of limestone breccia"); and E (brown sandstone) [28] (pp. 47–48). Most of Stock's control marks on the cave wall (indicating the 5 ft levels) were preserved and used in correlations between his work and the current study.

During the 1990 excavation, over 28 stratigraphic units were identified, ranging in thickness from 1 to 17 cm (Figure 3). A number have abrupt boundaries indicating a change in sedimentation while others have irregular or wavy boundaries perhaps indicating some sort of deformation before resumption of deposition.

The first and only radiometric controls for the cave sediments and its faunas have been obtained through radiocarbon assays targeting organic sediments. Radiocarbon dating using bone would not be reliable as shown by the results of the amino acid content assays (T.W. Stafford, personal communication, 1993). Six dates now are available for the cave stratigraphy from six strata within the upper and lowermost parts of the described profile [45]. They are critical to unraveling the history of investigations, Stock's specimen provenience, the stratigraphy, the paleoecology, and the age of the cave record. The deposits date between 28,000 and 45,000 ^{14}C years BP (Table 1). A problem in terms of reversed dates is noted for the lower excavated strata. With respect to the lowermost sample, INAH-940 (Table 1), it cross-cuts several strata and, therefore, represents a mixture of organic sediments from different ages. The incongruous age most likely is due to the long exposure of the profile within Stock's pit and lack of providing a fresh, uncontaminated exposure for sampling. This type of situation is known to yield falsely young ages. The date is discounted. The reversal between samples SMU-2590 and SMU-2532 (Table 1) are a different matter and which date should be discounted has not been resolved. All ages cited in the text are given as radiocarbon years before present (^{14}C BP). Corresponding calendar years are provided in Table 1.

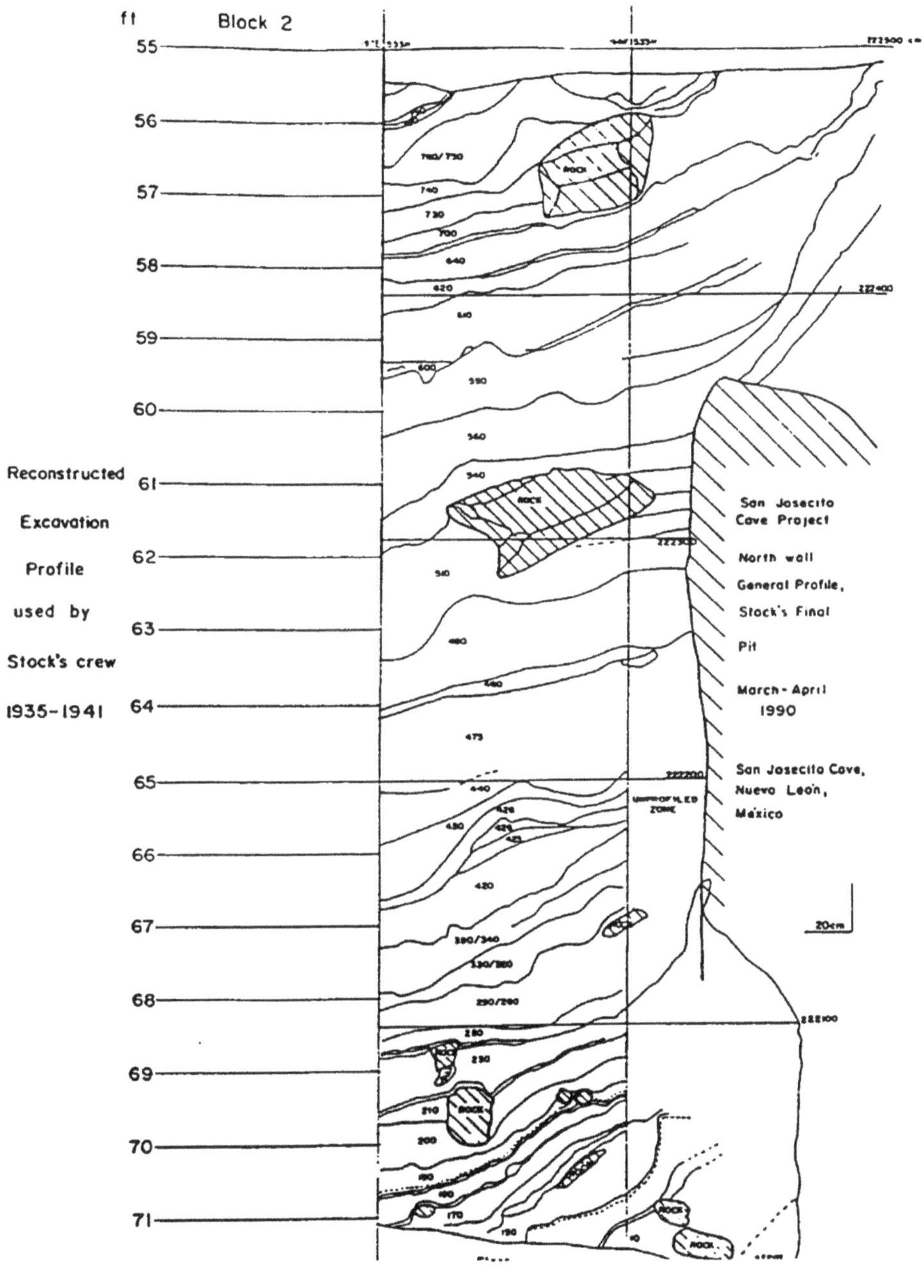

Figure 3. Stock's block 2 and excavated levels by foot increments superimposed over the 1990 stratigraphic profile of the north wall of Stock's pit [28].

Table 1. Radiocarbon ages from San Josecito Cave based on organic sediments [45].

Lab Number [1]	Elevation (cm)	Stratum	Uncalibrated Radiocarbon Age	±	δ13C	Corrected Radiocarbon Age [2]	Calendar Years [3]	±
INAH-943A	222,434–222,425	720	28,005	1035		34,892	32,942	4077
SMU-2518	222,425–222,422	710	31,290	1270	−16.9	39,390	37,440	4967
SMU-2589	222,415–222,413	670	34,830	470	−15.8	39,592	37,642	1536
SMU-2590	222,406–222,402	630	44,520	2590	−15.5	48,507	46,557	3751
SMU-2532	222,371–222,363	595	37,975	700	−14.4	42,684	40,734	1590
INAH-940	222,203–222,162	420–340	33,700	1365		43,974	42,024	6367

[1] INAH = Instituto Nacional de Antropología e Historia, SMU = Southern Methodist University. [2] uncalibrated but isotopic fractionation-corrected. [3] Calibration to calendar years at 1 σ (68.2% probability) was performed with OxCal v4.4 [46–49] using calibration dataset IntCal20 [50].

Based on the radiocarbon ages, the removed sediments from Stock's excavations above the 55 ft level represent a time span between 25,000 and 11,000 ^{14}C years BP, including the last glacial maximum (LGM) and the Pleistocene-Holocene boundary. This timespan corroborates that proposed by Harris [11] for the age of the general San Josecito Cave fauna. Holocene sediments may have been present, but their occurrence is not clear from the available documentation.

Bones recovered from Stock's pit from the 57 to 58 ft level overlap in age with the materials from stratum 720. In general, the renewed excavation and stratigraphic profile overlaps with Stock's 55 through 71 ft levels (Figure 3). This overlap is important as it permits a correlation of some of Stock's materials with the 1990 material and creates more representative local faunas for the period 45,000 to 28,000 ^{14}C years BP.

4.3. Fauna

Stock (personal correspondence with Francis D. Bode, 1935–1936) points out that below the 65–70 ft level, large bones are absent. The cave deposits thus can be divided into two sections: an upper section with abundant megafaunal remains; and a lower, tested section (Stock's pit), apparently lacking large bones but having abundant microfauna. The lower section of San Josecito Cave has a great potential for providing a complete record for the time span it represents (i.e., older than 45,000 ^{14}C years BP).

Originally, 53 species of mammals have been identified from Stock's [26] excavations (Table 2) [29,51]. The overall San Josecito Cave fauna is best represented with materials found within Stock's 55–60 ft level range that is within Stock's pit [29]. In the lower deposits, the reduction in number of taxa may indicate a faunal change. Stock (personal correspondence with Harold Sebenius, 1940–1941) notes this reduction as the end of the fossil deposits. The lowermost deposits, however, are the least excavated and bone numbers may be low since because those deposits were sampled less intensively. Furthermore, the main focus during Stock's excavations was on large animals. The crew collected bones of smaller animals only occasionally (as noted in several letters back and forth between Stock and crew members).

Table 2. Mammal species identified from paleontological excavations in San Josecito Cave, Nuevo León, Mexico from the original excavations (1935–1941) and the 1990 excavation (MNI are provided for bones from stratum 720). Nomenclature follows that proposed by [52]. An asterisk denotes extinct taxa (*).

Taxa	Original Excavations	1990 Excavation
Didelphidae		
Didelphis virginiana		X
Megatheriidae		
Megalonyx jeffersonii *	X	
Nothrotheriops shastensis *	X	X
Soricidae		
Cryptotis mexicana	X	X
Sorex cinereus or *S. milleri*	X	X
Sorex altoensis		9
Mormoopidae		
Mormoops megalophylla		X
Phyllostomidae		
Choeronycteris mexicana		X
Desmodus stocki *	X	X
Leptonycteris nivalis	X	4
Vespertilionidae		
Corynorhinus townsendii	X	
Eptesicus fuscus	X	
Lasiurus cinereus	X	
Myotis californicus		1
Myotis thysanodes		2
Canidae		
Aenocyon dirus *	X	
Canis latrans	X	
Canis lupus	X	
Cuon alpinus *	X	
Urocyon cinereoargenteus	X	
Felidae		
Herpailurus yagouaroundi	X	
Lynx rufus	X	
Panthera atrox *	X	
Panthera onca	X	
Puma concolor	X	
Smilodon fatalis *	X	
Mephitidae		
Conepatus mesoleucus	X	
Mephitis mephitis	X	
Spilogale angustifrons/gracilis	X	
Mustelidae		
Mustela frenata	X	
Taxidea taxus	X	
Procyonidae		
Bassariscus astutus	X	
Ursidae		
Tremarctos floridanus *	X	
Ursus americanus	X	
Equidae		
Equus conversidens *	X	1
Tapiridae		
Tapirus haysii *		1
Tayassuidae		
Platygonus *	X	
Camelidae		
Camelops *	X	

Table 2. *Cont.*

Taxa	Original Excavations	1990 Excavation
Cervidae		
Navahoceros fricki *	X	
Odocoileus	X	
Antilocapridae		
Stockoceros conklingi *	X	
Bovidae		
Euceratherium *	X	
Oreamnos harringtoni *	X	
Sciuridae		
Marmota flaviventris	X	
Parasciurus alleni	X	X
Xerospermophilus spilosoma	X	
Geomyidae		
Cratogeomys castanops	X	
Orthogeomys onerous *	X	
Thomomys umbrinus	X	X
Heteromyidae		
Heteromys irroratus	X	X
Cricetidae		
Microtus mexicanus	X	44
Neotoma albigula	X	4
Neotoma mexicana	X	9
Peromyscus difficilis		8
Peromyscus labecula		9
Peromyscus levipes	X	16
Reithrodontomy megalotis	X	16
Sigmodon hispidus	X	X
Synaptomys cooperi	X	X
Erethizontidae		
Erethizon dorsatum	X	X
Leporidae		
Sylvilagus floridanus	X	2
Sylvilagus leonensis *	X	X

4.4. Mammals

Subsequent analyses of the mammalian remains recovered during the 1990 excavation and those housed at LACM [53,54] brought the total taxa number to 62 (Table 2). Overall, 9 species and 4 genera had not been recorded previously from the cave. Also, six species and one genus (*Sorex altoensis*, *Corynorthinus towsendii*, *Equus conversidens*, *Marmota flaviventris*, *Parasciurus alleni*, *Peromyscus labecula*, and *Peromyscus levipes*) were reported under other names and now are reassigned to different species or genus [53,55–57]. Of these taxa, *Equus conversidens*, *Parasciurus alleni*, *Peromyscus labecula*, and *Peromyscus levipes* were present in the composite stratum 720 deposits. All the extant species occurred in the region (i.e., Sierra Madre Oriental) where the cave is located, and some even inhabited the cave.

No new extinct taxa were added to the cave mammal fauna, but four extralimital taxa (*Sorex milleri*, *S. altoensis* (previously reported as *S. saussurei*), *Marmota flaviventris*, and *Synaptomys cooperi*) were confirmed [53,54]. Of these extralimital taxa, only *S. altoensis* was present in the stratum 720 deposits. *Sorex milleri* is a minimally extralimital taxon relative to San Josecito Cave. The cave is less than 100 km south of the southernmost range of *S. milleri* [58]. *Marmota flaviventris* and *Synaptomys cooperi* came from deposits above Stock's 55 ft level (i.e., from upper deposits younger than stratum 720). These two taxa experienced range changes of at least 1000 km north of the cave [58], with *M. flaviventris* occurring today in the western US into Canada and *S. cooperi* in the northern to northeastern US into Canada.

Endemic species from other than stratum 720 consisted of *Cryptotis mexicanus*, *S. milleri*, and *Sylvilagus leonensis*. Based on a morphometric analysis, *C. mexicana* from San Josecito Cave was significantly larger than the extant form while *S. milleri* from the cave is similar in size to its extant form [54]. The endemic *Sylvilagus leonensis* (recovered in deposits above stratum 720) was an extinct dwarf rabbit known only from San Josecito Cave (type locality) and La Presita Cave, San Luis Potosí [59].

Eight hundred eighty-four mammalian cranial elements have been identified to species from stratum 720. The 13 species identified include 6 species new for San Josecito Cave (Table 3). Among the mammals are another two species endemic to Mexico. The endemic species are composed of a shrew (*Sorex altoensis*) and a white-footed mouse (*Peromyscus levipes*). Based on a morphometric analysis, *S. altoensis* from San Josecito Cave is similar in size to its extant form [54]. The modern ranges for the shrew and white-footed mouse are sympatric in the Sierra Madre Oriental, although they do not co-occur in the vicinity of San Josecito Cave. They have restricted and differing ranges in Mexico [58,60,61].

Table 3. Mammal species assigned to the composite stratum 720 and Stock's 57–58 ft level (~28,000 ^{14}C years BP). Nomenclature follows that proposed by [52]. An asterisk denotes extinct taxa (*).

Taxa	Stock's 57–58 ft Level	Stratum 720 (MNI)
Megatheriidae		
Nothrotheriops shastensis *	X	
Soricidae		
Sorex altoensis		9
Phyllostomidae		
Desmodus stocki *	X	
Leptonycteris nivalis	X	4
Vespertilionidae		
Myotis californicus		1
Myotis thysanodes		2
Canidae		
Aenocyon dirus *	X	
Canis latrans	X	
Felidae		
Panthera atrox *	X	
Panthera onca	X	
Puma concolor	X	
Smilodon fatalis *	X	
Ursidae		
Tremarctos floridanus *	X	
Ursus americanus	X	
Equidae		
Equus conversidens *	X	1
Cervidae		
Navahoceros fricki *	X	
Antilocapridae		
Stockoceros conklingi *	X	
Bovidae		
Euceratherium *	X	
Sciuridae		
Parasciurus alleni	X	
Cricetidae		
Microtus mexicanus	X	44
Neotoma albigula	X	4
Neotoma mexicana	X	9
Peromyscus difficilis		8
Peromyscus labecula		9
Peromyscus levipes	X	16
Reithrodontomy megalotis	X	16
Leporidae		
Sylvilagus floridanus	X	2

The occurrences of *Parasciurus alleni* and *Peromyscus levipes* at San Josecito Cave are the first known Quaternary fossil record for this squirrel [62] and mouse [61]. Both have very limited geographic ranges that are centered on the Sierra Madre Oriental. *Sorex altoensis* is a minimally extralimital taxon relative to San Josecito Cave. The cave is about 50 km north of the northernmost range of *S. altoensis* [28].

Based on the overlap of the 1990 stratigraphic profile and Stock's levels (Figure 3), when the mammalian species composition from stratum 720 is combined with the mammalian fauna from Stock's 57–58 ft level, then 27 species are recorded from this stratigraphic provenience (Table 3). That number represents 43.5% of the total cave mammalian fauna.

4.5. Birds

Miller [24] reported 43 fossil bird taxa from the more than 2100 specimens recovered from Stock's [26] excavations (Table 4). He segregated the original bird fauna into three groups. Group 1 were those preyed upon by raptors, composed of gallinaceous birds, parrots, doves, woodpeckers, ducks, shore birds, herons, rails, poor wills, and small passerines. Group 2 were those that nested in the cave mouth, represented by *Corvus corax, Coragyps atratus, Bubo virginianus, Tyto alba, Falco mexicanus*, and *Caracara plancus*. Group 3 were those that may have roosted outside the cave and senile or injured individuals that had sought refuge in the cave.

Table 4. Bird species identified from paleontological excavations in San Josecito Cave, Nuevo León, Mexico, from the original excavations (1935–1941) and the 1990 excavation. An asterisk denotes extinct taxa (*).

Taxa	Original Excavations	1990 Excavation (NISP)
Podicipedidade		
Podilymbus podiceps	X	
Ardeidae		
Nycticorax nycticorax	X	
Anatidae		
Aix cf. *A. sponsa*	X	
Anas sp.	X	
cf. Histrionicus sp.	X	
Oxyura cf. *O. jamaicensis*	X	
Ciconiidae		
Ciconia sp. or *Mycteria* sp.		1
Teratornithidae		
Teratornis merriami *	X	
Vulturidae		
Gymnogyps californianus	X	
Coragyps atratus	X	
Accipitridae		
Elanus leucurus	X	
Parabuteo unicinctus	X	
Buteo cf. *B. regalis*	X	
Buteo jamaicensis	X	1
Buteo nitidus	X	
Aquila chrysaetos	X	
Spizaetus grinnelli *	X	
Buteogallus daggetti *	X	
Neogyps errans *	X	
Neophrontops americanus	X	
Circus cyaneus	X	
Falconidae		
Caracara plancus	X	
Falco mexicanus	X	
Falco sparverius	X	20
Falco sp. *	X	
Phasianidae		
Cyrtonyx montezumae	X	85

Table 4. *Cont.*

Taxa	Original Excavations	1990 Excavation (NISP)
Dendrortyx sp.	X	
Meleagris crassipes *	X	
Rallidae		
Rallus elegans/longirostris	X	
Rallus limicola		2
Fulica americana	X	
Charadriidae		
Pluvialis cf. *P. squatarola*	X	1
Scolopacidae		
Scolopax minor		2
Numenius cf. *N. americanus*	X	
Burhinidae		
Burhinus cf. *B. bistriatus*	X	
Columbidae		
Patagioenas fasciata	X	10
Zenaida macroura	X	
Psittacidae		
Rhynchopsitta terrisi	X	
Rhynchopsitta pachyrhyncha	X	
Rhynchopsitta phillipsi *	X	
Cuculidae		
Geococcyx californianus conklingi *	X	41
Tytonidae		
Tyto alba	X	
Strigidae		
Asio flammeus	X	1
Asio otus	X	107
Megascops asio/kennicotti	X	
Megascops trichopsis	X	
Bubo virginianus	X	
Bubo sp.	X	
Glaucidium gnoma	X	
Micrathene cf. *M. whitneyi*	X	
Strix occidentalis	X	
Aegolius acadicus	X	
Psiloscops flammeolus		12
Ciccaba virgata	X	
Caprimulgidae		
Phalaenoptilus nuttallii	X	2
Picidae		
Colaptes auratus	X	
Hirundinidae		
Petrochelidon pyrrhonota		2
Corvidae		
Gymnorhinus cyanocephalus		1
Aphelocoma coerulescens		1
Corvus corax	X	1051
Muscicapidae		
Turdus migratorius		2
Icteridae		
Sturnella sp.		3
Small-medium passerines	X	X

Subsequent examination of the 2100 specimens [29,63–66] increased the taxa count to 54 for the overall bird fauna (Table 4). Steadman et al. [67], based on 90 bird bones from the 1990 excavation, identified 18 species (6 passerines and 12 non-passerines), with nine new species for the overall cave fauna. The new taxa were of temperate rather than tropical affinities. These new data increased the overall cave avifauna from 54 to 63 species (Table 4) and represents in general one of the largest and most diverse Late Pleistocene avifaunas in North America.

That overall avifauna was dominated by scavengers and predators and reflected three distribution categories. One category was that of birds widespread in both temperate and tropical settings. Another category represented birds restricted to temperate settings while the third category had birds restricted to tropical settings. In terms of broad habitat categories, a montane woodland or forest habitat was the most frequent represented in the overall avifauna, followed very closely by a grassland or grassy savannah habitat [67]. The overall San Josecito Cave bird fauna was a time-transgressive (i.e., not contemporaneous) mixture of species that today prefer coniferous or pine-oak forests/woodlands, grasslands/savannahs, and wetlands [24,67].

The 90 bones from the 1990 excavation yielded 39 identifiable elements to genus or species that came from two distinct periods. Of the 39 identifiable elements, 32 came from the unit in the bone alcove that corresponds to Stock's 35 ft level (Table 4). One of the taxa likely represented an undescribed extinct stork (*Ciconia* sp. or *Mycteria* sp.). Six others were extant birds new to the overall cave avifauna. Most notable were the presence of *Scolopax minor* (American woodcock) and *Aphelocoma coerulescens* (scrub jay). *Scolopax minor* was an extralimital, today ranging 500 to 800 km northeast of the cave and southern Nuevo Leon (winter and breeding ranges) [68] and with preferred habitats of brushy swamps or moist woodlands. *Aphelocoma coerulescens* would be minimally extralimital, where today they occur in Nuevo Leon only in the westernmost mountains of the Sierra Madre Oriental range in a preferred habitat of pinyon-juniper woodlands [68,69].

The remaining seven identifiable bones come from stratum 720. Despite so few bones, they represent six taxa (Table 5), one of which is new to the overall cave avifauna. *Gymnorhinus cyanocephalus* (pinyon jay) is another extralimital form, breeding today in the mountains of northern Baja California, with accidentals or non-breeding records known for the northern Sierra Madre Occidental [70,71]. This jay also has a preferred habitat of pinyon-juniper woodlands. In addition, small to medium passerines are represented within the unidentified elements (10 bones).

Table 5. Bird species assigned to the composite stratum 720 and Stock's 57–58 ft level (~28,000 ^{14}C years BP).

Taxa	Original Excavations	Stratum 720 (NISP)
Accipitridae		
Aquila chrysaetos	X	75
Phasianidae		
Cyrtonyx montezumae		85
Rallidae		
Rallus limicola		2
Charadriidae		
Pluvialis cf. *P. squatarola*	X	1
Strigidae		
Megascops asio/kennicottii		7
Asio otus	X	107
Ciccaba virgata	X	20
Corvidae		
Gymnorhinus cyanocephalus		1
Corvus corax	X	1051
Icteridae		
Sturnella sp.		3
Small-medium passerines	X	X

The inclusion of the birds from Stock's 57–58 ft level brings the avifauna to 10 taxa plus the small to medium passerines (Table 5). Of interest is the referred *Pluvialis* cf. *squatarola* (black-bellied plover), a wading shore bird. In the Americas, this globally occurring bird nests in the Arctic and winters as far south as along the coasts of Argentina and Chile. It primarily is a coastal form in Mexico, where they feed in tidal flats, beaches, salt marshes, and meadows. Winter habitats can include inland shallow water and grassland [72,73]. Sightings today in the Sierra Madre Oriental are rare to accidental occurrences [74] (p. 247). With only one specimen and a referred identification, the significance of this bird's presence in the composite stratum 720 fauna (fortuitous incorporation of a vagrant or actual range change) is speculative.

4.6. Squamate Reptiles

Brattstrom [75] identified two species of reptiles from Stock's excavation, that of a new extinct species *Phrynosoma josecitensis* and the extant species *P. orbiculare*. *Phrynosoma josecitensis* (Table 6) was known only from San Josecito Cave and is poorly understood due to its rarity. No remains were recovered from the 1990 excavation and the status of the taxon needs to be verified [76].

Table 6. Squamate reptiles identified from paleontological excavations in San Josecito Cave, Nuevo León, Mexico, from the original excavations (1935–1941) and the 1990 excavation (all MNI from stratum 720 [~28,000 ^{14}C years BP]). An asterisk denotes extinct taxa (*).

Taxa	Original Excavations	Stratum 720 (1990 Excavation)
Anguidae		
Barisia ciliaris		23
Barisia imbricata		8
Phrynosomatidae		
Phrynosoma modestum		1
Phrynosoma josecitensis *	X	
Phrynosoma orbiculare	X	2
Sceloporus spp.		4
Colubridae		
Tantilla sp.		1
Heterodon cf. *H. simus*		1
Hypsiglena		1
Storeria sp.		1
Viperidae		
Crotalus spp.		1

Mead et al. [76] recognized three taxa from stratum 720 deposits, all three of which occur in the region today around the cave. These were the endemic Mexican Plateau horned lizard (*Phrynosoma orbiculare*), spiny lizard (*Sceloporus* sp.) and the endemic Transvolcanic alligator lizard (*Barisia* cf. *B. imbricata*) (Table 6). Subsequently, the subspecies of *B. imbricata* was elevated to species level [77], and other herpetologists have accepted that change [78]. Bryson and Riddle [79] used mtDNA and phylogenic analysis to examine diversification across the genus and timing. Diversification within *B. imbricata* and *B. ciliaris* appeared to be around the Pliocene-Pleistocene border and into the Pleistocene. While not addressing the taxonomic status, they concluded that each of the two taxa seemed to be a composite of three different lineages [79] (p. 387). Unrecognized or cryptic species may be within the *B. imbricata* complex with further investigation required for clarification. Until that time, the taxa as used here followed Smith et al. [77].

Cruz-Silva [80] identified *B. imbricata* among the stratum 720 remains as well as the endemic *B. ciliaris* (Sierra alligator lizard) and additional reptile species, seven of which are new to the overall San Josecito Cave fauna (Table 6). Snakes constituted the majority (n = 5; 71.4%) of the new taxa. *Barisia ciliaris*, *Phrynosoma modestum* (roundtail horn lizard), *Heterodon simus* (Southern hognose snake), *Tantilla* (blackhead snake), and *Storeria* (brown snake) were the first records for the Pleistocene in Mexico (Table 6).

These new data increased the overall cave squamate reptiles to 11 species (Table 6). Of the 10 identified squamates from stratum 720, all are extant but three do not occur in the region today. In the Late Pleistocene, the squamate diversity represented in stratum 720 is composed of elements from three different modern biogegraphical regions. *Heterodon simus* ranges throughout the southeastern USA [81]. *Barisia imbricata* is from the Neovolcanic Transmexican Belt [82]. *Phrynosoma modestum* inhabits the Mexican Plateau into western Texas and New Mexico [83]. San Josecito Cave contains a relatively large number of taxa for localities in Mexico. The record is exceeded only by those from Santa Cruz Nuevo in Puebla [84], Rancho La Brisca in Sonora [85], and Loltun Cave in Yucatán [86].

Sometime after 28,000 ^{14}C years BP, retraction occurs in the distribution of *Heterodon simus* from the Sierra Madre Oriental to the southeastern USA. Some mammals, such as *Synaptomys cooperi* (Southern bog lemming) and *Microtus pennsylvanicus* (meadow vole) follow similar retraction patterns but to the northern and northeastern USA respectively [21,22]. The timing of the retraction for those rodents, however, is much later than 28,000 ^{14}C years BP. On the US Southern Plains, the timing for those two rodents is between 10,000 and 8500 ^{14}C years BP [8,14].

Barisia imbricata also shows a retraction in its Late Pleistocene distribution from the Sierra Madre Oriental south to the Mexican Transvolcanic Belt at present. Molecular data for *Ambystoma, Crotalus*, and *Phrynosoma* indicate a strong phylogenetic relation between the Mexican Transvolcanic Belt and the Sierra Madre Oriental [87–89]. The presence of *B. imbricata* (endemic to the Transvolcanic Belt) and *B. ciliaris* in the same geographic area has only been recognized at present in the State of Hidalgo [82,90]. This co-occurrence in the Late Pleistocene of San Josecito Cave indicates that sympatry extended as far as 450 km north to the State of Nuevo León.

Although *Phrynosoma modestum* is found in the State of Nuevo León, it does not inhabit the pine-oak forest around the cave. It inhabits the open shrub areas in arid and semiarid watersheds [83], and *Juniperus* forest in the State of Nuevo Leon [91,92]. The distributional pattern linking the Sierra Madre Oriental with the Mexican Plateau represented by *P. modestum* is reinforced by the presence of *P. orbiculare* that today occurs primarily on the Mexican Plateau. *Phrynosoma orbiculare* inhabits semiarid shrubland as well as montane woodlands and forests [83]. This pattern also is repeated by the spiny pocket mouse *Liomys irroratus* [21,22]. This relationship supports Martin's [93] proposal that a strong connection existed between the Sierra Madre Oriental and Mexican Plateau during the Late Pleistocene. That suggestion is based on biogeographic analysis of amphibians and reptiles, climate, vegetation types, and historical geology in the Gomez Farias Region. This pattern also is inferred for the genus *Sceloporus* [94,95].

Shifts in the distribution of taxa appear related to environmental changes associated with changes in global weather. Distributional changes found in studies such as phylogeography demonstrate the expansion and contraction of amphibian and reptile ranges during glacial and interglacials [96–99]. These studies support the biological dynamism during the Quaternary and indicate that the Pleistocene was not a steady state for herptiles as proposed by Holman [100].

The Late Pleistocene (i.e., stratum 720; ~28,000 ^{14}C years BP) vegetation surrounding San Josecito Cave based on the reptile record is a mosaic principally of an open forest with savanna, dominated by *Pinus* and *Quercus*. A xeric scrub with *Juniperus* also is present in the cave area. This reconstruction differs from previous environmental inferences for the cave area based on mammals [29–31,101], birds [67], and lizards [76]. Those studies indicate the presence of an oak-pine forest similar to the current setting but within a subhumid climate, more marked winter season, and drier conditions than today. The expanded palaeovegetation reconstruction based on vertebrates is in agreement with the pollen data recovered from the Late Pleistocene in the northern part of Mexico. Those data indicate a *Juniperus* forest into a xeric scrub was widespread in the Late Pleistocene. A *Juniperus* forest around the cave is 400 km south from its current distribution [102–105]. This situation underscores that shifts in vegetation distributions also were occurring in the Late Pleistocene.

4.7. *Taphonomy*

4.7.1. Stratum 720

The taphonomic data collected focused on refitting, minimum number of individuals (MNI), and bone modification for the stratum 720 small mammalian assemblage. Refitting was accomplished only in part as most postcranial material lacked specific identification. Cranial matching was conducted in terms of teeth-mandible, and tooth-tooth for specimens from the same taphonomic box. The findings were not considered valid for individual quantification as little confidence was placed on some of the refits (a more detailed consideration of age and size is required).

Minimum number of individuals is an import taphonomic feature in quantifying the fauna. Large differences have been found with absolute values (Figure 4), but when dealing with relative values for the results from the different counting methods, no significant differences were found (G-test, $F = 49.4$; $p = 0.20$). Nonetheless, as the MNI is used most frequently with cave faunas due to low dispersal of bone remains [106], it is considered here as the most reliable for a study involving many disarticulated bone elements from small vertebrates.

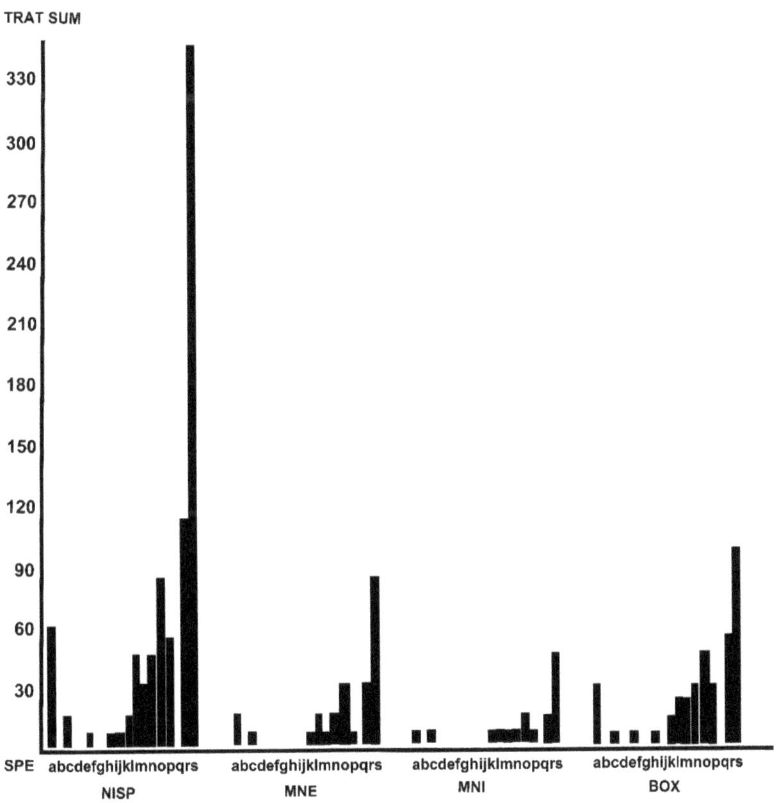

Figure 4. The plotting of the absolute values obtained for the different quantification methods used in the taphonomic study. Plots represent the utilized method, as follow: 1—NISP; 2—MNE; 3—MNI; 4—BOX. Species are represented by the following abbreviations: (a) *Sorex altoensis*; (b) Chiroptera; (c) *Leptonycteris nivalis*; (d) *Desmodus stocki*; (e) *Myotis californicus*; (f) *Myotis thysanodes*; (g) *Myotis* sp.; (h) *Sylvilagus floridanus*; (i) *Sylvilagus* sp.; (j) *Parasciurus alleni*; (k) *Neotoma leucodon*; (l) *Neotoma mexicana*; (m) *Neotoma* sp.; (n) *Peromyscus difficilis*; (o) *Peromyscus levipes*; (p) *Peromyscus labecula*; (q) *Peromyscus* sp.; (r) *Reithrodontomys megalotis*; (s) *Microtus mexicanus*.

Breakage patterns for the small mammal bones were examined using both cranial and postcranial elements [28] (pp. 129–133). A total of 815 postcranial elements were studied, corresponding to long bones. From those, less than 16% ($n = 130$) were complete bones, 36.9% ($n = 301$) proximal sections, 24.5% ($n = 200$) shafts, and 22.6% ($n = 184$) distal sections. Identified bones included 173 humeri (21.2%), 165 tibiae (20.2%), 133 femora (16.3%), 129 ulnae (15.8%), 125 pelves (15.3%), 49 scapulae (6.0%), and 41 radii (5.0%). The analysis of the overall breakage patterns showed that the proximal regions of the femur, ulna, scapula, and radius, the distal portions of the humerus and tibia, and the pelvic shaft (partial illium and acetabulum) are usually the less-damaged bone parts.

This pattern, however, is modified when looking at taxonomic groups. The breakage patterns differ from one group to another. Only the ones with a large number of specimens are close to the pattern shown by the overall counts, while those groups with fewer numbers are biased. For example, Eulipotyphla has mostly complete individuals; other groups are composed mainly of shafts or section fragments.

Comparisons with other cave faunas had to rely on the element frequencies and breakage pattern for the overall stratum 720 mammalian assemblage. Those patterns were similar to the ones found for some levels from Gran Dolina, Spain [107]. This similarity indicated that owls most likely were the accumulating agents for the stratum 720 small remains. Further studies using SEM should test this hypothesis by looking at the cortical surface damage [108].

Spatial distribution analysis (Figure 5) of the small mammal bones was pursued to learn the taphonomic agents affecting materials prior to burial, as well as to test the usefulness of the excavation methodology approach (taphonomic boxes). A statistically significant correlation (Mantel statistic, $z = 5865.7$, $p = 0.028$) was obtained between the distribution of individuals in taphonomic boxes and the spatial position of such boxes. Such correlation not only was derived due to physical causes (gravity), but other abiotic and biotic factors that may have affected the final position of the bones. Biotic factors may have included trampling and bone transportation by other small mammals. Those processes, however, were not recorded clearly on the bones themselves.

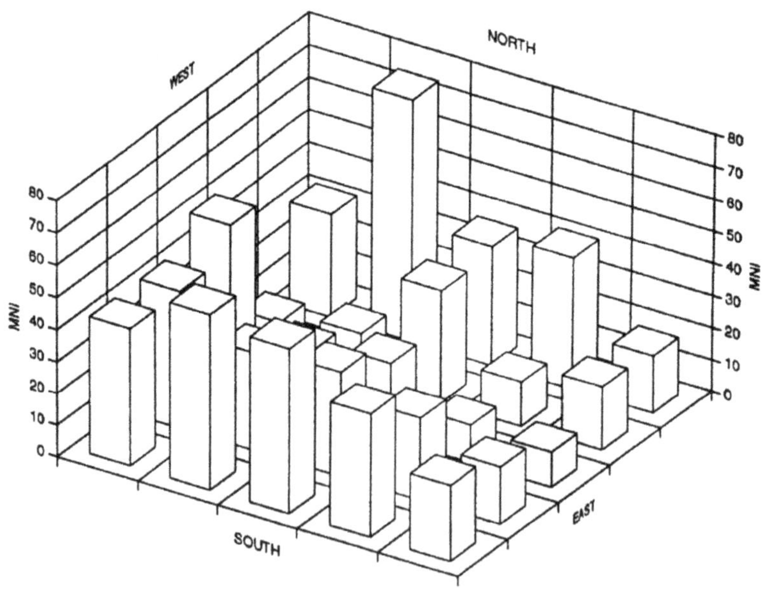

Figure 5. The distribution of the overall number of individuals (based on MNI) by taphonomic box for stratum 720 in excavation unit 534N197E, San Josecito Cave, Nuevo León, Mexico.

The overall taphonomic analysis for stratum 720 indicates that small mammal bone accumulation most likely was due primarily to owls. After deposition of the pellets, bones from the pellets may have been displaced either by gravity or moved by woodrats or other agents to their final position. Breakage seems to correspond to the normal effect while animals are contained in an owl's gizzard [108].

4.7.2. Stratum 770

Four of the identified strata from the 1990 excavation contained blue and gray colored bone. The diagenetic processes causing the colorations were investigated with bone from stratum 770 [101,109,110]. Results from energy-dispersive X-ray (EDAX) fluorescence and chemical analyses by instrumental neutron activation (INAA) indicated the presence of various trace elements including copper (Cu). Robles et al. [109] (p. 128) concluded that the color was due to impurities within the transition trace metal ions and most likely by copper. The source of the copper, however, was unknown.

Further research involved structural and elemental analyses and spectroscopic investigations. The spectroscopic data were the most informative for understanding the blue and gray colorations. The data revealed a heat-induced modification (apatite crystals of different shapes at least two orders of magnitude larger than those from modern bone) and traces of manganese (Mn) within the typical bone constituents. Different cations of manganese formed by the heat treatment were responsible for the two colorations. For the blue bones, heating took place under oxidizing conditions. Temperature range for color formation was estimated to have been above 500 °C and below 600 °C. The source of the heat, however, was not identified [110].

4.8. *Paleoecology*

Many studies have explored the relationship between ecological function and species diversity [111], including faunal comparisons and a community structural analysis (methods for describing the evenness in relative abundance among species). Two approaches were used to examine paleoecological influences based on the stratum 720 mammalian assemblage (Table 3). Measures of species number and the distribution of their relative abundances do not require taxonomic similarity as does the taxonomic uniformitarian approach. Dominance-diversity measures, however, do not convey anything about the kinds of ecological roles occupied by species or the interactions between them [112].

4.8.1. Faunal Comparisons

Twenty-one mammal faunas chosen for comparisons with the composite stratum 720 mammals from San Josecito Cave encompassed an approximate range of ages from 21,500 to 38,700 ^{14}C years BP and ee values from 3648 through 5540 (these values were available only for nine sites). For most of these localities, specific fauna for a dated stratum was not available. Comparisons were made based on the overall fauna at each locality.

The dendrogram obtained from the cluster analysis (Figure 6) shows that the San Josecito Cave composite stratum 720 mammalian fauna broadly groups with the faunas from southwestern New Mexico (Dry and U-bar caves), Arizona (Rampart Cave), and Nevada (Smith Creek Cave, Nevada). The age for San Josecito Cave stratum 720 (~28,000 ^{14}C years BP) is within the age ranges provided for the New Mexican [113] (25,160–33,590 ^{14}C years BP) and Arizonan [114,115] (23,540–36,200 ^{14}C years BP) caves, and close to the age of the Nevadan cave fauna [116] (28,650 ^{14}C years BP). Based on a re-evaluation of the arvicoline rodents, however, the age of the Smith Creek Cave fauna (Nevada) appears much older (from 820,000 to 146,000) than the original age and considerably older than stratum 720 [117].

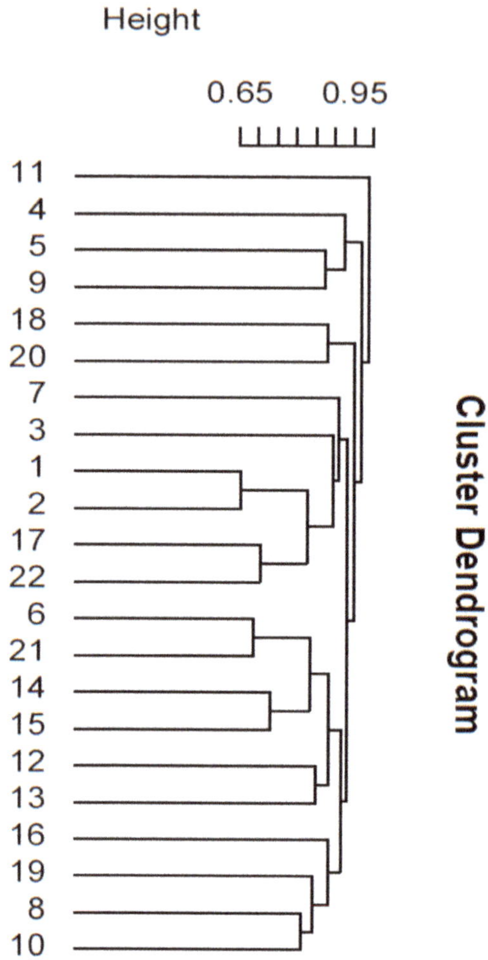

Figure 6. Dendrogram obtained from the UPGMA analysis of Pre-LGM faunas from different localities. Site number is located before the level value; those numbers correspond to the ones utilized Arroyo-Cabrales ([28], Table 28). (1) Dry Cave, New Mexico (25,610–33,590 ^{14}C years BP), (2) U-Bar Cave, New Mexico (26,150–35,890 ^{14}C years BP), (3) San Josecito Cave, Nuevo Leon, Mexico (27,970 ^{14}C years BP), (4) Hueco Mountains. New Mexico (>33,000 ^{14}C years BP), (5) Vulture Cave, Arizona (29,000–30,000 ^{14}C years BP), (6) Ambystoma Faunule, Jones Local Fauna, Meade Co., Kansas (26,700–29,000 ^{14}C years BP), (7) Rampart Cave, Arizona (23,540–36,200 ^{14}C years BP), (8) Hot Springs, South Dakota (26,075 ^{14}C years BP), (9) El Cedral, San Luis Potosi, Mexico (31,800 ^{14}C years BP), (10) Lewisville, Denton Co., Texas (>37,000 ^{14}C years BP), (11) Channel Islands, Santa Barbara Co., California (29,650 ^{14}C years BP), (12) Easley Ranch Local Fauna, Foard Co., Texas (>35,000 ^{14}C years BP), (13) Quitaque Local Fauna, Motley Co., Texas (>35,000 ^{14}C years BP), (14) Carrol Creek Local Fauna, Donley Co., Texas (32,400 ^{14}C years BP), (15) Canadian Local Fauna, Hemphill Co., Texas (34,900 ^{14}C years BP), (16) Rainbow Beach Local Fauna, Power Co., Idaho (21,500–31,300 ^{14}C years BP), (17) Smith Creek Cave, Nevada (28,650 ^{14}C years BP), (18) Fort Qu'appelle, Saskatchewan (>30,000 ^{14}C years BP), (19) Medicine Hat Area, Alberta (37,900–38,700 ^{14}C years BP), (20) Gold Run Creek, Yukon Territory (22,200–32,350 ^{14}C years BP), (21) Selby-Dutton Fauna (29,000 ^{14}C years BP), (22) Tesequite Canyon, Cimarron Co., Oklahoma (31,360 ^{14}C years BP).

The Rampart Cave fauna has a very low number of species (n = 7). Two of those species are shared with composite stratum 720 at San Josecito Cave, i.e., the Pleistocene vampire bat and the mountain lion. The sharing of one of the few tropical elements (the vampire bat) may be important in grouping these two caves. If so, the locality grouping is not only based on number of species, but also on the actual species represented.

The New Mexican and Nevadan caves are the localities with the most abundant number of species. The number of shared taxa with San Josecito Cave composite stratum 720, however, is not the same. Dry Cave (New Mexico) and Smith Creek Cave (Nevada) only share five taxa each (mostly carnivores) with composite stratum 720, while U-Bar Cave (New Mexico) shares 12 taxa (bats, carnivores, horses, ungulates, and rodents).

Harris [11] concluded that the U-Bar Cave fauna indicated more equitable conditions than today. Absence of high temperatures together with greater effective moisture during the spring allowed the occurrence of now-distant plants and animals favored by cool summers and moderate year-round moisture. Although this scenario could have occurred around San Josecito Cave, it was not very likely as most of the extant mammal taxa found in stratum 720 occur in the region today but not necessarily in the vicinity of the cave.

4.8.2. Community Structural Analysis

Food habits and habitat preference are the two measures used in the community structural analysis for the composite stratum 720 mammalian assemblage. Food habits provide an understanding of the different environments contributing certain species, and an explanation for bone accumulation in the cave [118]. Affecting the utility of this measure is the lack of knowledge of the detailed food habits of some recent species, and even more so for the fossil ones. Following a methodological approach proposed by Mares and Willig [119], determining habitat preference is based on the general vegetational associations proposed by Barbour and Billings [120]. Data for the different species are easy to obtain for the general associations based on the literature, but a specific association was precluded due to the lack of knowledge of habitat preferences for several extant as well as extinct species.

4.8.3. Food Habits

A total of 1261 specimens represent 27 species. Of the specimens, 976 individuals (77.4%) pertain to the herbivore niche, 201 are carnivores (15.9%), 69 insectivores (5.5%), 12 pollen-eaters (1.0%), 2 omnivores (0.2%), and one blood-eater (0.1%). In terms of food habits by species, percentages are: herbivores (50%), carnivores (26.7%), insectivore (10%), omnivores (6.7%), pollenophage (3.3%), and sanguinivore (3.3%). In comparison, the modern mammalian fauna collected in the area of San Josecito Cave in 1990 is composed of (based on number of taxa in Table 4 of [26] (pp. 57–59): 27 herbivores (45%), 18 insectivores (30%), 9 carnivores (15%), 3 omnivores (5%), 2 pollenophages (3.3%), and one sanguinivore (1.7%). Proportions in the fossil assemblage seem to deviate toward a larger number of carnivores and a smaller number of insectivores than in the modern mammalian fauna.

When comparing the percentage of carnivores present in the composite stratum 720 mammalian assemblage (Table 3) with those provided by White et al. [118] for several North American caves, the San Josecito Cave percentage is greater than the ones for most of those caves. Jaguar, Moonshiner, and Middle Butte in Idaho, however, have higher percentages of carnivores. White et al. [118] proposed two different hypotheses for those high percentages, and those hypotheses are used to evaluate the findings in San Josecito Cave.

One explanation is that San Josecito Cave could have been used as a denning site (as proposed by [15]). A carnivore denning site, however, implies a number of chewed bones, as well as young carnivore remains [118]. Neither is the case for the stratum 720 carnivore assemblage.

The San Josecito Cave carnivore percentage is not as high and disproportionate as those for the Idaho caves [118] that were considered carnivore traps. The great number of large mammals, including carnivores, collected in the overall deposits from San Josecito Cave may suggest that the cave was functioning as a trap throughout the Late Pleistocene. Carnivores may well have gone inside to scavenge the available meat from herbivores that fell down and themselves became entrapped [121]. A similar scenario has been proposed for Rancho La Brea in the Los Angeles basin (southern California) with an assemblage dominated by carnivores and herbivores [122,123]. The available evidence (physiography and taphonomy) indicates the cave functioned as a pit trap in general and the carnivores were present due to entrapment.

4.8.4. Habitat Preferences

The habitat preferences for extant and extinct taxa have been examined. For extinct taxa, past preferences have been inferred from the available literature based on modern analogs and isotope studies. All but one of the species with available data (21) occur in the Rocky Mountain biome. That biome is composed of several kinds of vegetational associations. Although the white-throated wood rat, *Neotoma albigula*, is not considered within that biome, it is known to occupy the slopes of the hills where the cave is located [124]. Other vegetation regimes where most taxa could occur are: intermountain deserts and woodlands (11 species), deserts (9 species), chaparral (8 species), and grassland (8 species).

Several species are stenotypic (adapted to a narrow range of environmental conditions) and these are the most valuable as paleoenvironmental indicators [119]. Three extant species (*Parasciurus alleni*, *Peromyscus difficilis*, *P. levipes*) presently are known to occur only in the Rocky Mountain forests [125,126]. Three extinct taxa (*Desmodus stocki*, *Navahoceros fricki*, and *Euceratherium*) are thought to occur only in such a vegetational association [15,127]. In addition, two extant species (*Sorex altoensis* and *Microtus mexicanus*) occupy only two biomes, that of the Rocky Mountains and Alpine forests [125,128].

While the results of the mammal analysis point to the Rocky Mountain biome as being present around 28,000 ^{14}C years BP, the specific vegetational association most likely is not the same as today for the San Josecito Cave area. In order to define more clearly the type of vegetational community that existed, a paleoenvironmental analysis has been undertaken to include all composite stratum 720 vertebrate taxa.

4.8.5. Paleoenvironmental Reconstruction

Several taxa provide additional environmental indications. The extinct Pleistocene vampire bat (*Desmodus stocki*) is considered to have required a mild climate with winter temperatures warmer than at present [129] and the genus *Desmodus* is considered to be highly sensitive to environmental temperature [130]. The association of the extinct vampire bat and the Shasta ground sloth (*Nothrotheriops shastensis*) in the composite stratum 720 at 2250 masl suggests a similar climatic restriction for both species associated with winter temperatures [131]. The presence of the Virginia rail (*Rallus limicola*) and black-bellied plover (*Pluvialis squatarola*) indicate a wetlands habitat (freshwater to brackish marshes) and grassland not available in the vicinity of the cave today [72,73,132].

The presence of two *Barisia* species (Alligator lizards) in the same geographic area is uncommon. Today, *B. ciliaris* and *B. imbricata* are sympatric only in Atotonilco el Grande, northern State of Hidalgo [82,90]. Assuming niche conservatism, climatic conditions around 28,000 ^{14}C years BP for the San Josecito Cave area would have been similar to that 450 km north. Data from 1951 to 2010 in the Atotonilco el Grande weather station show an average annual temperature of 16.3 °C and an annual precipitation of 787.3 mm (smn.cna.gob.mx). In contrast, San Francisco (the closest weather station to the cave area) has an annual average temperature of 17.3 °C and annual precipitation of 319.9 mm (smn.cna.gob.mx). During the Late Pleistocene, then, the annual average temperature in the cave area may have been 1 °C lower and the annual precipitation as much as 467.4

mm greater than today. Such an increase in annual precipitation would support wetlands in the area as indicated by the Virginia rail and black-bellied plover.

No differences occurred in the results using the habitat weighting method for composite stratum 720 with only the presence of the organism in different habitats or using the presence and the relative abundance (MNI or NISP) of organisms. The small mammals (Table 3) indicated a coniferous forest. The birds (Table 5) indicated a grassland. The squamate reptiles (Table 6) further indicated a coniferous forest (Tables 7–9). All vertebrates indicated a combined coniferous and oak forest and grassland at ~28,000 ^{14}C years BP for the San Josecito Cave area.

Table 7. Habitat weighting values for composite stratum 720 (~28,000 ^{14}C years BP) from San Josecito Cave using small mammals and habitat preferences.

Taxa		Coniferous Forest	Oak Forest	Xeric Scrub	Grassland	Tropical Forest
		Habitat weighting using presence data				
Sorex altoensis		1	0	0	0	0
Desmondus stocki		0	0	0.33	0.33	0.33
Leptonycteris nivalis		0.25	0.25	0.25	0	0.25
Myotis californicus		0.25	0.25	0.25	0	0.25
Myotis thysanodes		0.2	0.2	0.2	0.2	0.2
Parasciurus alleni		0.5	0.5	0	0	0
Microtus mexicanus		0.33	0.33	0	0.33	0
Neotoma albigula		0.25	0.25	0.25	0.25	0
Neotoma mexicana		0.2	0.2	0.2	0.2	0.2
Peromyscus difficilis		0.25	0.25	0.25	0.25	0
Peromyscus levipes		0.25	0.25	0.25	0	0.25
Peromyscus labecula		0.25	0.25	0.25	0	0.25
Reithrodontomys megalotis		0.33	0.33	0	0.33	0
Sylvilagus floridanus		0.2	0.2	0.2	0.2	0.2
TOTAL		**4.26**	3.26	2.43	2.09	1.93
		Habitat weighting using presence and MNI data				
Taxa	**MNI**	**Coniferous Forest**	**Oak Forest**	**Xeric Scrub**	**Grassland**	**Tropical Forest**
Sorex altoensis	9	9	0	0	0	0
Desmondus stocki	1	0	0	0.33	0.33	0.33
Leptonycteris nivalis	4	1	1	1	0	1
Myotis californicus	1	0.25	0.25	0.25	0	0.25
Myotis thysanodes	2	0.4	0.4	0.4	0.4	0.4
Parasciurus alleni	2	1	1	0	0	0
Microtus mexicanus	44	14.52	14.52	0	14.52	0
Neotoma albigula	4	1	1	1	1	0
Neotoma mexicana	9	1.8	1.8	1.8	1.8	1.8
Peromyscus difficilis	8	2	2	2	2	0
Peromyscus levipes	16	4	4	4	0	4
Peromyscus labecula	9	2.25	2.25	2.25	0	2.25
Reithrodontomys megalotis	16	5.28	5.28	0	5.28	0
Sylvilagus floridanus	2	0.4	0.4	0.4	0.4	0.4
TOTAL	127	**42.9**	33.9	13.43	25.73	10.43
		Habitat weighting using presence and NISP data				
Taxa	**NISP**	**Coniferous Forest**	**Oak Forest**	**Xeric Scrub**	**Grassland**	**Tropical Forest**
Sorex altoensis	60	60	0	0	0	0
Desmondus stocki	1	0	0	0.33	0.33	0.33
Leptonycteris nivalis	12	3	3	3	0	3
Myotis californicus	2	0.5	0.5	0.5	0	0.5
Myotis thysanodes	7	1.4	1.4	1.4	1.4	1.4
Parasciurus alleni	3	1.5	1.5	0	0	0
Microtus mexicanus	342	112.86	112.86	0	112.86	0
Neotoma albigula	16	4	4	4	4	0
Neotoma mexicana	44	8.8	8.8	8.8	8.8	8.8
Peromyscus difficilis	46	11.5	11.5	11.5	11.5	0
Peromyscus levipes	83	20.75	20.75	20.75	0	20.75
Peromyscus labecula	53	13.25	13.25	13.25	0	13.25
Reithrodontomys megalotis	110	36.3	36.3	0	36.3	0
Sylvilagus floridanus	3	0.6	0.6	0.6	0.6	0.6
TOTAL	782	**274.46**	214.46	64.13	175.79	48.63

Table 8. Habitat weighting values for composite stratum 720 (~28,000 ^{14}C years BP) from San Josecito Cave using birds and habitat preferences.

Taxa	NISP	Coniferous Forest	Oak Forest	Xeric Scrub	Grassland	Tropical Forest
		Habitat weighting using presence data				
Aquila chrysaetos		0.2	0.2	0.2	0.2	0.2
Asio otus		0.5	0	0	0.5	0
Ciccaba virgata		0.33	0.33	0	0	0.33
Corvus corax		0.2	0.2	0.2	0.2	0.2
Cyrtonyx montezumae		0.33	0.33	0	0.33	0
Gymnorhinus cyanocephalus		1	0	0	0	0
Megascops asio/kennicottii		0.2	0.2	0.2	0.2	0.2
Pluvialis cf. *P. squatarola*		0	0	0	1	0
Rallus limicola		0.2	0.2	0.2	0.2	0.2
Sturnella sp.		0	0	0	1	0
TOTAL		2.96	1.46	0.8	**3.63**	1.13
		Habitat weighting using presence and NISP data				
Taxa	**NISP**	**Coniferous Forest**	**Oak Forest**	**Xeric Scrub**	**Grassland**	**Tropical Forest**
Aquila chrysaetos	75	15	15	15	15	15
Asio otus	107	0.5	0	0	0.5	0
Ciccaba virgata	20	6.6	6.6	0	0	6.6
Corvus corax	1051	15	15	15	15	15
Cyrtonyx montezumae	85	0.33	0.33	0	0.33	0
Gymnorhinus cyanocephalus	1	1	0	0	0	0
Megascops asio/kennicottii	7	0.2	0.2	0.2	0.2	0.2
Pluvialis cf. *P. squatarola*	1	0	0	0	1	0
Rallus limicola	2	0.2	0.2	0.2	0.2	0.2
Sturnella sp.	3	0	0	0	20	0
TOTAL	1352	38.83	37.33	30.4	**52.23**	37.0

Table 9. Habitat weighting values for composite stratum 720 (~28,000 ^{14}C years BP) from San Josecito Cave using squamate reptiles and habitat preferences.

Taxa	Coniferous Forest	Oak Forest	Xeric Scrub	Grassland	Tropical Forest
	Habitat weighting using presence data				
Barisia imbricata	0.33	0.33	0	0.33	0
Barisia ciliaris	0.33	0.33	0	0.33	0
Phrynosoma modestum	0	0	1	0	0
Phrynosoma orbiculare	0.5	0.5	0	0	0
Sceloporus spp.	0.2	0.2	0.2	0.2	0.2
Tantilla sp.	0.25	0.25	0.25	0	0.25
Heterodon cf. *H. simus*	0.5	0	0	0.5	0
Hypsiglena sp.	0.25	0.25	0.25	0.25	0
Storeria sp.	0.5	0.5	0	0	0
Crotalus sp.	0.2	0.2	0.2	0.2	0.2
TOTAL	**3.06**	2.56	1.9	1.81	0.65
	Habitat weighting using presence and MNI data				

Table 9. *Cont.*

Taxa	MNI	Coniferous Forest	Oak Forest	Xeric Scrub	Grassland	Tropical Forest
Barisia imbricata	8	2.64	2.64	0	2.64	0
Barisia ciliaris	23	7.59	7.59	0	7.59	0
Phrynosoma modestum	1	0	0	1	0	0
Phrynosoma orbiculare	2	1	1	0	0	0
Sceloporus spp.	4	0.8	0.8	0.8	0.8	0.8
Tantilla sp.	1	0.25	0.25	0.25	0	0.25
Heterodon cf. *H. simus*	1	0.5	0	0	0.5	0
Hypsiglena sp.	1	0.25	0.25	0.25	0.25	0
Storeria sp.	1	0.5	0.5	0	0	0
Crotalus sp.	1	0.2	0.2	0.2	0.2	0.2
TOTAL	43	**13.73**	13.23	2.5	11.98	1.25
Habitat weighting using presence and NISP data						
Taxa	**NISP**	**Coniferous Forest**	**Oak Forest**	**Xeric Scrub**	**Grassland**	**Tropical Forest**
Barisia imbricata	32	10.56	10.56	0	10.56	0
Barisia ciliaris	75	24.75	24.75	0	24.75	0
Phrynosoma modestum	1	0	0	1	0	0
Phrynosoma orbiculare	6	3	3	0	0	0
Sceloporus spp.	14	2.8	2.8	2.8	2.8	2.8
Tantilla sp.	3	0.75	0.75	0.75	0	0.75
Heterodon cf. *H. simus*	1	0.5	0	0	0.5	0
Hypsiglena sp.	6	1.5	1.5	1.5	1.5	0
Storeria sp.	61	30.5	30.5	0	0	0
Crotalus sp.	41	8.2	8.2	8.2	8.2	8.2
TOTAL	240	**82.56**	82.06	14.25	48.31	11.75

The canonical correlations analysis (CCA) compared the correlation between the presence of the small mammals, birds, squamate reptiles, and all vertebrates in different vegetation types, and the habitat weighting values with relative abundance or only presence. The correlation was regularized and the values of the cross validation (CV-scores) is high in all cases.

The CCA for small mammals infers a coniferous-oak forest (Figure 7), where species such as *Sorex altoensis*, *Parasciururs alleni*, *Peromyscus difícilis*, and *Neotoma albigula* support high cross correlations values. The CCA for birds indicates that the habitat weighting analysis infer a grassland supported by *Sturnella* and *P. squatarola* (Figure 7). The CCA for squamate reptiles infers a coniferous-oak forest (Figure 7), where taxa such as *Hypsiglena*, *Barisia imbricata*, *B. ciliaris*, *Phrynosoma orbiculare*, and *Storeria* support high cross correlation values. The CCA for all vertebrates indicates a coniferous-oak forest (Figure 7).

The habitat weighting and CCA analyses, with different groups or all vertebrates, indicate in common the presence of coniferous-oak forest in the vicinity of San Josecito Cave at ~28,000 ^{14}C years BP. This result is different than the oak-pine forest present today. The mean annual temperature would have been colder, at least by two degrees in the past [133]. The presence of *Barisia imbricata* and *B. ciliaris* in sympatry during this time in San Josecito Cave mirrors their current sympatry at Atotonilco el Grande, Hidalgo [82]. This location has colder and more humid conditions than those of the San Josecito Cave area today.

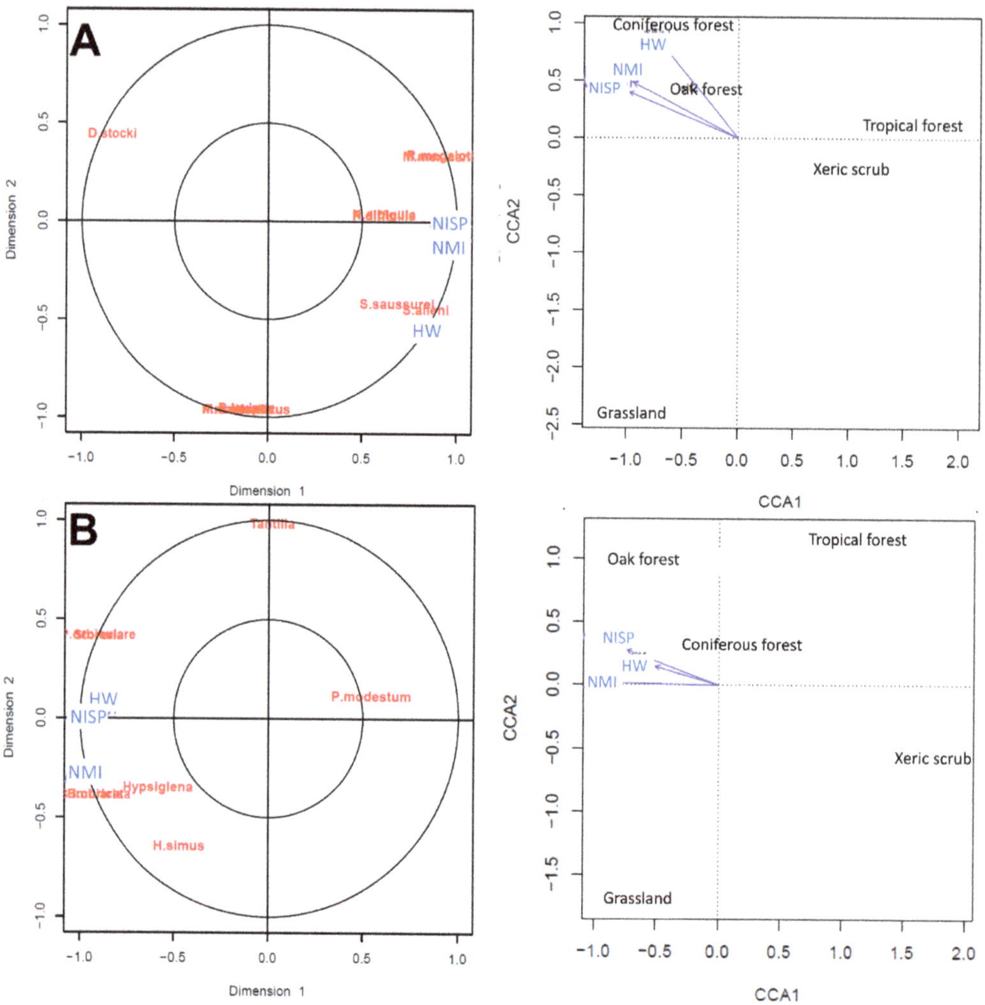

Figure 7. *Cont.*

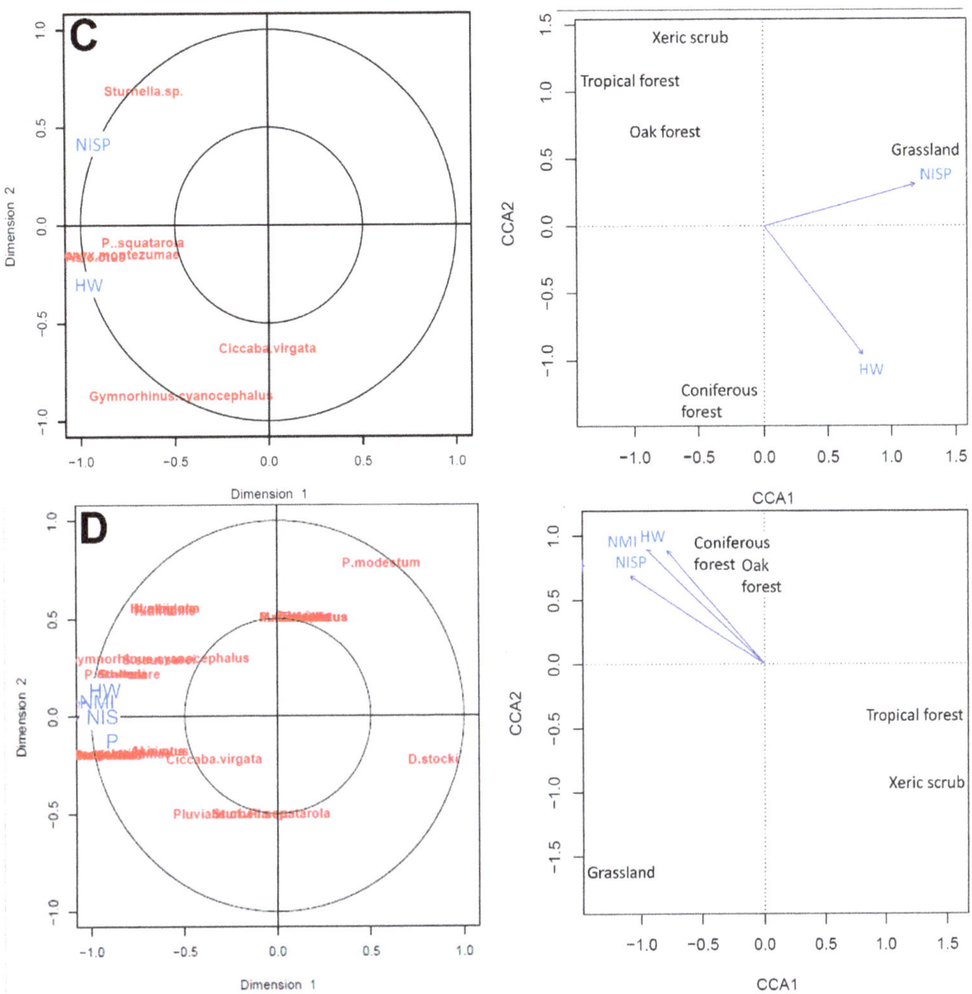

Figure 7. Correspondence canonical analysis (CCA) in different taxonomic groups from composite stratum 720, using the habitat weighting with presence species (HW) and habitat weighting with relative abundance employing number of individual specimens (NISP) and number of individuals (MNI). (**A**) mammals, (**B**) reptiles, (**C**) birds, and (**D**) all vertebrates.

5. Discussion

Results from the taphonomic studies of the 1990 strata 720 and 770 materials provide insights into the accumulation of the small vertebrates and the cave dynamics. The bone breakage patterns for the stratum 720 vertebrates indicate that owls most likely were the accumulating agents. Three species of owls occur in the composite stratum 720 fauna (Table 5). All three have a varied diet focused on small mammals (e.g., shrews, microtines, mice, bats), birds (particularly small passerines), snakes, and other small vertebrates and insects. This range of prey vertebrates is mirrored in the stratum 720 fauna. The most frequent remains are of *Asio otus* (long-eared owl) [67]. These owls generally hunt within their home range which can be up to 980 ha (~2422 ac). Home range use changes through the year, with the smallest use area being during the breeding season [134–136]. Assuming this behavior was occurring in the Late Pleistocene, home range use by this owl provides a view of the small vertebrate composition within a fairly localized area around San Josecito Cave.

The taphonomic boxes provided an enhanced estimate of abundance and MNI for the stratum 720 fauna. Bones were not distributed randomly across the 720 surfaces. Spatial distribution by taphonomic box indicted a differential distribution of bones across an excavation unit based on the abundance of elements in each box. The MNI by taphonomic box for the stratum 720 mammals (Table 10) increased almost three times from a general count for the overall stratum. The most abundant mammals were a vole (*Microtus mexicanus*) and mouse (*Reithrodontomy megalotis*) with the vole more than three times as abundant as the mouse. The accuracy of the individual counts was dependent upon the unit size used [137,138].

Table 10. Mammal species assigned to stratum 720 (~28,000 ^{14}C years BP) and minimum number of individuals (MNI) based on taphonomic box occurrences. Nomenclature follows that proposed by [52]. An asterisk denotes extinct taxa (*).

Taxa	MNI
Sorex altoensis	60
Leptonycteris nivalis	12
Myotis californicus	2
Myotis thysanodes	7
Equus conversidens *	1
Microtus mexicanus	342
Neotoma albigula	16
Neotoma mexicana	44
Peromyscus difficilis	46
Peromyscus labecula	53
Peromyscus levipes	83
Reithrodontomy megalotis	110
Sylvilagus floridanus	3
Total MNI	779

Bone discoloration that produced blue and gray bones was due to at least two factors, that of heat and of environmentally available manganese. These physical and chemical processes were occurring within the cave rather than from outside conditions. The heat-induced modifications were caused by an unknown source in the cave (combustible guano deposits may be a likely source) that reached temperatures above 500 °C but below 600 °C. Two different cations of manganese were responsible for the gray and blue colors. Although the cause of the heating events was unknown, these incidents occurred at least four times over a 10,000-year period (from ~35,000 to 25,000 ^{14}C years BP).

The squamate reptiles indicate a strong relationship between the Sierra Madre Oriental and Mexican Transvolcanic Belt regions during the Late Pleistocene. Sympatric taxa found in San Josecito Cave at ~28,000 ^{14}C years BP now range in one or the other region. The retraction pattern is to the south, whereas other taxa retreated north. The relationship between the Sierra Madre Oriental and the Mexican Transvolcanic Belt regions has not yet been examined for mammals experiencing range changes by the end of the Pleistocene [21,22]. In addition, future climatic studies with the taxa found in the composite stratum 720 local fauna may provide more information on the paleoclimatic conditions in the San Josecito Cave area and Sierra Madre Oriental that allowed co-existence of the taxa and community structure development around 28,000 ^{14}C years BP.

The overall fauna of San Josecito Cave now comprises 136 species and prevails as one of the most important Late Pleistocene faunas for North America. The great possibility exists of more unreported taxa within the other strata from the 1990 excavation as well as the non-cataloged materials housed at LACM. Two points stand out. First, paleontological excavations must be undertaken within a field methodology that allows for a very detailed recovery and data recording, with precise spatial (horizontal and vertical) and radiometric controls. Second, general paleoenvironmental conditions can be inferred following the approaches used here. For more detailed results, however, that lead to a better understanding of the Quaternary ecosystems and biotic communities, a greater knowledge of the biology of the recent species and their ecomorphological traits is needed.

The question is raised about the concept of a cave fauna as a single local fauna. Other strata have very different species composition. The term fauna seems better applied to a group of organisms sharing a suite of common species and representing its maximum geographic and temporal limits [139] such as the Late Pleistocene fauna of Mexico. On the other hand, local fauna refers to aggregates of species that have a defined distribution in time and space, based on the record from a restricted geographic area.

The term local fauna could be applied to the overall San Josecito Cave assemblage if all of the materials came from a single stratigraphic level. Stock's [26] excavations, however, crossed several stratigraphic levels that encompassed many thousands of years. Local faunas should be defined according to stratigraphic provenience and geochronology [8,12–14]. Most of Stock's materials cannot be allocated to a specific local fauna due to their lack of provenience data. The overall vertebrate assemblage reflects many, successive local faunas. Currently, then, a San Josecito Cave local fauna refers only to the composite stratum 720 and Stock's 57–58 ft level. That local fauna is composed of 47 taxa (Table 11), and dates at ~28,000 ^{14}C year BP.

It is on these grounds that it is considered both inappropriate and inadequate to treat the San Josecito Cave fauna as a single entity representing a disharmonious (i.e., non-analog) fauna [17]. The lack of exact provenience of most of Stock's collection does not allow ascertaining the relationships of the elements from different biomes within one or several temporal communities. The disharmony in the composite stratum 720 local fauna is limited. The level of disharmony within the local fauna is variable, with most non-analog species being minimally extralimital. A bird (*Gymnorhinus cyanocephalus*) is extralimital by several hundred km. A snake (*Heterodon simus*) has the most extensive extralimital distribution. Without knowing the provenience of most of the non-analog elements within the overall cave fauna, the trajectory of disharmony is difficult to track. What can be said is that a minimal non-analog fauna existed at ~28,000 ^{14}C years BP. Whether the composite stratum 720 is the first non-analog local fauna for the San Josecito Cave record is unknown until the vertebrate remains from the earlier strata are identified and evaluated. An important point, however, is that the non-analog nature of a fauna should be evaluated on all the vertebrate classes represented rather than just one class (typically mammals). Each faunal assemblage needs to be defined not just in terms of faunal composition, but also in terms of the stratigraphic position of the fauna and associated radiometric data. This situation points to the need of stratigraphically-controlled excavations to access and evaluate the actual components of a locality's overall fauna.

A shortcoming of the community structure analysis is the level of knowledge available for specific relationships between the ecological parameters and the morphological traits. Nevertheless, the approaches taken reveal different paleoenvironmental aspects. First, the Rocky Mountains biome is the general biome for the cave local fauna and regionally at ~28,000 ^{14}C years BP similar to that of today. The difference is in the specific vegetational association that appears distinct from today. Second, the vegetational association is a mosaic of coniferous-oak forest and grasslands (different from the pine-oak forest of today). This association in and of itself is a non-analog one. It is not represented in any of the current vegetational associations within the Rocky Mountains biome.

Table 11. The San Josecito Cave composite 720 local fauna, dated at ~28,000 ^{14}C years BP (MNI are provided for bones from stratum 720). Extinct taxa are denoted by an asterisk (*).

Taxa	Stock's 57–58 ft Level	Stratum 720
Nothrotheriops shastensis *	X	
Sorex altoensis		9
Desmodus stocki *	X	
Leptonycteris nivalis	X	4
Myotis californicus		1
Myotis thysanodes		2
Aenocyon dirus *	X	
Canis latrans	X	
Panthera atrox *	X	
Panthera onca	X	
Puma concolor	X	
Smilodon fatalis *	X	
Tremarctos floridanus *	X	
Ursus americanus	X	
Equus conversidens *	X	1
Navahoceros fricki *	X	
Stockoceros conklingi *	X	
Euceratherium *	X	
Parasciurus alleni	X	
Microtus mexicanus	X	44
Neotoma albigula	X	4
Neotoma mexicana	X	9
Peromyscus difficilis		8
Peromyscus labecula		9
Peromyscus levipes	X	16
Reithrodontomy megalotis	X	16
Sylvilagus floridanus	X	2
Aquila chrysaetos	X	
Cyrtonyx montezumae		85
Rallus limicola		2
Pluvialis cf. *squatarola.*	X	
Megascops asio/kennicottii		7
Asio otus	X	107
Ciccaba virgata	X	
Gymnorhinus cyanocephalus		1
Corvus corax	X	
Sturnella sp.		3
Barisia ciliaris		23
Barisia imbricata		8
Phrynosoma modestum		1
Phrynosoma orbiculare	X	2
Sceloporus sp.		4
Tantilla sp.		1
Heterodon cf. *simus*		1
Hypsiglena sp.		1
Storeria sp.		1
Crotalus spp.		1

6. Conclusions

The current synthesis examined a paleontological cave deposit and the significance of its faunal data for contributing to the interpretation of Quaternary ecosystems. The hypothesis was that no drastic climatic change occurred during the deposition of the studied strata at the mid-Wisconsin period (40,000 to 25,000 ^{14}C years BP). This hypothesis was tested by comparing the independently-derived data sets from the faunal and geochronological results focusing on stratum 720. The focus on one unique stratum decreased the general inferences that could be made regarding the overall late Quaternary ecosystems of northeastern Mexico. This approach, however, assured both the best knowledge of the San Josecito Cave local fauna and the most supported paleoenvironmental and paleoclimatic reconstruction.

Although the hypothesis of no change originally proved valid based only on mammals [28], the broader and more recent analyses have demonstrated the hypothesis to be false. In general, the Rocky Mountains biome is represented in the vicinity of the cave around 28,000 ^{14}C years BP as it is today. But, the vegetational association is a non-analog mosaic of coniferous-oak forest and grasslands. The current interpretation points to the importance of detailed paleoecological studies of all vertebrate taxa for each locality as well as within a stratigraphic and geochronological framework.

The continued study of the San Josecito Cave faunal assemblages is a building block toward the understanding of Late Pleistocene environments, climate, and faunal communities on the Mexican Plateau and the Southern Plains in general. The great potential of the sediments for a detailed fossil record for the mid-Wisconsin period exemplified by this synthesis indicates the importance of such research. The San Josecito Cave composite stratum 720 local fauna provides the first inferences of the paleoecology and paleoenvironment at the mid-Wisconsin of northeastern Mexico. Non-analog faunas are beginning to appear by at least ~28,000 ^{14}C years BP in the cave record. Equally significant, the local fauna indicates a non-analog vegetational community.

This synthesis of San Josecito Cave and its faunal data has demonstrated that the previous view of the assemblage as a single local fauna is erroneous and instead, several successive local faunas are present within a stratigraphic framework. The overall San Josecito Cave fauna (sensu [17,18]) is a time-transgressive one covering many thousands of years. Treating it as a single fauna masks changes within the faunal community, its paleoecology, and climatic conditions that may have existed.

The study also underscores the need for detailed studies of single localities in building paleoenvironmental models. Furthermore, the recognition of San Josecito Cave as an important Late Pleistocene vertebrate paleontological locality is enhanced with the consideration of its faunal data for paleoenvironment reconstruction and possible contribution to Quaternary paleoclimatic modeling. In addition, the field and analytical methodologies demonstrate the importance of very detailed paleontological excavations, with precise spatial and temporal controls, to assess the taphonomic history of a locality and infer the paleoecological conditions during the time span considered.

Author Contributions: Conceptualization, J.A.-C. and E.J.; methodology, J.A.-C. and E.J.; validation, E.J.; formal analysis, J.A.-C., E.J. and J.A.C.; investigation, J.A.-C. and E.J.; resources, J.A.-C. and E.J.; data curation, J.A.-C.; writing—original draft preparation, J.A.-C. and E.J.; writing—review and editing, J.A.-C., E.J. and J.A.C.; project administration, J.A.-C. and E.J.; funding acquisition, J.A.-C. and E.J. All authors have read and agreed to the published version of the manuscript.

Funding: External funding was received from the National Geographic Society (E.J.), National Speleological Society (E.J. and J.A.-C.), American Society of Mammologists (J.A.-C.), Cave Research Foundation (J.A.-C.), Geological Society of America (J.A.-C.), and the American Museum of Natural History Theodore Roosevelt Memorial Fund (J.A.-C.).

Institutional Review Board Statement: Not applicable.

Informed Consent Statement: Not applicable.

Data Availability Statement: All data are housed at the Archaeozoology Lab, Instituto Nacional de Antropología e Historía, Mexico City, Archaeozoology Lab, National Institute of Anthropology and History, Mexico City, and available on request to Joaquín Arroyo-Cabrales.

Acknowledgments: This manuscript represents part of the ongoing Lubbock Lake Landmark regional research program on the Quaternary paleoecology and taphonomy of the Southern Plains.

Conflicts of Interest: The authors declare no conflict of interest.

References

1. Nolan, C.; Overpeck, J.T.; Allen, J.R.M.; Anderson, P.M.; Betancourt, J.L.; Binney, H.A.; Brewer, S.; Bush, M.B.; Chase, B.M.; Cheddadi, R.; et al. Past and future global transformation of terrestrial ecosystems under climate change. *Science* **2018**, *361*, 920–923. [CrossRef]
2. Past Global Changes Project. Proposed implementation plans for research activities. *IGBP Glob. Chang. Rep.* **1992**, *19*, 1–110.
3. Blum, M.D.; Toomey, R.S., III; Valastro, S., Jr. Fluvial response to late Quaternary climatic and environmental change, Edwards Plateau, Texas. *Palaeogeogr. Palaeoclimatol. Palaeoecol.* **1994**, *108*, 1–21. [CrossRef]
4. Werdelin, L.; von Koenigswald, W. Climate and fauna: A database of the Quaternary mammals of Europe. *Eur. Mam. Newsl.* **1994**, *1*, 9–15.
5. Behrensmeyer, A.K.; Hook, R.W. Paleoenvironmental contexts and taphonomic modes. In *Terrestrial Ecosystems through Time. Evolutionary Paleoecology of Terrestrial Plants and Animals*; Behrensmeyer, A.K., Damuth, J.D., DiMichele, W.A., Potts, R., Sues, H.-D., Wing, S.L., Eds.; The University of Chicago Press: Chicago, IL, USA, 1992; pp. 15–136.
6. Behrensmeyer, A.K.; Kidwell, S.M.; Gastaldo, R.A. Taphonomy and paleobiology. *Paleobiology* **2000**, *26*, 103–147. [CrossRef]
7. Martin, R.E. *Taphonomy: A Process Approach*; Cambridge University Press: Cambridge, MA, USA, 1999; p. 508.
8. Johnson, E. *Lubbock Lake: Late Quaternary Studies on the Southern High Plains*; Texas A&M University Press: College Station, TX, USA, 1987; pp. xii + 179.
9. Pérez-Crespo, V.A.; Arroyo-Cabrales, J.; Benammi, M.; Johnson, E.; Polaco, O.J.; Santos-Moreno, A.; Morales-Puente, P.; Cienfuegos-Alvarado, E. Geographic Variation of Diet and Habitat of the Mexican Populations of Columbian Mammoth (*Mammuthus columbi*). *Quat. Int.* **2012**, *276–277*, 8–16. [CrossRef]
10. Pérez-Crespo, V.A.; Prado, J.L.; Alberdi, M.T.; Arroyo-Cabrales, J.; Johnson, E. Diet and Habitat for Six American Pleistocene Proboscidean Species Using Carbon and Oxygen Stable Isotopes. *Ameghiniana* **2016**, *53*, 39–51. [CrossRef]
11. Harris, A.H. Wisconsin age environments in the northern Chihuahuan Desert: Evidence from the higher vertebrates. In *Transactions of the Symposium on the Biological Resources of the Chihuahuan Desert Region*; National Park Service Transactions and Proceedings Series; U.S. Department of the Interior: Chihuahua, CH, USA, 1977; Volume 3, pp. 23–52.
12. Johnson, E. Late Pleistocene and early Holocene vertebrates and paleoenvironments on the southern High Plains (USA). *Geogr. Phys. Quat.* **1986**, *40*, 249–261.
13. Johnson, J.D.; Wilson, L.D.; Mata-Silva, V.; García-Padilla, E.; DeSantis, D.L. The Endemic Herpetofauna of Mexico: Organisms of Global Significance in Severe Peril. *Mesoam. Herpetol.* **2017**, *4*, 544–620.
14. Moretti, J.A. The Vertebrate Fauna of Macy Locality 100: Exploring Late Pleistocene Community Composition in Non-Analog North America. Master's Thesis, Texas Tech University, Lubbock, TX, USA, 2018.
15. Kurten, B.; Anderson, E. *Pleistocene Mammals of North America*; Columbia University Press: New York, NY, USA, 1980; pp. xvii + 442.
16. Álvarez, T.; Ocaña, M.A.; Arroyo-Cabrales, J. Restos de mamífero. In *Rancho "La Amapola", Cedral. Un Sitio Arqueológico-Paleontológico Pleistocénico-Holocénico con Restos de Actividad Humana*; México, D.F., Mirambell, L.E., Eds.; Colección Interdisciplinaria-Serie Memorias; INAH: México City, Mexico, 2012; pp. 147–194.
17. Graham, R.W. Response of mammalian communities to environmental changes during the late Quaternary. In *Community Ecology*; Diamond, J., Case, T.J., Eds.; Harper & Row: New York, NY, USA, 1986; pp. 300–313.
18. Graham, R.W.; Lundelius, E.L., Jr. Coevolutionary disequilibrium and Pleistocene extinctions. In *Quaternary Extinctions*; Martin, P.S., Klein, R.G., Eds.; University of Arizona Press: Tucson, AZ, USA, 1984; pp. 223–249.
19. Semken, H.A., Jr. Micromammal distribution and migration during the Holocene I. In *3rd Annual Meeting Abstracts*; Knox, J.C., Mickelson, D.M., Eds.; American Quaternary Association: Madison, WI, USA, 1974; p. 25.
20. Arroyo-Cabrales, J.; Johnson, E. Catálogo de los ejemplares tipo procedentes de la Cueva de San Josecito, Nuevo León, México. *Rev. Mex. Cienc. Geol.* **2003**, *20*, 79–93.
21. Ceballos, G.; Arroyo-Cabrales, J.; Ponce, E. Effects of Pleistocene environmental changes on the distribution and community structure of the mammalian fauna of Mexico. *Quat. Res.* **2010**, *73*, 464–473. [CrossRef]
22. Ferrusquía-Villafranca, I.; Arroyo-Cabrales, J.; Martínez-Hernández, E.; Gama-Castro, J.; Ruíz-González, J.; Polaco, O.J.; Johnson, E. Pleistocene mammals of Mexico: A critical review of regional chronofaunas, climate change response and biogeographic provinciality. *Quat. Int.* **2010**, *217*, 53–104. [CrossRef]
23. Graham, R.W.; Mead, J.I. Environmental fluctuations and evolution of mammalian faunas during the last deglaciation in North America. In *The Geology of North America: North America and Adjacent Oceans during the Last Deglaciation*; Geological Society of America: Boulder, CO, USA, 1987; Volume K-3, pp. 371–402.
24. Miller, L. The Pleistocene Birds of San Josecito Cavern, Mexico. *Univ. Calif. Publ. Zool.* **1943**, *47*, 143–167.
25. Naples, V.L. Morphological changes in the facial region and a model of dental growth and wear pattern in *Nothrotheriops shastensis*. *J. Vertebr. Paleontol.* **1990**, *10*, 372–389. [CrossRef]
26. Stock, C. The Cave of San Josecito, Mexico. *Calif. Inst. Technol. Balch Grad. Sch. Geol. Sci. Contrib.* **1943**, *361*, 1–5.
27. Arroyo-Cabrales, J. New Studies on San Josecito Cave, Nuevo Leon, Mexico. *Cave Res. Found. Rep.* **1991**, *1990*, 43–45.
28. Arroyo-Cabrales, J. Taphonomy and Paleoecology of San Josecito Cave, Nuevo León, México. Ph.D. Thesis, Department of Biological Sciences, Texas Tech University, Lubbock, TX, USA, 1994.

29. Arroyo-Cabrales, J.; Johnson, E. Arroyo-Cabrales, J.; Johnson, E. A Reappraisal of Fossil Vertebrates From San Josecito Cave, Nuevo Leon, Mexico. In *Ancient Peoples and Landscapes*; Johnson, E., Ed.; Museum of Texas Tech University: Lubbock, TX, USA, 1995; pp. 217–231.
30. Arroyo-Cabrales, J.; Johnson, E.; Ralph, R.W. An initial reappraisal of San Josecito Cave and Its Late Pleistocene fauna. *Curr. Res. Pleistocene* **1989**, *6*, 63–65.
31. Arroyo-Cabrales, J.; Johnson, E.; Ralph, R.W. New excavations at San Josecito Cave, Nuevo León, México. *Curr. Res. Pleistocene* **1993**, *10*, 91–94.
32. Hibbard, C.W. Techniques of Collecting Microvertebrate Fossils. *Univ. Mich. Contrib. Mus. Paleontol.* **1949**, *8*, 7–19.
33. Ralph, R.W.; Arroyo-Cabrales, J.; Johnson, E. San Josecito Cave. Recent Explorations into the Past. *NSS News* **1994**, *52*, 230–235.
34. Jones, J.K., Jr. Pleistocene bats from San Josecito Cave, Nuevo León, México. *Univ. Kans. Mus. Nat. Hist.* **1958**, *9*, 389–396.
35. Russell, R.J. Pleistocene pocket gophers from San Josecito Cave, Nuevo León, México. *Kans. Mus. Nat. Hist.* **1960**, *9*, 539–548.
36. Webb, S.D. A cranium of Navahoceros and its phylogenetic place among New World Cervidae. *Ann. Zool. Fenn.* **1992**, *28*, 401–410.
37. Lyman, R.L. *Quantitative Paleozoology*; Cambridge University Press: New York, NY, USA, 2008; p. 348.
38. Faith, J.T.; Lyman, R.L. *Paleozoology and Paleoenvironments: Fundamentals, Assumptions, Techniques*; Cambridge University Press: New York, NY, USA, 2019; p. 398.
39. Blain, H.A.; Bailon, S.; Cuenca-Bescós, G. The Early–Middle Pleistocene palaeoenvironmental change based on the squamate reptile and amphibian proxies at the Gran Dolina site, Atapuerca, Spain. *Palaeogeogr. Palaeoclimatol. Palaeoecol.* **2008**, *261*, 177–192. [CrossRef]
40. González, I.; Déjean, S.; Martin, P.; Baccini, A. CCA: An R package to extend canonical correlation analysis. *J. Stat. Softw.* **2008**, *23*, 1–14. [CrossRef]
41. Oksanen, J.; Blanchet, F.G.; Friendly, M.; Kindt, R.; Legendre, P.; McGlinn, D.; Minchin, P.R.; O'Hara, R.B.; Simpson, G.L.; Solymos, P.; et al. Package Vegan: Community Ecology Package. 2016. Available online: https://CRAN.R-project.org/package=vegan (accessed on 10 August 2020).
42. Rolong, N.A.; Arroyo-Cabrales, J.; Allen, B.L.; Johnson, E.; Flores-Díaz, A. Sediments Properties from San Josecito Cave, Nuevo León, México. In Proceedings of the Transactions of the 15th World Congress of Soil Science, Commission V: Symposia, Acapulco, Mexico, 10–16 July 1994; Etchevers, B., Ed.; Sociedad Mexicana de la Ciencia del Suelo, International Society of Soil Science: Acapulco, México, 1994; Volume 6a, pp. 129–141.
43. Stock, C. The succession of mammalian forms within the period in which human remains are known to occur in America. *Am. Nat.* **1936**, *70*, 324–331. [CrossRef]
44. Simpson, G.G. Chester Stock 1892–1950. *Natl. Acad. Biogr. Mem.* **1952**, *27*, 334–362.
45. Arroyo-Cabrales, J.; Johnson, E.; Hass, H.; de los Ríos-Paredes, M.; Ralph, R.W.; Hartwell, W.T. First Radiocarbon Dates for San Josecito Cave, Nuevo León, México. *Quat. Res.* **1995**, *43*, 255–258. [CrossRef]
46. Bronk-Ramsey, C. Radiocarbon calibration and analysis of stratigraphy: The OxCal program. *Radiocarbon* **1995**, *37*, 425–430. [CrossRef]
47. Bronk-Ramsey, C. Bayesian analysis of radiocarbon dates. *Radiocarbon* **2009**, *51*, 337–360. [CrossRef]
48. Bronk-Ramsey, C.; Dee, M.; Lee, S.; Nakagawa, T.; Staff, R. Developments in the calibration and modelling of radiocarbon dates. *Radiocarbon* **2010**, *52*, 953–961. [CrossRef]
49. Bronk-Ramsey, C. Methods for Summarizing Radiocarbon Datasets. *Radiocarbon* **2017**, *59*, 1809–1833. [CrossRef]
50. Reimer, P.; Austin, W.; Bard, E.; Bayliss, A.; Blackwell, P.; Bronk-Ramsey, C.; Butzin, M.; Cheng, H.; Edwards, R.; Friedrich, M.; et al. The IntCal20 Northern Hemisphere radiocarbon age calibration curve (0–55 cal kBP). *Radiocarbon* **2020**, *62*, 725–757. [CrossRef]
51. Jakway, G.E. Pleistocene Lagomorpha and Rodentia from the San Josecito Cave, Nuevo Leon, Mexico. *Trans. Kans. Acad. Sci.* **1958**, *61*, 313–327. [CrossRef]
52. Mammal Diversity Database. Version 1.6. 2021. Available online: http://doi.org/10.5281/zenodo.4139818 (accessed on 12 May 2021). [CrossRef]
53. Arroyo-Cabrales, J.; Johnson, E. Mammalian additions to the faunal assemblages from San Josecito Cave, Nuevo León, México. In *Avances en el Estudio de los Mamíferos de México II*; Lorenzo, C., Espinoza, E., Ortega, J., Eds.; Asociación Mexicana de Mastozoología Asociación Civi: San Cristóbal de las Casas, CS, Mexico, 2008; pp. 65–87.
54. Esteva, M.; Arroyo-Cabrales, J.; Flores Martínez, A.; Johnson, E.; Polaco, O.J. Fossil Shrews from San Josecito Cave, Nuevo Leon, Mexico. In *Advances in the Biology of Shrews II*; Merritt, J.F., Churchfield, S., Hutterer, R., Sheftel, B.L., Eds.; Special Publication of the International Society of Shrew Biologists: New York, NY, USA, 2005; Volume 1, pp. 31–43.
55. Woodman, N. American Recent Eulipotyphla: Nesophontids, Solenodons, Moles, and Shrews in the New World. *Smithson. Contr. Zool.* **2018**, *650*, 1–107. [CrossRef]
56. Greenbaum, I.F.; Honeycutt, R.L.; Chirhart, S.E. Taxonomy and Phylogenetics of the *Peromyscus maniculatus* Species Group. *Spec. Publ. Mus. Tex. Tech. Univ.* **2019**, *71*, 559–575.
57. de Abreu, E.F., Jr.; Pavan, S.E.; Tsuchiya, M.T.N.; Wilson, D.E.; Percequillo, A.R.; Maldonado, J.E. Museomics of Tree Squirrels: A Dense Taxon Sampling of Mitogenomes Reveals Hidden Diversity, Phenotypic Convergence, and the Need of a Taxonomic Overhaul. *BMC Evol. Biol.* **2020**, *20*, 77. [CrossRef]

58. Hall, E.R. *The Mammals of North America*; John Wiley & Sons: New York, NY, USA, 1981; p. 1300.
59. Polaco, J.O.; Butrón, M.L. Mamíferos pleistocénicos de la cueva La Presita, San Luis Potosí. México. In *Homenaje al Profesor Ticul Álvarez*; Colección Científica; Arroyo-Cabrales, J., Polaco, J.O., Eds.; Instituto Nacional de Antropología e Historia: Mexico City, Mexico, 1997; pp. 279–296.
60. Chávez-Tovar, J.C. Peromyscus levipes Merriam, 1898. In *Los Mamíferos Silvestres de México*; Ceballos, G., Oliva, G., Eds.; Comisíon Nacional para el Conocimiento y Uso de la Biodiversidad, Fondo de Cultura Económica: México City, Mexico, 2005; pp. 745–746.
61. Álvarez-Castañeda, S.T.; González-Ruiz, N. *Peromyscus levipes* (Rodentia: Cricetidae). *Mamm. Species* **2009**, *824*, 1–6. [CrossRef]
62. Best, T.L. Sciurus alleni. *Mamm. Species* **1995**, *501*, 1–4.
63. Howard, H. Quaternary avian remains from Dark Canyon Cave, New Mexico. *Condor* **1971**, *73*, 237–240. [CrossRef]
64. Olson, S.L. The Pleistocene rails of North America. *Condor* **1974**, *76*, 169–175. [CrossRef]
65. Olson, S.L. An evaluation of the fossil curlew *Palnumenius victima* L. Miller (Aves: Scolopacidae). In *Contributions in Quaternary Vertebrate Paleontology: A Volume in Memorial to John E. Guilday*; Genoways, H.H., Dawson, M.R., Eds.; Special Publication; Carnegie Museum of Natural History: Pittsburgh, PA, USA, 1984; pp. 53–55.
66. Rea, A.M. The Indeterminate Parrot of Nuevo León. In *The Era of Allan R. Phillips: A Festschrift*; Dickerman, R.W., Ed.; Horizon Communications: Albuquerque, NM, USA, 1991; pp. 167–176.
67. Steadman, D.W.; Arroyo-Cabrales, J.; Johnson, E.; Guzmán, A.F. New information on the Late Pleistocene birds from San Josecito Cave, Nuevo León, México. *Condor* **1994**, *96*, 577–589. [CrossRef]
68. American Ornithologists' Union (AOU). *Check-List of North American Birds*, 6th ed.; Allen Press: Lawrence, KS, USA, 1983; p. 877.
69. Phillips, A.R. *The Known Birds of North and Middle America. Part I. Hirundinidae to Mimidae, Certhiidae*; UNKNO: Denver, CO, USA, 1986; p. 320.
70. Moore, R.T. Records of two North American corvids in Mexico. *Condor* **1951**, *53*, 101.
71. Selander, R.K. Additional record of the Pinon Jay in Mexico. *Condor* **1956**, *58*, 239.
72. Hayman, P.; Marchant, J.; Prater, T. *Shorebirds. An Identification Guide to the Wader of the World*; Croom Helm: London, UK, 1986; p. 412.
73. Johnsgard, P.A. *The Plovers, Sandpipers and Snipes of the World*; University of Nebraska Press: Lincoln, NE, USA, 1981; p. 519.
74. Van Perlo, B. *Birds of Mexico and Central America*; Princeton University Press: Princeton, NJ, USA, 2006; p. 336.
75. Brattstrom, B.H. Pleistocene lizards from San Josecito Cavern, Mexico, with description of new species. *Copeia* **1955**, *1955*, 133–134. [CrossRef]
76. Mead, J.I.; Arroyo-Cabrales, J.; Johnson, E. Pleistocene lizards (Reptilia: Squamata) from San Josecito cave, Nuevo Leon, Mexico. *Copeia* **1999**, *1999*, 163–173. [CrossRef]
77. Smith, H.M.; Burg, T.M.; Chiszar, D. Evolutionary speciation in the alligator lizard of the genus *Barisia*. *Bull. Md. Herpetol. Soc.* **2002**, *38*, 23–26.
78. Cruz-Sáenz, D.; Muñoz-Nolasco, F.J.; Mata-Silva, V.; Johnson, J.D.; García-Padilla, E.; Wilson, L.D. The Herpetofauna of Jalisco, Mexico: Composition, Distribution, and Conservation Status. *Mesoam. Herpetol.* **2017**, *4*, 23–118.
79. Bryson, R.W., Jr.; Riddle, B.R. Tracing the Origins of Widespread Highland Species: A Case of Neogene Diversification Across the Mexican Sierras in an Endemic Lizard. *Biol. J. Linn. Soc.* **2012**, *105*, 382–394. [CrossRef]
80. Cruz-Silva, J.A. Reconstrucción paleoambiental del Pleistoceno final usando modelos de nicho ecológico inferidos con vertebrados: El caso de la Cueva de San Josecito, Nuevo León. Master's Thesis, Instituto de Biología, Universidad Nacional Autónoma de México, México City, Mexico, 2012; p. 132.
81. Conant, R. *Collins, J.T. A Field Guide to Reptiles and Amphibians of Eastern/Central North America*, 3rd ed.; Houghton Mifflin: Boston, MA, USA; New York, NY, USA, 1991; 450p.
82. Zaldivar-Riveron, A.; Nieto-Montes de Oca, A.; Laclette, J.P. Phylogeny and evolution of dorsal pattern in the Mexican endemic lizard genus *Barisia* (Anguidae: Gerrhonotinae). *J. Zool. Syst. Evol. Res.* **2005**, *43*, 243–257. [CrossRef]
83. Sherbrooke, W.C. *Introduction to Horned Lizards of North America*; California Natural History Guides; University of California Press: Berkeley, CA, USA, 2003; p. 191.
84. Tovar-Liceaga, R.E.; Montellano-Ballesteros, M. Pleistocene microvertebrates from Santa Cruz Nuevo, Puebla, Mexico. *Curr. Res. Pleistocene* **2006**, *23*, 72–74.
85. Van Devender, T.R.; Rea, A.M.; Smith, M.L. The Sangamon interglacial vertebrate fauna from Rancho La Brisca, Sonora, Mexico. *Trans. San Diego Soc. Nat. Hist.* **1985**, *21*, 23–55. [CrossRef]
86. Cruz, J.A.; Arroyo-Cabrales, J.; Reynoso, V.H. Reconstructing the paleoenvironment of Loltún Cave, Yucatán, México, with Pleistocene amphibians and reptiles and their paleobiogeographic implications. *Rev. Mex. Cienc. Geol.* **2016**, *33*, 342–354.
87. Shaffer, H.B.; McKnight, M.L. The polytypic species revisited: Genetic differentiation and molecular phylogenetics of the tiger salamander *Ambystoma tigrinum* (Amphibia: Caudata) complex. *Evolution* **1996**, *50*, 417–433. [CrossRef]
88. Bryson, R.W., Jr.; Murphy, R.W.; Graham, M.R.; Lathrop, A.; Lozano, D. Ephemeral Pleistocene woodlands connect the dots for highland rattlesnakes of the *Crotalus intermedius* group. *J. Biogeogr.* **2011**, *38*, 2299–2310. [CrossRef]
89. Bryson, R.W., Jr.; García-Vasquez, U.O.; Riddle, B.R. Diversification in the mexican horned lizard *Phrynosoma orbiculare* across a dynamic landscape. *Mol. Phylogenet. Evol.* **2012**, *62*, 87–96. [CrossRef]
90. Guillette, L.J., Jr.; Smith, H.M. A review of the Mexican lizard *Barisia imbricata*, and the description of a new subspecies. *Trans. Kans. Acad. Sci.* **1982**, *85*, 13–33. [CrossRef]

91. Canseco-Márquez, L.; Mendoza-Quijano, F.; Gutiérrez-Mayen, M.G. Análisis de la distribución de la herpetofauna. In *Biodiversidad de la Sierra Madre Oriental*; Luna, I., Morrone, J.J., Espinosa, D., Eds.; Las Prensas de Ciencias: México City, Mexico, 2004; pp. 417–437.
92. Lazcano, D.; Sánchez-Almazán, A.; García-de la Peña, C.; Castañeda, G.; Contreras-Balderas, A.J. Notes on Mexican herpetofauna 9: Herpetofauna of a fragmented *Juniperus* forest in the state Natural Protected Area of San Juan y Puentes, Arramberri, Nuevo León, México. *Bull. Chic. Herpetol. Soc.* **2007**, *42*, 21–26.
93. Martin, P.S. A biogeography of reptiles and amphibians in the Gomez Farias region, Tamaulipas, Mexico. *Misc. Publ. Mus. Zool. Univ. Mich.* **1958**, *101*, 1–102.
94. Gadsden, H.; Rodríguez-Romero, F.J.; Méndez-de la Cruz, F.R.; Gil-Martínez, R. Ciclo reproductor de *Sceloporus poinsettii* Baird y Girard 1852 (Squamata: Phrynosomatidae) en el centro del desierto chihuahuense, México. *Acta Zool. Mex.* **2005**, *21*, 93–107.
95. Méndez-de la Cruz, F.R.; Villagrán-Santa Cruz, M.; Cuellar, O. Geographic variation of spermatogenesis in the Mexican viviparous lizard *Sceloporus mucronatus*. *Biogeographica* **1994**, *70*, 59–67.
96. Douglas, M.E.; Douglas, M.R.; Schuett, G.W.; Porras, L.W. Evolution of rattlesnakes (Viperidae; *Crotalus*) in the warm deserts of western North America shaped by Neogene vicariance and Quaternary climate change. *Mol. Ecol.* **2006**, *15*, 3353–3374. [CrossRef]
97. Hutchison, D.W.; Malcomber, S.T.; Pletscher, L.S. A multidisciplinary investigation of the applicability of the Pleistocene herpetofaunal stability model to collared lizards (*Crotaphytus collaris*). *Herpetol. Monogr.* **1999**, *13*, 81–141. [CrossRef]
98. McGuire, J.A.; Linkem, C.W.; Koo, M.S.; Hutchison, D.W.; Lappin, A.K.; Orange, D.I.; Lemos-Espinal, J.; Riddle, B.R.; Jaeger, J.R. Mitochondrial introgression and incomplete lineage sorting through space and time: Phylogenetics of crotaphytid lizards. *Evolution* **2007**, *61*, 2879–2897. [CrossRef] [PubMed]
99. Waltari, E.; Hijmans, R.J.; Townsend-Peterson, A.; Nyári, A.S.; Perkins, S.L.; Guralnick, R.P. Locating Pleistocene refugia: Comparing phylogeography and ecological niche model predictions. *PLoS ONE* **2007**, *2*, e563. [CrossRef] [PubMed]
100. Holman, J.A. *Pleistocene Amphibians and Reptiles in North America*; Oxford University Press: New York, NY, USA, 1995; p. 256.
101. Arroyo-Cabrales, J.; Johnson, E. Preliminary analysis of bone from San Josecito Cave, Nuevo Leon, Mexico. In *Proceedings of the 1993 Bone Modification Conference*; Occasional Papers; Hannus, L.A., Rossum, L., Winham, R.P., Eds.; Archeology Lab Augustana College: Hot Springs, SD, USA, 1997; pp. 50–64.
102. Caballero, M.; Lozano-García, S.; Vázquez-Selem, L.; Ortega, B. Evidencias de cambio climático y ambiental en registros glaciales y en cuencas lacustres del centro de México durante el último máximo glacial. *Boletín Soc. Geol. Mex.* **2010**, *62*, 359–377. [CrossRef]
103. Metcalfe, S.E.; O'Hara, S.L.; Caballero, M.; Davies, S.J. Records of Late Pleistocene-Holocene climatic change in Mexico: A review. *Quat. Sci. Rev.* **2000**, *19*, 699–721. [CrossRef]
104. Metcalfe, S.E. Late Quaternary environments of the northern deserts and Central Transvolcanic Belt of Mexico. *Ann. Mol. Bot. Gard.* **2006**, *93*, 258–273. [CrossRef]
105. Van Devender, T.R.; Mead, J.I. Early Holocene and Late Pleistocene Amphibians and Reptiles in Sonoran Desert Packrat Middens. *Copeia* **1978**, *1978*, 464–475. [CrossRef]
106. Badgley, C. Counting individuals in mammalian fossil assemblages from fluvial environments. *Palaios* **1986**, *1*, 328–338. [CrossRef]
107. Fernández-Jalvo, Y.; Andrews, P. Small mammal taphonomy of Gran Dolina, Atapuerca (Burgos), Spain. *J. Archaeol. Sci.* **1992**, *19*, 407–428. [CrossRef]
108. Andrews, P. *Owls, Caves, and Fossils*, 1st ed.; The University of Chicago Press and British Museum (Natural History): Chicago, IL, USA, 1990; p. 231.
109. Robles, J.; Arroyo-Cabrales, J.; Johnson, E.; Allen, B.L.; Izquierdo, G. Blue Bone Analyses as a Contribution to the Study of Bone Taphonomy in San Josecito Cave, Nuevo León, México. *J. Cave Karst Stud.* **2002**, *64*, 147–152.
110. Chadefaux, C.; Vignaud, C.; Chalmin, E.; Robles-Camacho, J.; Arroyo-Cabrales, J.; Johnson, E.; Reiche, I. Colour Origin and Heat Evidence of Paleontological Bones: Case Study of Blue and Gray Bones from San Josecito Cave, México. *Am. Mineral.* **2009**, *94*, 27–33. [CrossRef]
111. Balvanera, P.; Kremen, C.; Martínez-Ramos, M. Applying community structure analysis to ecosystem function: Examples from pollination and carbon storage. *Ecol. Appl.* **2005**, *15*, 360–375. [CrossRef]
112. Wing, S.L.; Sues, H.-D.; Potts, R.; DiMichele, W.A.; Behrensmeyer, A.K. Evolutionary Paleoecology. In *Terrestrial Ecosystems through Time. Evolutionary Paleoecology of Terrestrial Plants and Animals*; Behrensmeyer, A.K., Damuth, J.D., DiMichele, W.A., Potts, R., Sues, H.-D., Wing, S.L., Eds.; The University of Chicago Press: Chicago, IL, USA, 1992; pp. 1–13.
113. Harris, A.H. Reconstruction of mid-Wisconsin environments in southern New Mexico. *Natl. Geogr. Res.* **1987**, *3*, 142–151.
114. Meltzer, D.J.; Mead, J.I. Dating Late Pleistocene extinctions: Theoretical issues, analytical bias, and substantive results. In *Environments and Extinctions: Man in Late Glacial North America*; Peopling of the Americas Series; Mead, J.I., Meltzer, D.J., Eds.; Center for the Study of Early Man, University of Maine: Orono, MI, USA, 1985; pp. 145–173.
115. Wilson, R.W. Preliminary study of the fauna of Rampart Cave, Arizona. *Publ. Carnegie Inst. Wash.* **1942**, *530*, 169–185.
116. Miller, S.J. The archaeological fauna of four sites in Smith Creek Canyon. In *The Archaeology of Smith Creek Canyon, Eastern Nevada*; Anthropological Papers; Tuohy, D.R., Randall, D.L., Eds.; Nevada State Museum: Carson City, Nevada, 1979; Volume 17, pp. 271–329.
117. Jass, C. Caves, Arvicoline Rodents, and Chronologic Resolution. *Palaeontol. Electron.* **2011**, *14*, 1–21.

118. White, J.A.; McDonald, H.G.; Anderson, E.; Soiset, J.M. Lava blisters as carnivore 25s. In *Contributions in Quaternary Vertebrate Paleontology: A Volume in Memorial of John E. Guilday*; Genoways, H.H., Dawson, M.R., Eds.; Carnegie Museum of Natural History: Carnegie, PA, USA, 1984; Volume 8, pp. 241–256.
119. Mares, M.A.; Willig, M.R. Inferring biome associations of recent mammals from samples of temperate and tropical faunas: Paleoecological considerations. *Hist. Biol.* **1994**, *8*, 31–48. [CrossRef]
120. Barbour, M.G.; Billings, W.D. *North American Terrestrial Vegetation*; Cambridge University Press: New York, NY, USA, 1988; p. 434.
121. Sutcliffe, A.J. *On the Track of the Ice Age Mammals*; British Museum (Natural History): London, UK, 1985; p. 224.
122. Carbone, C.; Maddox, T.; Funston, P.J.; Mills, M.G.L.; Grether, G.F.; Van Valkenburgh, B. Parallels between Playbacks and Pleistocene Tar Seeps Suggest Sociality in an Extinct Sabretooth cat, Smilodon. *Biol. Lett.* **2009**, *5*, 81–85. [CrossRef]
123. DeSantis, L.R.G.; Crites, J.M.; Feranec, R.S.; Fox-Dobbs, K.; Farrell, A.B.; Harris, J.M.; Takeuchi, G.T.; Cerling, T.E. Causes and Consequences of Pleistocene Megafaunal Extinctions as Revealed from Rancho La Brea Mammals. *Curr. Biol.* **2019**, *29*, 2488–2495. [CrossRef] [PubMed]
124. Zepeda, G.J. Los Mamíferos Recientes de la Región de la Cueva de San Josecito, Nuevo León, México. Unpublished Bachelor's Thesis, Escuela Nacional de Ciencias Biológicas, IPN, México City, Mexico, 2003.
125. Baker, R.H. Mammals of Coahuila, Mexico. *Univ. Kans. Publ. Kans. Mus. Nat. Hist.* **1956**, *9*, 125–335.
126. Schmidly, D.J. *The Mammals of Trans-Pecos Texas*; Texas A&M University: College Station, TX, USA, 1977; p. 225.
127. Ray, C.E.; Linares, O.J.; Morgan, G.S. Paleontology. In *Natural History of Vampire Bats*; Greenhall, A.M., Schmidt, U., Eds.; CRC Press, Inc.: Boca Raton, FL, USA, 1988; pp. 19–29.
128. Koestner, E.J. An annotated list of mammals collected in Nuevo Leon, Mexico, in 1938. *Great Basin Nat.* **1941**, *2*, 9–15.
129. Morgan, G.S. Neotropical Chiroptera from the Pliocene and Pleistocene of Florida. *Bull. Am. Mus. Nat. Hist.* **1991**, *206*, 176–213.
130. McNab, B.K. Energetics and the distribution of vampires. *J. Mammal.* **1973**, *54*, 131–144. [CrossRef]
131. Grayson, D.K. Minimum Numbers and Sample Size in Vertebrate Faunal Analysis. *Am. Antiq.* **1978**, *43*, 53–65. [CrossRef]
132. Sibley, D.A. *The Sibley Guide to Birds*; Alfred A. Knopf: New York, NY, USA, 2000; p. 624.
133. Rzedowski, J. *Vegetación de México*; CONABIO: México City, Mexico, 2006; p. 505.
134. Craig, E.H.; Craig, T.H.; Powers, L.R. Patterns and Home-Range Use of Nesting Long-Eared Owls. *Wilson J. Ornithol.* **1988**, *100*, 204–213.
135. Henrioux, F. Home Range and Habitat Use by the Long-Eared Owl in Northwestern Switzerland. *J. Raptor Res.* **2000**, *34*, 93–101.
136. Lövy, M.; Riegert, J. Home Range and Land Use of Urban Long-Eared Owls. *Condor* **2013**, *115*, 551–557. [CrossRef]
137. Lyman, R.L. Quantitative Units and Terminology in Zooarchaeology. *Am. Antiq.* **1994**, *59*, 36–71. [CrossRef]
138. McDonald, H.G.; Jefferson, G.T. Distribution and habitat of *Nothrotheriops* (Xenarthra, Nothrotheridae) in the Pleistocene of North America. In *Geology and Vertebrate Paleontology of Western and Southern North America, Contributions in Honor of David P. Whistler*; Wang, X., Barnes, L.G., Eds.; Natural History Museum of Los Angeles County Science Series: Los Angeles, CA, USA, 2008; No. 41; pp. 313–331.
139. Tedford, R.H. Principles and Practices of Mammalian Geochronology in North America. In Proceedings of the North American Paleontological Convention: Field Museum of Natural History, Chicago, IL, USA, 5–7 September 1969; Yochelson, E.L., Ed.; pp. 666–703.

quaternary

Article

Middle Pleistocene Variations in the Diet of *Equus* in the South of France and Its Morphometric Adaptations to Local Environments

Antigone Uzunidis [1,2]

[1] Institut Català de Paleoecologia Humana i Evolució Social (IPHES-CERCA), Zona Educacional 4, Campus Sescelades URV (Edifici W3), 43007 Tarragona, Spain; antigone.uzunidis@wanadoo.fr
[2] UMR 7269 LAMPEA, CNRS, Aix Marseille Université, Ministry Culture, 13190 Aix-en-Provence, France

Abstract: *Equus* is a very sensitive genus which has expanded over a large area and lived in Europe despite the climatic instability of the Pleistocene. Its persistence and abundance are helpful in understanding and describing environmental and climatic regional parameters. In this study, we present the result of dental mesowear and microwear analysis and post-cranial skeleton biometry on *Equus* populations located in two regions in the South of France from ten sites, corresponding to twelve assemblages dated from MIS 12 to MIS 5. The areas refer to two major climatic zones: the oceanic or subcontinental climate for the South West of France, and the Mediterranean for the South East. The first objective of this study is to integrate and compare biometric data, dental wear, and other already-published environmental proxies. The goal is to discuss the validity of horse body shape adaptations on a small geographical scale. The second objective is to describe the impact of environmental features on the horse population through time in the two regions. We observe that the *Equus* diet was quite diverse, according to microwear analysis which shows adaptations according to seasonal variations. However, they remained mostly grazers over a long period of time. Estimated body mass of *Equus* in the localities studied here varies from a mean of 468 up to a mean of 570 kg, but these variations failed to be correlated with the diet, the climatic period, or the geographical position of the horse population, probably because of the sample size or the restricted time-span or geographical scale. However, the conformation of the metapodials and the width of the third phalanges may have been linked with environmental and behavioural parameters. The width of the third phalange may be correlated with the recurrence of the snow cover, while the robustness of the metapodial co-occurs with a humid climate. Also, diet may influence the conformation of the bones, since the tall and slender horses seem to be preferentially grazers all year long and seasonally browser horses are tall and robust. Seasonally mixed-feeder horses, all coming from the Mediterranean area, were found to be smaller, perhaps in relation to a less productive environment. The correspondence of the dietary and morphometrical data could suggest high pressure on the horse population, which caused rapid body adaptation. Thus, the combination of these different proxies allows us to suggest more accurate large mammal paleoenvironmental reconstructions.

Keywords: *Equus*; Pleistocene; paleontology; South of France; mesowear analysis; microwear analysis; body mass; morphometry; ecology; ecomorphology

1. Introduction

This paper addresses *Equus* from the caballoid lineage populations from the end of the Middle Pleistocene to the beginning of the Late Pleistocene succession in the South of France. This area can be divided into two major climatic zones: the South East, subject to a Mediterranean climate, and the South West, with an oceanic to subcontinental climate, depending on distance from the sea [1]. In addition, the two zones are characterised by different hydrographic basins and reliefs. These parameters, along with latitude and longitude, affect temperature, precipitation [2], and the composition of the plant communities [3]. The South East is known for its relative climatic stability across time, allowing the persistence of temperate plant species during glacial periods [4]. The South West climate is much more sensitive to global variations [5–7]. The interglacial stages covered by this paper are isotopic stages 9, 7 and 5. The first two of these are slightly cooler than today, and the last one is warmer, with a mean of 5 Celsius degrees more than today [8]. Despite the fact that temperatures vary a lot during the phases, the vegetation cover generally remains similar to current cover [9]. The main vegetal cover in the South West is a continental acidophilous oak woodland. In the South East, the vegetation is mainly medio-Mediterranean vegetation (with pine, evergreen oak, rosemary, rock rose and lavender) [3]. The glacial stages relevant to this paper are isotopic stages 12, 10 and 6. Temperatures have been modelled for the Saale Glaciation (MIS 6) in particular and the means show differences between the two regions: the South East is warmer (approximately 15 °C in summer and −1 °C in winter) than the South West (apparently 10 °C in summer and −10 °C in winter) [10]. The vegetation assemblages have also been modelled, and assume many differences between the two regions. The South East vegetation may have been dominated by a temperate savannah and the South West vegetation may have been part of a taiga Forest and part of a mammoth steppe biotope [11]. The South of France is a contrasted region with distinct environments; thus, it may be a key region for understanding the impact of local environmental conditions on ecological, ethological and morphological adaptations of herbivorous megafauna. During these periods and in these regions, Pleistocene horses were able to occupy these various biotopes, which suggests behavioural flexibility.

Many studies have already shown that *Equus* is a flexible genus that can live in a wide range of environments and occasionally adopt a diversified diet (e.g., [12–18]). Some populations are characterised by a high amount of browsing in their diet, for example at Schöningen or Taubach [14,16]. Other horse assemblages, located in the Fairbanks area of Alaska, for example, suggest that the *Equus* diet is more variable than other ungulates from the same area [13]. While, as a general trend, the size of *Equus* decreased continuously through the Late Pleistocene [19–21], during the Middle Pleistocene horses present various morphometrical features, including size, that vary according to environmental conditions [12]. However, for now, the rate of evolutionary adaptation or the phenotypic plasticity of the skeleton is unknown, and so is the geographical and chronological accuracy of these observations.

In order to discuss this observation, in this paper we describe the morphological features of twelve populations of horses from the caballoid lineage from the South of France. We also reconstruct the dietary habits of those populations with the help of meso- and microwear dental analysis. This will allow us: (1) to determine if interpretations of horses' biometrical adaptations are still valid for a restricted geographical scale; and (2) to document the change of those features through time in these regions.

2. Paleoenvironmental Parameters of the Study Sites

We have selected ten sites located in South West and South East France, corresponding to twelve fossil assemblages (Figure 1, two sites—Rameaux and Pech de l'Azé II—present different layers). The dates of the studied deposits range between MIS 12 (ca. 478-424 ky) and MIS 5 (ca. 130-71 ky) (dating from [22]) (Table 1), covering the period from the end of the Middle Pleistocene to the beginning of the Late Pleistocene. These sites correspond to different climatic phases, from glacial (Camp-de-Peyre, La Micoque, Rigabe, Suard, Pech de l'Azé II-l.9-6 and Coudoulous II) and interglacial contexts (Rameaux, Lunel-Viel, Payre, Pech de l'Azé II.l.4-2 and Peyre). They are rich in horse fossil material with well-preserved bones and teeth. In this paper, we have referred to the Latin terminology *E. caballus* rather than *E. ferus* following the demonstration of J.-L. Guadelli [23]. Moreover, we referred to several species and subspecies described for the caballoid lineage in each site through the latest works on this subject (see below).

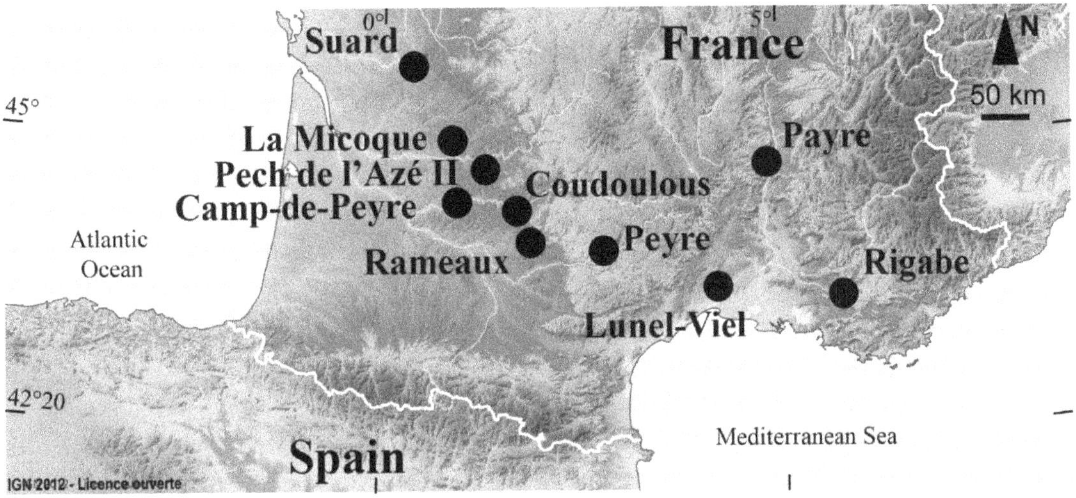

Figure 1. Location of the sites from the South West and South East of France selected for this study: Camp-de-Peyre, La Micoque, Rameaux, Lunel-Viel, Payre, Rigabe, Suard, Pech de l'Azé II, Coudoulous II and Peyre. Map: IGN 2012_Licence ouverte, modified A. Uzunidis (http://education.ign.fr/primaire/fonds-de-cartes (accessed on 20 May 2018)).

Table 1. Localities cited in this study with dating, location, location of the collection and enumeration of the bones and teeth studied. MMSH: Maison Méditérranéenne des Sciences de l'Homme; MNP: Musée National de Préhistoire. MIS 12: ca 478-424 ky; MIS 10: ca 374-337 ky; MIS9: ca 337-300 ky; MIS 7: ca 243-191 ky; MIS6: ca 191-130 ky; MIS 5: ca 130-71 ky (age interval for each isotopic stage in [22]).

Locality	Region	Age	Reference for Dating	Location	*Equus* Specie	Reference for Palaeontological Determination	Number of Studied Bones	Number of Studied Teeth
Peyre	South-West	MIS 5	[24]	MMSH	*Equus caballus*	[25]	138	27
Pech de l'Azé II 1.4-2	South-West	MIS 5	[26]	Les Eyzies MNP	*Equus caballus*	[19,25]	18	31
Coudoulous II 1.9	South-West	MIS 6	[27,28]	MMSH	*Equus mosbachensis*	[25,28]	115	25
Pech de l'Azé II l. 9-6	South-West	MIS 6	[26]	Les Eyzies MNP	*Equus mosbachensis*	[19,25]	6	63
Suard 1.53-50	South-West	MIS 6	[29–31]	Angoulême museum	*Equus caballus piveteaui*	[25,32]	190	116
Rigabe	South-East	MIS 6	[33]	Les Eyzies MNP	*Equus mosbachensis palustris*	[25]	29	18
Payre l. F	South-East	MIS 7	[34]	Orgnac museum	*Equus mosbachensis*	[25]	26	8
Lunel-Viel	South-East	MIS 7	[35]	Les Eyzies MNP	*Equus mosbachensis palustris*	[25,36]	1109	65
Rameaux-Amont	South-West	MIS 9	[25,37,38]	MMSH	*Equus mosbachensis*	[25]	247	151
Rameaux-Aval	South West	MIS 9	[25,37,38]	MMSH	*Equus mosbachensis*	[25]	190	8
La Micoque l. E, H, J	South-West	MIS 10	[39]	Les Eyzies MNP	*Equus mosbachensis*	[25,40,41]	38	34
Camp-de-Peyre	South-West	MIS 12	[42]	Les Eyzies MNP	*Equus mosbachensis campdepeyri*	[25,42]	164	38

Local environmental parameters of some of the sites have already been investigated through various proxies (sediment, pollen, microfauna, etc). However, not all of these proxies were used in the study of all the sites. Thus, the palaeoenvironmental data available for some sites are not as complete as for others.

The site of Camp-de-Peyre accumulated abundant elements such as *Rangifer* sp., *Dicrostonyx torquatus* and *Microtus gregalis*, which are clearly indicative of the cold climatic phase [42,43]. However, the presence of some taxa, such as *Apodemus sylvaticus* and *Clethrionomys glareolus*, nonetheless suggests the existence of sheltered places. According to these compositions of the mammal species and sedimentological studies [44], the landscape must have been a steppe with some wooded patches. The horse described in this site is *Equus mosbachensis campdepeyri* [25,42].

At La Micoque, we studied the levels E, J and H. The faunal association is largely dominated by *Equus mosbachensis* materials, which represent 75% of the remains [40, 41]. The study of sediment suggests that the median ensemble (from which all of the studied bones and teeth come) is the result of river transport [45,46]. Occasionally across the year, the flow was very strong, which is typical of arid or semi-arid climate conditions [45]. Furthermore, some sediment lenses correspond to ice block ersatz. Overall, these observations strongly suggest a very cold and arid climate.

As at La Micoque, *Equus mosbachensis* is the most abundant taxon in the two parts of the Igue des Rameaux (Amont and Aval), with at least 25 individuals [47]. The other herbivores are: *Palaeoloxodon antiquus, Coelodonta antiquitatis, Equus hydruntinus, Haploidoceros mediterraneus, Cervus elaphus, Hemitragus* sp. and *Bos primigenius* [37]. Microfauna studies show differences between the paleoenvironment of the two parts of the Igue des Rameaux [48]. For the Aval ensemble, which is a bit more recent than that of the Amont [25,38], the climate must have been temperate-cool, with more precipitation but less snow cover than in the Amont ensemble, and the forest cover was quite significant (25% of the vegetal composition). For the Amont ensemble, the climate was semi-continental and the steppe more significant. In the two parts of the site the horse was described as *E. mosbachensis* [25].

At Lunel-Viel, the climate must have been temperate, according to fauna associations (*Cervus elaphus, Capreolus* cf. *süssenbornensis, Sus* sp., *Equus hydruntinus*, etc.); furthermore, some birds such as *Hieraaetus fasciatus* and *Falco naumanni* are related to the Mediterranean area [35,49–51]. The study of the horses' skeleton suggests a very humid environment, described as a swamp, similar to the current Camargue [36]. The tooth wear study made it possible to rule out this last hypothesis and suggest a rich biotope instead, both in monocot and dicot plants, allowing strong resource partitioning between bovids and equids, which is not consistent with a swamp biotope [52]. The horse was described as *Equus mosbachensis palustris* [25,36].

At Payre, the herbivore assemblage is rich and diverse (*Palaeoloxodon antiquus, Stephanorhinus kirchbergensis, S. hemitoechus, Bos primigenius, Bison priscus, Hemitragus* sp., *Rupicapra* sp., *Cervus elaphus, Megaloceros* sp., *Dama dama, Capreolus capreolus*, and *Sus scrofa*), probably owing to its geographical position. Indeed, it is located in the Rhône Valley at the crossroads of various biotopes, mainly the plateau and the bottom of the valley. For level F (studied here), microfauna studies show that the climate must have been cool and dry, but not enough to indicate a glacial episode [53]. The isotopic analysis indicates that the horses lived in the bottom of the valley, rather than on the top of the plateau [54,55]. The horse is described as Equus mosbachensis in the layer F [25].

The faunal association of the site of Rigabe's Rissian levels (X to J) is dominated by *Equus* sp., *Cervus elaphus, Capreolus capreolus* and *Sus scrofa* which suggests a forested area [56]. Moreover, the sedimentary analysis suggests a cold climate [33]. The horse was attributed to *E. mosbachensis palustris* [25].

At Suard, the faunal association is dominated by *Equus caballus piveteaui* [25,32], *Rangifer tarandus* and *Coelodonta antiquitatis*, and the presence of *Mammuthus primigenius* and *Saiga tatarica* is signaled. The pollen and faunal analyses of the Suard cave show that the climate must have been cold and dry, with a mosaic landscape dominated by steppes, with some wooded habitats [57–59].

With *Equus mosbachensis* in layers 9 to 6 and *E. caballus* in layer 4 [19,25], several herbivore species have been described in the different layers of Pech de l'Azé II: for example,

Stephanorhinus kirchbergensis (layer 9 to 7) replaced by *Coelodonta antiquitatis* (in layer 6), *Bison priscus* (l. 9 to 5), *Bos primigenius* (l. 9 to 4), *Cervus elaphus* (l.9 to 4), *Megaloceros* sp. (l.9), and *Capreolus* sp. (l.4). The pollen of the sequence of the site of Pech de l'Azé II has been studied [60] but the sedimentary analysis shows a "ruissellement" (overland flow) phenomenon [46,61], which interferes with layer integrities [62]. Nonetheless, the presence of cryoclastic complexes in layers 9 to 6 suggests very cold conditions, whereas layer 4 corresponds to a temperate climate.

Cold-adapted taxa have been described in layer 9 of Coudoulous II, such as *Mammuthus intermedius, Coelodonta antiquitatis* and *Rangifer tarandus*. Moreover, the microfauna analysis indicates a cold period, relatively dry, with a mixed-landscape dominated by a steppe [27,28,63]. The horse was attributed to *E. mosbachensis* [25,28].

Some taxa present at Peyre are typical of cold environments such as *Mammuthus primigenius* and *Coelodonta antiquitatis* [64] but the faunal association is dominated by temperate taxa such as *Cervus elaphus, Equus hydruntinus, Bos primigenius, Sus scrofa* and *Hystrix* sp., which more likely to indicate a temperate climate [24,65–68]. Moreover, the study of the Microtidae, the Muridae and the gastropods suggests a dry climate [24]. The horse was described as *E. caballus* [25].

3. Biometric and Dental Wear Methods

3.1. Osteologic Measurement Protocols and Taphonomic Limitations

The morphometric characteristics observed on horse long bones (humerus, radius, femur and tibia) provide information on locomotor adaptations [69], and some features also depend on environmental parameters (e.g., [12,70–73]). As already shown, very few differences between adult horse bones are linked to age or sex [74,75]; only the fully fused long bone epiphyses were selected. Limb bone measurements were taken in accordance with the works of V. Eisenmann [76–78]. All of the long bones were measured with a calliper: humerus, radius, femur, tibia, metapodials and phalanges (I, II and III).

Although all of the selected sites are rich in bone remains, their conservation differs and constrains this study. Indeed, some of the sites considered here correspond to accumulations in caves of natural origin or linked to the activity of carnivores (Camp-de-Peyre, Les Rameaux, Lunel-Viel, Coudoulous II and Peyre). The bones are mostly very well preserved and numerous morphometric observations have been possible. On the other hand, the accumulation of other sites is due to human activities (La Micoque, Payre I.F, Suard, Pech de l'Azé II and Rigabe). In these sites, bone fragmentation is much more significant and has limited the observations. The most striking example is the La Micoque assemblage which, in addition to being of anthropogenic origin, is also a cliff foot site. A minimum of twenty individuals have been counted there [40] and no whole long bones are known. Thus, although the selected assemblages are rich, their origin and preservation conditions are factors that may have limited our observations.

3.2. Body Mass Estimation and Limb Proportions

Body mass was estimated following the recommendations of Saarinen et al. [18]. They established that the better equations to estimate body mass were the ones published by Scott [79] using the metapodials. In this paper, we use measurements of the minimum mid-shaft width, maximum proximal width, and maximum distal width of third metacarpals and third metatarsals in order to reconstruct body mass.

The relative proportions of the limb bones were analysed using log-ratio diagrams [80]. This graphic allows us to compare the differences in size and conformation between different populations based on measurements standardised on a referenced population. In this paper, the standard used for comparison is *E. hemionus*, published by V. Eisenmann on her website [76]. Along with the fossil sample, a sample for *Equus przewalskii* [81] (p. 26) is included for comparative purposes. The maximal length of the long bones (humerus, femur, radius, tibia, third metacarpal and metatarsal, first anterior and posterior phalanges) and the greatest width of the third phalanges of all the studied populations were compared.

Comparing the proportions of the different limb segments allows the development of one segment to be compared to another (e.g., length of radius to metacarpals) and therefore provides information on locomotion [12]. Comparing the greatest width of the third phalanges allows us to compare ground hardness adaptation in several populations [72].

3.3. *Mesowear and Microwear Methods*

The horses' diet in each site was studied using dental mesowear and microwear analysis. These two techniques provide information about the diet on two different timescales: mesowear indicates the average diet during the individual's life or its last months or years [82,83], and microwear reveals the diet of the last few days to months of an animal's life [84–88]. Combined together, these two methods allow us to understand the evolution of the animal's diet in relation to seasonal variations over a period of several months. All the meso- and microwear data were collected by the author except for those from Payre, studied by F. Rivals [14,89].

3.3.1. Dental Mesowear

Mesowear analysis is a method based on the observation of wear patterns on ungulate molar cusps that indicate the diet of an individual animal [83,90]. The sharpness and the morphology of cusp apices are correlated with relative attritive and abrasive dental wear. Thus, a diet with low abrasion (and high attrition) shows very sharp and high relief molar buccal cusps and, on the contrary, a diet with high abrasion (due especially to the consumption of siliceous grass [91]) will result in rounded, blunted and low relief cusps.

We selected only upper intact adult teeth. As for the bones, taphonomic processes constrained data collection since the slightest alteration (e.g., breaks or polish) in cusp integrity led to its exclusion from selection. Observation of the cusp shape on horse teeth is easier on upper ones, which decreases the chance of potential mistakes. Unworn (and marginally worn) teeth as well as extremely worn ones are excluded from the following analysis [83,89,90].

For this study, the preference was for observation of the paracone of the second upper molar, as recommended by [83], but we also extended the analysis to the upper fourth premolar and first and third molar, as no differences were detected between the teeth [92], in order to increase the number of samples. Furthermore, only medium-age teeth were selected and juvenile or senescent teeth were excluded, as it was observed in [90] that they were less sensitive to diet.

In this study, we employed the standardised method proposed by the authors of [93] and modified by the authors of [94], already used in other works (e. g. [95,96]). This method categorises dental wear into seven groups (numbered from 0 to 6), according to their shape and topography (0 = high and sharp; 6 = blunt with no relief). The average value of mesowear data from a single sample corresponds to the "mesowear score" noted here "MWS" [93].

3.3.2. Dental Microwear

The microwear analysis study followed the protocol established by [97,98]. The occlusal surface of each tooth was cleaned using acetone and 96% alcohol. Then the surface was moulded with a high-resolution silicone and casts were made using clear epoxy resin. The transparent casts were then observed with a stereomicroscope at magnifications of ×35. Observations were restricted to a standard surface of 0.16 mm^2 (using an ocular reticule), localised preferably on the upper tooth paracone and the lower tooth protoconid.

We selected adult molars and premolars, as it was shown that both provide consistent data for dietary studies in Artiodactyla and Perissodactyla, and because the intertooth variation in the microwear pattern is not significant [99].

Teeth with taphonomic alterations (e.g., wide and short striae or pits with irregular morphology) were excluded from this analysis [100].

The observed micro-traces, scratches, and pits in particular are left on the occlusal surfaces during mastication [101]. The variability of the density of these traces is indicative of various diets: grazer, mixed-feeder, and browser. We observed various features following the classification of [97,98] (Figure 2): pits (small and large), scratches (fine, coarse and hypercoarse) and gouges. The scratch width score (SWS) is also calculated with a score of '0' for teeth with predominantly fine scratches per tooth surface, '1' for those with mixed fine and coarse scratches on the tooth surface, and '2' for those with predominantly coarse scratches. In addition, we used the method of [102], based on the observations of [97], which separates the three diet categories according to the number of scratches between 0 and 17 in the 0.16 mm^2 area (0–17%). For extant ungulates, the diet categories match the percentages as follows: grazers (0.0–22.2% of scratches between 0 and 17), mixed feeders (20.9–70.0% of scratches between 0 and 17), and browsers (72.7–100% of scratches between 0 and 17).

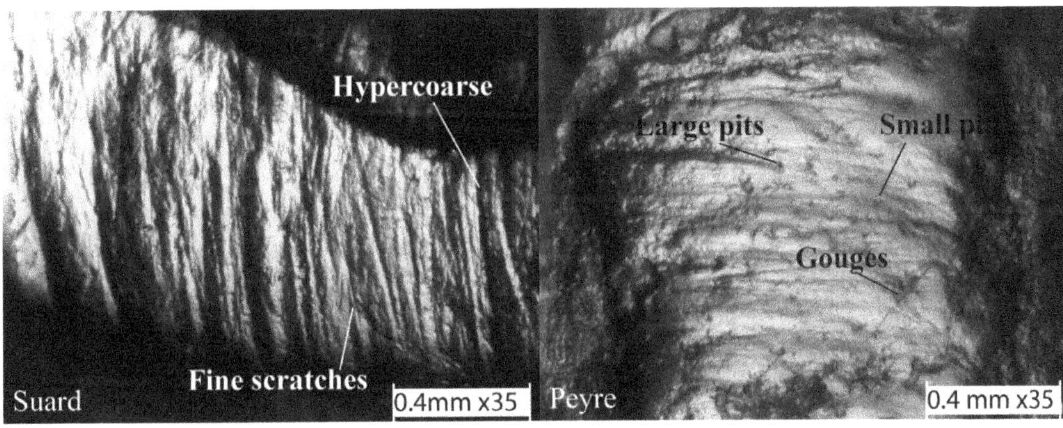

Figure 2. Photomicrographs of horse teeth from Suard (n° 52-K14-50) and Peyre (n° E9-347) with scratches and pits. Magnification ×35.

3.4. Statistical Analysis

Several statistical analyses were performed in order to fully investigate ecological and morphometrical horse adaptations. We use principal component analysis (PCA) to observe the morphometrical variability of the horses metapodials. We used the Pearson correlation coefficient to construct the PCA. The variables used were tested using the Kaiser-Meyer-Olkin (KMO) index, which indicates the degree to which each variable in a set is predicted without error by the other variables. Every value below 0.5 was rejected. The PCA does not accept missing data, which led to the rejection of any incomplete sample in this analysis.

The Kruskal-Wallis test was used to test statistical differences between several groups with the Steel-Dwass-Critchlow-Fligner method of multiple comparisons for the post-hoc test. We used the Mann-Whitney test when we were comparing only two samples. In both cases, since the sample size is quite small, we used the exact p-value.

The Spearman correlation test was used to evaluate the liaison intensity between the diet categories defined by enamel wear studies and the size of the different horse populations. All the statistics were carried out using the Xlstat v. 2014.5.03 software with a risk threshold of 5%.

4. Results

4.1. Horse Size and Ecomorphological Variation through Time

In order to understand the impact of the environment on the morphology of horses from the South of France, we compared the ratio of limb bone proportions, and the body size of the different populations. First, we checked the homogeneity of the morphometric characteristics of the bones in the different levels of a same site. For the length of the limb bones and the greatest width of the phalanx III, only the Lunel-Viel site contained a sufficient number of specimens in each level to perform a Kruskal-Wallis test. The statistical results are non-significant for all bones and level of Lunel-Viel. For the La Micoque and Suard sites (which also contain several levels), we assumed that the bone dimensions used for this study are homogenous as this is the case for all the other bones and measurements (see [25]).

Based on the metapodials, and following the recommendations of [18], we estimated the body mass for each cohort (Figure 3). The average body mass within late Middle Pleistocene and early Late Pleistocene horse populations in southern France varied from 468 kg at Pech de l'Azé II l. 6-4 to 571 kg at Coudoulous II l.9. They therefore fall within the 'large' (450–550 kg) and 'very large' (>550 kg) weight classes defined by the authors of [18]. The variability of the weights of these populations, of the order of 100 kg, is therefore quite low.

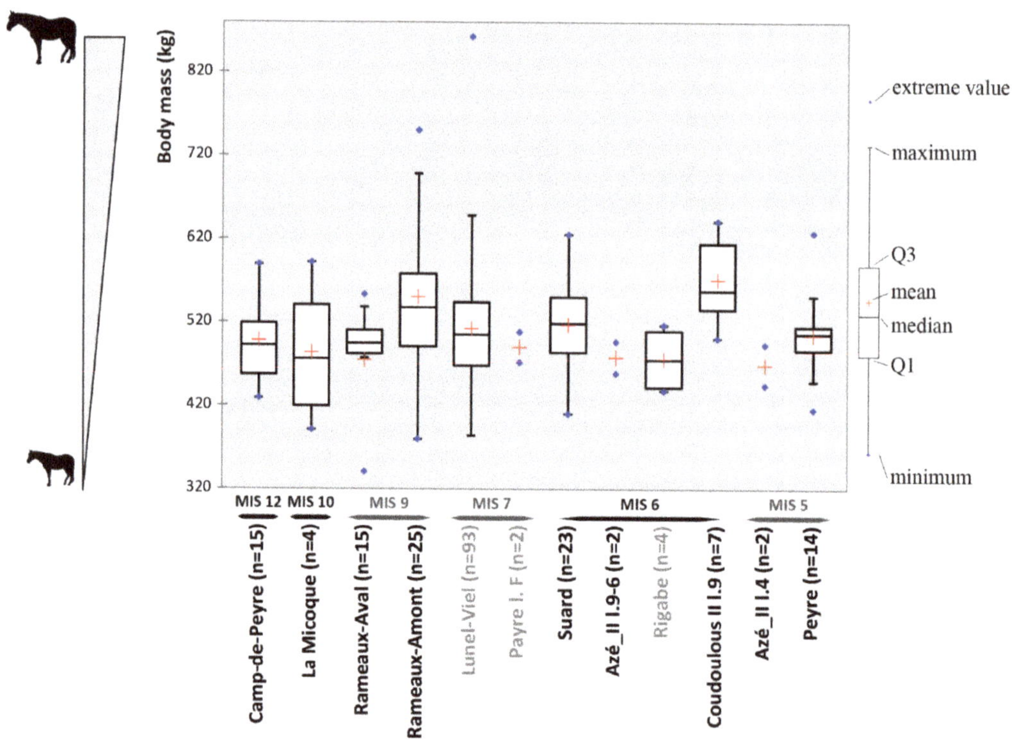

Figure 3. Box plot of the estimated weight in kilograms of Southern France Middle and Late Pleistocene caballoid horses based on metacarpal and metatarsal measurements (MC2, MC4, and MC6, MT2, MT4 and MT6 of Scott [79]). Camp-de-Peyre: *E. m. campdepeyri*; La Micoque: *E. mosbachensis*; Rameaux-Aval and Amont: *E. mosbachensis*; Lunel-Viel: *E. m. palustris*; Payre l.F: *E. mosbachensis*; Suard: *E. caballus piveteaui*; Pech de l'Azé II l.9-6: *E. mosbachensis*; Rigabe: *E. m. palustris*; Coudoulous II l.9: *E. mosbachensis*; Pech de l'Azé II l.4: *E. caballus*; Peyre: *E. caballus*. Vertically: in grey, South East sites, in black South West sites. Horizontally: in grey temperate isotopic stages, in black glacial isotopic stages.

The comparison of the proportion of horse limb length in the Pleistocene South of France populations is still preliminary due to the sample size (Table 2). The missing data and the low sample size prevent further statistical analysis. Nonetheless, the log ratio diagram (Figure 4) gives preliminary indications. It indicates that all of the samples have proportionally long limb bones compared to metapodials and phalanges, which is not the case with *E. prezwalskii*. The main difference between these cohorts of horses is in the width of the third phalanges. Those for which this data is available can be divided into two groups: (1) the populations with narrow phalanges composed of the horses of Lunel-Viel, La Micoque, Rameaux-Aval, and Pech de l'Azé l.8 (the two extremes of this group are the populations of Lunel-Viel with an average of 78.78 mm and Pech de l'Azé l.8 with a phalanx of 72.81 mm wide); and (2) the populations with wide phalanges composed of the horses of Peyre, Coudoulous II l.9, Rameaux-Amont, Suard and Camp-de-Peyre (the two extremes of this group are the populations of Peyre with an average of 90.26 mm and Camp-de-Peyre with phalanges III of 85.03 mm wide on average). The difference between these two groups is statistically significant (Mann-Whitney test: p-value = 0.0001). This measurement was not possible for the horse population of Rigabe and Pech de l'Azé II l. 6-4.

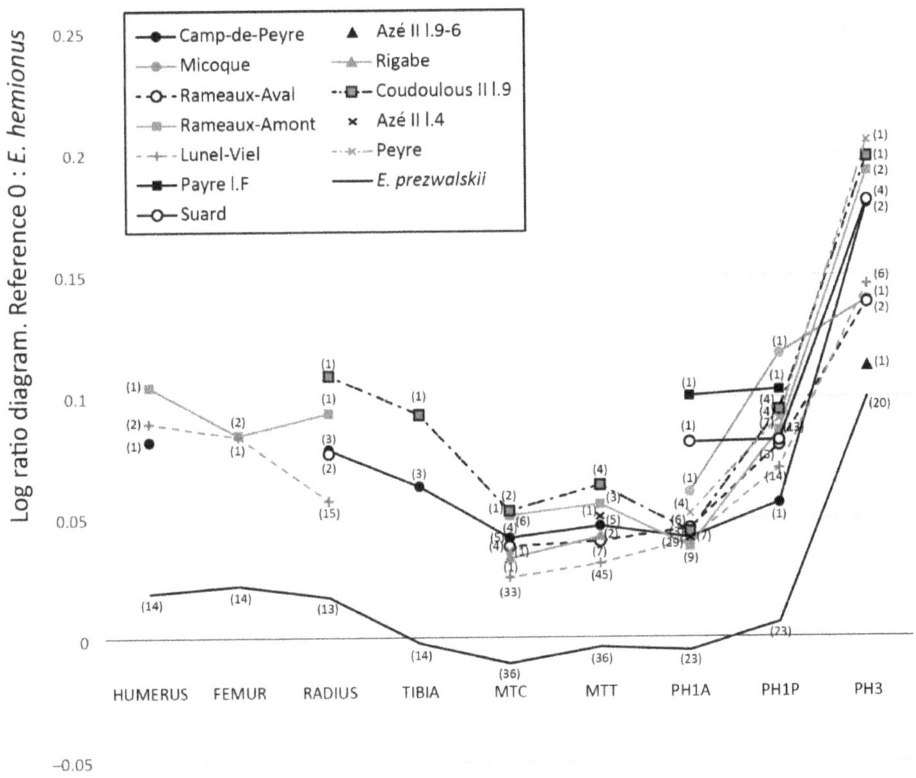

Figure 4. Log ratio diagram of humerus, femur, radius, tibia, third metacarpal and metatarsal, first anterior and posterior phalanges greatest length and third phalanx greatest width for caballoid horses from the South of France. Reference *E. hemionus* [78]. The number of specimens is indicated for each population. Camp-de-Peyre: *E. m. campdepeyri*; La Micoque: *E. mosbachensis*; Rameaux-Aval and Amont: *E. mosbachensis*; Lunel-Viel: *E. m. palustris*; Payre l.F: *E. mosbachensis*; Suard: *E. caballus piveteaui*; Pech de l'Azé II l.9-6: *E. mosbachensis*; Rigabe: *E. m. palustris*; Coudoulous II l.9: *E. mosbachensis*; Pech de l'Azé II l.4: *E. caballus*; Peyre: *E. caballus*.

Table 2. Measurements in millimetres of the greatest length of humerus, radius, femur, tibia, metapodials, first anterior phalanges (Phalanges I A), first posterior phalanges (Phalanges I P) and of the greatest width of the third phalange (Phalange III) of caballoid horses of the Middle and Late Pleistocene from France. n = number of specimen; m = mean. Camp-de-Peyre: *E. m. campdepeyri*; La Micoque: *E. mosbachensis*; Rameaux-Aval and Amont: *E. mosbachensis*; Lunel-Viel: *E. m. palustris*; Payre l.F: *E. mosbachensis*; Suard: *E. caballus piveteaui*; Pech de l'Azé II l.9-6: *E. mosbachensis*; Rigabe: *E. m. palustris*; Coudoulous II l.9: *E. mosbachensis*; Pech de l'Azé II l.4: *E. caballus*; Peyre: *E. caballus*.

Site/Bone	Humerus		Radius		Femur		Tibia		Metacarpal		Metatarsal		Phalange I A		Phalange I P		Phalange III	
	n	m	n	m	n	m	n	m	n	m	n	m	n	m	n	m	n	m
Camp-de-Peyre	1	300.15	3	355.43			3	373.27	4	242.69	5	287.78	7	86.9	1	82.65	2	85.03
La Micoque													1	90.73	1	95.18	1	77.65
Rameaux-Aval			2	354.08					5	240.75	7	283.65	6	87.68	3	87.28	2	77.45
Rameaux-Amont	1	316.14	1	368	2	410.44			6	247.89	3	293.79	9	86.16	7	88.51	2	87.7
Lunel-Viel	2	305.51	15	338.48	1	409.74			33	233.81	45	277.75	29	86.7	14	85.4	6	78.78
Payre l.F									1	249			1	99.4			1	92
Suard													3	95.14	13	87.7	4	85.27
Pech de l'Azé II l.9-6																	1	72.81
Rigabe									1	237.99	2	284.88						
Coudoulous II l.9			1	381.19			1	399.68	2	249	4	299.175	3	87.4	4	90.25	1	88.9
Pech de l'Azé II l.4											2	290.43						
Peyre									4	239.71	5	285.09	4	88.9	4	89.52	1	90.26

Principal component analyses were performed on the metapodials. Figure 5 shows the comparison of the conformation of the third metacarpal of the populations studied based on the measurement summarised in Table 3. It was constructed from the maximum length of the bone (M. 1), the width and thickness of the distal (M. 11 and 14) and proximal (M. 5 and 6) epiphyses, and the diaphysis of the bone (M. 3 and 4). The sagittal crest thickness (M. 12) and distal supra-articular width (M. 10) were excluded from the analysis because they were poorly represented on PCA axes 1 and 2. In this way the analysis presents 74.39% of the variability of the populations on these two axes. The third metacarpals of the Lunel-Viel, Rameaux-Aval and Peyre cohorts are located at the top of the graph, and are characterised in particular by the robustness of their diaphysis and the short overall length of the bone. The KMO index displays a value of 0.8354, which confirms the adequacy of the chosen variables. At the bottom left, those of Camp-de-Peyre are characterised by a greater length and a much slenderer appearance at the diaphysis and the distal and proximal ends. In Payre l.F, the metacarpal is long, slender at the diaphysis, but more developed at the extremities. Finally, the third metacarpals of Coudoulous II l.9 and the Rameaux-Amont, located at the bottom right of the graph, are both very long and very robust compared to the other populations.

Figure 6 presents a comparison of the same type of data for the third metatarsals (summarised in Table 4). For this PCA, the data used were: maximum bone length (M. 1), width and thickness of the diaphysis (M. 3 and 4) and distal end (M. 11 and 14), and width of the proximal end (M. 5). In order to keep only well-represented data on axes 1 and 2, we had to exclude the distal sagittal crest (M. 12) and proximal end (M. 6) thickness and the distal supra-articular width (M. 10). Thus, the PCA represents 70.86% of the variability of the third metatarsals in the study populations. The KMO index displays a value of 0.6650, indicating that the chosen variables are suitable for the analysis. As for the third metacarpals, the cohorts at the top of the graph corresponding to the Lunel-Viel, Peyre and Rameaux-Aval are short and robust. The third metatarsals at the bottom left of the graph from the Pech de l'Azé l.6-4, Camp-de-Peyre and Rigabe cohorts are longer and slender, particularly at the diaphysis. Finally, the cohorts of Coudoulous II l.9 and Rameaux-Amont are characterised by long and very robust bones.

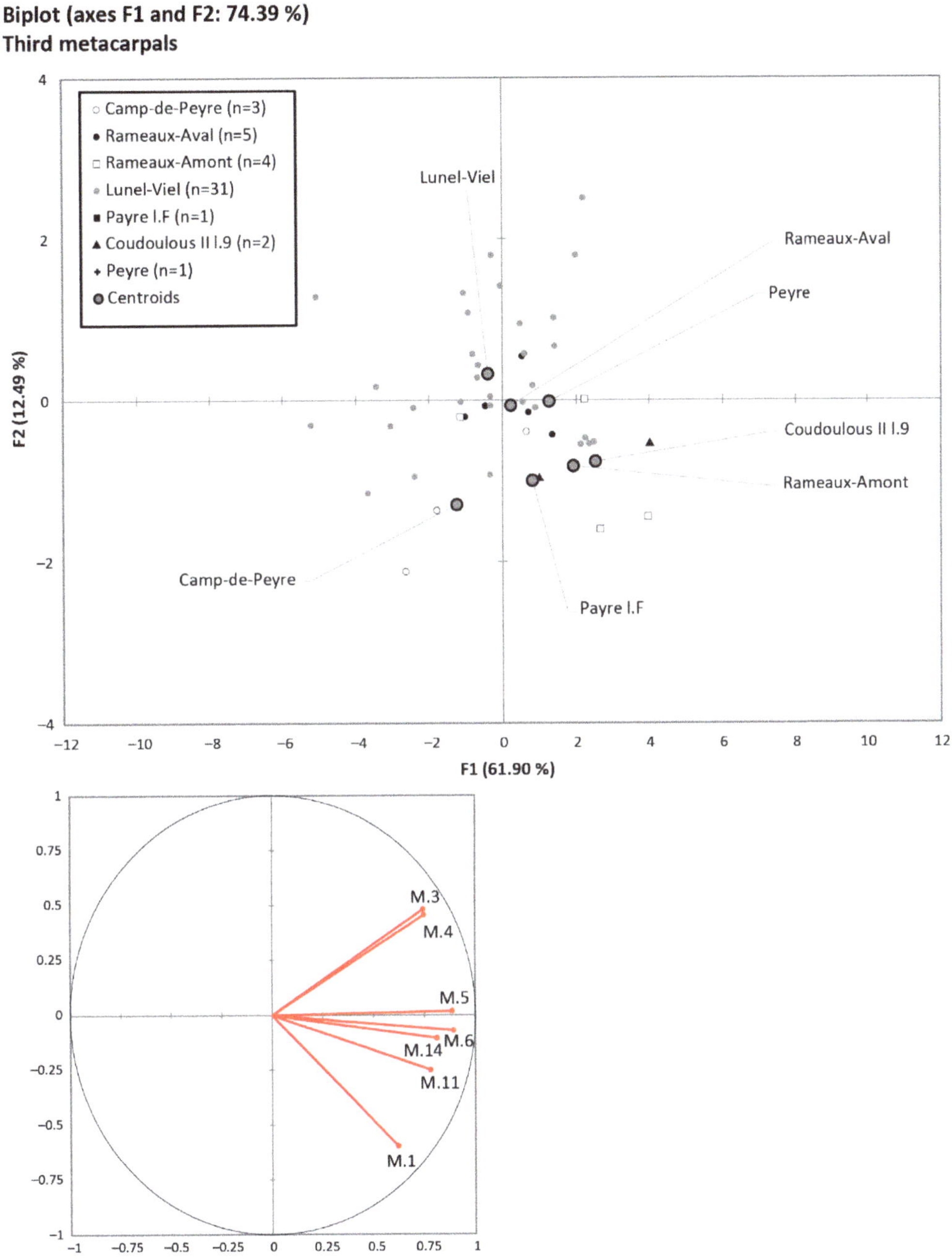

Figure 5. Principal component analysis (PCA) with measurements of the third metacarpal of Southern Middle and Late Pleistocene horses from the caballoid lineage. The correlation circle illustrates the contribution of each variable. M. 1: Maximal length, M. 3: Breadth at the middle of the diaphysis; M. 4: Depth of the diaphysis at the same level; M. 5: Proximal breadth; M. 6: Proximal depth; M. 11: Distal articular breadth; M.14: Greatest depth of the medial condyle. The large dots correspond to the centroid of each cohort. Camp-de-Peyre: *E. m. campdepeyri*; Rameaux-Aval and Amont: *E. mosbachensis*; Lunel-Viel: *E. m. palustris*; Payre I.F: *E. mosbachensis*; Coudoulous II I.9: *E. mosbachensis*; Peyre: *E. caballus*.

Table 3. Measurements in millimetres of the third metacarpals of horses from the caballoid lineage. 1: Maximal length; 3: Breadth at the middle of the diaphysis; 4: Depth of the diaphysis at the same level; 5: Proximal breadth; 6: Proximal depth; 10: Distal supra-articular breadth; 11: Distal articular breadth; 12: Depth of the sagittal crest; 14: Greatest depth of the medial condyle (measurements from V. Eisenmann [76]). n = number of specimens; m = mean; s = standard deviation; min-max: minimal and maximal values. Camp-de-Peyre: *E. m. campdepeyri*; La Micoque: *E. mosbachensis*; Rameaux-Aval and Amont: *E. mosbachensis*; Lunel-Viel: *E. m. palustris*; Payre l.F: *E. mosbachensis*; Suard: *E. caballus piveteaui*; Pech de l'Azé II l.9-6: *E. mosbachensis*; Rigabe: *E. m. palustris*; Coudoulous II l.9: *E. mosbachensis*; Peyre: *E. caballus*.

Site		1	3	4	5	6	10	11	12	14
Camp-de-Peyre	n	*4*	*7*	*7*	*4*	*5*	*4*	*4*	*4*	*4*
	m	**242.70**	**37.92**	**29.42**	**54.29**	**36.95**	**53.33**	**52.97**	**38.82**	**32.99**
	s	4.43	1.98	0.99	2.16	1.40	3.07	2.00	0.42	1.97
	min	238.25	35.98	28.18	51.34	34.55	51.05	50.60	38.21	30.47
	max	248.16	40.28	30.83	56.30	37.99	57.86	55.26	39.18	34.84
Micoque	n		*1*	*1*	*3*	*5*	*1*	*1*	*1*	*1*
	m		**43.65**	**26.49**	**55.19**	**35.88**	**52.03**	**53.03**	**40.60**	**34.18**
	s				3.04	1.21				
	min				52.61	34.50				
	max				58.54	37.67				
Rameaux-Aval	n	*5*	*8*	*8*	*7*	*7*	*8*	*6*	*6*	*6*
	m	**240.76**	**38.36**	**28.78**	**55.26**	**36.15**	**51.74**	**52.83**	**39.55**	**32.90**
	s	4.08	4.69	2.69	1.97	2.11	2.66	0.97	1.18	0.96
	min	237.20	29.85	24.18	52.31	33.31	47.26	51.70	37.69	31.71
	max	247.40	42.71	31.18	57.66	38.86	54.10	54.19	40.72	33.80
Rameaux-Amont	n	*6*	*9*	*9*	*8*	*7*	*7*	*6*	*5*	*7*
	m	**247.89**	**40.23**	**30.54**	**55.86**	**37.60**	**54.62**	**54.42**	**40.59**	**33.05**
	s	12.18	2.03	0.88	2.43	2.05	3.06	2.55	1.98	1.88
	min	235.99	37.60	29.25	52.13	34.74	51.44	51.10	38.17	29.59
	max	267.10	42.77	31.75	60.13	40.83	59.53	56.97	42.57	35.01
Lunel-Viel	n	*33*	*43*	*44*	*37*	*38*	*37*	*37*	*34*	*37*
	m	**233.81**	**39.54**	**30.26**	**54.13**	**36.35**	**51.77**	**52.07**	**40.47**	**32.90**
	s	5.92	3.52	2.41	3.26	2.56	2.33	2.46	1.74	1.89
	min	221.20	26.16	24.26	46.82	31.34	47.17	45.51	37.53	27.79
	max	242.90	45.48	40.69	58.80	39.73	58.15	57.65	46.46	39.81
Payre l.F	n	*1*	*1*	*1*	*1*	*4*	*2*	*2*	*2*	*2*
	m	**249.00**	**41.50**	**29.30**	**54.40**	**37.43**	**52.45**	**53.30**	**41.25**	**34.35**
	s					2.06	0.35	0.14	1.20	0.07
	min					35.00	52.20	53.20	40.40	34.30
	max					40.00	52.70	53.40	42.10	34.40
Suard	n		*5*	*6*	*13*	*15*	*2*	*2*	*2*	*2*
	m		**42.87**	**31.02**	**56.30**	**36.81**	**56.28**	**55.42**	**41.23**	**34.26**
	s		1.40	0.61	2.01	1.15	0.34	0.21	0.57	0.30
	min		41.39	30.20	53.67	33.88	56.04	55.27	40.83	34.05
	max		45.15	32.02	60.05	38.61	56.52	55.56	41.63	34.47
Pech de l'Azé II	n			*1*	*1*	*2*				
l. 9-6	m			**28.74**	**57.49**	**37.88**				
	s					0.71				
	min					37.38				
	max					38.38				
Rigabe	n	*1*	*1*	*1*		*1*	*1*	*1*	*1*	*1*
	m	**237.99**	**37.33**	**28.67**		**34.16**	**49.36**	**49.48**	**38.65**	**31.36**
Coudoulous II l.9	n	*2*	*2*	*2*	*2*	*2*	*2*	*2*	*2*	*2*
	m	**249.00**	**41.32**	**31.10**	**58.30**	**38.70**	**54.90**	**55.60**	**39.20**	**34.26**
	s	1.41	2.91	0.42	2.83	2.12	1.56	2.83	1.70	0.21
	min	248.00	39.26	30.80	56.30	37.20	53.80	53.60	38.00	34.11
	max	250.00	43.37	31.40	60.30	40.20	56.00	57.60	40.40	34.40
Peyre	n	*4*	*4*	*4*	*2*	*2*	*4*	*4*	*3*	*4*
	m	**239.72**	**39.75**	**30.45**	**57.54**	**37.12**	**50.24**	**50.90**	**38.92**	**32.40**
	s	7.05	0.46	1.42	0.66	0.09	2.03	1.73	0.67	1.21
	min	233.59	39.22	29.08	57.07	37.05	47.22	48.36	38.19	30.64
	max	249.32	40.22	32.34	58.00	37.18	51.53	52.08	39.49	33.37

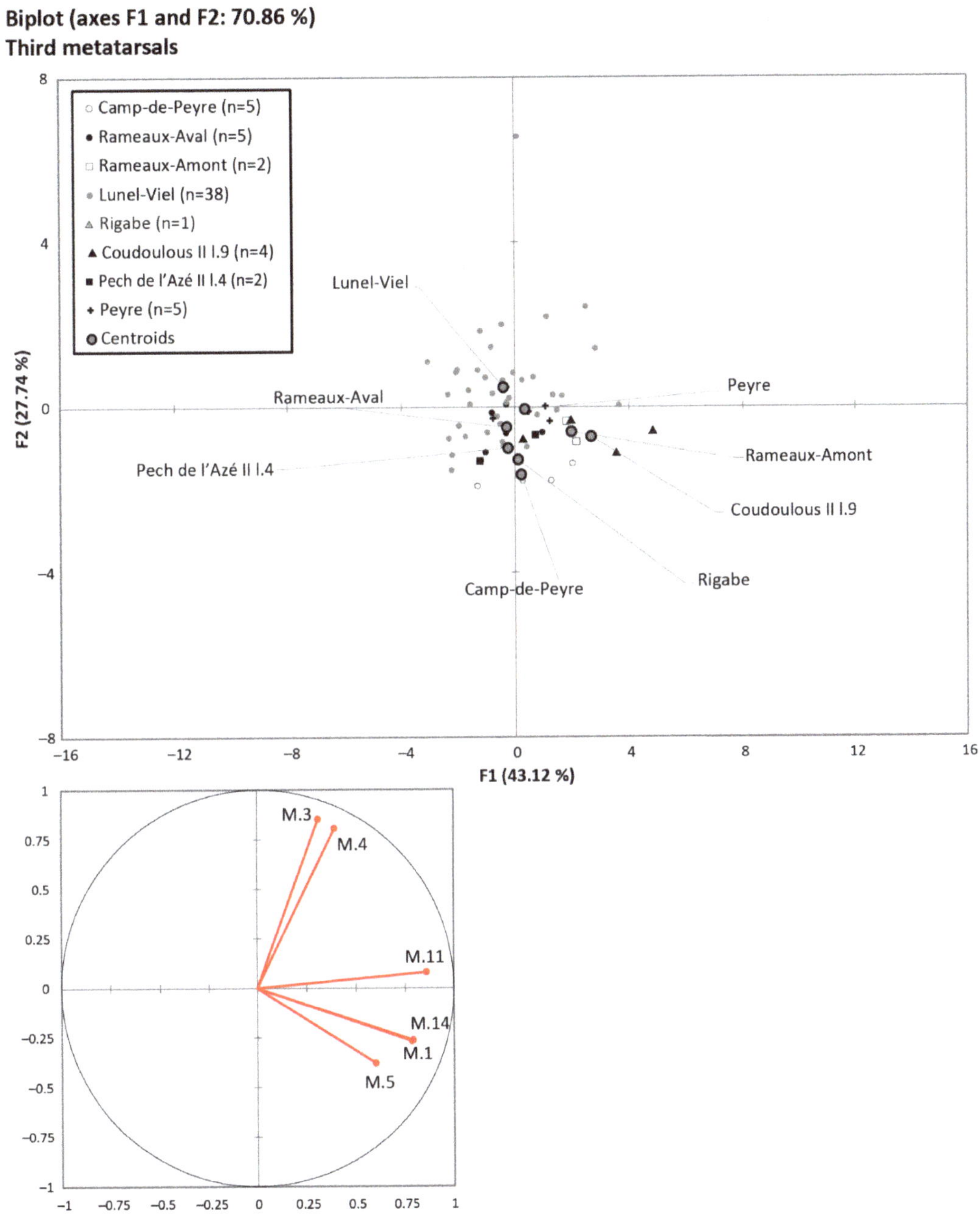

Figure 6. Principal component analysis (PCA) with measurements of the third metatarsals of southern Middle and Late Pleistocene horses from the caballoid lineage. The correlation circle illustrates the contribution of each variable. M. 1: Maximal length, M. 3: Breadth at the middle of the diaphysis; M. 4: Depth of the diaphysis at the same level; M. 5: Proximal breadth; M. 11: Distal articular breadth; M.14: Greatest depth of the medial condyle. The large dots correspond to the centroid of each cohort. Camp-de-Peyre: *E. m. campdepeyri*; Rameaux-Aval and Amont: *E. mosbachensis*; Lunel-Viel: *E. m. palustris*; Rigabe: *E. m. palustris*; Coudoulous II l.9: *E. mosbachensis*; Pech de l'Azé II l.4: *E. caballus*; Peyre: *E. caballus*.

Table 4. Measurements in millimetres of the third metatarsals of horses from the caballoid lineage. 1: Maximal length; 3: Breadth at the middle of the diaphysis; 4: Depth of the diaphysis at the same level; 5: Proximal breadth; 6: Proximal depth; 10: Distal supra-articular breadth; 11: Distal articular breadth; 12: Depth of the sagittal crest; 14: Greatest depth of the medial condyle (measurements from V. Eisenmann [76]). n= number of specimens; m=mean; s=standard deviation; min-max: minimal and maximal values. Camp-de-Peyre: *E. m. campdepeyri*; La Micoque: *E. mosbachensis*; Rameaux-Aval and Amont: *E. mosbachensis*; Lunel-Viel: *E. m. palustris*; Suard: *E. caballus piveteaui*; Pech de l'Azé II l.9-6: *E. mosbachensis*; Rigabe: *E. m. palustris*; Coudoulous II l.9: *E. mosbachensis*; Pech de l'Azé II l.4: *E. caballus*; Peyre: *E. caballus*.

Site		1	3	4	5	6	10	11	12	14
Camp-de-Peyre	n	*5*	*8*	*8*	*6*	*5*	*7*	*7*	*7*	*7*
	m	**287.78**	**37.91**	**36.07**	**56.91**	**48.62**	**54.81**	**54.19**	**42.17**	**35.76**
	s	7.59	1.52	1.17	2.96	2.27	2.14	1.98	1.69	1.29
	min	282.58	35.25	34.53	51.79	46.00	52.42	51.57	39.37	33.95
	max	301.08	40.35	37.85	59.46	51.09	58.65	57.01	43.95	37.15
Micoque	n			*2*						
	m			**38.52**						
	s			3.71						
	min			35.89						
	max			41.14						
Rameaux-Aval	n	*7*	*7*	*7*	*5*	*5*	*7*	*7*	*6*	*7*
	m	**283.65**	**37.17**	**36.12**	**55.96**	**49.60**	**54.31**	**52.69**	**39.76**	**33.48**
	s	5.96	3.04	2.74	1.08	2.49	1.64	2.00	1.18	1.07
	min	274.11	30.55	30.98	54.72	47.03	51.57	48.57	37.84	31.40
	max	290.80	39.80	38.56	57.36	53.71	55.82	54.99	41.32	34.90
Rameaux-Amont	n	*3*	*12*	*13*	*5*	*4*	*6*	*6*	*6*	*6*
	m	**293.79**	**40.31**	**35.35**	**56.00**	**49.40**	**54.94**	**54.28**	**41.83**	**34.71**
	s	13.38	1.83	4.71	2.14	2.70	2.88	1.23	2.07	1.29
	min	278.79	37.41	28.98	53.71	45.85	50.13	52.90	38.89	32.79
	max	304.50	43.13	44.71	58.41	52.38	58.38	55.86	43.87	36.07
Lunel-Viel	n	*44*	*51*	*50*	*42*	*41*	*47*	*48*	*45*	*48*
	m	**277.63**	**41.00**	**37.73**	**55.59**	**49.45**	**54.37**	**53.79**	**42.12**	**33.75**
	s	7.63	3.07	2.46	3.65	3.24	2.66	2.45	4.76	1.42
	min	262.61	36.24	33.61	40.47	38.32	50.30	48.43	38.84	30.89
	max	294.75	50.28	47.27	65.34	56.30	60.87	61.46	71.77	37.58
Suard	n		*3*	*2*	*6*	*6*	*2*	*2*	*2*	*2*
	m		**40.41**	**35.54**	**58.11**	**48.14**	**55.28**	**57.59**	**44.06**	**35.86**
	s		1.84	4.07	1.54	2.12	0.34	0.23	1.22	0.28
	min		38.39	32.66	55.19	45.65	55.04	57.42	43.20	35.66
	max		42.00	38.41	59.84	51.45	55.52	57.75	44.92	36.05
Pech de l'Azé II	n				*1*	*1*				
l. 9-6	m				**56.84**	**52.37**				
Rigabe	n	*2*	*3*	*3*	*1*	*2*	*3*	*3*	*2*	*2*
	m	**284.88**	**39.05**	**35.53**	**56.22**	**47.58**	**53.14**	**50.82**	**43.89**	**35.61**
	s	1.00	1.61	1.94		0.90	1.98	2.15	1.12	1.07
	min	284.17	37.23	34.39		46.94	51.31	48.38	43.10	34.85
	max	285.59	40.28	37.77		48.21	55.24	52.44	44.68	36.36
Coudoulous II l.9	n	*4*	*5*	*5*	*5*	*5*	*5*	*5*	*5*	*5*
	m	**299.18**	**40.86**	**37.21**	**59.45**	**49.45**	**57.42**	**57.76**	**44.86**	**37.37**
	s	9.06	0.63	1.33	3.90	1.03	2.56	3.11	1.84	1.83
	min	292.20	40.00	35.59	56.44	48.18	54.20	53.16	42.30	34.68
	max	311.80	41.62	39.30	65.44	50.90	60.31	60.42	46.70	39.34
Pech de l'Azé II	n	*2*	*2*	*2*	*2*	*2*	*2*	*2*	*2*	*2*
l.4	m	**290.44**	**38.16**	**36.36**	**53.93**	**46.33**	**55.16**	**52.66**	**41.41**	**34.97**
	s	4.87	0.32	2.32	2.86	0.62	0.64	2.07	0.22	0.06
	min	286.99	37.93	34.72	51.90	45.89	54.70	51.19	41.25	34.93
	max	293.88	38.38	38.00	55.95	46.77	55.61	54.12	41.56	35.01
Peyre	n	*5*	*7*	*7*	*7*	*6*	*6*	*6*	*4*	*6*
	m	**285.10**	**40.34**	**37.40**	**56.27**	**46.78**	**55.65**	**53.10**	**42.16**	**34.94**
	s	9.90	0.72	1.51	1.30	1.77	2.05	0.92	1.01	1.63
	min	268.84	38.90	35.54	54.55	44.57	52.68	52.03	40.72	32.10
	max	292.30	40.93	38.91	58.75	49.12	58.74	54.59	42.93	36.48

4.2. Dietary Traits of Horses

The horses from Coudoulous II l.9, Suard and La Micoque display the highest mesowear values, in particular at Coudoulous II l.9, with a score of 5.36 (Table 5; Figure 7). The lowest values come from Rigabe, Payre l.F, Peyre, Pech de l'Azé II l.4-2, Rameaux-Aval and Rameaux-Amont, with a very low score at Payre l.F (MWS = 3.13). The populations from Lunel-Viel (MWS = 4.88) and Camp-de-Peyre (MWS = 4.18) are in the middle between the two extremes, Coudoulous II l.9 and Payre l.F. Nonetheless, the populations' error bars (Figure 7) overlap and show that horses from the South of France remained preferential grazers according to mesowear analysis.

Table 5. Summary of the mesowear and microwear data for Pleistocene caballoid *Equus* from the South of France. Abbreviations: N = number of specimens; MWS = mesowear score; NP = average number of pits; NS = average number of scratches; LP% = percentage of specimens with large pits; G% = percentage of specimens with gouges; SWS = scratch width score; HC% = percentage of specimens with hypercoarse scratches; 0–17% = percentage of specimens with 17 or fewer scratches per counting area. M = average; SD = standard deviation. * Comm pers. F. Rivals. Camp-de-Peyre: *E. m. campdepeyri*; La Micoque l.E, H, J: *E. mosbachensis*; Rameaux-Aval and Amont: *E. mosbachensis*; Lunel-Viel: *E. m. palustris*; Payre l.F: *E. mosbachensis*; Suard l.53-50: *E. caballus piveteaui*; Pech de l'Azé II l.9-6: *E. mosbachensis*; Rigabe: *E. m. palustris*; Coudoulous II l.9: *E. mosbachensis*; Pech de l'Azé II l.4: *E. caballus*; Peyre: *E. caballus*.

Site		Mesowear		Microwear							
		N	MWS	N	NP	NS	LP%	G%	SWS	HC%	0–17%
Peyre	M	27	3.33	14	25.71	15.21	50	28.57	0.92	7.14	85.71
	SD		0.73		5.89	1.40					
Pech de l'Azé II l.4-2	M	31	4.87	18	24.50	20.86	0	11.11	1	27.78	11.11
	SD		0.72		5.94	3.67					
Coudoulous II l.9	M	25	5.36	11	24.05	21.86	63.63	0	1	0	0
	SD		0.76		5.07	3.17					
Pech de l'Azé II l. 9-6	M	63	4.65	42	18.13	20.21	0	23.81	1	27.78	16.67
	SD		0.9		4.25	2.63					
Suard l. 50	M	12	5.17	7	19.57	17.93	57.14	14.28	1	0	42.86
	SD		0.83		4.14	2.17					
Suard l. 51	M	70	5.01	28	32.04	26.05	3.57	0	1	3.57	0
	SD		0.84		5.66	3.42					
Suard l. 52	M	30	4.80	26	29.83	22.35	3.84	3.84	0.92	0	0
	SD		0.92		4.66	1.62					
Suard l. 53	M	4	5.25	9	23.72	19.89	0	0	0.89	0	0
	SD		0.96		3.91	1.34					
Rigabe	M	18	4.94	12	15.21	15.71	75	50	0.5	0	66.67
	SD		0.73		3.44	3.95					
Payre l. F *	M	8	3.13	10	20.80	19.65	20	10	1.1	63.63	30
	SD		1.13		2.84	3.64					
Lunel-Viel	M	65	4.88	36	15.85	19.61	31.43	8.33	1	2.78	22.22
	SD		0.93		4.17	2.65					
Rameaux-Amont	M	151	4.37	24	18.88	13.77	37.5	12.5	1.04	4.17	87.5
	SD		0.97		7.19	2.79					
Rameaux-Aval	M	8	3.50	3	16.83	13.67	66.66	33.33	1	0	100
	SD		1.07		2.84	2.02					
La Micoque l. J	M	21	4.48	19	21.89	21.13	94.74	42.10	1.05	10.53	10.53
	SD		0.93		4.82	2.99					
La Micoque l. H	M	8	3.88	15	22.40	23.33	46.66	0	1	0	0
	SD		0.99		3.52	3.19					
La Micoque l. E	M	5	5	22	18.34	24.55	0	0	0.72	0	4.55
	SD		0		5.41	4.27					
Camp-de-Peyre	M	38	4.18	7	28.57	22.00	0	0	0.71	14.28	0
	SD		0.51		4.17	2.47					

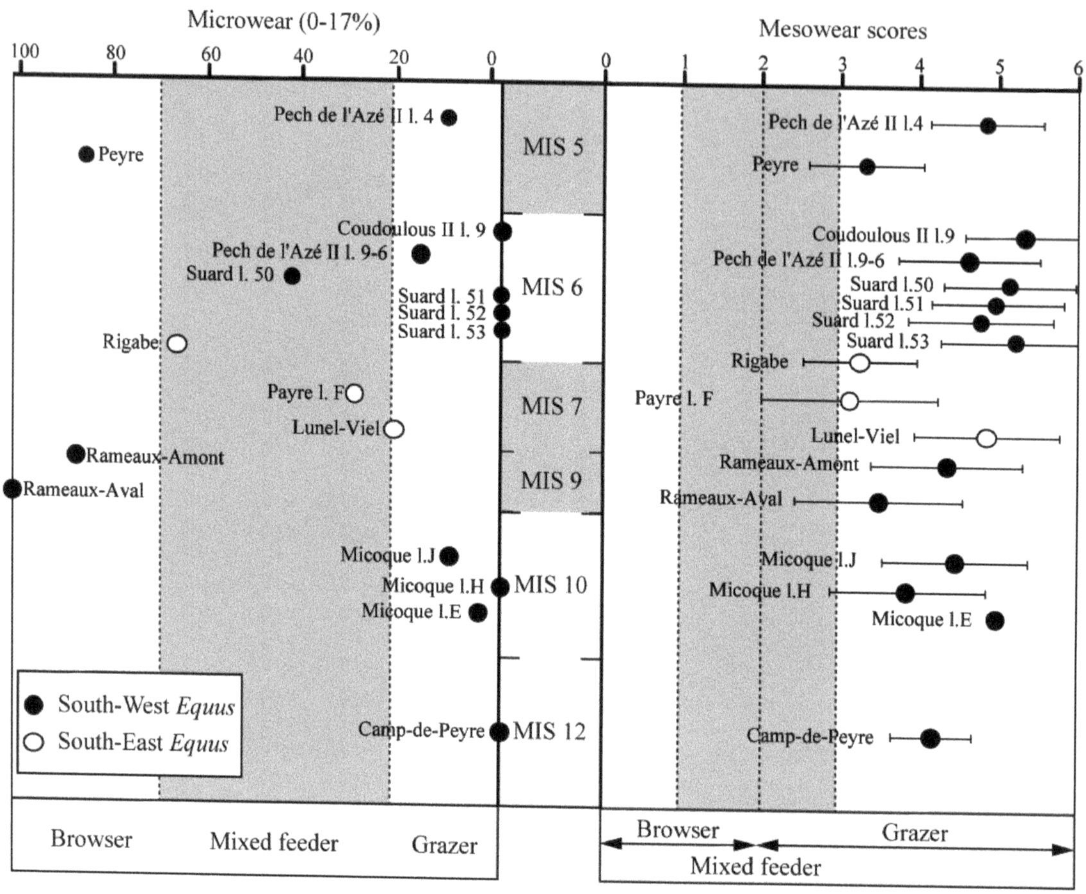

Figure 7. Low scratch percentages (0%–17%) and mesowear scores for each cohort through time. The grey areas correspond to the range of values for extant mixed feeders. Camp-de-Peyre: *E. m. campdepeyri*; La Micoque l.E, H, J: *E. mosbachensis*; Rameaux-Aval and Amont: *E. mosbachensis*; Lunel-Viel: *E. m. palustris*; Payre l.F: *E. mosbachensis*; Suard l. 53-50: *E. caballus piveteaui*; Pech de l'Azé II l.9-6: *E. mosbachensis*; Rigabe: *E. m. palustris*; Coudoulous II l.9: *E. mosbachensis*; Pech de l'Azé II l.4: *E. caballus*; Peyre: *E. caballus*.

The preservation of teeth on a microscopic scale is quite similar from one site to another. At Camp-de-Peyre, 75% of the cast teeth were suitable for study, and this value was 81.8% at La Micoque l.E, 86.6% at La Micoque l.H, 57.8% at La Micoque l.J, 100% at Rameaux-Aval, 85.7% at Rameaux-Amont, 81.8% at Lunel-Viel, 100% at Payre l.F, 66.6% at Rigabe, 100% at Suard l.53, 89.6% at Suard l.52, 100% at Suard l. 51 and l.50, 65.6% at Pech de l'Azé II l.9-6, 100% at Coudoulous II l.9, 60% at Pech de l'Azé l.4, and 87.5% at Peyre. Thus, on the whole, the teeth are well preserved on a microscopic scale, with the exception of the Rigabe and Pech de l'Azé II sites, where about 40% had to be excluded from the analysis.

The microwear patterns of some of the horse populations show an abundant number of teeth with a lot of scratches, for example at: Camp-de-Peyre, La Micoque, Suard l.53-52-51, Pech de l'Azé II and Coudoulous II l.9, whereas this number at Lunel-Viel, Payre l.F, Rigabe and Suard l.50 is lower. It is lowest for the populations of les Rameaux and Peyre. (Figures 7 and 8). This suggests that the first group corresponds to grazer feeding habits, the second to mixed-feeder ones, and the third to a browser diet. Some populations show a relatively high mean number of pits on occlusal tooth surfaces (Figure 8), such as Suard

l. 51-52 or Camp-de-Peyre. This suggests the consumption of a larger amount of dusty grass than the other populations [102]. The scratch width score of the Rigabe horses is very low (=0.5), which could indicate the occasional consumption of leaves [103], in addition to a more abrasive diet. On the contrary, the scratch width score of Payre l. F is the highest (=1.1). Moreover, the high percentage of teeth with hypercoarse scratches (63.63%) could indicate the consumption of bark and twigs [102]. Thus, the microwear analysis results show high variations between the different populations.

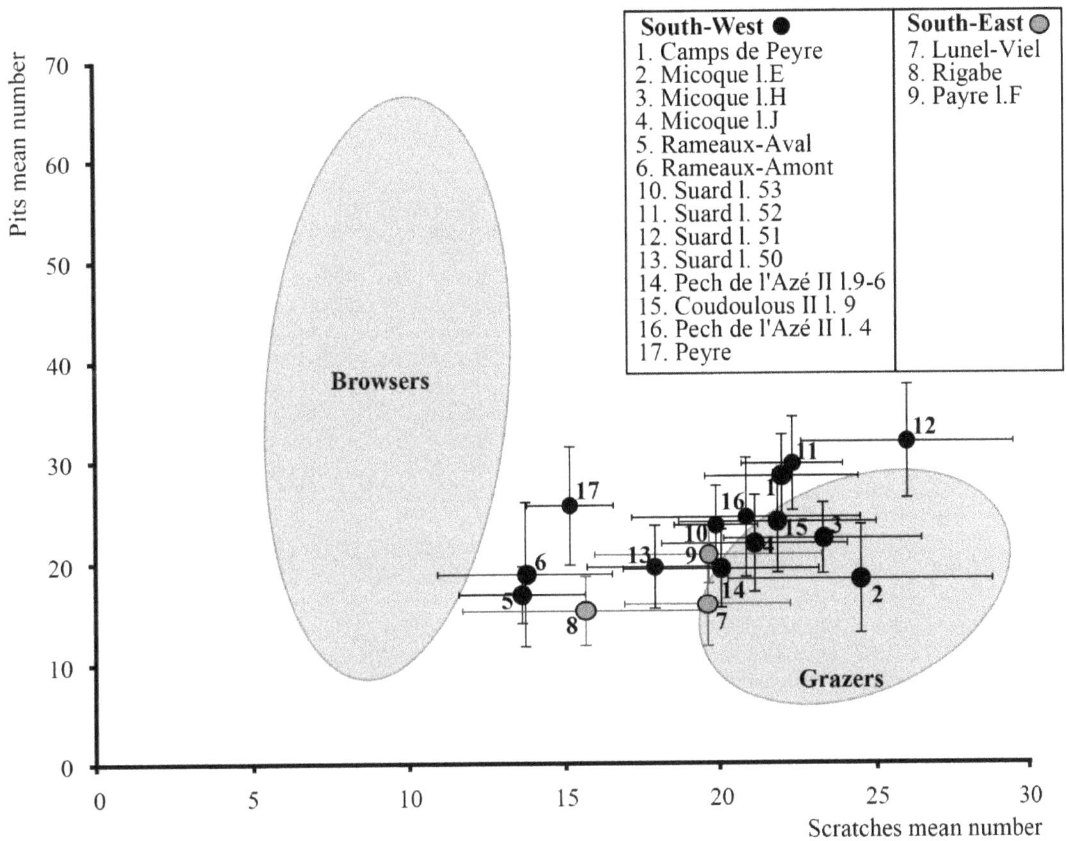

Figure 8. Bivariate plot of the average numbers of pits and scratches in horse cohorts from the South of France. Error bars correspond to standard deviations for the fossil samples. Plain ellipses correspond to Gaussian confidence ellipses (p = 0.95) on the centroid for extant leaf browsers and grazers from Solounias and Semprebon [97]. Camp-de-Peyre: *E. m. campdepeyri*; La Micoque l.E, H, J: *E. mosbachensis*; Rameaux-Aval and Amont: *E. mosbachensis*; Lunel-Viel: *E. m. palustris*; Payre l.F: *E. mosbachensis*; Suard l. 53-50: *E. caballus piveteaui*; Pech de l'Azé II l.9-6: *E. mosbachensis*; Rigabe: *E. m. palustris*; Coudoulous II l.9: *E. mosbachensis*; Pech de l'Azé II l.4: *E. caballus*; Peyre: *E. caballus*.

5. Discussion

5.1. Body Mass and Morphometrical Adaptations to Local Environments

The body mass of late Middle Pleistocene and early Late Pleistocene horses in southern France varied from one population to another by a margin of about 100 kg. On average, horses from south-western France were slightly heavier than those from the South East, and their body mass was similar in the isotopic stages of cold and temperate climates. In the South East, horses in the temperate isotopic stages were on average heavier than those in the cold isotopic stages, but this observation is not well based as only one site represents a cold isotopic stage (Rigabe). Statistically, the variation in body mass of these cohorts cannot be explained by the geographical position of the sites (West/East) or by the general climatic ambience of the associated isotopic stages (Cold/Temperate) (Table 6). Moreover, our data, which are still preliminary, do not allow us to propose a chronological scheme for the evolution of horse size over this period of time.

Table 6. Comparison of the mean (m) body mass with the standard deviation (s) in kg of the cohorts (n = number) of caballoid horses from the western and eastern part of France and from cold and temperate isotopic stages (IS) using the Mann-Withney statistical test.

Group	n	m	s	*p*-Values
Western cohorts	9	504.91	35.67	0.8636
Eastern cohorts	3	492.88	17.89	
Cold IS	6	502.22	37.91	0.9372
Temperate IS	6	501.59	28.13	

The Pleistocene Equus paleopopulations from France analysed in this study all have elongated upper limb bones and proportionally short metapodials compared to *E. przewalskii*. This observation is similar to that made by the authors of [12] regarding Pleistocene horses from Britain and Germany. The elongation of distal limb bones has been linked to the occupation of arid and resource-poor habitats by Prezwalski's horses and their significant mobility related to their living conditions [12,104]. During the Pleistocene, horses in southern France would have occupied environments richer in plant resources, allowing a less mobile lifestyle, similar to that of German or British populations with seasonal moves, which is also supported by microwear analysis [105]. The width of the third phalanx is much more variable in the cohorts studied. We have observed two groups: the first including La Micoque, Rameaux-Aval, Lunel-Viel and Pech de l'Azé II l. 8 with narrow third phalanges; and the second including Camp-de-Peyre, Rameaux-Amont, Suard, Coudoulous II l.9 and Peyre with wider third phalanges similar in proportion to *E. prezwalskii*. The width of the third phalanges is considered an adaptation to the hardness of the substrate. Narrow phalanges are thought to be an adaptation to hard ground [72]. In general, the phalanges of caballoid equids are wider than those of stenonian, asinian and zebra equids, and thus adapted to softer or more humid ground than the other ones, which are adapted to harder and more arid substrate [12,71,106,107]. The largest third phalanges are identified in a majority of cohorts from the glacial isotopic stages: Camp-de-Peyre (MIS 12), Suard and Coudoulous II l. 9 (MIS 6), which could indicate a regular snow cover. This is not systematic since two temperate isotopic stage cohorts also have very large third phalanges: Rameaux-Amont (MIS 9) and Peyre (MIS 5). At Rameaux-Amont, the study of the microfauna describes a cold interstage and an increase of the snow cover [48], which could be consistent with the increase of the width of the third phalange from Rameaux-Aval to Rameaux-Amont. At Peyre, not enough information about the paleoenvironment is available to discuss the width of the third phalange with regard to the significance of the snow cover or the softness of the substrate. On the other hand, narrower third phalanges can be found in both glacial (La Micoque, MIS 10 and Pech de l'Azé l. 9-6, MIS 6) and temperate isotopic stages (Rameaux-Aval, MIS 9 and Lunel-Viel, MIS 7). At La Micoque, the climate was suspected to be really dry [45], which could indicate a smaller period of snow cover compared to Camp-de-Peyre, for example, and be consistent with the morphology of the third phalange. At Pech de l'Azé II l.9-6, paleoenvironmental information is too limited to establish any correlation. At Lunel-Viel, based on the morphology of the horses, the paleoenvironment was first suspected to be very humid and similar to today's Camargue wetland [36]. The comparison of the dietary habits of ungulates from the site and ones from Camargue allows us to discard this hypothesis and suggest a mosaic environment rich in monocot and dicot plants [52].

The morphology of the third phalanx of the horse from Lunel-Viel could indicate a harder substrate than horses confronted with significant snow cover or simply soft soils.

The morphology of the metapodials also shows a different pattern in the studied cohorts. The PCAs (Figures 5 and 6) show that the limb bones of the horses from Lunel-Viel and Rameaux-Aval are short and robust, while the ones from Pech de l'Azé II l.6-4, Camp-de-Peyre and Rigabe are longer and slenderer, especially at the diaphysis. Finally, the metapodials of the horses from Coudoulous II l.9 and Rameaux-Amont are longer and more robust than the ones from the other populations. Limb length was correlated with various variables: those from a cold climate and/or a forested biotope could have short limbs [12,15,108], while long limb bones could indicate a rich environment with diversified dietary resources, as in mosaic steppe habitats [12], or strong seasonal variations [109]. The robustness of the bones can be interpreted as a wet climate indicator, according to the authors of [70,71], but [12] shows that, during glacial periods, North European horses tend to always be more robust than those from temperate periods. Moreover, temperate North European horses seem to be slenderer under oceanic conditions related to increasing humidity. Some temperate populations living in a mosaic steppe habitat can grow to a larger size, according to their limb bones length and robusticity, owing to a very suitable environment.

The variations in the slenderness of horse metapodials from the South of France are neither linked to the isotopic stage nor site location. Indeed, cold isotopic stage horses from Camp-de-Peyre (South West, MIS 12) and Rigabe (South East, MIS 6) are slenderer, while the horses from Coudoulous II l.9 (South West, MIS 6) are much more robust. During temperate stages, some populations are slender, as in Payre (South East, MIS 7), Pech de l'Azé II l.4, and Peyre (South West, MIS 5), and others have very robust limb bones, even superior to some glacial period populations, as at Rameaux-Amont and Rameaux-Aval (South West, MIS 9) and Lunel-Viel (South East, MIS 7). The slenderness of the horses from Camp-de-Peyre, Payre and Peyre is associated with a dry climate [24,34,43,110,111]. On the contrary, robust horses from Rameaux-Aval and Rameaux-Amont and Lunel-Viel are associated with a humid climate [36,48,52]. While the correlation between humidity and robusticity of the bones is repeated in our data in accordance with [70,71], it is not always the case. The metapodials of the horse from Coudoulous II l.9 are very robust and this is associated with humid conditions [27]. This feature may be related to very cold temperature adaptations associated with a diverse steppe vegetation [12].

5.2. Horse Dietary Traits at the End of Middle Pleistocene and the Beginning of Late Pleistocene in the South of France

Mesowear analysis shows that the horses from the South of France usually ate abrasive plants like grass during the year, which corresponds to the general trend of this genus [112], even if mixed-feeder populations have been investigated [113–115]. Microwear analysis, on the other hand, shows that their diet varies a lot over a shorter time scale. The flexibility of horses' diet and their ability to incorporate grass and browse in their feeding or rely exclusively on browse has already been shown in several Pleistocene populations, such as Schöningen (MIS 9) [16] and Taubach (MIS 5) [89] in Germany. In a landscape where both browse and grass are available, they seem to prefer to select the latter [52,116], but in the context of the mammoth-steppe landscape, their diet seems to be less exclusive than that of extant horses [115].

In this study, often we found that horses from the South West of France from glacial periods, such as those from Camp-de-Peyre, La Micoque, Suard l.53-52-51, Pech de l'Azé II and Coudoulous II l.9, mainly ate grass all year long (with the notable exception of Suard l.50), whereas horses from the South East, from glacial and interglacial periods, were mixed-feeders shortly before their death. Finally, some horses from the South West from interglacial periods (Rameaux-Amont, Rameaux-Aval and Peyre) were browsers at the time of their death. In addition, in Suard l.51-52 and Camp-de-Peyre, located in the South West and characterised by a particularly cold and rigorous climate [42,43,58,59,117], the horses probably ate dusty or short grass. This behaviour seems to have been quite rare among horses from the South of France and corresponds to a specific local context. Likewise, south-eastern horses display dietary particularities that may be related to local environmental conditions. Indeed, at Rigabe, horses may have occasionally eaten leaves to supplement a more abrasive diet. At Payre l. F, microwear analysis shows the presence of bark and twigs in the horses' diet, which is consistent with their habitat in the forested bottom of the valley below the site [55].

While Middle to Late Pleistocene horses from the South of France remained preferentially grazers, their diet was strongly impacted by seasonal changes in the environment. They displayed a variety of behaviour from grazers to browsers in accordance with local conditions. In the South West, the influence of the Atlantic Ocean induced colder conditions during the glacial isotopic stages (MIS 12, 10 and 6) [10] which may have favoured open areas and steppes and allowed horse cohorts to remain grazers all year long. In this general scheme, exceptions still exist, like at Suard l.50, where south-western glacial isotopic stage horses could incorporate more browse in their diet. Such behaviour may be the result of competition with other species, as at Canyars with *Bos primigenius* [96] or the indicator of the development of a forest during the accumulation of this layer. The interglacial periods in this region were more forested, perhaps similar to today's vegetation with continental acidophilous oak woodland [3], which lead the horses to seasonally eat browse.

In the South East region, in each isotopic stage (MIS 7 and 6), the horses were mixed feeders at the time of their death. This behaviour may be related to the influence of the Mediterranean climate. This region is considered a refuge area where the impact of major global climate changes was buffered by the influence of the Mediterranean Sea [4]. The succession of the different isotopic stages was therefore less marked than in the South West, allowing the horses to maintain the same behaviour from one stage to the next. Their mixed-feeder diet at the time of their death illustrates the composition of the Mediterranean vegetation, rich in both grass and browse, with one rarely taking precedence over the other [3].

5.3. *Relations between Horse Dietary Habits, Body Mass, and Morphometrical Parameters*

The relationship between body mass, morphometrical skeleton features, and dietary habits can be difficult to establish, as they do not necessarily correspond to the same time span. Indeed, microwear analysis displays information about the final days to months of life [85–88], and mesowear indicates a general or monthly dietary signal [82,83]. The importance of phenotypic plasticity, and the rate of body adaptation, is largely unknown and may depend on the living condition of each cohort. However, several authors have found fairly large variations in body mass and bone conformation over relatively short time scales within the isotopic stage chronology [12,71,118].

In this study, the variability of the body mass of Middle to Late Pleistocene horses from the South of France was found to be quite low. Probably because of the size of the sample, no trend related to the isotopic stage or the region is apparent. Over larger time-spans and geographical areas, observations have already been made of underlying relationships between horses' body mass and environmental factors, diet and behaviour. The growth of a large body mass has been suggested to be associated with seasonally highly-productive and high-quality plant communities [119–121], or with environments that have a seasonal shortage of dietary resources [109,121,122]. Also, a large body mass can be related to small group size and population densities [18,118,123]. Here, the Spearman correlation test between the mean weight and the MWS of each cohort shows a very light and non-significant positive relationship between those two data ($n = 12$; p-value = 0.4368; rho = 0.2448) in contrast to the studies described in [18] and [91]. This difference can be explained by the different chronological and geographical scales studied. This work focuses on a shorter period and a smaller area. The variation in body mass of horses according to their diet in the dental mesowear would therefore not be noticeable here. The impact of seasonal variations on horse body mass was also investigated by statistically comparing the body mass of the three dietary categories (G-G: grazer all year long; G-MF: generally, grazer, mixed-feeder at time of death; G-B: generally, grazer, browser at time of death). Each diet-body mass related group was not statistically different one from the other (Table 7).

Table 7. Comparison of the mean (m) body mass with the standard deviation (s) in kg of the cohorts (n = number) of caballoid horses according to their diet using a Kruskal-Wallis statistical test. G-G: grazer according to meso- and microwear; G-MF: grazer according to mesowear analysis and mixed-feeder at the time of death (microwear analysis); G-B: grazer according to mesowear analysis and Browser at the time of death (microwear analysis).

Group	n	m	s
G-G	6	502.64	37.57
G-MF	3	492.88	17.89
G-B	3	509.45	38.91
	Kruskal-Wallis Test *p*-Values		
	G-G/G-MF	G-G/G-B	G-MF/G-B
	−0.3651	0.3651	−0.3086

On the other hand, morphometric analysis concerning metapodials and third phalanges underline a variability between the studied cohorts. Figure 9 displays the centroid of the diet categories according to meso- and microwear data (cf. supra) in relation to metacarpal and metatarsal morphometrics. In the two tests, the G-MF diet was shown to be related to the shortest bones, while the G-G diet was related to the longest. The G-B diet is more represented by the long and robust bones. Open areas rich with grass allow horses to grow tall, since there are no closed-area restrictions. Also, tall and slender metapodials in relation to proximal limb bones indicate cursoriality and an increased ability to escape predators by efficient running in open environments. In more closed-areas, small size may facilitate movement and predator avoidance [108]. However, G-B populations are also tall, but with a more robust conformation. This type of diet indicates strong seasonal variations in the plant composition in a humid climate, allowing the development of wooded areas. The humidity of the climate is one of the strongest factors explaining the growth in robusticity of the horses (cf. supra). The large size of the individuals may be explained by the small group size induced, at least seasonally, by the expansion of the forest cover, as demonstrated by the authors of [18]. The smaller horses in terms of height were shown, in this set of data, to be G-MF and were all located in the South East of France. This effect may be a consequence of the low productivity and quality of Mediterranean plants [124], which may have restrained the horses' size. On the other hand, the seasonal variability under a Mediterranean climate is less pronounced than under an Atlantic one, which may not have led to a selection of tall horses [109].

In a small timespan and geographical range, the body mass of the horses does not vary enough to describe patterns and adaptations to environmental features. However, the conformation of the bones, here the metapodials, seems to be partly driven by the diet and environmental conditions, which themselves partly conditioned the diet. While body mass may be a good indicator of a larger time span and geographical range, bone conformation seems to be a more sensitive proxy. The relationship between it and the dietary analysis could indicate rapid skeleton adaptations for horses during the end of the Middle Pleistocene and the beginning of the Late Pleistocene in the South of France. While it is not possible to establish the rate of anatomical changes, some must have been very rapid, as illustrated at Les Rameaux. Within the same isotopic stage (MIS 9), the height of the horses and the width of their third phalanges increased in order to face more humid conditions, while their diet remained similar.

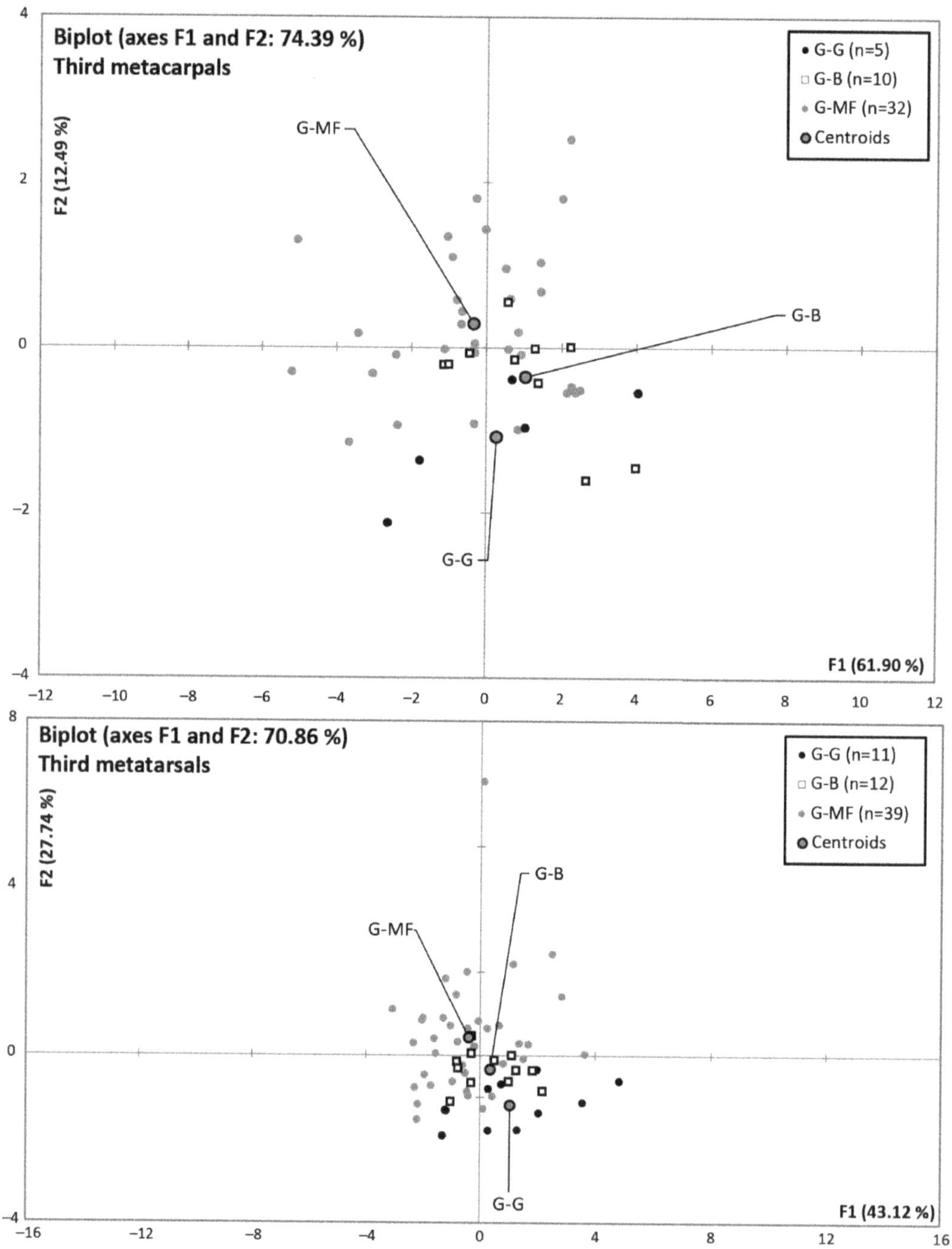

Figure 9. Principal component analysis (PCA) with measurements of the third metacarpals and metatarsals of Southern Middle and Late Pleistocene caballoid horses with regards to the dietary categories displayed by meso- and microwear analysis. G-G: grazer all year long; G-MF: generally, grazer, mixed-feeder at the time of death; G-B: generally, grazer, browser at the time of death. The large dots correspond to the centroid of each dietary category. For the correlation circles, see Figure 5 for metacarpals and 6 for metatarsals.

6. Conclusions

This study confirms the significant flexibility of horses and their capacity to live in a wide range of habitats. Very common in archaeological and paleontological sites from the end of the Middle and the beginning of the Late Pleistocene in the South of France, they adapted to various environmental and climatic conditions. While at these scales the variation of their body mass did not seem to be a good indicator of their adaptations, metapodials and third phalanges show rapid changes in their conformation in relation to the environment. The width of the third phalanges have been, in a majority of cases, linked with the significance of the snow cover, while the length and robusticity of the metapodials have been linked with the humidity of the climate, the size of the forest cover, the productivity of the environment, and the social and dietary behaviour of the horses.

While the mesowear analysis indicates that the horses remained preferentially grazers, their diet could be impacted by strong seasonal variation. Thus, several populations remained grazers (Camp-de-Peyre, La Micoque, Suard l.53-51, Pech de l'Azé l. 9-4 and Coudoulous II l.9), others could, at their time of death, have been mixed-feeders (Payre l. F, Lunel-Viel, Suard l.50 and Rigabe) or browsers (Rameaux-Aval, Rameaux-Amont and Peyre). Their diet, strongly related to the vegetation, may also have influenced the conformation of their metapodial. Thus, tall horses were grazers all year long and must have inhabited open landscape like the steppe. Tall and robust horses were more likely to be seasonal browsers, with their morphometrical and dietary features explained by the more sizeable forest cover which induces smaller groups. Finally, smaller horses were found in the South East of France, under the Mediterranean climate; their size may be explained by low-quality food resources.

All of the cohorts from this corpus also have similarities. In particular, the ratio of the limb proportions shows elongated upper limb bones compared to the autopodial, which indicates general low-mobility behaviour.

In the case of horses, morphometrical and dental wear studies seem to be good proxies in order to investigate their past behaviour and adaptations in relation to their environment. While the rate of anatomical adaptation cannot be truly established, the concordance between the results of the two proxies could indicate rapid anatomical changes and validate their integration at a paleontological study scale. Moreover, their combination could be a good tool to reconstruct their palaeoenvironment and their eco-ethology.

Funding: This research received no external funding.

Institutional Review Board Statement: Not applicable.

Informed Consent Statement: Not applicable.

Data Availability Statement: The data presented in this study are available in this paper.

Acknowledgments: I am particularly grateful to Jean-Philip Brugal who directed my PhD, from where this study is extracted. I would also like to thank LAMPEA, ED 355, and Aix-Marseille University for the help and funding they provided during this work. I also would like thank F. Rivals for sharing the teeth wear data from Payre. Finally, I would like to thanks the five reviewers and the academic editors, Juan Rofes, Janine Ochoa and Emmanuelle Stoetzel who help to greatly improve this manuscript.

Conflicts of Interest: The author declares no conflict of interest.

References

1. Ozenda, P. *La Végétation Du Continent Européen*; Delachaux et Niestlé: Lausanne, Switzerland, 1994; p. 271.
2. Joannin, S. Changements Climatiques en Méditerranée à la Transition Pléistocène Inférieur-Moyen: Pollens, Isotopes Stables et Cyclostratigraphie. Ph.D. Thesis, Claude Bernard—Lyon I University, Lyon, France, 2007.
3. Ozenda, P.; Borel, J.-L. An Ecological Map of Europe: Why and How? *Compt. Rendus Acad. Sci. Ser. III Sci. Vie* **2000**, 323, 983–994. [CrossRef]

4. Petit, R.J.; Aguinagalde, I.; de Beaulieu, J.-L.; Bittkau, C.; Brewer, S.; Cheddadi, R.; Ennos, R.; Fineschi, S.; Grivet, D.; Lascoux, M.; et al. Glacial Refugia: Hotspots But Not Melting Pots of Genetic Diversity. *Science* **2003**, *300*, 1563–1565. [CrossRef]
5. Couchoud, I. Etude Pétrographique et Isotopique Des Spéléothèmes Du Sud-Ouest de La France Formés En Contexte Archéologique: Contribution à La Connaissance Des Paléoclimats Régionaux Du Stade Isotopique 5. Ph.D. Thesis, Bordeaux I University, Bordeaux, France, 2006.
6. Wainer, K.; Genty, D.; Blamart, D.; Hoffmann, D.; Couchoud, I. A New Stage 3 Millennial Climatic Variability Record from a SW France Speleothem. *Palaeogeogr. Palaeoclimatol. Palaeoecol.* **2009**, *271*, 130–139. [CrossRef]
7. Couchoud, I.; Genty, D.; Hoffmann, D.; Drysdale, R.; Blamart, D. Millennial-Scale Climate Variability during the Last Interglacial Recorded in a Speleothem from South-Western France. *Quat. Sci. Rev.* **2009**, *28*, 3263–3274. [CrossRef]
8. Masson-Delmotte, V. Les grandes oscillations du climat depuis 800,000 ans. In *Des Climats et des Hommes*; Berger, J.-F., Ed.; La Découverte: Paris, France, 2012; pp. 57–72.
9. Magri, D.; Palombo, M.R. Early to Middle Pleistocene Dynamics of Plant and Mammal Communities in South West Europe. *Quat. Int.* **2013**, *288*, 63–72. [CrossRef]
10. Colleoni, F.; Krinner, G.; Jakobsson, M.; Peyaud, V.; Ritz, C. Influence of Regional Parameters on the Surface Mass Balance of the Eurasian Ice Sheet during the Peak Saalian (140 Kya). *Glob. Planet. Chang.* **2009**, *68*, 132–148. [CrossRef]
11. Colleoni, F.; Krinner, G.; Jakobsson, M. Sensitivity of the Late Saalian (140 Kyrs BP) and LGM (21 Kyrs BP) Eurasian Ice Sheet Surface Mass Balance to Vegetation Feedbacks. *Geophys. Res. Lett.* **2009**, *36*, 1–5. [CrossRef]
12. van Asperen, E.N. Ecomorphological Adaptations to Climate and Substrate in Late Middle Pleistocene Caballoid Horses. *Palaeogeogr. Palaeoclimatol. Palaeoecol.* **2010**, *297*, 584–596. [CrossRef]
13. Rivals, F.; Mihlbachler, M.C.; Solounias, N.; Mol, D.; Semprebon, G.M.; de Vos, J.; Kalthoff, D.C. Palaeoecology of the Mammoth Steppe Fauna from the Late Pleistocene of the North Sea and Alaska: Separating Species Preferences from Geographic Influence in Paleoecological Dental Wear Analysis. *Palaeogeogr. Palaeoclimatol. Palaeoecol.* **2010**, *286*, 42–54. [CrossRef]
14. Moncel, M.-H.; Rivals, F. On the Question of Short-Term Neanderthal Site Occupations: Payre, France (MIS 8-7), and Taubach/Weimar, Germany (MIS 5). *J. Anthropol. Res.* **2011**, *67*, 47–75. [CrossRef]
15. Van Asperen, E.N. Late Middle Pleistocene Horse Fossils from Northwestern Europe as Biostratigraphic Indicators. *J. Archaeol. Sci.* **2012**, *39*, 1974–1983. [CrossRef]
16. Rivals, F.; Julien, M.-A.; Kuitems, M.; Van Kolfschoten, T.; Serangeli, J.; Drucker, D.G.; Bocherens, H.; Conard, N.J. Investigation of Equid Paleodiet from Schöningen 13 II-4 through Dental Wear and Isotopic Analyses: Archaeological Implications. *J. Hum. Evol.* **2015**, *89*, 129–137. [CrossRef] [PubMed]
17. Rivals, F.; Lister, A.M. Dietary Flexibility and Niche Partitioning of Large Herbivores through the Pleistocene of Britain. *Quat. Sci. Rev.* **2016**, *146*, 116–133. [CrossRef]
18. Saarinen, J.; Cirilli, O.; Strani, F.; Meshida, K.; Bernor, R.L. Testing Equid Body Mass Estimate Equations on Modern Zebras—With Implications to Understanding the Relationship of Body Size, Diet, and Habitats of *Equus* in the Pleistocene of Europe. *Front. Ecol. Evol.* **2021**, *9*. [CrossRef]
19. Prat, F. Recherches sur les Equidés Pléistocènes en France. Ph.D. Thesis, Bordeaux Faculty of Sciences, Bordeaux, France, 1968.
20. Forsten, A. Size Decrease in Late Pleistocene-Holocene Caballoid Horses (Genus *Equus*), Intra-or Interspecific Evolution? A Discussion of Alternatives. *Quat. Int.* **1993**, *19*, 71–75. [CrossRef]
21. Magniez, P.; Boulbes, N. Environment during the Middle to Late Palaeolithic Transition in Southern France: The Archaeological Sequence of Tournal Cave (Bize-Minervois, France). *Quat. Int.* **2014**, *337*, 43–63. [CrossRef]
22. Lisiecki, L.E.; Raymo, M.E. A Pliocene-Pleistocene Stack of 57 Globally Distributed Benthic Δ18O Records. *Paleoceanography* **2005**, *20*, PA1003. [CrossRef]
23. Brugal, J.-P.; Argant, A.; Boudadi-Maligne, M.; Crégut-Bonnoure, E.; Croitor, R.; Fernandez, P.; Fourvel, J.-B.; Fosse, P.; Guadelli, J.-L.; Labe, B.; et al. Pleistocene Herbivores and Carnivores from France: An Updated Overview of the Literature, Sites and Taxonomy. *Ann. Paléontol.* **2020**, *106*, 102384. [CrossRef]
24. Vernet, J.-L.; Mercier, N.; Bazile, F.; Brugal, J.-P. Travertins et terrasses de la moyenne vallée du Tarn à Millau (Sud du Massif Central, Aveyron, France). *Quat. Rev. Assoc. Fr. Pour Létude Quat.* **2008**, *19*, 3–10. [CrossRef]
25. Uzunidis-Boutillier, A. Grands Herbivores de La Fin Du Pléistocène Moyen Au Début Du Pléistocène Supérieur Dans Le Sud de La France: Implications Anthropologiques Pour La Lignée Néandertalienne. Ph.D. Thesis, Aix-Marseille University, Aix-en-Provence, France, 2017.
26. Jacobs, Z.; Jankowski, N.R.; Dibble, H.L.; Goldberg, P.; McPherron, S.J.P.; Sandgathe, D.; Soressi, M. The Age of Three Middle Palaeolithic Sites: Single-Grain Optically Stimulated Luminescence Chronologies for Pech de l'Azé I, II and IV in France. *J. Hum. Evol.* **2016**, *95*, 80–103. [CrossRef]
27. Brugal, J.-P.; Aouadi, N.; Mai, B.T.; Cochard, D.; Costamagno, S.; Ellwood, B.; Girard, M.; Jaubert, J.; Jeannet, M.; Kervazo, B.; et al. *La Grotte de Coudoulous II (Tour-de-Faure, Lot)*; Unpublished Excavation Report, Bilan 1997—2000; Pgm P1 Min. de la Culture, SRA Midi-Pyrénées, 2000; 138p.
28. Uzunidis, A.; Brugal, J.-P. Les grands herbivores (Bovinés, Équidés, Rhinocérotidés, Proboscidiens) de la fin du Pléistocène Moyen: La couche 9 de Coudoulous II (Lot, Quercy, Sud-Ouest France). *PALEO* **2018**, *29*, 223–249. [CrossRef]
29. Schvoerer, M.; Rouanet, J.-F.; Navailles, H.; Debénath, A. Datation Absolue Par Thermoluminescence Des Restes Humains Antéwürmiens de l'abri Suard à La Chaise-de-Vouthon (Charente). *Comptes Rendus Acad. Sci. Paris* **1977**, *284*, 1979–1982.

30. Schwarcz, H.P.; Debénath, A. Datations Absolues Des Restes Humains de La Chaise-de-Vouthon (Charente) Au Moyen Des Déséquilibres Des Séries. *Comptes Rendus Acad. Sci. Paris* **1979**, *288*, 1155–1157.
31. Blackwell, B.; Schwarcz, H.P.; Debénath, A. Absolute Dating of Hominids and Palaeolithic Artifacts of the Cave of La Chaise-de-Vouthon (Charente), France. *J. Archaeol. Sci.* **1983**, *10*, 493–513. [CrossRef]
32. David, P.; Prat, F. Considérations Sur Les Faunes de La Chaise (Commune de Vouthon, Charente), Abris Suard et Bourgeois-Delaunay. *Bull. Assoc. Fr. Pour Létude Quat.* **1965**, *2*, 222–232. [CrossRef]
33. Bonifay, E. Etude Géologique de La Grotte de Rigabe, Artigues (Var). In Proceedings of the Congrès Préhistorique de Monaco, XVIe Session, Monaco, 28 August–5 September 1959; Imprimerie Monnoyer: Le Mans, France, 1959; pp. 265–291.
34. Moncel, M.-H. Les différentes phases d'occupations humaines à Payre et hypothèses sur les types d'occupation. In *Le Site de Payre: Occupations Humaines dans la Vallée du Rhône à la fin du Pléistocène Moyen et au début du Pléistocène Supérieur;* Patou-Mathis, M., Moncel, M.-H., Eds.; Société Préhistorique Française: Paris, France, 2008; pp. 309–314.
35. Brugal, J.-P.; Giuliani, C.; Fosse, P.; Fourvel, J.-B.; Magniez, P.; Pelletier, M.; Uzunidis, A. Preliminary Data on the Middle Pleistocene Site of Lunel Viel I (Hérault, France). *Alp. Mediterr. Quat.* **2021**, *34*, 1–13. [CrossRef]
36. Bonifay, M.-F. Le Cheval du pléistocène moyen des grottes de Lunel-Viel (Hérault). *Equus mosbachensis palustris* n. ssp. *Gall. Préhistoire* **1980**, *23*, 233–281. [CrossRef]
37. Rouzaud, F.; Soulier, M.; Brugal, J.-P.; Jaubert, J. L'igue Des Rameaux (Saint-Antonin-Noble-Val, Tarn-et-Garonne). Un Nouveau Gisement Du Pléistocène Moyen. Premiers Résultats. *PALEO* **1990**, *2*, 89–106. [CrossRef]
38. Argant, A.; Brugal, J.-P. The Cave Lion *Panthera (Leo) Spelaea* and Its Evolution: *Panthera Spelaea Intermedia* Nov. Subspecies. *Acta Zool. Crac.* **2017**, *60*, 58–103. [CrossRef]
39. Falguères, C.; Bahain, J.-J.; Saleki, H. U-Series and ESR Dating of Teeth from Acheulian and Mousterian Levels at La Micoque (Dordogne, France). *J. Archaeol. Sci.* **1997**, *24*, 537–545. [CrossRef]
40. Langlois, A. Au Sujet du Cheval de La Micoque (Dordogne) et des Comportements Humains de Subsistance au Pléistocène Moyen Dans le Nord-est de l'Aquitaine. Ph.D. Thesis, Bordeaux I University, Bordeaux, France, 2004.
41. Langlois, A. Le Cheval du gisement Pléistocène moyen de La Micoque (Les Eyzies-de-Tayac, Dordogne): *Equus mosbachensis micoquii* nov. ssp. *PALEO* **2005**, *17*, 73–110. [CrossRef]
42. Guadelli, J.-L.; Prat, F. Le Cheval Du Gisement Pléistocène Moyen de Camp-de-Peyre (Sauveterre-La-Lémance, Lot-et-Garonne): *Equus mosbachensis campdepeyri* n. ssp. *PALEO* **1995**, *7*, 85–121. [CrossRef]
43. Marquet, J.-C. Paléoenvironnement et Chronologie Des Sites Du Domaine Atlantique Fraçais d'âge Pléistocene Moyen et Supérieur d'après l'étude Des Rongeurs. Ph.D. Thesis, Dijon University, Dijon, France, 1993.
44. Le Tensorer, J.-M. *Le Paléolithique de l'Agenais;* Cahiers du Quaternaire; Editions du Centre National de la Recherche Scientifique: Paris, France, 1981; 526p.
45. Texier, J.-P.; Bertran, P. Nouvelle Interprétation Paléoenvironnementale et Chronostratigraphique Du Site Paléolithique de La Micoque (Dordogne): Implications Archéologiques. *Comptes Rendus Acad. Sci. Paris* **1993**, *316*, 1611–1617.
46. Texier, J.-P. *Histoire Géologique de Sites Préhistoriques Classiques Du Périgord: Une Vision Actualisée La Micoque, La Grotte Vaufrey, Le Pech de l'Azé I et II, La Ferrassie, l'abri Castanet, Le Flageolet, Laugerie Haute;* Documents Préhistoriques; Comité des Travaux Historiques et Scientifiques: Paris, France, 2009; p. 193.
47. Coumont, M.-P. Taphonomie Préhistorique: Mammifères Fossiles en Contexte d'avens Pièges. Apport Pour L'étude des Archeofaunes. Ph.D. Thesis, Aix-Marseille I University, Aix-en-Provence, France, 2006.
48. Jeannet, M.; Mein, P. Les Muridae (Mammalia, Rodentia) Du Pléistocène Moyen de l'Igue Des Rameaux (Tarn-et-Garonne, France). *PALEO* **2016**, *27*, 177–205. [CrossRef]
49. Fosse, P. L'hyène Des Cavernes Comme Agent d'accumulation Des Ossements à Lunel-Viel 1 (Hérault, France): Observations Préliminaires. In *Outillage peu Élaboré en os et bois de Cervidés;* CERDAC: Treignes, Belgium, 1994; pp. 91–96.
50. Fosse, P. Taphonomie Paléolithique: Les Grands Mammifères de Soleilhac (Haute-Loire) et de Lunel-Viel 1 (Hérault). Ph.D. Thesis, Provence University Aix-Marseille I, Marseille, France, 1994.
51. Fosse, P. La Grotte N° 1 de Lunel-Viel (Hérault, France): Repaire d'hyènes Du Pléistocène Moyen. Etude Taphonomique Du Matériel Osseux. *PALEO* **1996**, *8*, 47–79. [CrossRef]
52. Uzunidis, A. Dental Wear Analyses of Middle Pleistocene Site of Lunel-Viel (Hérault, France): Did *Equus* and *Bos* Live in a Wetland? *Quat. Int.* **2020**, *557*, 39–46. [CrossRef]
53. Desclaux, E.; El Hazzazi, N.; Vilette, P.; Dubar, M. Le contexte environnemental des occupations humaines. L'apport de la microfaune, des restes aviaires et de la malacofaune. In *Le site de Payre. Occupations Humaines dans la Vallée du Rhône à la fin du Pléistocène Moyen et au Début du Pléistocène Supérieur;* Moncel, M.-H., Ed.; Mémoires de la Société Préhistorique Française; Société Préhistorique Française: Paris, France, 2008; pp. 91–105.
54. Ecker, M.; Bocherens, H.; Julien, M.-A.; Rivals, F.; Raynal, J.-P.; Moncel, M.-H. Middle Pleistocene Ecology and Neanderthal Subsistence: Insights from Stable Isotope Analyses in Payre (Ardèche, Southeastern France). *J. Hum. Evol.* **2013**, *65*, 363–373. [CrossRef] [PubMed]
55. Bocherens, H.; Díaz-Zorita Bonilla, M.; Daujeard, C.; Fernandes, P.; Raynal, J.-P.; Moncel, M.-H. Direct Isotopic Evidence for Subsistence Variability in Middle Pleistocene Neanderthals (Payre, Southeastern France). *Quat. Sci. Rev.* **2016**, *154*, 226–236. [CrossRef]

56. Dessart, C. *Etude de La Faune de La Grotte de Rigabe (Var) (Fouilles E. Bonifay)*; Post-Graduate Work, Aix-Marseille I-Université de Provence: Marseille, France, 2001.
57. Paquereau, M.-M. Flore Des Deux Derniers Interglaciaires Dans Le Sud-Ouest de La France. *L'Anthropologie* **1976**, *80*, 201–228.
58. Griggo, C. Significations Environnementales Des Communautés Animales Pléistocènes Reconnues à l'Abri Suard (Charente) et à La Grotte Du Bois-Ragot (Vienne). Ph.D. Thesis, Bordeaux 1 University, Bordeaux, France, 1995.
59. Griggo, C. Etablissement de Courbes Climatiques Quantifiées à Partir Des Communautés Animales Pléistocènes Suivi d'une Application Aux Gisements de l'Abri Suard (Charente) et La Grotte de Bois-Ragot (Vienne). *PALEO* **1996**, *8*, 81–97. [CrossRef]
60. Paquereau, M.-M. Etude Palynologique Du Wurm I Du Pech de l'Azé (Dordogne). *Quaternaria* **1969**, *XI*, 227–233.
61. Goldberg, P. Micromorphology of Pech-de-l'Azé II Sediments. *J. Archaeol. Sci.* **1979**, *6*, 17–47. [CrossRef]
62. Lenoble, A. Le Rôle Du Ruissellement Dans La Formation Des Sites Préhistoriques: Approche Expérimentale. Ph.D. Thesis, Bordeaux 1 University, Bordeaux, France, 2003.
63. Uzunidis, A. *Les Grands Herbivores de La Couche 9 (Fin Pléistocène Moyen) de Coudoulous II (Tour-de-Faure, Lot): Paléontologie, Paléoécologie, Taphonomie*; Post-Graduate Work, Aix-Marseille University: Aix-en-Provence, France, 2013.
64. Astre, G. Brèche Pléistocène de Peyre (Aveyron), Avec *Equus hydruntinus* et *Megaceros*. *Bull. Soc. Hist. Nat. Toulouse* **1963**, *98*, 321–331.
65. Brugal, J.-P. Recherches Préliminaires Sur Les Faunes Des Grands Causses et Vallées Périphériques. *Paléobiol. Cont.* **1981**, *12*, 145–158.
66. Vernet, J.-L.; Ambert, P.; Andre, J.; Brugal, J.-P.; Casanova, J.; Michaux, J.; Semah, F. Les Travertins Pléistocènes de La Vallée Du Tarn, Dans Le Sud Du Massif Central: Intérêts Paléoécologiques. In Proceedings of the Paléoécologie, Lyon, France, 18–23 July 1983.
67. Brugal, J.-P.; Jaubert, J. Les Gisements Paléontologiques Pléistocènes à Indice de Fréquentation Humaine: Un Nouveau Type de Comportement Prédateur ? *PALEO* **1991**, *3*, 15–41. [CrossRef]
68. Brugal, J.-P. Un repaire de hyène des cavernes et de porc-épic à Peyre (Comprégnac, Aveyron). In *Les Hommes et leurs Environnements Quaternaires dans les Causses de l'Aveyron*; Ambert, P., Vernet, J.-L., Eds.; Association des amis du Musée de Millau: Millau, France, 1993; pp. 47–48.
69. Sondaar, P.Y. *The Osteology of the Manus of Fossil and Recent Equidae with Special Reference to Phylogeny and Function*; Verhandelingen der Koninklijke Nederlandse Akademie van Wetenschappen, Afd. Natuurkunde; N.V. Noord-Hollandsche Utgevers Maatschappij: Amsterdam, The Netherlands, 1968.
70. Gromova, V. Le Genre "Equus", Premiere Partie: Revue et Description des Formes. In *Annales du Centre d'Etudes et de Documentation Paléontologiques, 1955*; n° 13; Pietresson de, S.-A., Translator; Travauxde l'Institut Paléontologique, Académie des Sciences de l'URSS: Paris, France, 1949; Volume 1, t. 17.
71. Eisenmann, V. Sur Quelques Caractères Adaptatifs Du Squelette d'*Equus* et Leurs Implications Paléoécologiques. *Bull. Muséum Hist. Nat.* **1984**, *6*, 185–195.
72. Kuzmina, I.E. Horses of North Eurasia from the Pliocene till the present time. In *Proceedings of the Zoological Institute*; Vereschagin, N.K., Ed.; Russian Academy of Sciences: Saint-Petersbourg, Russia, 1997; Volume 273.
73. Cramer, B. Morphometrische Untersuchungen an Quartären Pferden in Mitteleuropa. Ph.D. Thesis, Geowissenschaftliche Fakultät, Eberhard-Karls-Universität, Tübingen, Germany, 2002.
74. Levine, M.A. The use of crown height measurements and eruption-wear sequences to age horse teeth. In *Ageing and Sexing of Animal Bones from Archaeological Sites*; Wilson, B., Grigson, C., Payne, S., Eds.; BAR British Series; BAR Publishing: Oxford, UK, 1982; pp. 223–250.
75. Van Asperen, E.N. Implications of Age Variation and Sexual Dimorphism in Modern Equids for Middle Pleistocene Equid Taxonomy. *Int. J. Osteoarchaeol.* **2013**, *23*, 1–12. [CrossRef]
76. Eisenmann, V. Vera Eisenmann. Available online: https://vera-eisenmann.com/ (accessed on 20 May 2021).
77. Eisenmann, V.; Alberdi, M.T.; De Giuli, C.; Staesche, U. Methodology. In *Studying Fossil Horses*; Woodburne, M.-O., Sondaar, P.Y., Eds.; E. J. Brill: Leiden, The Netherlands, 1988; Volume 1, pp. 1–71.
78. Eisenmann, V. Les métapodes d'*Equus* sensu lato (Mammalia, Perissodactyla). *Géobios* **1979**, *12*, 863–886. [CrossRef]
79. Scott, K.M. Postcranial dimensions of ungulates as predictors of body mass. In *Body Size in Mammalian Palaeobiology—Estimation and Biological Implications*; Damuth, J., MacFadden, B.J., Eds.; Cambridge University Press: New York, NY, USA, 1990; pp. 301–355.
80. Simpson, G.G. *Large Pleistocene Felines of North America*; American Museum Novitates; American Museum of Natural History: New York, NY, USA, 1941.
81. Eisenmann, V. Proportions Squelettiques de Chevaux Quaternaires et Actuels. *Géobios* **1991**, *13*, 25–32. [CrossRef]
82. Ackermans, N.L.; Martin, L.F.; Codron, D.; Hummel, J.; Kircher, P.R.; Richter, H.; Kaiser, T.M.; Clauss, M.; Hatt, J.-M. Mesowear Represents a Lifetime Signal in Sheep (*Ovis Aries*) within a Long-Term Feeding Experiment. *Palaeogeogr. Palaeoclimatol. Palaeoecol.* **2020**, *553*, 109793. [CrossRef]
83. Fortelius, M.; Solounias, N. Functional Characterization of Ungulate Molars Using the Abrasion-Attrition Wear Gradient: A New Method for Reconstructing Paleodiets. *Am. Mus. Novit.* **2000**, *2000*, 1–36. [CrossRef]
84. Grine, F.E. Dental Evidence for Dietary Differences in *Australopithecus* and *Paranthropus*: A Quantitative Analysis of Permanent Molar Microwear. *J. Hum. Evol.* **1986**, *15*, 783–822. [CrossRef]

85. Hoffman, J.M.; Fraser, D.; Clementz, M.T. Controlled Feeding Trials with Ungulates: A New Application of in Vivo Dental Molding to Assess the Abrasive Factors of Microwear. *J. Exp. Biol.* **2015**, *218*, 1538–1547. [CrossRef]
86. Merceron, G.; Ramdarshan, A.; Blondel, C.; Boisserie, J.-R.; Brunetiere, N.; Francisco, A.; Gautier, D.; Milhet, X.; Novello, A.; Pret, D. Untangling the Environmental from the Dietary: Dust Does Not Matter. *Proc. Biol. Sci.* **2016**, *283*. [CrossRef]
87. Teaford, M.F. A Review of Dental Microwear and Diet in Modern Mammals. *Scanning Microsc.* **1988**, *2*, 1149–1166. [PubMed]
88. Winkler, D.E.; Schulz-Kornas, E.; Kaiser, T.M.; Codron, D.; Leichliter, J.; Hummel, J.; Martin, L.F.; Clauss, M.; Tütken, T. The Turnover of Dental Microwear Texture: Testing the "Last Supper" Effect in Small Mammals in a Controlled Feeding Experiment. *Palaeogeogr. Palaeoclimatol. Palaeoecol.* **2020**, *557*, 109930. [CrossRef]
89. Rivals, F.; Moncel, M.-H.; Patou-Mathis, M. Seasonality and Intra-Site Variation of Neanderthal Occupations in the Middle Palaeolithic Locality of Payre (Ardèche, France) Using Dental Wear Analyses. *J. Archaeol. Sci.* **2009**, *36*, 1070–1078. [CrossRef]
90. Rivals, F.; Mihlbachler, M.C.; Solounias, N. Effect of Ontogenetic-Age Distribution in Fossil and Modern Samples on the Interpretation of Ungulate Paleodiets Using the Mesowear Method. *J. Vertebr. Paleontol.* **2007**, *27*, 763–767. [CrossRef]
91. Saarinen, J.; Lister, A.M. Dental Mesowear Reflects Local Vegetation and Niche Separation in Pleistocene Proboscideans from Britain. *J. Quat. Sci.* **2016**, *31*, 799–808. [CrossRef]
92. Kaiser, T.M.; Solounias, N. Extending the Tooth Mesowear Method to Extinct and Extant Equids. *Geodiversitas* **2003**, *25*, 321–345.
93. Mihlbachler, M.C.; Rivals, F.; Solounias, N.; Semprebon, G.M. Dietary Change and Evolution of Horses in North America. *Science* **2011**, *331*, 1178–1181. [CrossRef] [PubMed]
94. Rivals, F.; Rindel, D.; Belardi, J.B. Dietary Ecology of Extant Guanaco (*Lama Guanicoe*) from Southern Patagonia: Seasonal Leaf Browsing and Its Archaeological Implications. *J. Archaeol. Sci.* **2013**, *40*, 2971–2980. [CrossRef]
95. Uzunidis, A.; Rivals, F.; Brugal, J.-P. Relation between Morphology and Dietary Traits in Horse Jugal Upper Teeth during the Middle Pleistocene in Southern France. *Quat. Rev. Assoc. Fr. Pour Létude Quat.* **2017**, *28*, 303–312. [CrossRef]
96. Rivals, F.; Uzunidis, A.; Sanz, M.; Daura, J. Faunal Dietary Response to the Heinrich Event 4 in Southwestern Europe. *Palaeogeogr. Palaeoclimatol. Palaeoecol.* **2017**, *473*, 123–130. [CrossRef]
97. Solounias, N.; Semprebon, G. Advances in the Reconstruction of Ungulate Ecomorphology with Application to Early Fossil Equids. *Am. Mus. Novit.* **2002**, *3366*, 49. [CrossRef]
98. Semprebon, G.; Godfrey, L.R.; Solounias, N.; Sutherland, M.R.; Jungers, W.L. Can Low-Magnification Stereomicroscopy Reveal Diet? *J. Hum. Evol.* **2004**, 115–144. [CrossRef]
99. Xafis, A.; Nagel, D.; Bastl, K. Which Tooth to Sample? A Methodological Study of the Utility of Premolar/Non-Carnassial Teeth in the Microwear Analysis of Mammals. *Palaeogeogr. Palaeoclimatol. Palaeoecol.* **2017**, *487*, 229–240. [CrossRef]
100. Uzunidis, A.; Pineda, A.; Jiménez-Manchón, S.; Xafis, A.; Ollivier, V.; Rivals, F. The Impact of Sediment Abrasion on Tooth Microwear Analysis: An Experimental Study. *Archaeol. Anthropol. Sci.* **2021**, *13*, 134. [CrossRef]
101. Walker, A.; Hoeck, H.N.; Perez, L. Mecrowear of Mammalian Teeth as an Indicator of Diet. *Science* **1978**, *201*, 908–910. [CrossRef]
102. Semprebon, G.M.; Rivals, F. Was Grass More Prevalent in the Pronghorn Past? An Assessment of the Dietary Adaptations of Miocene to Recent Antilocapridae (Mammalia: Artiodactyla). *Palaeogeogr. Palaeoclimatol. Palaeoecol.* **2007**, *253*, 332–347. [CrossRef]
103. Semprebon, G.M.; Sise, P.J.; Coombs, M.C. Potential Bark and Fruit Browsing as Revealed by Stereomicrowear Analysis of the Peculiar Clawed Herbivores Known as Chalicotheres (Perissodactyla, Chalicotherioidea). *J. Mamm. Evol.* **2011**, *18*, 33–55. [CrossRef]
104. Kaczensky, P.; Ganbaatar, O.; Wehrden, H.V.; Walzer, C. Resource Selection by Sympatric Wild Equids in the Mongolian Gobi. *J. Appl. Ecol.* **2008**, *45*, 1762–1769. [CrossRef]
105. Uzunidis, A. Large Ungulates Mobility and Neanderthal Subsistence Behaviours: A Preliminary Tooth Microwear Analysis. *J. Archaeol. Sci. Rep.* **2020**, *29*, 102084. [CrossRef]
106. Foronova, I. Late Quaternary Equids (Genus Equus) of South-Western and Southcentral Siberia. In *Proceedings of the Equids in Time and Space: Papers in Honour of Véra Eisenmann: Proceedings of the 9th Conference of the International Council of Archaeozoology, Durham, UK, 23–28 August 2002*; Mashkour, M., Ed.; Août 2002; Oxbow Books: Oxford, UK, 2006; pp. 20–30.
107. Forsten, A. Middle Pleistocene Replacement of Stenonoid Horses by Caballoid Horses—Ecological Implications. *Palaeogeogr. Palaeoclimatol. Palaeoecol. Elsevier Sci. Publ. BV* **1988**, *65*, 23–33. [CrossRef]
108. Bro-Jørgensen, J. Dense Habitats Selecting for Small Body Size: A Comparative Study on Bovids. *Oikos* **2008**, *117*, 729–737. [CrossRef]
109. Lindstedt, S.L.; Boyce, M.S. Seasonality, Fasting Endurance, and Body Size in Mammals. *Am. Nat.* **1985**, *125*, 873–878. [CrossRef]
110. El Hazzazi, N. Apport Du Genre Arvicola (Rodentia, Mammalia) Dans La Connaissance de La Chronostratigraphie Des Sites Pléistocène Moyen et Supérieur: Orgnac 3, Payre 2 et l'abri Des Pêcheurs (Ardèche, France). In *Proceedings of the Theoretical and Methodological Problems—Archaeometry—Paleoecology, Forli, Italy, 8–14 September 1996*; Arias, C., Bietti, A., Castelletti, L., Peretto, C., Eds.; A.B.A.C.O. Edizioni: Forli, Italy, 1998; Volume 1, pp. 125–133.
111. Moncel, M.-H.; Debard, E.; Desclaux, E.; Dubois, J.-M.; Lamarque, F.; Patou-Mathis, M.; Villette, P. Le Cadre de Vie Des Hommes Du Paléolithique Moyen (Stades Isotopiques 6 et 5) Dans Le Site de Payre (Rompon, Ardèche): D'une Grotte à Un Abri Sous Roche Effondré. *Bull. Soc. Préhistorique Fr.* **2002**, *99*, 249–273. [CrossRef]
112. Schulz, E.; Kaiser, T.M. Historical Distribution, Habitat Requirements and Feeding Ecology of the Genus *Equus* (Perissodactyla). *Mammal Rev.* **2013**, *43*, 111–123. [CrossRef]

113. Kaiser, T.M.; Franz-Odendaal, T.A. A Mixed-Feeding *Equus* Species from the Middle Pleistocene of South Africa. *Quat. Res.* **2004**, *62*, 316–323. [CrossRef]
114. Valli, A.M.F.; Palombo, M.R.; Alberdi, M.T. How Homogeneous Are Microwear Patterns on a Fossil Horse Tooth? Preliminary Test on a Premolar of *Equus Altidens* from Barranco Leon 5 (Spain). *Alp. Mediterr. Quat.* **2012**, *25*, 25–33.
115. Kelly, A.; Miller, J.H.; Wooller, M.J.; Seaton, C.T.; Druckenmiller, P.; DeSantis, L. Dietary Paleoecology of Bison and Horses on the Mammoth Steppe of Eastern Beringia Based on Dental Microwear and Mesowear Analyses. *Palaeogeogr. Palaeoclimatol. Palaeoecol.* **2021**, *572*, 110394. [CrossRef]
116. Rivals, F.; Kitagawa, K.; Julien, M.-A.; Patou-Mathis, M.; Bessudnov, A.A.; Bessudnov, A.N. Straight from the Horse's Mouth: High-Resolution Proxies for the Study of Horse Diet and Its Relation to the Seasonal Occupation Patterns at Divnogor'ye 9 (Middle Don, Central Russia). *Quat. Int.* **2018**, *474*, 146–155. [CrossRef]
117. Delpech, F.; Le Tensorer, J.-M.; Pineda, R.; Prat, F. Un Nouveau Gisement Du Pleistocène Moyen: Camp-de-Peyre à Sauveterre-La-Lémance (Lot-et-Garonne). *Comptes Rendus Acad. Sci. Paris D* **1978**, *286*, 1101–1103.
118. Saarinen, J.; Eronen, J.; Fortelius, M.; Seppä, H.; Lister, A.M. Patterns of Diet and Body Mass of Large Ungulates from the Pleistocene of Western Europe, and Their Relation to Vegetation. *Palaeontol. Electron.* **2016**, *19.3.32A*, 1–58. [CrossRef]
119. Geist, V. *Deer of the World: Their Evolution, Behaviour, and Ecology*; Stackpole Books: Mechanicsburg, PA, USA, 1998; ISBN 978-0-8117-0496-0.
120. Guthrie, R.D. Origin and Causes of the Mammoth Steppe: A Story of Cloud Cover, Woolly Mammal Tooth Pits, Buckles, and inside-out Beringia. *Quat. Sci. Rev.* **2001**, *20*, 549–574. [CrossRef]
121. Saarinen, J. Ecometrics of Large Herbivorous Land Mammals in Relation to Climatic and Environmental Changes during the Pleistocene. Ph.D. Thesis, Department of Geosciences and Geography, University of Helsinki, Helsinki, Finland, 2014.
122. Blackburn, T.M.; Gaston, K.J.; Loder, N. Geographic Gradients in Body Size: A Clarification of Bergmann's Rule. *Divers. Distrib.* **1999**, *5*, 165–174. [CrossRef]
123. Boulbes, N.; van Asperen, E.N. Biostratigraphy and Palaeoecology of European Equus. *Front. Ecol. Evol.* **2019**, *7*. [CrossRef]
124. Richerson, P.J.; Boyd, R.; Bettinger, R.L. Was Agriculture Impossible during the Pleistocene but Mandatory during the Holocene? A Climate Change Hypothesis. *Am. Antiq.* **2001**, *66*, 387–411. [CrossRef]

MDPI AG
Grosspeteranlage 5
4052 Basel
Switzerland
Tel.: +41 61 683 77 34

Quaternary Editorial Office
E-mail: quaternary@mdpi.com
www.mdpi.com/journal/quaternary

www.ingramcontent.com/pod-product-compliance
Lightning Source LLC
Chambersburg PA
CBHW061555220926
51850CB00005B/341
* 9 7 8 3 7 2 5 8 4 6 7 7 1 *